Emerging Nanotechnologies and their Applications

Emerging Nanotechnologies and their Applications

Edited by **Rich Falcon**

Willford Press
New York

Published by Willford Press,
118-35 Queens Blvd., Suite 400,
Forest Hills, NY 11375, USA
www.willfordpress.com

Emerging Nanotechnologies and their Applications
Edited by Rich Falcon

International Standard Book Number: 978-1-68285-178-4 (Hardback)

Printed in China.

Contents

Preface

Over the recent decade, advancements and applications have progressed exponentially. This has led to the increased interest in this field and projects are being conducted to enhance knowledge. The main objective of this book is to present some of the critical challenges and provide insights into possible solutions. This book will answer the varied questions that arise in the field and also provide an increased scope for furthering studies.

The recent developments in the field of nanotechnology and nanomaterials has brought massive shift in various engineering and scientific disciplines. This book includes diverse topics related to the field of nanotechnology such as diagnosis and drug delivery, nanosensors, water treatment, nanoelectronics and nanodevices. It also includes chapters on crucial aspects of nanotechnology and nanoscience like properties and synthesis of nanomaterials, structure of nanodevices, modeling and simulation of different nanostructures, etc. A number of latest studies and examples have been included to keep the readers up-to-date with the significant concepts in this area of study.

I hope that this book, with its visionary approach, will be a valuable addition and will promote interest among readers. Each of the authors has provided their extraordinary competence in their specific fields by providing different perspectives as they come from diverse nations and regions. I thank them for their contributions.

Editor

1

A facile and rapid method for the black pepper leaf mediated green synthesis of silver nanoparticles and the antimicrobial study

Robin Augustine · Nandakumar Kalarikkal · Sabu Thomas

Abstract Green synthesis of nanoparticles is widely accepted due to the less toxicity in comparison with chemical methods. But there are certain drawbacks like slow formation of nanoparticles, difficulty to control particle size and shape make them less convenient. Here we report a novel cost-effective and eco-friendly method for the rapid green synthesis of silver nanoparticles using leaf extracts of *Piper nigrum.* Our results suggest that this method can be used for obtaining silver nanoparticles with controllable size within a few minutes. The fabricated nanoparticles possessed excellent antibacterial property against both Gram-positive and Gram-negative bacteria.

Keywords Biosynthesis · Silver nanoparticles · *Piper nigrum* · Reduction · Organic capping

R. Augustine · N. Kalarikkal (✉) · S. Thomas (✉)
International and Interuniversity Centre for Nanoscience and Nanotechnology, Mahatma Gandhi University, Priyadarshini Hills P O, Kottayam 686 560, Kerala, India
e-mail: nkkalarikkal@mgu.ac.in

S. Thomas
e-mail: sabuchathukulam@yahoo.co.uk

N. Kalarikkal
School of Pure and Applied Physics, Mahatma Gandhi University, Priyadarshini Hills P O, Kottayam 686 560 Kerala, India

S. Thomas
School of Chemical Sciences, Mahatma Gandhi University, Priyadarshini Hills P O, Kottayam 686 560, Kerala, India

Introduction

Reducing the particle size of materials is an efficient tool for improving their bioactivity. A number of preparation methods have been described for the synthesis of metallic nanoparticles (Pal et al. 2007a; Rosemary and Pradeep 2003); such as, reverse micelles process (Xie et al. 2006; Maillard et al. 2002), salt reduction (Pillai and Kamat 2004), microwave dielectric heating reduction (Patel et al. 2005), ultrasonic irradiation (Salkar et al. 1999), radiolysis (Karim et al. 2007; Remita et al. 2007), solvothermal synthesis (Starowicz et al. 2006), electrochemical synthesis (Zhu et al. 2001; Liu et al. 2001) and biological techniques (Naik et al. 2002). The most widespread method of synthesis of metallic nanoparticles is based on the chemical reduction of metal salt solution by a reducing agent (Szczepanowicz et al. 2010). But the chemicals used in such processes are quite often toxic and flammable which make them unsuitable for many applications including biological applications like impregnation in wound dressings (Maneerung et al. 2008; Vivekanandhan et al. 2012) and sutures (Augustine and Rajarathinam 2012).

Biological syntheses of nanoparticles are not only a good way to fabricate benign nanostructure materials, but also to reduce the use or generation of hazardous substances to human health and the environment (Justin Packia Jacob et al. 2012). These biological methods are regarded as safe, cost-effective, sustainable and environment friendly as well as they do not require any special culture preparation and isolation techniques (Gardea et al. 2003). Biosynthesis of silver nanoparticles using microorganisms like bacteria (Saifuddin et al. 2009), fungi (Duran et al. 2009) and yeast (Kowshik et al. 2003) are already reported. However, exploitation of the plant extracts as the potential agents for the biosynthesis of nanoparticles has opened a new way—the green synthesis of nanoparticles. Further, synthesis of silver nanoparticles using extracts of various plants like *Aloe vera* (Chandran et al. 2006), *Cinnamon zeylanicum* (Sathishkumar et al. 2009), *Stevia rebaudiana* (Varshney et al. 2010), *Papaya* (Jain et al. 2009), Geranium (Shiv Shankar et al. 2008), *Cassia angustifolia* (Peter Amaladhas et al. 2012), *Ocimum tenuiflorum* (Patil et al. 2012), lemon (Vankar and Shukla 2012), carob (Awwad et al. 2013), *Coleus aromaticus* (Vanaja and Annadurai 2012), were also reported.

Piper nigrum Linn. or Black pepper is a perennial woody climbing liana belonging to the family Piperaceae. It is native to India, Indonesia, Malaysia, South America and West Indies but is also widely cultivated in the tropical regions. It is considered as the 'King of Spices' (Srinivasan 2007; Mathew et al. 2001). The presence of a pungent alkaloid (Tripathi et al. 1996) piperine ($C_{17}H_{19}NO_3$) attributes a spicy taste to the seeds, leaves and other parts of *Piper* (Khajuria et al. 2002). It also contains small amounts of safrole ($C_{10}H_{10}O_2$), pinene ($C_{10}H_{16}$), sabinene ($C_{10}H_{16}$), limonene ($C_{10}H_{16}$), caryophyllene ($C_{15}H_{24}$) and Linalool ($C_{10}H_{18}O$) compound. The fruits have variety of activities including CNS depressant, antipyretic, analgesic, hepatoprotective (Lee et al. 1984), bioavailability enhancer (Zhao et al. 2007; Kasibhatta and Naidu 2007), antioxidant (Khajuria et al. 1997) and anti-inflammatory (Majumdar et al. 1990).

Antioxidants are compounds that can delay or inhibit the oxidation of lipids or other molecules by inhibiting the initiation or propagation of oxidative chain reaction (Velioglu et al. 1998; Chanwitheesuk et al. 2005). From the *P. nigram*, a number of such compounds have been isolated and characterized. Alkaloids such as pellitorine ($C_{14}H_{25}NO$), (E)-1-[3',4'-(methylenedioxy)cinnamoyl] piperidine ($C_{15}H_{17}NO_3$), 2,4-tetradecadienoic acid isobutyl amide ($C_{16}H_{25}NO$), piperine ($C_{17}H_{19}NO_3$), cepharadione A ($C_{18}H_{11}NO_4$), piperolactam D ($C_{17}H_{13}NO_4$) and paprazine ($C_{17}H_{17}NO_3$) have been isolated from black pepper (Ee et al. 2009). Wei et al. (2004) isolated some amide-containing alkaloids from *P. nigrum.* They are Pipercycliamide ($C_{14}H_{26}NO_3$), (±)-*erythro*-1-(1-oxo-4,5-dihydroxy-2E-decaenyl)piperidine ($C_{15}H_{28}NO_3$), (±)-*threo*-1-(1-oxo-4,5-dihydroxy-2E-decaenyl)piperidine ($C_{15}H_{28}NO_3$), (±)-*threo*-*N*-isobutyl-4,5-dihydroxy-2E-octanamide ($C_{12}H_{23}NO_3Na$), 1-(1,6-dioxo-2E,4E-decadienyl)piperidine ($C_{15}H_{23}NO_2$), 1-1-oxo-3(3,4-methylenedioxy-5-methoxyphenyl)-2Z-propenylpiperidine ($C_{16}H_{19}NO_4$), and 1-1-oxo-5(3,4-methylenedioxyphenyl)-2Z,4E-pentadienylpyrrolidine ($C_{16}H_{17}$–NO_3) (Wei et al. 2004). A number of novel dimeric amide alkaloids possessing a cyclobutane or a cyclohexene ring termed as nigramides have also been isolated from the roots of *P. nigrum* (Wei et al. 2005). Shanmugapriya et al. (2012) extracted alkaloids, phenolic compounds, saponins and flavonoids from the leaves of black pepper. They have also shown the antioxidant potential of these compounds. These alkaloids may act as reducing agents, free radical scavengers or potential complexers of pro-oxidant metals (Hudson 1990). Thus as in the chemical reduction process, these natural reducing agents may act upon silver nitrate and reduce it into metallic silver nanoparticles.

Even though the biological synthesis of nanoparticles is advantageous in many aspects, the long duration needed for the formation of nanoparticles makes it inconvenient. Many workers tried to develop a method for the rapid biosynthesis of various kinds of nanoparticles but all these attempts took more than 15 min to get nanoparticles (Darroudi et al. 2010; Shenya et al. 2012; Philip 2010; Praveen Kumar et al. 2011; Bar et al. 2009). Further the controllability of the size of the nanoparticles was poor in previous reports. Thus in the present study, we report a green method for the synthesis of silver nanoparticles using an aqueous leaf extract of *P. nigram* and no toxic chemicals are used as reducing and stabilizing agents during the synthesis.

Materials and methods

Materials

Silver nitrate of analytical grade purchased from Sigma Aldrich was used as a starting material without further purification. The *P. nigram* leaves were freshly collected from the agricultural field near Kottayam, Kerala, India.

Preparation of the leaf extract

Around 5 g of freshly collected *P. nigram* leaves were grinded well using mortar and pestle. 100 ml of deionized water was added into the slurry and filtered through a cheese cloth. The filtrate was again filtered through Whatman No. 1 filter paper (pore size 25 μm). The fresh filtrate was used for the present study.

Synthesis of silver nanoparticles

1–5 mM aqueous solutions of silver nitrate ($AgNO_3$) was prepared in double distilled water and used for the synthesis of silver nanoparticles. 100 ml aqueous solution of the silver nitrate solution was taken in a round bottom flask and heated to boiling with a magnetic stirrer. The flask was covered with aluminum foil to prevent light-mediated reduction of silver nitrate. Then around 5 ml of the *P. nigram* leaf extract was added drop wise into the silver nitrate solution. During this process solution was mixed vigorously. Within 1 min the color change was evident (pale red). Then it was removed from the heating mantle and stirred until cooled to room temperature. A portion of the obtained suspensions were used for UV–Visible spectroscopic analysis. Remaining part was centrifuged at 12,000 rpm several times in distilled water and finally in ethanol to get pure silver nanoparticles.

UV–Vis spectra analysis

The optical properties (absorbance) of colloidal solution were evaluated using a Shimadzu double-beam spectrophotometer (model 1700) at a resolution of 1 nm, between 200 and 600 nm. The reduction of pure Ag^+ ions was monitored by measuring the UV–Vis spectrum of the reaction medium after diluting a small aliquot of the sample into distilled water.

FTIR (Fourier-transform IR) analysis

FTIR studies on the samples were carried out using Perkin Elmer, Spectrum 400 to ensure the formation of silver nanoparticles. FTIR measurements help to identify the possible interactions between silver and bioactive molecules, which may be responsible for formation and stabilization (capping material) of silver nanoparticles.

TEM analysis of silver nanoparticles

To understand the morphology and size distribution, the synthesized nanoparticles were imaged using JEOL JEM 2100 high resolution transmission electron microscope. The sample was prepared by air-drying drops of diluted solutions of the preparations on carbon films supported by copper grids. The particle size was measured using ImageJ software and the average particle size was determined.

XRD analysis

XRD was recorded in the 2θ range of 30–80° using D8-Advance of Bruker (Germany), of CuKα radiation, the energy of which was 8.04 keV and wavelength was 1.54 Å. The applied voltage was 40 kV and current was 25 mA.

Antimicrobial activity

The disc-diffusion assay Kirby–Bauer method (Pal et al. 2007b) was used by many workers to determine the growth inhibition of bacteria by the silver nanoparticles (Sadeghi et al. 2010; Augustine and Rajarathinam 2012; Amany and El-Rab 2012). The bacterial strains *Escherichia coli* (ATCC 12228) and *Staphylococcus aureus* (ATCC6538-P) were used as representatives of Gram-negative and Gram-positive bacteria, respectively. The bacteria were cultured in Mueller–Hinton Broth (MHB) at 37 °C and prepared to the turbidity equivalent to 0.5 McFarland standards (McFarland 1907). Then 100 µl of the bacterial suspension was spread on the nutrient agar. Sterile blank discs with 6 mm diameter (HiMedia, Mumbai) were impregnated with silver nanoparticles synthesized at different silver nitrate concentrations so that to get a final nanoparticle concentration of 10 µg/disc. These were then placed on the surface of the test plate. A standard antibiotic disc was used as positive control (Ciprofloxacin, 10 µg/disc). The culture plates were incubated for overnight in an incubator at 37 °C. The diameters of the inhibition zones were measured in millimeters (mm). The experiment was repeated for three times to get an average value.

Results and discussion

Color change

From the visual color change we can get preliminary information regarding the formation of silver nanoparticles. As the silver nanoparticles are formed, the color of the solution changes from white to pale yellow to brick red which is an indication of the presence of silver nanoparticles (Fig. 1). The variation of the color was due to the change in surface plasmon resonance of silver nanoparticles during the formation. At a silver nitrate concentration of 1 mM a very pale yellow color was obtained. While increasing the concentration of silver nitrate up to 5 mM, there is an increase in the redness of the solution. This is due to the fact that as the concentration of silver nitrate increases, aggregation of formed silver ions occurs and which leads to the formation of larger sized silver nanoparticles.

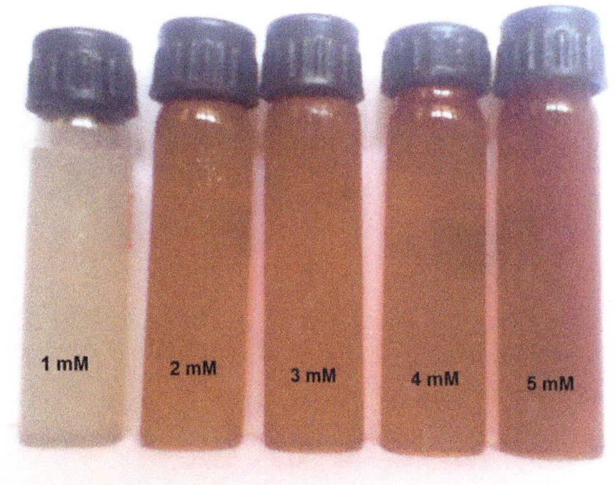

Fig. 1 Visible color change during the formation of silver nanoparticles at varying molar concentrations of silver nitrate

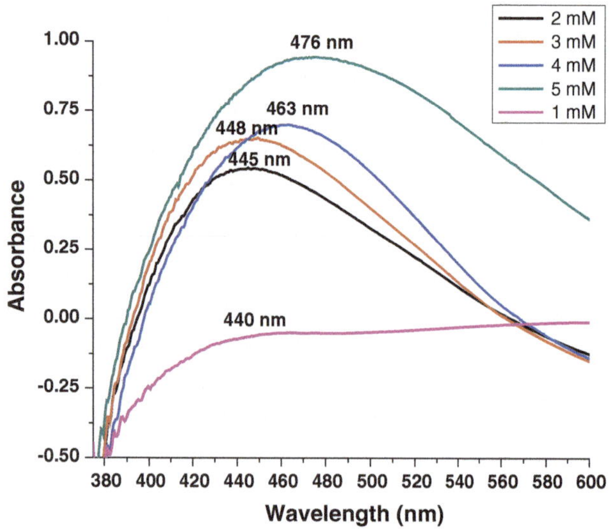

Fig. 2 UV–Visible absorption spectra of synthesized silver nanoparticles at varying silver nitrate concentrations

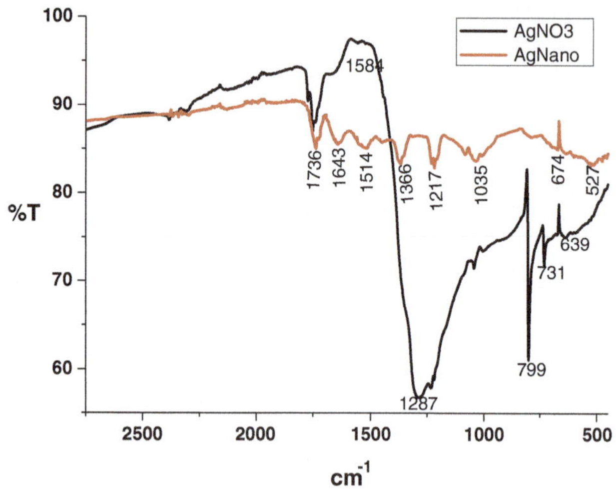

Fig. 3 FTIR spectra of silver nitrate and synthesized silver nanoparticles

UV–Visible spectrum

Characteristic absorption peaks of silver nanoparticles can be seen at around 400–475 nm. Figure 2 shows the UV–Visible spectra which are recorded after the completion of the reaction. For 1 mM solution, silver nanoparticles has shown absorbance maxima at 440 nm, 2 mM solution has peak at 445 nm, 2 mM solution has at 445 nm, 3 mM has at 448 nm, 4 mM has at 463 nm, and 5 mM solution has at 476 nm. Electrons are limited to specific vibrations modes by the particle's size and shape. The frequency and width of the surface plasmon absorption depend on the size and shape of the metal nanoparticles (Wang et al. 2007) as well as on the dielectric constant of the metal itself and the surrounding medium (Rai et al. 2006). Here the medium dielectric constant and temperature can be considered as the same but the mean diameter and morphology of the nanoparticles strongly affect the SPR band in aqueous solution (Wiley et al. 2006). The spectrum shows the red shift with increasing the molar concentration of silver nitrate. It indicates the increase of mean diameter of the silver nanoparticles as the concentration increases (Rai et al. 2006; Song and Kim 2009; Fayaz et al. 2009).

FTIR

The dried nanoparticle samples were analyzed in FTIR to identify the possible bio molecules responsible for the reduction of the Ag^{+} ions by *Piper* leaf extract. The FTIR spectrum is presented in Fig. 3.

The FTIR absorption spectra of unreduced silver nitrate and silver nanoparticles after the reduction and capping by *Piper* extract were taken. While comparing the FTIR spectra of unreduced silver nitrate and biosynthesized silver nanoparticles, there is a notable reduction in certain peaks and appearance of certain new peaks was observed. Absorbance bands in pure silver nitrate were observed in the region of 450–1,750 cm^{-1} and are 1,584, 1,287, 799, 731 and 639 cm^{-1}. These absorbance bands are associated with the nitro compounds. A broad peak at 1,287 cm^{-1} which is present in the spectrum of silver nitrate is not found in the spectrum of silver nanoparticles. It indicates the loss of nitro group from silver species during the reduction process. Molecules containing NO_2 groups, such as nitro compounds, nitrates, and nitramines, commonly exhibit asymmetric and symmetric stretching vibrations of the NO_2 group at 1,660–1,500 and 1,390–1,260 cm^{-1} region. Apart from these losses of functional moieties, a number of new peaks found in the case of biosynthesized silver nanoparticles.

The FTIR spectrum of Ag nanoparticles showed distinct peaks 1,643 cm^{-1}, which represent the involvement of C–N in plane vibrations, 1,035 and 1,217 cm^{-1} shows the involvement of C–N in plane vibrations of aliphatic amines. The peak 1,514 cm^{-1} arises from N–H in-plane bending and C–N stretching modes. It may be due to the C–N bond present in the amide-containing alkaloids attached to the silver nanoparticles, which make the synthesized nanoparticles stable. Peak at 1,736 cm^{-1} is an indication of C=O Stretch and probably due to ketones. Further the presence of C–O Stretch in between 1,300 and 1,000 cm^{-1} may be due to the covalent linking of ester or ether groups to the nanoparticle. Esters show their carbonyl C=O stretch at 1,750–1,735 cm^{-1}, but also exhibit their characteristic absorption at 1,300–1,000 cm^{-1} from the couplings of C–O and C–C stretches. Thus, the peak at 1,035 cm^{-1} indicates the coupling of C–O and C–C stretches and 1,736 and

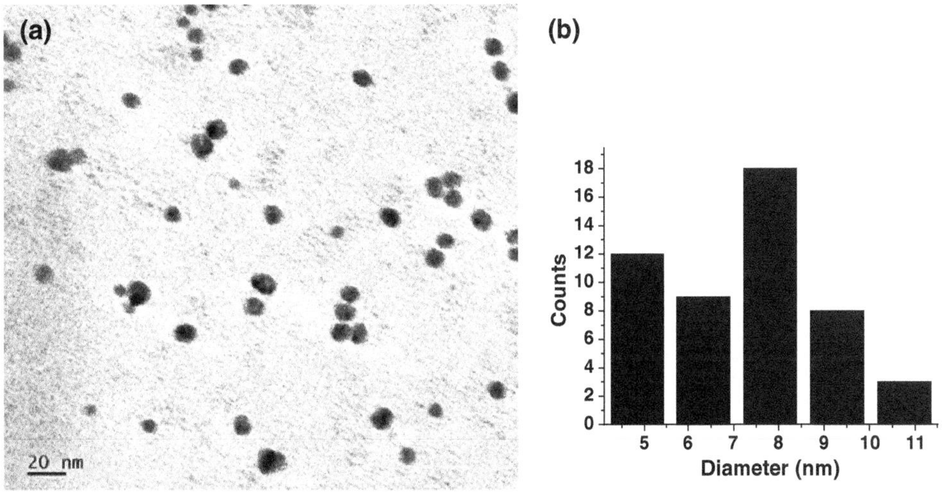

Fig. 4 TEM image of silver nanoparticles formed at 1 mM silver nitrate concentration (**a**) and the particle size distribution (**b**)

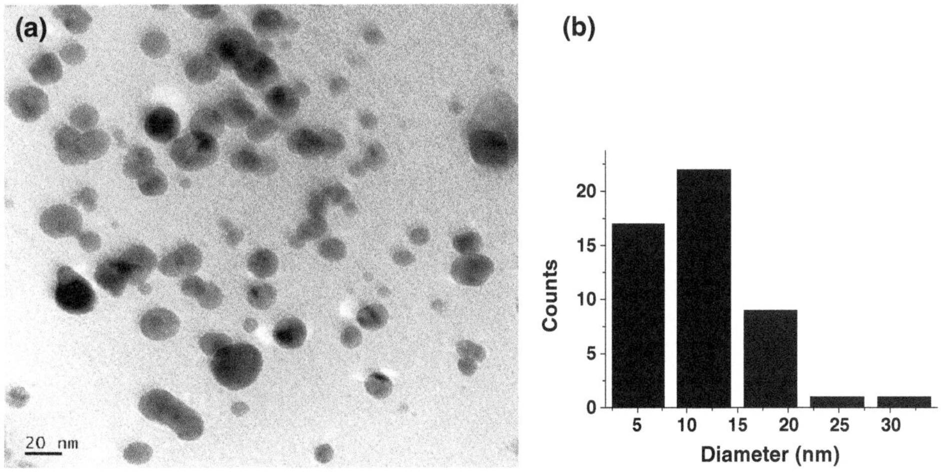

Fig. 5 TEM image of silver nanoparticles formed at 2 mM silver nitrate concentration (**a**) and the particle size distribution (**b**)

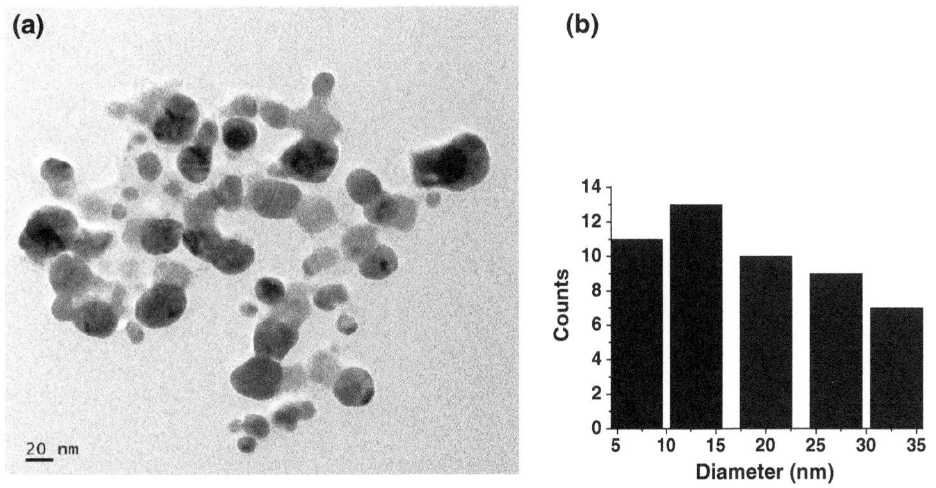

Fig. 6 TEM image of silver nanoparticles formed at 3 mM silver nitrate concentration (**a**) and the particle size distribution (**b**)

1,643 cm^{-1} indicates C=O stretch. Hence, there may be an ester link with the silver nanoparticle. These observations indicate the presence and binding of certain amide-containing compounds with silver nanoparticles. It may be due to the binding of one or more amide-containing alkaloids which is present in *Piper* like nigramides, *N*-isobutyl-4-hexanoyl-4-hydroxypyrrolidin-1-one, (±)-*erythro*-1-(1-oxo-4,5-dihydroxy-2E-decaenyl)piperidine, (±)-*threo*-1-(1-oxo-4,5-dihydroxy-2E-decaenyl)piperidine, (±)-*threo*-*N*-isobutyl-4,5-dihydroxy-2E-octanamide, 1-(1,6-dioxo-

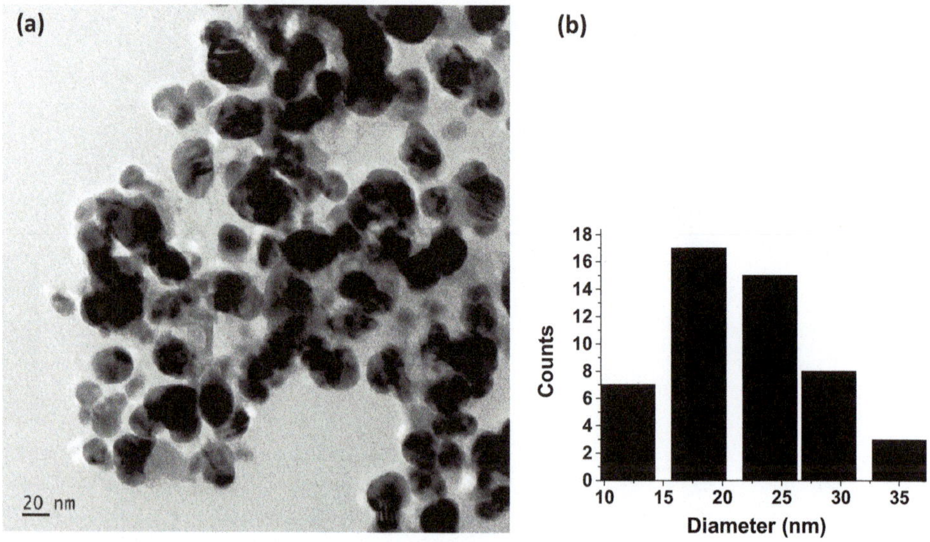

Fig. 7 TEM image of silver nanoparticles formed at 4 mM silver nitrate concentration (**a**) and the particle size distribution (**b**)

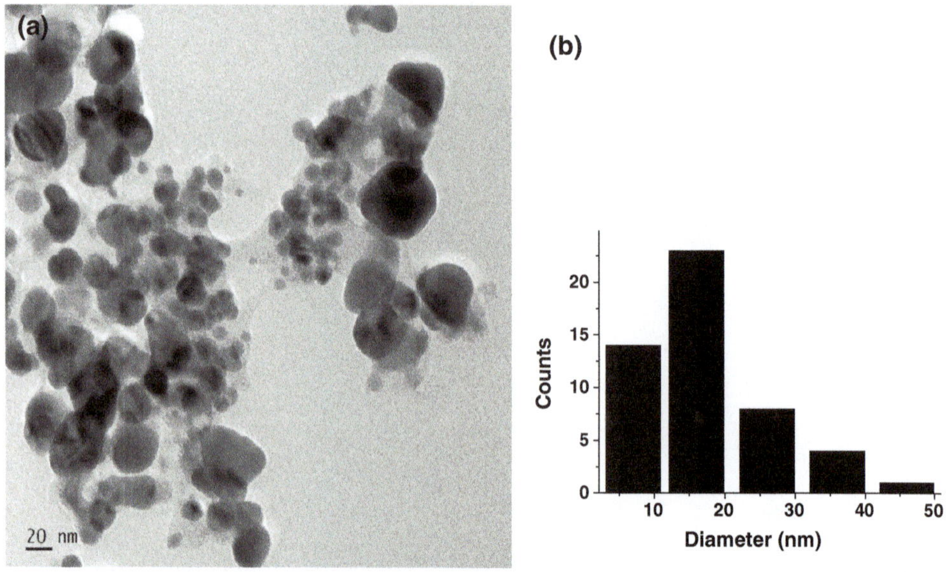

Fig. 8 TEM image of silver nanoparticles formed at 5 mM silver nitrate concentration (**a**) and the particle size distribution (**b**)

2E,4E-decadienyl)piperidine, 1-1-oxo-3(3,4-methylenedioxy-5-methoxyphenyl)-2Z-propenylpiperidine, 1-1-oxo-5(3,4-methylenedioxyphenyl)-2Z, 4E-pentadienylpyrrolidine, etc., to the synthesized silver nanoparticles. This may be the possible reason for the stability of synthesized silver nanoparticles.

TEM

Transmission electron microscopy provided a clear idea about the morphology and size of the synthesized silver nanoparticles. Comparison of TEM results has shown that the diameter of prepared nanoparticles in the solution was about 5–50 nm. Figures 4, 5, 6, 7 and 8 show the transmission electron micrograph of silver nanoparticles obtained from the proposed bioreduction method at various silver nitrate concentrations. At a silver nitrate concentration of 1 mM, silver nanoparticles with spherical morphology is obtained (Fig. 4). Each nanoparticle is individually remained without forming any agglomeration. The average particle size at this silver nitrate concentration is 8 nm. At a silver nitrate concentration of 2 mM almost similar morphology obtained as in the case of 1 mM concentration but the average particle diameter increased to 12 nm (Fig. 5). When the silver nitrate concentration was further increased, the size of the nanoparticles tends to decrease. Agglomeration of nanoparticles also observed at higher silver nitrate concentrations. At 3 mM silver nitrate concentration nanoparticles with almost spherical morphology were obtained with an average particle size of 18 nm (Fig. 6). The heterogeneity in particle size can be observed here. At 4 and 5 mM silver nitrate concentrations, the morphology of the nanoparticles becomes irregular higher tendency for agglomeration could be observed

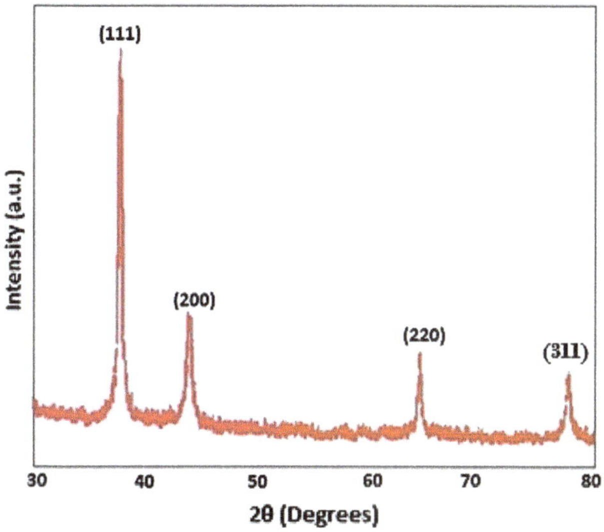

Fig. 9 Representative XRD pattern of synthesized silver nanoparticles at 2 mM silver nitrate concentration

(Figs. 7, 8). At 5 mM silver nitrate concentration the control over the particles size seemed to be completely lost and nanoparticles with varying diameters in the range of 5–50 nm obtained.

XRD

The XRD result (Fig. 9) shows that the silver nanoparticles formed are crystalline. A number of strong Bragg reflections can be seen which correspond to the (111), (200), (220), (311) reflections of face-centered cubic (fcc) silver. This is in agreement with the results obtained by other workers (Parikh et al. 2011; Theivasanthi and Alagar 2011; Martinez-Castanon et al. 2008). A sharp and strong diffraction peak centered at 38° was appeared, which can be indexed to the (1 1 1) reflection of the metallic silver with fcc structure (JCPDS File No. 04-0783). Comparably weak diffraction peaks at 44°, 66° and 77° in the pattern agree well with the (2 0 0), (2 2 0) and (3 1 1) reflections, respectively. The broadening of the peaks clearly indicates that the particles formed are in the nanoscale.

Table 1 Inhibitory zone diameter of disc diffusion assay

Samples (mM)	Inhibitory zone diameter (mm)	
	E. Coli	*S. aureus*
1	10.863	12.12
2	10.26	10.52
3	8.61	9.12
4	8.23	8.65
5	8.15	8.24
Positive control (Ciplofloxacin)	27.53	28.4

Antimicrobial activity

The disc diffusion assay or Kirby–Bauer method (Pal et al. 2007b), which is employed to assess preliminary antimicrobial activity of antimicrobial agents, was used to evaluate the antibacterial activity of silver nanoparticles synthesized at different silver nitrate concentrations. Silver nanoparticles synthesized at all the silver nitrate concentrations displayed antibacterial activity against both *E. coli* and *S. aureus* (Fig. 10). The diameter of inhibition zones around the disc containing silver nanoparticles in *E. coli* and *S. aureus* are given in Table 1. Nanoparticles prepared using 1 mM silver nitrate was the most efficient to inhibit both *E. coli* and *S. aureus*. Then there is a decrease in antibacterial activity as the concentration of silver nitrate increased. This can be correlated to the increase in particle size as the concentration of silver nitrate increased during synthesis. Antibacterial activity of silver nanoparticles was found to be dependent on the size of silver particles and as the size increases the antibacterial activity decreases (Panáček et al. 2006). All the synthesized nanoparticles have shown more antibacterial activity against *S. aureus*

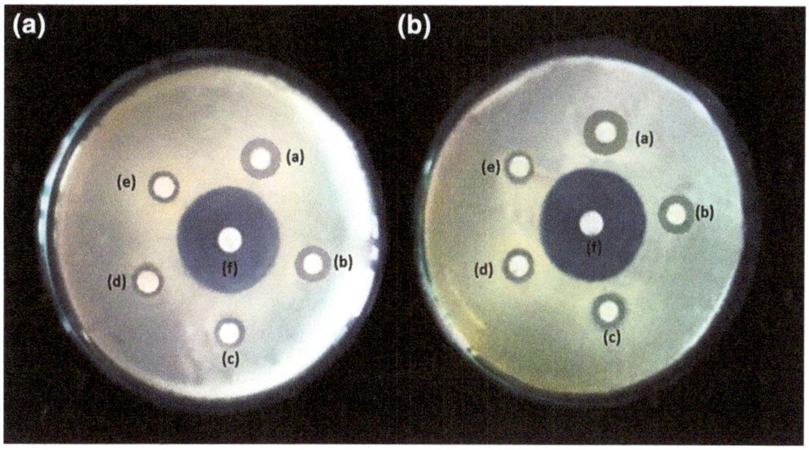

Fig. 10 Antimicrobial activity of synthesized silver nanoparticles against *E. coli* (plate **a**) and *S. aureus* (plate **b**). In both cases nanoparticles synthesized at 1, 2, 3, 4 and 5 mM silver nitrate concentrations are denoted by *a*, *b*, *c*, *d* and *e*, respectively. In both plates f denotes standard antibiotic disc (Ciplofloxacin)

than *E. coli* as reported by other workers (Sadeghi et al. 2010; Shrivastava et al. 2007), probably due to the difference in cell walls between Gram-positive and Gram-negative bacteria. Moreover, the synthesized nanoparticles enhance the therapeutic efficacy and strengthen the medicinal values of this plant. Antibacterial activity of black pepper (*P. nigrum* Linn.) and its mode of action on both Gram-negative and Gram-positive bacteria were already reported (Karsha and Lakshmi 2010).

Conclusions

A simple and facile process has been described for the preparation of silver nanoparticles starting from silver nitrate as a precursor and leaf extract of *P. nigram* as reducing as well as stabilizing agent. The characterization from UV–Visible, FTIR and TEM data supports the stability of the biosynthesized nanoparticles. The average particle size and morphology of the synthesized nanoparticles has been found to depend on the molar concentration of silver nitrate solution used. The FTIR studies indicate the capping of certain amide-containing compounds on the synthesized nanoparticles. The XRD analysis of the powder indicates that the synthesized nanoparticles are crystalline with fcc structure. The results obtained in this study are comparable with that obtained while chemical reducing agents like citrate of sodium or sodium borohydrate are used.

Acknowledgments The authors acknowledge Department of Science and Technology (DST), Government of India, New Delhi for the financial support through Nanomission and PURSE programs. Funding from UGC SAP, DST-FIST are also gratefully acknowledged.

References

Amaladhas Peter, Sivagami S, Akkini Devi T, Ananthi N, Priya Velammal S (2012) Biogenic synthesis of silver nanoparticles by leaf extract of *Cassia angustifolia*. Adv Nat Sci Nanosci Nanotechnol 3(4):045006

Amany A, El-Rab SFG (2012) Effect of reducing and protecting agents on size of silver nanoparticles and their anti-bacterial activity. Der Pharma Chemica 4(1):53–65

Augustine R, Rajarathinam K (2012) Synthesis and characterization of silver nanoparticles and its immobilization on alginate coated sutures for the prevention of surgical wound infections and the in vitro release studies. Int J Nano Dimens 2(3):205–212

Awwad AM, Salem NM, Abdeen AO (2013) Green synthesis of silver nanoparticles using carob leaf extract and its antibacterial activity. Int J Ind Chem 4(1):1–6

Bar H, Bhui DK, Sahoo GP, Sarkar P, Pyne S, Misra A (2009) Green synthesis of silver nanoparticles using seed extract of *Jatropha curcas*. Colloids Surf A Physicochem Eng Aspects 348:212–216

Chandran SP, Chaudhary M, Pasricha R, Ahmad A, Sastry M (2006) Synthesis of gold nanotriangles and silver nanoparticles using *Aloe vera* plant extract. Biotechnol Prog 22:577–583

Chanwitheesuk A, Teerawutgulrag A, Rakariyatham N (2005) Screening of antioxidant activity and antioxidant compounds of some edible plants of Thailand. Food Chem 92:491–497

Darroudi M, Ahmad MB, Abdullah AH, Ibrahim NA, Shameli K (2010) Effect of accelerator in green synthesis of silver nanoparticles. Int J Mol Sci 11:3898–3905

Duran N, Marcato PD, De Souza GIH (2009) Antibacterial effect of silver nanoparticles produced by fungal process on textile fabrics and their effluent treatment. J Biomed Nanotechnol 5:247–253

Ee GCL, Lim CM, Lim CK, Rahmani M, Shaari K, Bong CFJ (2009) Alkaloids from *Piper sarmentosum* and *Piper nigrum*. Nat Prod Res 23:1416–1423

Fayaz AM, Balaji K, Kalaichelvan PT, Venkatesan R (2009) Fungal based synthesis of silver nanoparticles—an effect of temperature on the size of particles. Colloids Surf B 74:123–126

Gardea-Torresdey JL, Gomez E, Peralta-Videa J, Parsons JG, Troiani HE, Jose-Yacaman M (2003) *Alfalfa sprouts*: a natural source for the synthesis of silver nanoparticles. Langmuir 19:357–1361

Hudson BJF (1990) Food antioxidants. Elsevier Applied Science, London, pp 65–98

Jain D, Kumar Daima H, Kachhwaha S, Kothari SL (2009) Synthesis of plant-mediated silver nanoparticles using papaya fruit extract and evaluation of their antimicrobial activities. Dig J Nanomater Biostruct 4:557–563

Justin Packia Jacob S, Finub JS, Narayanan A (2012) Synthesis of silver nanoparticles using *Piper longum* leaf extracts and its cytotoxic activity against Hep-2 cell line. Colloids Surf B Biointerfaces 91:212–214

Karim MR, Lim KT, Lee CJ, Bhuiyan MTI, Kim HJ, Park L-S, Lee MS (2007) Synthesis of core-shell silver–polyaniline nanocomposites by gamma radiolysis method. J Polym Sci Part A Polym Chem 45:5741–5747

Karsha P, Lakshmi VOB (2010) Antibacterial activity of black pepper (*Piper nigrum* Linn.) with special reference to its mode of action on bacteria. Indian J Nat Prod Resour 1:213–215

Kasibhatta R, Naidu MU (2007) Influence of piperine on the pharmacokinetics of nevirapine under fasting conditions: a randomized, crossover placebo-controlled study. Drugs RD 8:83–139

Khajuria A, Thusus N, Zutshi U, Bedi KL (1997) Antioxidant potential of piperine on oxidant induced alterations in rat intestinal lumen. Indian Drugs 34:557–563

Khajuria A, Thusu N, Zutshi U (2002) Piperine modulates permeability characteristics of intestine by inducing alterations in membrane dynamics: influence on brush border membrane fluidity, ultra-structure and enzyme kinetics. Phytomedicine 9(3):224–231

Kowshik M, Ashtaputre S, Kharrazi S, Vogel W, Urban J, Kulkarni SK, Paknikar KM (2003) Extracellular synthesis of silver nanoparticles by a silver-tolerant yeast strain MKY3. Nanotechnology 14:95–100

Lee EB, Shin KH, Woo WS (1984) Central nervous system depressant and anti-inflammatory activity of piperine. Arch Pharmacol Res 7:127–132

Liu S, Chen S, Avivi S, Gendanken A (2001) Synthesis of X-ray amorphous silver nanoparticles by the pulse sonoelectrochemical method. J Non Cryst Solids 283:231–236

Maillard M, Giorgo S, Pileni MP (2002) Silver nanodisks. Adv Mater 14(15):1084–1086

Majumdar AM, Dhuley JN, Deshmukh VK, Raman PH, Naik SR (1990) Antiinflammatory activity of piperine. Jpn J Med Sc Biol 43:95–100

Maneerung T, Seiichi T, Ratana R (2008) Impregnation of silver nanoparticles into bacterial cellulose for antimicrobial wound dressing. Carbohydr polym 72(1):43–51

Martinez-Castanon GA, Nino-Martinez N, Martinez-Gutierrez F, Martinez-Mendoza JR, Ruiz F (2008) Synthesis and antibacterial activity of silver nanoparticles with different sizes. J Nanopart Res 10(8):1343–1348

Mathew PJ, Mathew PM, Kumar V (2001) Graph clustering of *Piper nigram* L. (black pepper). Euphytica 118:257–264

McFarland J (1907) The nephelometer: an instrument for estimating the number of bacteria in suspensions for calculating the opsonic index and vaccines. J Am Med Assoc 49:1176–1178

Naik RR, Stringer SJ, Agarwal G, Jones S, Stone MO (2002) Biomimetic synthesis and patterning of silver nanoparticles. Nat Mater 1(3):169–172

Pal A, Shah S, Devi S (2007a) Preparation of silver, gold and silver–gold bimetallic nanoparticles in w/o microemulsion containing TritonX-100, colloids and surfaces. Physicochem Eng Aspects 302:483–487

Pal S, Kyung Y, Song Myong (2007b) J Appl Environ Microbiol 73(6):1712–1720

Panáček A, Kvitek L, Prucek R, Kolar M, Vecerova R, Pizurova N, Sharma VK, Nevečná T, Zboril R (2006) Silver colloid nanoparticles: synthesis, characterization, and their antibacterial activity. J Phys Chem B 110(33):16248–16253

Parikh RY, Ramanathan R, Coloe PJ, Bhargava SK, Patole MS, Shouche YS, Bansal V (2011) Genus-wide physicochemical evidence of extracellular crystalline silver nanoparticles biosynthesis by *Morganella* spp. PLoS One 6(6):e21401

Patel K, Kapoor S, Dave DP, Mukherjee T (2005) Synthesis of nanosized silver colloids by microwave dielectric heating. J Chem Sci 117:53–60

Patil RS, Kokate MR, Kolekar SS (2012) Bioinspired synthesis of highly stabilized silver nanoparticles using *Ocimum tenuiflorum* leaf extract and their antibacterial activity. Spectrochimica Acta Part A Mol Biomol Spectrosc 91:234

Philip Daizy (2010) Rapid green synthesis of spherical gold nanoparticles using Mangiferaindica leaf. Spectrochimica Acta Part A Mol Biomol Spectrosc 77:807–810

Pillai ZS, Kamat PV (2004) What factors control the size and shape of silver nanoparticles in the citrate ion reduction method? J Phys Chem B 108:945–951

Praveen Kumar K, Paul W, Chandra Sharma P (2011) Green synthesis of gold nanoparticles with *Zingiber officinale* extract: characterization and blood compatibility. Process Biochem 46:2007–2013

Rai AK, Singh R, Singh KN, Singh VB (2006) FTIR, Raman spectra and ab initio calculations of 2-mercaptobenzothiazole. Spectrochim Acta A Mol Biomol Spectrosc 63(2):483–90

Remita S, Fontaine P, Lacaze E, Borensztein Y, Sellame H, Farha R, Rochas C, Goldmann M (2007) X-ray radiolysis induced formation of silver nano-particles: a SAXS and UV–Visible absorption spectroscopy study. Nucl Instrum Methods Phys Res B 263:436–440

Rosemary MJ, Pradeep T (2003) Solvothermal synthesis of silver nanoparticles from thiolates. J Colloid Interface Sci 268: 81–84

Sadeghi B, Jamali M, Kia S, Amini NA, Ghafari S (2010) Synthesis and characterization of silver nanoparticles for antibacterial activity. Int J Nano Dimens 1(2):119–124

Saifuddin N, Wong CW, Nur Yasumira AA (2009) Rapid biosynthesis of silver nanoparticles using culture supernatant of bacteria with microwave irradiation. Eur J Chem 6:61–70

Salkar RA, Jeevanandam P, Aruna ST, Koltypin Y, Gedanken A (1999) The sonochemical preparation of amorphous silver nanoparticles. J Mater Chem 9:1333–1335

Sathishkumar M, Sneha K, Won SW, Cho CW, Kim S, Yun YS (2009) *Cinnamon zeylanicum* bark extract and powder mediated green synthesis of nano-crystalline silver particles and its bactericidal activity. Colloids Surf B Biointerfaces 73:332–338

Shanmugapriya K, Saravana PS, Payal H, Mohammed P, Williams SB (2012) Antioxidant potential of pepper (*Piper nigrum* Linn.) leaves and its antimicrobial potential against some pathogenic microbes. Indian J Nat Prod Resour 3(4):570–577

Shenya DS, Philip Daizy, Mathew Joseph (2012) Rapid green synthesis of palladium nanoparticles using the dried leaf of *Anacardium occidentale*. Spectrochim Acta Part A Mol Biomol Spectrosc 91:35–38

Shiv Shankar S, Ahmad A, Sastry M (2008) Geranium leaf assisted biosynthesis of silver nanoparticles. Biotechnol Prog 19(6):1627–1631

Shrivastava S, Bera T, Roy A, Singh G, Ramachandrarao P, Dash D (2007) Characterization of enhanced antibacterial effects of novel silver nanoparticles. Nanotechnol 18(22):225103

Song JY, Kim BS (2009) Rapid biological synthesis of silver nanoparticles using plant leaf extracts. Bioprocess Biosyst Eng 32:79–84

Srinivasan K (2007) Black pepper and its pungent principle-piperine: a review of diverse physiological effects. Crit Rev Food Nutr 47:735–748

Starowicz M, Stypula B, Banaoe J (2006) Electrochemical synthesis of silver nanoparticles. Electrochem Commun 8:227–230

Szczepanowicz K, Stefańska J, Socha PR, Warszyński P (2010) Preparation of silver nanoparticles via chemical reduction and their antimicrobial activity. Physicochem Probl Miner Process 45:85–98

Theivasanthi T, Alagar M (2011) Electrolytic synthesis and characterizations of silver nanopowder (arXiv preprint, arXiv:1111.0260)

Tripathi AK, Jain DC, Kumar S (1996) Secondary metabolites and their biological and medical activities of *Piper* species plants. J Med Aromat Plant Sci 18:302–321

Vanaja M, Annadurai G (2012) *Coleus aromaticus* leaf extract mediated synthesis of silver nanoparticles and its bactericidal activity. Appl Nanosci 3:217–223

Vankar PS, Shukla D (2012) Biosynthesis of silver nanoparticles using lemon leaves extract and its application for antimicrobial finish on fabric. Appl Nanosci 2(2):163–168

Varshney R, Bhadauria S, Gaur MS (2010) Biogenic synthesis of silver nanocubes and nanorods using sundried *Stevia rebaudiana* leaves. Adv Mater Lett 1:232–237

Velioglu YS, Mazza G, Gao L, Oomah BD (1998) Antioxidant activity and total phenolics in selected fruits, vegetables, and grain products. J Agric Food Chem 46:4113–4117

Vivekanandhan S, Christensen L, Misra M, Mohanty AK (2012) Green process for impregnation of silver nanoparticles into microcrystalline cellulose and their antimicrobial bionanocomposite films. J Biomater Nanobiotechnol 3:371–376

Wang G, Shi Ch, Zhao N, Du X (2007) Synthesis and characterization of Ag nanoparticles assembled in ordered array pores of porous anodic alumina by chemical deposition. Mater Lett 61:3795–3797

Wei K, Li W, Koike K, Pei Y, Chen Y, Nikaido T (2004) New amide alkaloids from the roots of *Piper nigrum*. J Nat Prod 67:1005–1009

Wei K, Li W, Koike K, Chen Y, Nikaido T (2005) Nigramides A-S, dimeric amide alkaloids from the roots of *Piper nigrum*. J Org Chem 70:1164–1176

Wiley BJ, Im SH, Li ZY, McLellan J, Siekkinen A, Xia Y (2006) Maneuvering the surface plasmon resonance of silver nanostructures through shape-controlled synthesis. J Phys Chem B 110:15666–15675

Xie Y, Ye R, Liu H (2006) Synthesis of silver nanoparticles in reverse micelles stabilized by natural biosurfactant. Colloids Surf A 279:175–178

Zhao JQ, Du GZ, Xiong YC, Wen YF, Bhadauria M, Nirala SK (2007) Attenuation of beryllium induced hepatorenal dysfunction and oxidative stress in rodents by combined effect of gallic acid and piperine. Arch Pharm Res 30: 1575–1578

Zhu JJ, Liao XH, Zhao XN, Chen HY (2001) Preparation of silver nanorods by electrochemical methods. Mater Lett 49:91–95

2

Enhanced photodynamic efficacy of PLGA-encapsulated 5-ALA nanoparticles in mice bearing Ehrlich ascites carcinoma

Maryam N. Shaker · Heba S. Ramadan · Moustafa M. Mohamed · Ahmed M. El khatib · Gamal D. Roston

Abstract Nanoparticles (NPs) fabricated from the biodegradable copolymer poly(lactic-*co*-glycolic acid) (PLGA) were investigated as a drug delivery system to enhance the photodynamic efficacy of 5-aminolevulinic acid (5-ALA) in mice bearing Ehrlich ascites carcinoma. The PLGA-encapsulated 5-ALA NPs were prepared using binary organic solvent diffusion method and characterized in terms of shape and particle size. The in vivo photodynamic efficiency in Ehrlich ascites-bearing mice was studied. The obtained particles were uniform in size with spherical shape of mean size of 249.5 nm as obtained by particle size analyzer and the in vitro release studies demonstrated a controlled release profile of 5-ALA. Tumor-bearing mice injected with PLGA-encapsulated 5-ALA NPs exhibited significantly smaller mean tumor volume, increased tumor growth delay compared with the control group and the group injected with free 5-ALA during the time course of the experiment. Histopathological examination of tumor from mice treated with PLGA-encapsulated 5-ALA NPs showed regression of tumor cells, in contrast to those obtained from mice treated with free 5-ALA. The results indicate that PLGA-encapsulated 5-ALA NPs are a successful delivery system for improving photodynamic activity in the target tissue.

Keywords Photodynamic therapy (PDT) · Nanoparticles (NPs) · Photosensitizer (PS) · 5-Aminolevulinic acid (5-ALA) · Poly-lactide-*co*-glycolide (PLGA) · Percentage of drug entrapment (PDE) · Drug loading · Drug release

M. N. Shaker (✉) · A. M. El khatib · G. D. Roston
Physics Department, Faculty of Science, Alexandria University, Alexandria, Egypt
e-mail: maryam.nabil@alex-sci.edu.eg

H. S. Ramadan · M. M. Mohamed
Biophysics Department, Medical Research Institute, Alexandria University, Alexandria, Egypt

Introduction

Photodynamic therapy also known as PDT, photo-radiation therapy, phototherapy, or photo chemotherapy is a promising treatment for a variety of oncological, cardiovascular, dermatological and ophthalmic diseases. PDT is emerging as an alternative modality for cancer therapy for the treatment of light accessible tumors (Saxena 2005).

Photodynamic therapy incorporates a photosensitizer (PS), light source, and oxygen. Individually, the components are nontoxic. The activated (irradiated) form of the photosensitizer is toxic at the site of action (Issa and Manela-Azulay 2010; Yoo et al. 2012). Among various photosensitizers, 5-aminolevulinic acid (5-ALA) is a precursor of a strong photosensitizing agent, protoporphyrin IX (PpIX) and has been used extensively in clinical applications, due to its safety, feasibility as an effective endogenous photosensitizer for fluorescence diagnosis, and its efficacy in PDT. However, owing to the poor penetration of ALA into skin, ALA–PDT is inappropriate for difficult-to-treat deep skin neoplasias (Donnelly et al. 2010). As most photosensitizers, 5-ALA is characterized by a high lipophilicity. Although interesting with respect to the selectivity for neoplastic tissue, the resulting insolubility in physiologically acceptable media makes the systemic administration of conventional PS problematic and restricts their medical applications (Castano et al. 2005; Konan et al. 2002). To overcome this limitation, PS have been associated with colloidal carriers such as liposomes, emulsions, microparticles or nanoparticles as reviewed by Konan et al. (2002). The interest in nanocarriers such as polymeric nanoparticles (NPs) to enable the production of an injectable suspension of PS has been increasing over recent years. NPs made of a polymeric matrix are a very stable and biocompatible drug delivery system.

Polymeric NPs have the potential to dramatically improve the diagnosis and treatment of tumor (Lee et al. 2012; Qian et al. 2008). The ability to fine-tune the properties of nanoparticles by controlling their components in the core or the shell region renders them useful and optimal for pharmaceutical purposes (Lee et al. 2012; Ferrari 2005). Among them, Poly(D,L-lactide-co-glycolide) (PLGA) NPs have been demonstrated as excellent carriers for imaging and targeting of tumor cells and for carrying anti-tumor drugs to tumor cells, due to their unique features including biocompatibility, biodegradability, tailor-made functionality, and the large interior cargo volume (Lee et al. 2004, 2012; Zhang et al. 2007).

5-Aminolevulinic acid was simply conjugated to PLGA NPs for enhancing the nanoparticles stability in the PDT. The conjugation of 5-ALA to PLGA NPs was designed to perform photodynamic tumor therapy by generating reactive oxygen species (e.g., singlet oxygen: $^1O_2$) from light absorption. The released reactive oxygen species lead to irreversible damage to the treated tumor cells (Oh et al. 2012; Park et al. 2011).

In this study, we prepared PLGA NPs by emulsion solvent evaporation method compared with the binary organic solvent diffusion (BOSD) method, and studied the possibility of nanoparticles as drug carriers by using 5-ALA as a model of photosensitizer drug. The drug loading contents, loading efficiency, changes of particle size, and physicochemical properties of PLGA NPs after the drug is entrapped into the nanoparticles are investigated. Also the use of laser and PLGA-encapsulated 5-ALA NPs as possible dual-mode enhancement agents in mice bearing Ehrlich ascites carcinoma is explored.

Materials and methods

Materials

Poly (D,L-lactide-co-glycolide) (PLGA, lactide/glycolide molar ratio = 85/15, average Mw 50,000–75,000), 5-aminolevulinic acid hydrochloride, poly (vinyl alcohol), 87–89 % hydrolyzed, average Mw 13,000–23,000 all those was provided by Sigma Chemical Co. USA, and all other chemicals and solvents were of reagent grade.

PLGA nanoparticles preparation

PLGA NPs were prepared as follows.

Spontaneous emulsification solvent diffusion (SESD)

Briefly, 60 mg of PLGA (85:15) were first dissolved in 25 mL of acetone; this organic phase was poured at the rate of 0.4 mL/min into 30 mL phosphate buffer (pH 7.4, 5 mM) containing 100 mg PVA as the hydrophilic surfactant under moderate magnetic stirring of 2,000 rpm. Finally, the organic solvents were evaporated under reduced pressure at 58 °C, and the final volume of the aqueous suspension was adjusted to 10 mL (Lamprecht 1999).

But this method resulted in nanoparticles of large mean size of about 233 nm.

Binary organic solvent diffusion (BOSD) (Jiang et al. 2003)

1. 60 mg of PLGA (85:15) was first dissolved in 17 mL of acetone in a glass beaker then 8 mL of ethanol is added drop-wise using a 22-gauge micro-infusion syringe pump at rate of 0.4 mL/min into 30 mL double distilled water containing 100 mg PVA as the hydrophilic surfactant with stirring of 2,000 rpm with the help of a high-speed magnetic stirrer.
2. After addition of the organic solvent, the solution stirred on the high-speed magnetic stirrer for about one extra day.

Purification

3. The organic solvents of acetone and ethanol were evaporated under reduced pressure at 58 °C using rotary evaporator and then centrifuged at 12,000 rpm for 1 h.
4. The precipitate was washed with 25 mL distilled water. This centrifugation and washing process was repeated three times to remove the PVA and residual organic solvents.

Freeze-drying

5. The precipitate was frozen by vacuum freeze-drying machine (Lyophilizer) (Model/CRYODOS-50, Spain) at 230 V, 50 Hz, at a pressure of 26.5 Pa and 5 % saccharose was used as cytoprotectant. The lyophilized particles were then characterized.

Characterization of PLGA nanoparticles

Particle size (Thakkar et al. 2004)

The particle size distribution of the nanoparticles was determined by laser light scattering on a Beckman Coulter Particle Size Analyzer (N5 submicron particle size analyzer, Japan) equipped with a He–Ne Laser beam at a fixed scattering angle of 11.1°. The nanoparticles suspension were added to the sample dispersion unit containing the stirrer and stirred to reduce the aggregation between the particles, and laser obscuration range was maintained at 15–20 %. The mean particle size was measured after performing the experiment in triplicate.

Shape of PLGA nanoparticles (Gupta and Jabrail 2006)

The physical size and shape of the prepared nanoparticles were determined by transmission electron microscope (JEOL-JSM-6360LA, Japan). For this purpose, the particle suspension was diluted 10 times with distilled water and deposited drop-wise onto a 400-mesh copper grid coated with carbon film, and were allowed to dry in the air before studying by microscope.

Fourier-transform infrared spectroscopy (FTIR) (Bhumkar and Pokharkar 2006)

Fourier-transform infrared (FTIR) spectra of the blank PLGA NPs were obtained on a Shimadzu FTIR-8400S (Tokyo, Japan). The pellets were prepared by grinding 4–8 mg of samples with 200 mg potassium bromide and by a compression using Shimadzu compressor (Japan) tablets were prepared then fixed on holder to be examined. The spectra were scanned over the wave number range of 4,600–500 cm^{-1} in which the region of 4,600–600 cm^{-1} are called the functional group region.

Thermo-gravimetric analysis (TGA) (Bhumkar and Pokharkar 2006)

The compositions of the prepared nanoparticles were determined by thermo-gravimetric analysis (TGA) (TGA 50, Shimadzu, Japan). The weight of the dried nanoparticles for the TGA experiment was between 13 and 16 mg. The experiments were conducted using closed aluminum pans with a cover hole. The sample was examined under a nitrogen flow rate of 20 mL/min at a scan rate 10 °C/min with the range from 25 to 600 °C.

Determination of encapsulation efficiency (Thakkar et al. 2004)

The percentage of drug entrapment (PDE) was calculated according to Eq. (1) as following:

$$\text{PDE}\ (\%) = \frac{\text{Actual drug content}}{\text{Theoretical drug content}} \times 100\,\% \tag{1}$$

Theoretical drug content was determined assuming that the entire drug present in the PLGA solution used gets entrapped in nanoparticles and no loss occurs.

The actual drug content in the 5-ALA loaded PLGA nanoparticles was quantitatively determined by the following procedure:

First standard curve from known concentration of 5-ALA was constructed. Freeze-dried nanoparticles were weighted and suspended into methanol and vigorously stirred for 3 h and sonicated for 15 min. Resulting solution was centrifuged at 12,000 rpm for 20 min. The supernatant was taken for measurement of drug concentration using ultraviolet (UV)/visible spectrophotometer (new Jenway Model 6315) at 670 nm. The drug loading efficiency (DLE) was presented by Eq. (2) (Nah et al. 2000).

$$\text{DLE} = \frac{\text{Weight of drug in nanoparticles}}{\text{Weight of nanoparticles}} \tag{2}$$

In vitro drug release

The in vitro drug release study was performed in triplicate. Briefly, 15 mg freeze-dried drug-loaded nanoparticles was suspended in 50 mL phosphate buffer saline (PBS), pH 7.4, in a glass flask and sonicated using Bransonic Ultrasonic B-220 for 10 min to get a homogenous suspension. This suspension was kept in shaking on a mechanical shaking bath shaken horizontally at 100 cycles/min, with temperature adjusted to 37 °C. At different time intervals, a sample volume of 1.5 mL was transferred and centrifuged at 10,000 rpm for 15 min. A quantity of 1 mL of the supernatant was taken for laser analysis at 670 nm. 1 mL of fresh PBS buffer was added to the remaining 0.5 mL sample. 1.5 mL of the suspension was agitated vigorously by vortexing and replaced in the flasks. The cumulative percentage of 5-ALA released from the nanoparticles at a certain time point was obtained using Eq. (3).

$$\text{Cumulative } 5-\text{ALA released }\% = 100 - (5-\text{ALA remaining})\,\% \tag{3}$$

The same procedure was repeated where the sample were irradiated to a diode laser at 650 nm laser beam from a He–Ne laser (photon company) as measured by spectrometer, the power of each laser on the sample was 150 mW. During the experiments, the laser beam was delivered to the tubes containing samples with an irradiation spot of 7 mm diameter. The fluency rate was 3.90 mW/mm^3. Laser irradiation lasted for 11 min, and then a sample volume of 1.5 mL was transferred and centrifuged at 10,000 rpm for 15 min. A quantity of 1 mL of the supernatant was taken for laser analysis at 670 nm. 1 mL of fresh PBS buffer was added to the remaining

0.5 mL sample. 1.5 mL of the suspension was agitated vigorously by vortexing and replaced in the flasks.

Animal care

In vivo studies were done on a total of 80 male Swiss albino mice 6–8 weeks old, weighing 16–25 g, purchased from animal house at National Cancer Institute, Cairo University. Ten animals each were housed in plastic cages in a well-ventilated room (25 ± 2 °C) with a relative humidity of (43 ± 3 %) and maintained on a high protein basal diet.

PDT protocol

Animals were acclimated to the environment for at least 2 weeks before onset of the experiment. Ehrlich ascites carcinoma cells, 2×10^6 mammary in origin, diluted approximately (1–4) in 0.2 mL normal saline. The mouse's leg was shaved; the thigh skin was pulled up with forceps, and then inoculated subcutaneously.

The treatment protocol started with intralesional injection on the eighth day after tumor implantation. Animals were anesthetized with ketamine/xylazine/acepromazine and the dose of PLGA NPs administrated to each mouse was equivalent to 20 mg/kg which were suspended in saline and sonicated for 10 min before injection to get a homogenous suspension then directly injected into the tumor interstitial coordinates using insulin syringe. Within 2 min of injection to limit particle diffusion beyond the tumor boundaries, animal tumors were extra-corporeally exposed for 11 min to 670 nm visible diode laser at power of 150 mW (Casas et al. 1999) for each other day for 1 week (three times/week). During the experiment, laser beam was delivered to the tumor with an irradiation spot directly on the skin over the tumor of 7 mm diameter; fluency rate was 3.90 mW/mm^2 at room temperature (25 ± 2 °C). During the procedure, a homemade thermo sensitive sensor connected to computer is to monitor the thermal transient of tumor and measure its temperature using a across a software interface the circuit of the sensor.

Experimental design

The animals injected with tumor cells were subdivided into four main groups each comprising 20 animals as follows:

Group I mice injected with free 5-ALA at dose of 20 mg/kg.

Group II mice injected with PLGA-encapsulated 5-ALA NPs at dose equivalent of 20 mg/kg.

Group III mice injected with blank PLGA NPs.

Group IV mice injected with 50 μL normal saline and serve as the control untreated group.

Each group was divided into two subgroups each of 10 mice; subgroup A which did not receive laser irradiation. While subgroup B irradiated with laser beam for 11 min each other day (three times/week) for 1 week.

Inhibitory effect of 5-ALA loaded on nanospheres

To study the cytotoxic effect of 5-ALA loaded on PLGA nanospheres irradiated with laser on the tumor in different animal groups, the following parameters were estimated as follows.

Tumor volume

The tumor was measured every other day between the skin surface layers. Tumor volume was estimated using Eq. (4) as follows (Tang et al. 2007).

$$\text{Volume of tumor} = D_{\max} \times (D_{\min})^2/2. \tag{4}$$

where $D_{\max}$ maximum diameter, $D_{\min}$ minimum diameter

Tumor volume ratio (TVR) (Chen et al. 2004)

$$\text{TVR} = \frac{\text{Tumor volume after treatment}}{\text{tumor volume before treatment}} \tag{5}$$

Inhibition ratio of tumor (Tang et al. 2007)

$$\begin{aligned}\text{Inhibition ratio of tumor} = 1 - \{ &(\text{average tumor volume at beginning} - \text{average tumor}\\ &\text{volume at the end of treatment})/(\text{average tumor volume}\\ &\text{of control at beginning} - \text{average}\\ &\text{tumor volume of control at the end of treatment})\}\end{aligned} \tag{6}$$

Tumor growth relative rate (TGRR) (Venslauskas and Batiuskaite 2003)

The tumor growth at a given moment is dependent on the size of tumor cell population at the previous time interval and that the tumor growth rate at different time interval is not steady. On the other hand, under the influence of anticancer drug tumor growth could be stopped or reduced. These findings suggest proposing a new quantitative measure-tumor growth relative rate (TGRR) defined as:

$$r_t = \left(v_t - v_{(t-1)}\right)/\Delta t * \left(1/v_{(t-1)}\right) \tag{7}$$

where v_t is tumor volume at the present time interval, $v_{(t-1)}$ is the tumor volume at the previous time interval, Δt is the time interval between the measurements so r_t is the value of the tumor volume change per day expressed by part of the former volume.

Biochemical analysis

Blood samples were collected when the mice were killed. Sera were harvested by centrifugation at 3,500 rpm and stored in aliquots at −80 °C until assayed. Serum TNF-α were determined using enzyme linked immunosorbent assay (ELISA) kit for the quantitative detection of TNF-α (eBioscience, Europe) according the producer's protocol.

Histopathological study

Light microscopic examination On dissection, tumor samples were excised from animals of each group and preserved in a 10 % formalin solution for histopathological sections. The fixed tissues were embedded in paraffin; sections 3–5 μm thick were obtained, deparaffinized, dehydrated in ethanol (50–100 %), and cleared with xylene. The extent of induced cytotoxicity, including inflammation, cell apoptosis, cell necrosis, steatosis and others was evaluated by assessing the morphological changes in tumor sections stained with hematoxylin and eosin (H&E) under light microscope.(Bancroft et al. 1996)

Results and discussion

Characterization of blank PLGA and PLGA-encapsulated 5-ALA NPs

Particle size

Figure 1a shows the particle size distribution curve, as measured by particle size analyzer, for blank PLGA NPs prepared by SESD method, sharp distribution range of nanoparticles, with 90 % of spheres in size range of 130–290 nm with mean size of 233.1 nm and only 10 % were oversized at diffraction angle of 11.1° with baseline error of 0.01 % and poly-dispersity index of 0.856 (<1) which indicates homogeneous distribution of the particles. These results were in accordance with Murakami et al's. (1999) study regarding production of PLGA NPs suggested that the original SESD method sometimes causes a severe aggregation in the particle formation process, resulting in formation of nanoparticles with larger size. This indicates that the SESD method should be modified. While Fig. 1b shows the particle size distribution curve for blank PLGA NPs prepared by BOSD method, sharp distribution range of nanoparticles, with 90 % of spheres in size range of 100–140 nm with mean size of 124.6 nm and only 10 % were oversized at diffraction angle of 11.1° with baseline error of 0.01 % and poly-dispersity index of 0.784 which also indicates homogeneous distribution of the particles. The binary organic solvents used were acetone and ethanol at the volume fraction of (2/1). The sizes of PLGA (85:15) NPs changes from 233 to 124 nm. Since ethanol has larger affinity to water than acetone and accelerates the diffusion of PLGA during dispersion process. This benefits the fast formation of NPs and avoids the aggregation of small particles, which results in the production of NPs with larger size (Wu et al. 2011).

Figure 1c shows the particle size distribution curve for PLGA-encapsulated 5-ALA NPs, it shows distribution range of nanoparticles, with 90 % of spheres in size range of 160–360 nm with mean size of 249.5 nm and only 10 % were oversized at diffraction angle of 11.1° with baseline error of 0.02 % and poly-dispersity index of 18.480 (more than 1) which indicates inhomogeneous distribution of the particles. Here, the increase in size is due to the addition of 5-ALA in NPs.

Transmission electron microscope (TEM)

Figure 2a shows TEM examination for PLGA NPs prepared by SESD method; it revealed the presence of completely spherical PLGA NPs with smooth surfaces, and no agglomeration formed.

While Fig. 2b shows TEM examination for blank PLGA NPs prepared by BOSD method, it revealed the presence of

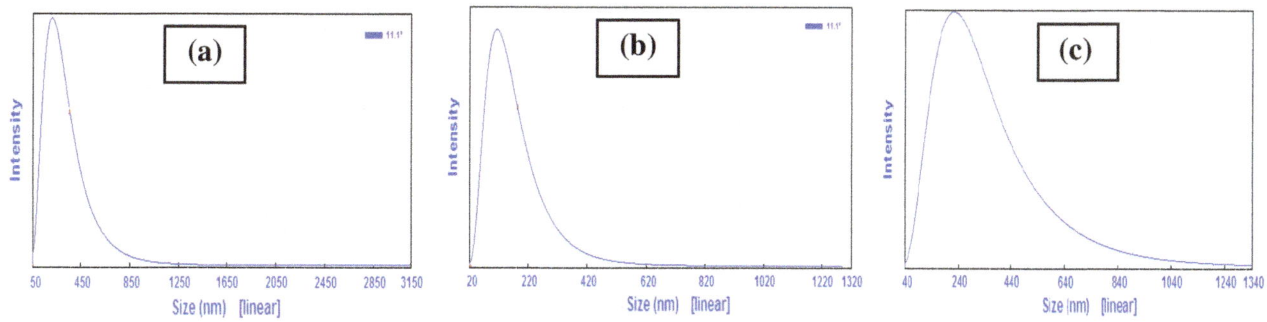

Fig. 1 The particle size distribution curve for **a** free PLGA NPs prepared by SESD method. **b** Free PLGA NPs prepared by BOSD method. **c** PLGA-encapsulated 5-ALA NPs prepared by BOSD method

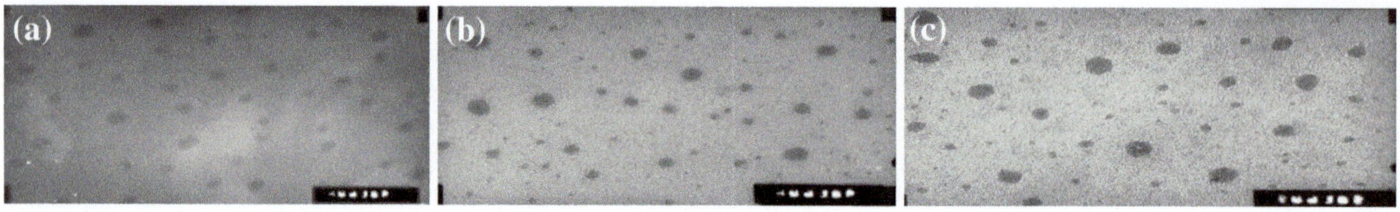

Fig. 2 TEM of **a** blank PLGA NPs prepared by SESD. **b** Blank PLGA NPs prepared by (BOSD). **c** TEM of PLGA-encapsulated 5-ALA NPs. All in ×40

Fig. 3 Fourier-transform infrared spectra of PLGA polymer, blank PLGA NPs and PLGA-encapsulated 5-ALA NPs

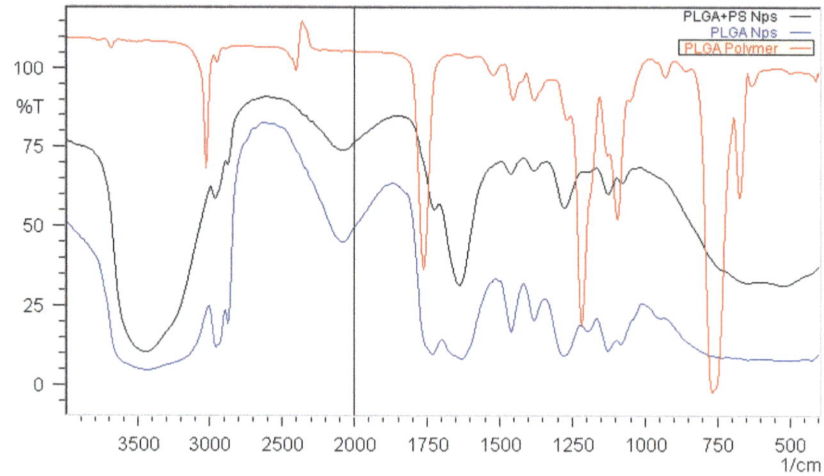

completely spherical PLGA NPs with low level of agglomeration or adhesion that occurred among particles and high level of uniformity. But for PLGA-encapsulated 5-ALA NPs, Fig. 2c showed increased in size compared to free PLGA NPs and no change in the spherical shape which mean that the spherical shape of blank PLGA NPs has not been compromised.

Fourier-transform infrared spectroscopy (FTIR)

For the IR spectrum of PLGA polymer, the characteristic absorption bands appeared at wave number 3,023 and 2,401 cm^{-1} (CH bend), 1,759 cm^{-1} (C=O ester), 1,520, 1,451, 1,378 cm^{-1} (CH_3), 1,215, 1,126, 1,093 cm^{-1} (C–O stretch), 748, 671 cm^{-1} (CH– bend) while the band at 4,214 and 3,685 cm^{-1} belongs to the OH group of the water molecule. While for blank PLGA NPs stabilized by PVA, it shows the characteristic absorption bands at wave number 3,423 cm^{-1} (OH group), 2,954 cm^{-1} (CH alkyl group), 2,866, 2,075 cm^{-1} (CH bend), 1,730 cm^{-1} (C=O ester), 1,631, 1,456, 1,379 cm^{-1} (CH_3), 1,276, 1,195, 1,126 cm^{-1} (C–O stretch), 1,080 cm^{-1} (CH_2), 945, 732 cm^{-1} (CH– bend). Also for PLGA-encapsulated 5-ALA NPs, it shows the same characteristic absorption bands at 3,438, 2,960, 2,867, 2,071, 1,724, 1,635, 1,458, 1,379, 1,274, 1,197, 1,126, 1,078, 642, 520 cm^{-1}, in addition to a new band at 1,033 cm^{-1} that C–O bond.

Comparing the obtained IR spectra for PLGA polymer, blank PLGA NPs and PLGA-encapsulated 5-ALA NPs as in Fig. 3 with the IR spectra characteristic for PLGA, PVA and 5-ALA shown in the literature, it confirmed that the obtained NPs are composed of PLGA and PVA with no interaction with 5-ALA. It was confirmed that there are no major shifting as well as no loss of functional peaks between the spectra of drug, polymer and drug-loaded NPs (3,438, 2,960, 2,867, 2,071, 1,724, 1,635, 1,458, 1,379, 1,274, 1,197, 1,126, 1,078, 642, 520 cm^{-1}). Besides the characteristic groups for PLGA, the large bands observed between 3,550 and 3,200 cm^{-1} are linked to the stretching O–H from the intermolecular and intramolecular hydrogen bonds. The vibrational band observed between 2,840 and 3,000 cm^{-1} refers to the stretching C–H from alkyl groups and the peaks between 1,750 and 1,735 cm^{-1} are due to the stretching C–O and C–O from acetate group remaining from PVA (Mansur et al. 2008; Andrade et al. 2008).

Our results coincide with the results obtained by dos Reis et al. Their results indicated that FTIR spectrum of pure PVA sample clearly reveals the major peaks associated with poly(vinyl alcohol) that can be observed C–H broad alkyl stretching band (ν = 2,850–3,000 cm^{-1}) and typical strong hydroxyl bands for free alcohol (nonbonded–OH stretching band at ν = 3,600–3,650 cm^{-1}), and hydrogen bonded band (ν = 3,200–3,570 cm^{-1}) (Budhian et al. 2007; Jain 2000). Intramolecular and intermolecular

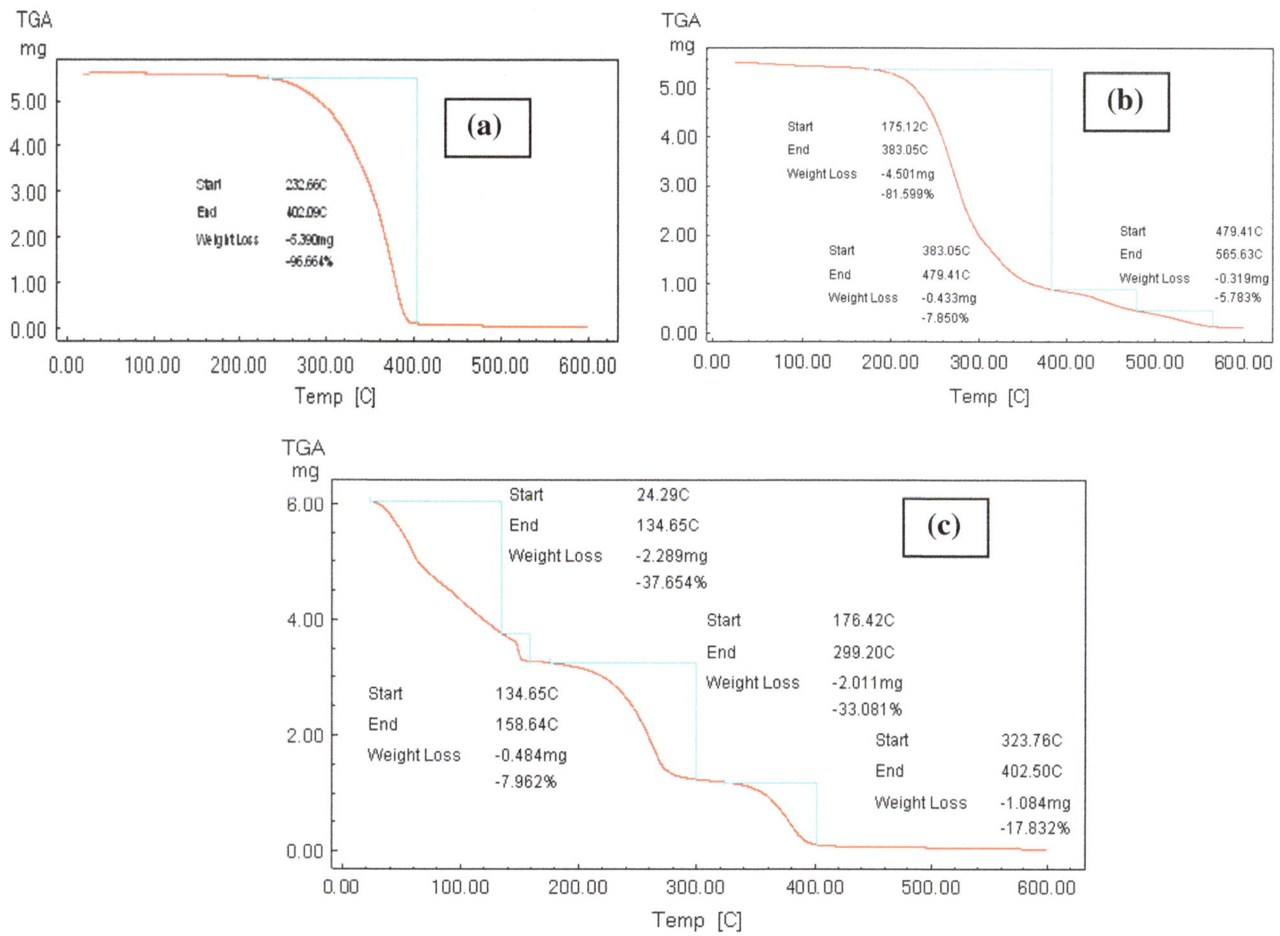

Fig. 4 TGA of **a** PLGA powders. **b** Blank PLGA NPs. **c** PLGA-encapsulated 5-ALA NPs

hydrogen bonding are expected to occur among PVA chains due to high hydrophilic forces. An important absorption peak was verified at $\nu = 1{,}142\ \mathrm{cm}^{-1}$. This band has been used as an assessment tool of PVA structure because it is a semi-crystalline synthetic polymer able to form some domains depending on several process parameters (John Coates and Meyers 2000; Elizabeth Fonseca dos Reis et al. 2006).

Our results are in agreement with the results obtained by Jawahar et al. They designed PLGA NPs for ramipril as anti-hypertensive agent confirmed no loss of functional groups of PLGA when processed into NPs loaded with ramipril (1,652, 1,701, 1,743, 2,866, 1,323 cm^{-1}) (Jawahar et al. 2009).

Thermo-gravimetric analysis (TGA)

Figure 4a shows TGA for PLGA powder which shows only one strong peak at 232–402 °C with 96.66 % loss of weight.

While Fig. 4b shows TGA for free PLGA NPs which shows three peaks, the larger one at 175–383 °C with 81.599 % loss of weight, next by shorter peak at 383–479 °C with 7.85 % loss of weight and the shortest peak at 479–566 °C with 5.783 % loss of weight.

While Fig. 4c shows TGA for PLGA-encapsulated 5-ALA NPs which shows four intermediate peaks at 24–135 °C with 37.654 % loss of weight, next by a very short peak at 135–159 °C with 7.962 % loss of weight and return to a stronger peak at 176–299 °C with 33.081 % loss of weight and ended by an intermediate peak at 324–402.5 °C with 17.832 % loss of weight.

Characterization of drug loading

According to Eqs. 1 and 2, PDE was 66.8 %, while DLE was 12.74 %. The drug load was 127.5 μg drug/mg nanoparticles for PLGA NPs containing 5-ALA as shown in Table 1.

These results confirm the result of Jawahar et al. They prepared PLGA NPs containing ramipril a hypertensive agent with high PDE of 68.28 % (2009), while Budhian et al. (2008) prepared PLGA NPs (200 nm) encapsulating hydrophilic drug haloperidol with drug content of 0.2 %.

Table 1 PDE and DLE for PLGA-encapsulated 5-ALA NPs

PLGA concentration (mg)	Initial concentration of 5-ALA (mg)	Encapsulated concentration of 5-ALA (mg)
26	5	3.34
PDE %	66.8 %	
DLE %	12.74 %	

The high 5-ALA incorporation efficiency in PLGA can be attributed to several factors:

Firstly, the hydrophobic nature of PLGA molecules makes it relatively easy to entrap hydrophobic 5-ALA into PLGA NPs.

Secondly, the hydrophobic nature of 5-ALA results in a minimum loss of the drug to the external aqueous phase during the interfacial deposition preparation process.

The high entrapment efficiency proves that the affinity between 5-ALA and PLGA matrix was far stronger than the affinity between 5-ALA and external aqueous phase.

Thirdly, the drug incorporation method and variables such as, incubation time, temperature, stirring, solvent, presence of surfactants, MW of PLGA, L:G ratio of PLGA, PLGA concentration, removal rate of organic solvent, etc., are also factors which may influence drug entrapment efficiency (Kader et al. 1998) and this in accordance with the explanation of Zeng and Sun (2009).

Figure 5a showed initial burst release profile of 54 % of 5-ALA from PLGA NPs after 24 h followed by a slower release rate that lasts for 12 days; this biphasic pattern is characterized by initial burst release followed by a slow and continuous release rate as shown in the figure. This pattern of release is still the same even after laser irradiation at power intensity of 150 mW/cm^2 for 11 min as in Fig. 5b which shows the initial fast release of 5-ALA was probably due to 5-ALA, which was adsorbed or close to the surface of the nanoparticles and diffused out of the nanoparticles into the released medium. The large surface-to-volume ratio of the nanoparticles geometry can be the contributing factor for the initial fast release pattern. The slow release of 5-ALA can be attributed to the diffusion of 5-ALA entrapped in the core of the nanoparticles into the aqueous media surrounding the nanoparticles. The release pattern observed in our study has been similar to those reported by Zeng and Sun (2009).

In vivo study

Temperature monitoring

At time of laser irradiation, all subgroups were monitored for temperature which revealed an increase in temperature of all subgroups compared with that before laser irradiation. While temperature in the subgroups that were injected with free 5-ALA and PLGA-encapsulated 5-ALA NPs increased significantly compared with the control group and the subgroup injected with blank PLGA NPs after laser irradiation (Fig. 6). This increased temperature confirmed by a necrosis occurs in the mice which was irradiated by laser and had taken free 5-ALA than that is not irradiated with laser where necrosis was found on histological examination of sections from each tumor. The areas of

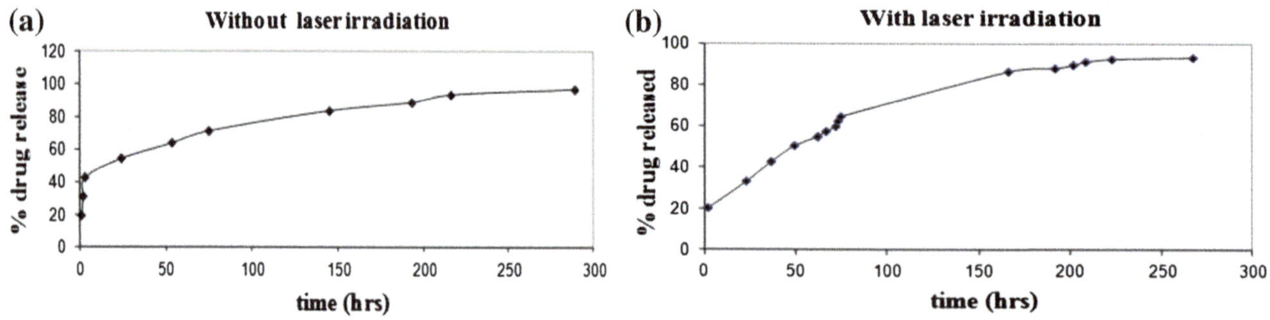

Fig. 5 Pattern of drug release of 5-ALA from PLGA NPs **a** without laser irradiation, **b** with laser beam for 11 min of power intensity 150 mW/cm^2 as a function of time

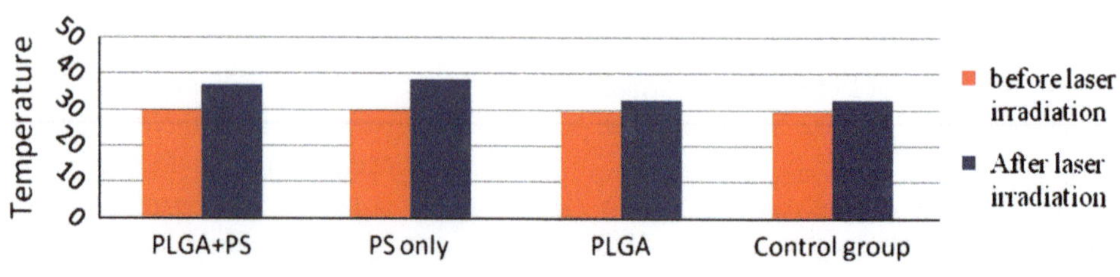

Fig. 6 Temperature change in the subgroups before and after irradiation with laser beam at 670 nm (150 mW/cm^2) for 11 min

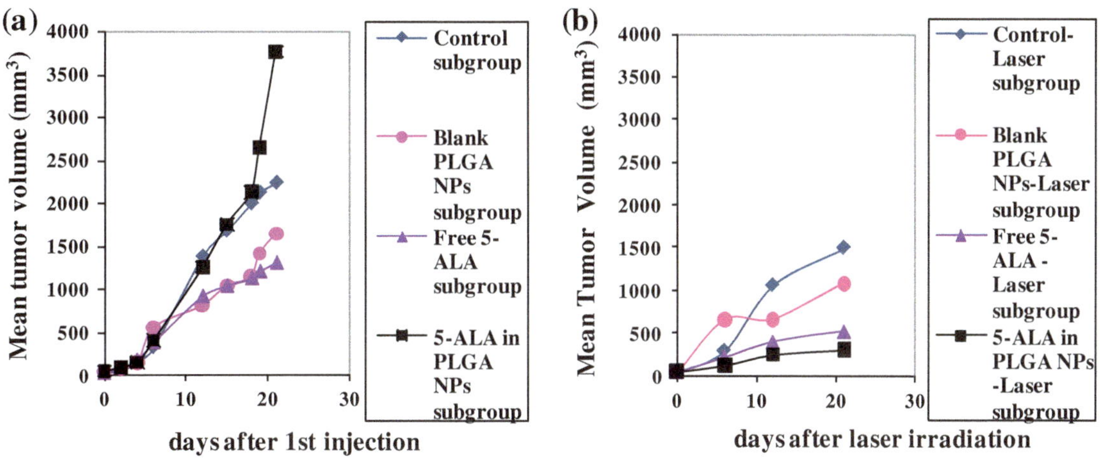

Fig. 7 Tumor growth curves for the studied subgroups **a** without laser irradiation. **b** With laser irradiation at 670 nm (150 mW/cm^2) for 11 min

necrosis in irradiated tumors were larger (higher rank) than areas of necrosis of non-irradiated tumors (Egger et al. 1997).

Tumor growth and growth rate

All tumor volumes were normalized to that at the first day of treatment. All the four studied subgroups that were not irradiated with laser showed rapid tumor growth as shown in Fig. 7a. Mice bearing Ehrlich ascites carcinoma injected with free 5-ALA and irradiated with laser beam for 11 min showed reduced tumor volume compared to mice injected saline (control group) Fig. 7b. A severe decrease in tumor volume were seen in the subgroup of mice bearing Ehrlich ascites carcinoma, injected with PLGA-encapsulated 5-ALA NPs and irradiated with laser beam Fig. 7b. These results are in accordance with the results obtained by Lee et al. (2012) and Fadel et al. (2010). They formulated nanoparticles from the biodegradable copolymer poly (lactic-co glycolic acid) (PLGA) for targeting of zinc (II) phthalocyanine (ZnPc) for PDT in Ehrlich ascites carcinoma bearing mice. Tumor-bearing mice injected with ZnPc NPs exhibited significantly smaller mean tumor volume, increased tumor growth delay when compared with the control group and the group injected with free ZnPc during the time course of the experiment. Histopathological examination of tumor from animals treated with PLGA-ZnPc showed regression of tumor cells, in contrast to those obtained from animals treated with free ZnPc (Fadel et al. 2010).

Tumor growth inhibition

No significant difference in tumor volume was found among subgroups before treatment. For all the studied subgroups with no laser irradiation, the tumor grew rapidly in mice of the control group, in which the mean tumor volume after 21 days was 60.99 times greater than that at the beginning as seen in Table 2.

For subgroups irradiated with laser beam, the tumor grew rapidly in mice of the control group, in which the mean tumor volume after 21 days was 39.72 times greater than that at the beginning as shown in Table 3. In contrast, tumor growth was less in volume in mice injected with free 5-ALA and PLGA-encapsulated 5-ALA NPs. The group treated with PLGA-encapsulated 5-ALA NPs showed a

Table 2 Tumor volume, tumor volume ratio (TVR) and inhibition ratio of tumor after treatment without laser irradiation

	Control	Blank PLGA	Free 5-ALA	5-ALA PLGA NPs
Tumor volume (mm^3)				
Before treatment	36.73	45.22	33.47	46.57
After 21 days of treatment	2,240.32	1,655.258	1,311.954	3,767.16
Tumor volume ratio (TVR)	60.99	36.60	39.19	80.89
Inhibition ratio of tumor after treatment (%)		26.9	42	−68

Table 3 Tumor volume, tumor volume ratio (TVR) and inhibition ratio of tumor after treatment in the presence of laser irradiation

	Control	Blank PLGA	Free 5-ALA	5-ALA PLGA NPs
Tumor volume (mm^3)				
Before treatment	37.58	39.75	44.21	45.74
After 21 days of treatment	1,492.92	1,082.7	530.76	302.3
Tumor volume ratio (TVR)	39.72	27.24	12	6.61
Inhibition ratio of tumor after treatment (%)		28.33	66.6	82.4

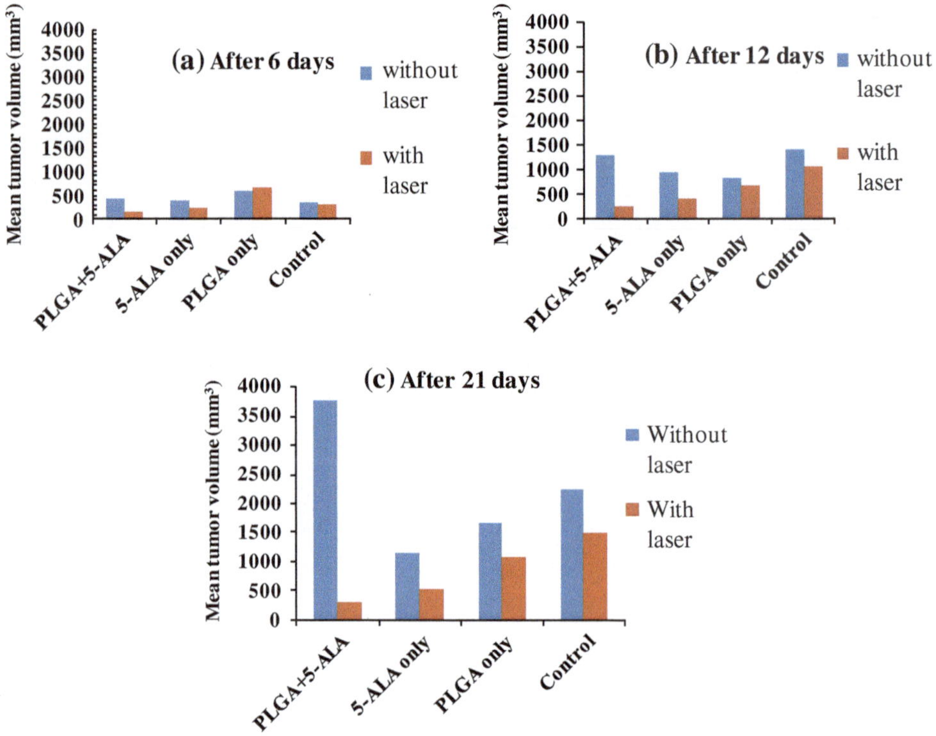

Fig. 8 Mean tumor volume for all studied subgroups in the absence and presence of laser irradiation. **a** After 6 days of treatment. **b** After 12 days. **c** After 21 days

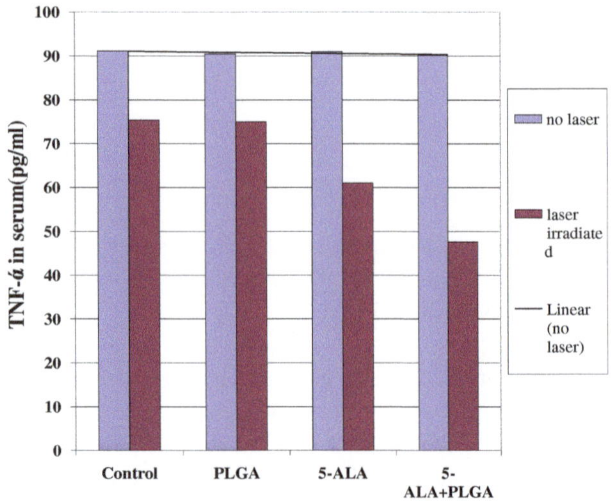

Fig. 9 Biochemical analysis by TNF-α in blood serum for all subgroups with and without laser irradiation

statistical significance difference in tumor volume compared to that treated with blank PLGA NPs and also with free 5-ALA as shown in Fig. 8.

Biochemical analysis

Figure 9 shows the mean serum levels of TNF-α in mice bearing Ehrlich ascites carcinoma with no laser irradiation were non-significantly changed ($P = 0.115$). While those injected with free 5-ALA and PLGA-encapsulated 5-ALA NPs and irradiated with laser showed significant reduction of serum levels of TNF-α ($P = 0.00$) compared with those injected with saline or blank PLGA NPs and irradiated with laser. These results confirmed the presence of necrosis and/or apoptosis which is confirmed by histopathological examination.

Macroscopic analysis and histological studies of paraffin section micrograph

Subgroups not treated with laser shows entirely viable tumors formed of sheets of polymorphic hyperchromatic malignant cells with bizarre giant forms and no evidence of necrosis was identified as in Fig. 10a, but tumor begins to be enhanced by exposing to laser beam 11 min 3 times per week for 1 week as it shows an increase number of large cells with visculated nuclei, reduction of multinucleate cell and no infiltrating lymphocytes cells Fig. 10b. For subgroups injected with PLGA NPs and irradiated to laser, it shows increase number of apoptotic cells characterized as pycnotic nuclei with basophilic cytoplasm other with hallow of vacuolated cytoplasm (Fig. 10c). For subgroups injected with 5-ALA only and irradiated with laser, it shows increase number of apoptotic cells regenerative cells characterized by vesicular nuclei and basophilic cytoplasm cells normal blood vessel large necrotic areas as shown in Fig. 10d. But for subgroup injected with PLGA-encapsulated 5-ALA NPs and irradiated with laser exhibited an improved in vivo PDT for the tumor site where it shows foci area of apoptotic cells surrounded by area of necrosis

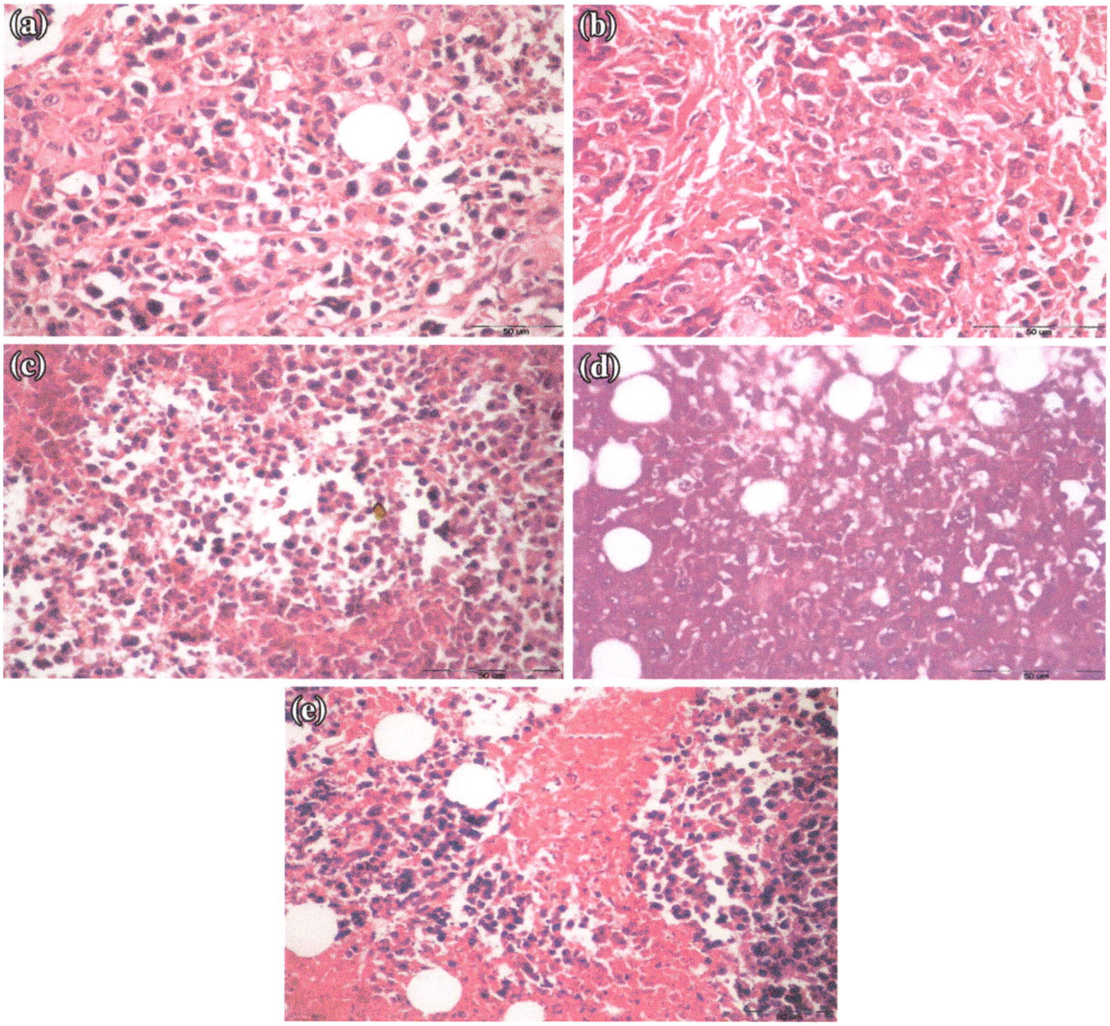

Fig. 10 Paraffin section micrograph for subgroups of Ehrlich ascites carcinoma **a** untreated. **b** Irradiated with laser beam. **c** Injected with free PLGA NPs and exposed to laser. **d** Injected free photosensitizer (5-ALA) and exposed to laser. **e** Injected PLGA-encapsulated 5-ALA NPs and exposed to laser

and regenerative eosinophilic cytoplasm cells with visculated nuclei (Fig. 10e).

The administration of PLGA NPs resulted in a significant growth inhibition of the KB tumor bearing nude mice injected only once with PLGA NPs encapsulating chlorine e6 (Ce6) (a commonly used photosensitizer) or free Ce6 where 24 h post-injection, tumor sites were locally illuminated at a light intensity of 5.2 mW/cm^2 using a 670 nm laser source for 40 min. The tumor volume in the nude mice treated with PLGA NPs was approximately 1.5 or 3 times smaller than those treated with free Ce6 or PBS (control). When compared to high-dose administration of free Ce6 (2.5 mg/kg), low-dose administration of NP1 NPs (equivalent Ce6 0.5 mg/kg body) were found to be more efficient for tumor regression (Lee et al. 2012). Also our results coincide with the results obtained by Fadel et al. (2010).

Conclusion

As well as this study reports that entirely spherical shaped PLGA nanospheres of (124 nm) mean size were efficiently prepared by BOSD method rather than Spontaneous Emulsification Diffusion (SESD) method, also the resultant PLGA NPs allowed structural stability upon photosensitizer loading expressed in the volumetric expansion indicated by the increase in the mean particle size from empty NPs to 5-ALA loaded ones. 5-ALA as a model of photosensitizer was loaded to PLGA NPs with no deviation in FTIR and TGA pattern in reference to PLGA polymer. 5-ALA showed bimodal release pattern from PLGA NPs which is not affected by laser irradiation. This work nominates PLGA-encapsulated 5-ALA NPs for in vivo application of PDT studies, strongly supporting that laser-irradiated PLGA-encapsulated 5-ALA NPs 5-ALA can effectively improve tumor PDT treatment options in terms of tumor volume reduction, growth regression and histopathology. We expect PLGA NPs to be tailor-made customizable for the desired functionality advancing the efficiency of anti-tumor management. So the feasibility of drug-encapsulated nanoparticles systems has been demonstrated and the advantages of such implanted systems are well known and include mainly: (1) minimizing systemic effects; (2) protecting the encapsulated drug from the physiologic medium; (3) obtaining high and sustained concentrations of the drug at the tumor site the particle chosen in this study was nanospheres. Indeed, if the particles are small enough, then they can be injected either through a needle into a discreet and functional area of the tissue or directly into a tumor that cannot be restricted because its location in turn will increase efficacy of PLGA-encapsulated 5-ALA nanoparticles in mice bearing Ehrlich ascites carcinoma.

References

Andrade G, Barbosa-Stancioli EF, Mansur AA, Vasconcelos W, Mansur HS (2008) Small-angle X-ray scattering and FTIR characterization of nanostructured poly (vinyl alcohol)/silicate hybrids for immunoassay applications. J Mater Sci 43:450–463

Bancroft JD, Stevans A, Tuener DR (1996) Theory and practice of histopathological techniques, 4th edn. Churchill Livingstone, Edinburgh

Bhumkar DR, Pokharkar VB (2006) Studies on effect of pH on cross-linking of Chitosan with sodium tripolyphosphate: a technical note. AAPS PharmSciTech 7(2):Article 50

Budhian A, Siegel SJ, Winey KI (2007) Haloperidol-loaded PLGA nanoparticles: systematic study of particle size and drug content. Int J Pharm 336:367–375

Budhian A, Siegel SJ, Winey KI (2008) Controlling the in vitro release profiles for a system of haloperidol-loaded PLGA nanoparticles. Int J Pharm 346(1–2):151–159

Casas A, Fukuda H, Meiss R, Batlle AM (1999) Topical and intratumoral photodynamic therapy with 5-aminolevulinic acid in a subcutaneous murine mammary adenocarcinoma. Cancer Lett 141(1-2):29–38

Castano AP, Demidova TN, Hamblin MR (2005) Mechanisms in photodynamic therapy: part three-photosensitizer pharmacokinetics, biodistribution, tumor localization and modes of tumor destruction. Photodiagn Photodyn Ther 2:91–106

Chen J, Ling R, Yao Q, Wang L, Ma Z, Li Y, Wang Z, Xu H (2004) Enhanced antitumor efficacy on hepatoma-bearing rats with adriamycin-loaded nanoparticles administered into hepatic artery. World J Gastroenterol 10:1989–1991

Coates J, Meyers RA (eds) (2000) Interpretation of infrared spectra, a practical approach. Encyclopedia of Analytical Chemistry. Wiley, Chichester, pp 10815–10837

Donnelly RF, Morrow DI, Fay F, Scott CJ, Abdelghany S, Singh RR, Garland MJ, Woolfson AD (2010) Microneedle-mediated intradermal nanoparticle delivery: potential for enhanced local administration of hydrophobic pre-formed photosensitisers. Photodiagnosis Photodyn Ther 7:222–231

Elizabeth Fonseca dos Reis, Campos FS, Lage AP, Leite RC, Heneine LG, Vasconcelos WL, Lobato ZIP, Mansur HS (2006) Synthesis and characterization of poly (Vinyl Alcohol) hydrogels and hybrids for rMPB70 protein adsorption. Mater Res 9(2):185–191

Egger NG, Schoenecker JA Jr, Gourley WK, Motamedi M, Anderson KE, Weinman SA (1997) Photosensitization of experimental hepatocellular carcinoma with protoporphyrin synthesized from administered delta-aminolevulinic acid: studies with cultured cells and implanted tumors. J Hepatol 26.4:913–920

Fadel M, Kassab K, Fadeel DA (2010) Zinc phthalocyanine-loaded PLGA biodegradable nanoparticles for photodynamic therapy in tumor-bearing mice. Lasers Med Sci 25(2):283–292

Ferrari M (2005) Cancer nanotechnology: opportunities and challenges. Nat Rev Cancer 5:161–171

Gupta KC, Jabrail FH (2006) Effects of degree of deacetylation and cross-linking on physical characteristics, swelling and release behavior of chitosan microspheres. Carbohydr Polym 66:43–54

Issa MC, Manela-Azulay M (2010) Photodynamic therapy: a review of the literature and image documentation. An Bras Dermatol 85(4):501–511

Jain RA (2000) The manufacturing techniques of various drug loaded biodegradable poly(lactide-co-glycolide) (PLGA) devices. Biomaterials 21:2475–2490

Jawahar N, Eagappanath T, Venkatesh N, Jubie S, Samanta MK (2009) Preparation and characterisation of PLGA-nanoparticles containing an anti-hypertensive agent. Int J Pharm Tech Res 1(2):390–393

Jiang XY, Zhou CS, Tang KW (2003) Preparation of PLA and PLGA nanoparticles by binary organic solvent diffusion method. J Central South Univ Technol 10(3):202–206

Kader A, Davis PJ, Kara M, Liu H (1998) Drug targeting using low density lipoprotein (LDL): physicochemical factors affecting drug loading into LDL particles. J Control Release 55:231–243

Konan YN, Gurny R, Allemann E (2002) State of the art in the delivery of photosensitizers for photodynamic therapy. J Photochem Photobiol B 66(2):89–106

Lamprecht A, Ubrich N, Hombreiro Perez M, Lehr C, Hoffman M, Maincent P (1999) Biodegradable monodispersed nanoparticles prepared by pressure homogenization-emulsification. Int J Pharm 184.1:97–105

Lee SJ, Jeong JR, Shin SC, Kim JC, Chang YH, Chang YM, Kim JD (2004) Nanoparticles of magnetic ferric oxides encapsulated with poly (d,l latide-coglycolide) and their applications to magnetic resonance imaging contrast agent. J Magn Magn Mater 272:2432–2433

Lee DJ, Park GY, Oh KT, Oh NM, Kwag DS, Youn YS, Oh YT, Park JW, Lee ES (2012) Multifunctional poly (lactide-co-glycolide) nanoparticles for luminescence/magnetic resonance imaging and photodynamic therapy. Int J Pharm 434(1–2):257–263

Mansur HS, Sadahira CM, Souza AN, Mansur AAP (2008) FTIR spectroscopy characterization of poly (vinyl alcohol) hydrogel with different hydrolysis degree and chemically cross linked with glutaraldehyde. Mater Sci Eng 28(4): 539–548

Murakami H, Kobayashi M, Takeuchi H, Kawashima Y (1999) Preparation of poly(DL-lactide-co-glycolide) nanoparticles by modified spontaneous emulsification solvent diffusion method. Int J Pharm 187(2):143–152

Nah J, Jeong Y, Koh J (2000) Drug release from nanoparticles of poly(DL-lactide-co-glycolide). Korean L Chem Eng 17(2): 230–236

Oh NM, Kwag DS, Oh KT, Youn YS, Lee ES (2012) Electrostatic charge conversion processes in engineered tumor-identifying polypeptides for targeted chemotherapy. Biomaterials 33:1884–1893

Park SY, Baik HJ, Oh YT, Oh KT, Youn YS, Lee ES (2011) A smart polysaccharide/drug conjugate for photodynamic therapy. An gew Chem Int Ed Engl 50:1644–1647

Qian X, Peng XH, Ansari DO, Yin-Goen Q, Chen GZ, Shin DM, Yang L, Young AN, Wang MD, Nie S (2008) In vivo tumor targeting and spectroscopic detection with surface-enhanced Raman nanoparticle tags. Nat Biotechnol 26:83–90

Saxena V (2005) Development of polymeric nanoparticle delivery system for indocyanine green. PhD thesis submitted to the faculty of Graduate Division, College of Pharmacy and Allied Health Professions, Saint John University, Jamaica

Tang T, Zheng J, Chen B, Li H, Li X, Xue K, Ai X, Zou S (2007) Effects of targeting magnetic drug nanoparticles on human cholangiocarcinoma xenografts in nude mice. Hepatobiliary Pancreat Dis Int 6:303–307

Thakkar H, Sharma RK, Mishra AK, Chuttani K, Murthy RSR (2004) Efficacy of chitosan microspheres for controlled intra-articular delivery of celecoxib in inflamed joints. JPP 56:1091–1099

Venslauskas MS, Batiuskaite D (2003) Irregular solid tumor growth: estimation of electrochemotherapy efficacy. Acta Med Lituanica 10:28–32

Wu L, Zhang J, Watanabe W (2011) Physical and chemical stability of drug nanoparticles. Adv Drug Deliv Rev 63(6):456–469

Yoo JJ, Kim C, Chung CW, Jeong Y, Kang DH (2012) 5-aminolevulinic acid-incorporated poly(vinyl alcohol) nanofiber-coated metal stent for application in photodynamic therapy. Int J Nanomed 7:1997–2005

Zeng QB, Sun M (2009) Poly (lactide-co-glycolide) nanoparticles as carriers for norcantharidin. Mater Sci Eng C 29(3):708–713

Zhang N, Chittasupho C, Duangrat C, Siahaan TJ, Berkland C (2007) PLGA nanoparticle-peptide conjugate effectively targets intercellular cell-adhesion molecule-1. Bioconjug Chem 19:145–152

3

Biosynthesis of silver nanoparticles from the marine seaweed *Sargassum wightii* and their antibacterial activity against some human pathogens

N. Shanmugam · P. Rajkamal · S. Cholan · N. Kannadasan · K. Sathishkumar · G. Viruthagiri · A. Sundaramanickam

Abstract In this paper, we have reported on biological synthesis of nano-sized silver and its antibacterial activity against human pathogens. The nanoparticles of silver were formed by the reduction of silver nitrate to aqueous silver metal ions during exposure to the extract of marine seaweed *Sargassum wightii*. The optical properties of the obtained silver nanoparticles were characterized using UV–visible absorption and room temperature photoluminescence. The X-ray diffraction results reveal that the synthesized silver nanoparticles are in the cubic phase. The existence of functional groups was identified using Fourier transform infrared spectroscopy. The morphology and size of the synthesized particles were studied with atomic force microscope and high-resolution transmission electron microscope measurements. The synthesized nanoparticles have an effective antibacterial activity against *S. aureus*, *K. pneumoniae*, and *S. typhi*.

Keywords Ag nanoparticles · *Sargassum wightii* · Antibacterial activity · Cubic phase · AFM

Introduction

In recent years the nanoparticles have consumed ample space in the field of marine bio-nanotechnology. It is believed that nanoparticles are clusters of atoms in the size range of 1–100 nm (Williams 2008). Nanoparticles are classified primarily into two types, viz organic and inorganic nanoparticles. The nanoparticles of carbon are called the organic nanoparticles. Magnetic nanoparticles, noble metal nanoparticles (platinum, gold and silver) and semiconductor nanoparticles (titanium dioxide, zinc oxide and zinc sulfide) are classified as inorganic nanoparticles (Kathiresan and Asmathunisha 2013). The metal nanoparticles have magnetic, electronic and optical properties, which make their usage in different fields like medicine, agriculture and electronics (Rai et al. 2012). Among the metal nanoparticles, silver nanoparticles have gained remarkable consideration owing to their physicochemical properties (Elechiguerra et al. 2005).

Nanoparticles of silver are widely used in water filters (Jain and Pradeep 2005), bio sensors (Chen et al. 2007), antibacterial activity (Venkatpurwar and Pokharkar 2011), anti-HIV activity (Elechiguerra et al. 2005) and in controlling plant pathogens (Krishnaraj et al. 2012). The biosynthetic method employing marine seaweed extract has received more attention as being simple, eco-friendly and less time consuming compared to usual chemical and physical methods.

In the present study, the extracellular synthesis of silver nanoparticles by the brown seaweed *Sargassum wightii* and their antibacterial effects against some selected human pathogens are reported.

Materials and methods

Sample collection

The green seaweed *S. wightii* was collected from Mandapam costal region (78°8′E, 9°17′N) in Gulf of manner, the southeast coast of India. Immediately after the collection, the

N. Shanmugam (✉) · P. Rajkamal · S. Cholan · N. Kannadasan · K. Sathishkumar · G. Viruthagiri
Department of Physics, Annamalai University, Annamalai Nagar, Chidambaram 608 002, Tamilnadu, India
e-mail: quantumgosh@rediffmail.com

A. Sundaramanickam
CAS in Marine Biology, Faculty of Marine Sciences, Annamalai University, Parangipettai, Tamil Nadu, India

samples were transferred to the laboratory in new plastic bags containing natural sea water to prevent desiccation.

Materials

Silver nitrate ($AgNO_3$) was purchased from Merck. The glass wares used in this experimental work were acid washed. Ultrapure water was used for all dilution and sample preparation.

Preparation of seaweeds extract

The seaweed was washed thoroughly thrice with distilled water and was shade dried for 10 days. Fine powder of the seaweed was obtained by using kitchen blender. The seaweed powder was sterilized at 120 °C for 15 min, 20 g of powder was taken and mixed with 100 ml of milli Q water and kept in boiling water bath at 60 °C for 20 min. The extracts were filtered with Whatman No. 1 filter paper and stored in refrigerator at 4 °C for further studies.

Synthesis of AgNPs

For the biosynthesis of silver nanoparticles, 5 ml of seaweed extract was mixed with 50 ml of $AgNO_3$ solution and incubated at 28 °C for 24 h. The bio-reduction of $AgNO_3$ into AgNps can be confirmed visually by the change in color from colorless to reddish brown (Fig. 1).

Characterization techniques

Characterization of nanoparticles is important to understand and control nanoparticles synthesis and applications (Al-Warthan et al. 2010). The formation of AgNPs was confirmed by sampling the reaction mixture at regular intervals and the absorption maximum was scanned by UV–visible spectra, in a range of wavelength between 400 and 550 nm using LAMDA 25 PERKIN ELMER spectrometer. The X-ray diffraction (XRD) patterns of the silver nanoparticles were recorded using X'PERT PRO diffractometer with Cu-Kα radiation ($\lambda = 1.5406$ Å). The crystallite size was estimated using the Scherer equation. Fluorescence measurements of the synthesized products with different incubation time were carried out using VARIAN Spectrometer. The presence of functional groups in seaweed extract and synthesized AgNPs was identified by SHIMADZU-8400 FTIR Spectrometer using KBr pellet technique. The morphology of the product was observed by Nano Surf Easy Scan 2 Atomic Force Microscope (AFM) measurement. High-resolution transmission electron microscopy (HRTEM) analysis was performed using JEDL 3010 HRTEM to study the morphology and size of the Ag nanoparticles.

Fig. 1 *A* Silver nitrate ($AgNO_3$) solution and *B* color change during the reduction of $AgNO_3$ into AgNPs by the extract of *S. wightii* after 24 h of incubation

Antibacterial activity

Silver is known for its antibacterial properties and has been used for years in the medical field for antimicrobial applications and even has shown to prevent HIV binding to host cells (Nino-Martinez et al. 2008; Alt et al. 2004; Russel and Hugo 1994; Lee et al. 2007; Jeong et al. 2005). In addition, silver has been used in water and air filters to eliminate microorganisms (Chou et al. 2005, Jin et al. 2007). In the present work, the synthesized Ag nanoparticles in aqueous were tested against bacterial pathogens by agar disc diffraction method (Karting et al. 1991). The pathogens namely *P. aeruginosa*, *V. cholerae*, *K. pneumoniae*, *S. aureus*, *E. coli*, *S. pneumoniae* and *S. typhi* were obtained from Raja Sir Muthaiya Medical College, Annamalai University. About 150 CFU/ml of inoculums was swabbed onto MH-agar plates uniformly and allowed to dry in a sterile environment. Sterile disc of 6 mm (HIMEDIA) were loaded with 30 µl of test solution ($AgNO_3$ and Ag nanoparticles). Ampicillin (10 mg in 1 ml) was used as positive control. The plates were incubated at 25 °C for 3 days to measure zone of inhibition. The mean was calculated by performing the experiments in triplicates.

Results and discussion

Crystal structure analysis and determination of crystallite size

Figure 2 shows the XRD pattern of silver nanoparticles synthesized after 120 h of incubation. The pattern has three main diffraction features corresponding to (1 1 1), (2 0 0) and

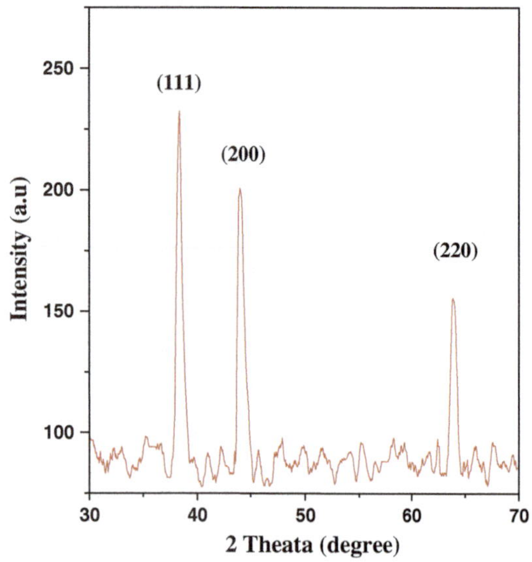

Fig. 2 XRD patterns of silver nanoparticles synthesized after 120 h of incubation

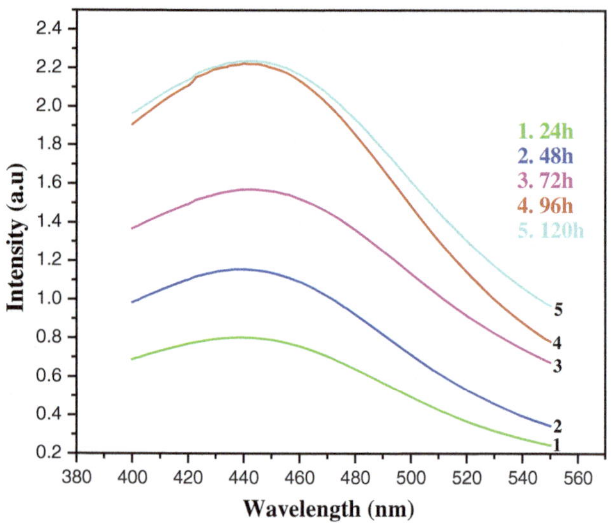

Fig. 3 UV–visible absorption spectra of silver nanoparticles with different incubation time

(2 2 0) planes and all the three peaks can be indexed to standard cubic phase of silver (JCPDS card No 04-0783). No reflection peaks corresponding to nitrate ions and other impurities were noted in this pattern, indicating the high purity of the final product. Moreover, the obtained reflections are sharp and high in intensity which reveals that the synthesized nanoparticles are well crystalline. In addition, the lattice constant of the synthesized cubic phase of silver is found to be '$a = 4.083$ Å'. This value is consistent with the JCPDS value ($a = 4.086$ Å). The size of the nanocrystals was determined from the Debye–Scherrer formula $D = 0.9\lambda/\beta\cos\theta$, where D is the crystallite size, λ is the wavelength of the incident X-ray (1.5406 Å), θ is the Bragg's angle and β is the full width at half maximum (FWHM). From the X-ray line broadening the crystallite size of the synthesized products is estimated around 17 nm ($r = \pm 10$ %).

Optical absorption studies

The optical properties of metallic nanoparticles depend on its surface plasmon resonance, where the plasmon refers to the collective oscillation of the free electrons within the metallic nanoparticles. It is obviously known that the plasmon resonant peaks and line widths are sensitive to the size and shape of the nanoparticles, the metallic species and the surrounding medium.

UV–visible absorption is one among the most important techniques to identify the formation of metal nanoparticles, provided surface plasmon resonance exists for metal (Binupriya et al. 2010). Figure 3 shows the UV–Vis absorption spectra of reaction mixtures studied at different levels of incubation time (24, 48, 72, 96 and 120 h). From the absorption spectra, it is noted that Ag surface plasmon resonance band occurs at about 439 nm and the intensity of absorption increases steadily as a function of incubation time. The interaction of electromagnetic radiation having wavelength smaller than the size of the Ag nanoparticles leads to polarization of the free conduction electrons with respect to the much heavier ionic core of Ag nanoparticles. Therefore, an electron dipolar orientation is set up and resulted in surface plasmon absorption band. The maximum absorption of 441 nm was obtained for the mixture with the incubation time of 120 h. The red shift of 2 nm may reduce the band gap and result in increased size of Ag nanoparticles.

The photoluminescence of silver metal and that of noble metal is generally attributed to electronic transition between the upper d band and conduction SP band (He et al. 2001). Room temperature PL of Ag nanoparticles was recorded using an excitation wavelength of 325 nm. As shown in Fig. 4, a broad emission peaking at 449 nm is observed in the emission spectrum due to metal–ligand charge transfer absorption (Xu et al. 2006). Furthermore, the intensity of the emission peak increases with increase of reaction time with a small red shift in peak position. This indicates that the concentration of silver increases with reaction time.

Functional group analysis

To understand the existence of different functional groups responsible for the bio-reduction of $AgNO_3$ into Ag Nanoparticles, FTIR studies were carried out for *S. wightii* and synthesized silver nanoparticles (Fig. 5a, b). As shown in Fig. 5a, FTIR spectrum of seaweed shows

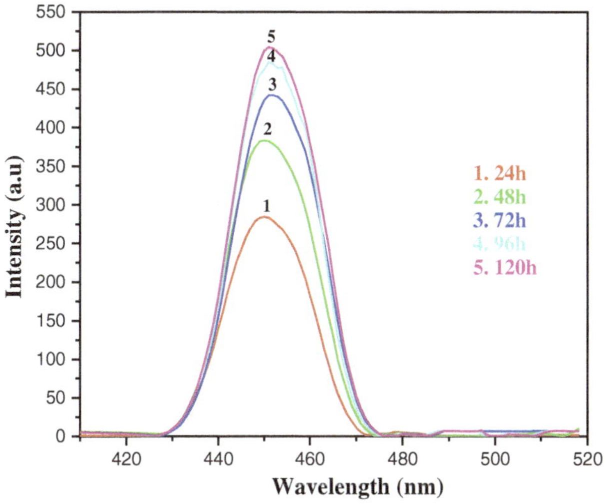

Fig. 4 PL emission spectra of silver nanoparticles with different incubation time

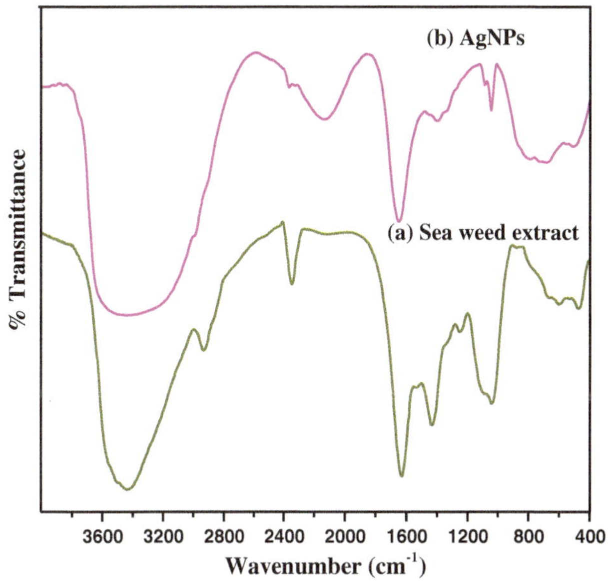

Fig. 5 FTIR spectra of *a* seaweed extract and *b* silver nanoparticles (AgNPs)

different major peaks positioned at 3,433, 2,931, 1,630, 1,087, 1,032, 661, 599 and 468 cm^{-1}. The presence of peak at 3,433 cm^{-1} could be due to O–H group in polyphenols or proteins/enzymes or polysaccharide (Song et al. 2009; Li et al. 2007; Susanto et al. 2009). A small peak observed at 2,931 cm^{-1} is due to CH-stretching of alkanes. A sharp and intense band observed at 1,630 cm^{-1} may be ascribed to the stretching vibration of the (NH)=O group. The bands positioned at 1,087 and 1,032 cm^{-1} are due to C–N stretching vibration of aliphatic amines. The observed bands ranging between 661 and 468 cm^{-1} are due to α-glucopyranose rings deformation of carbohydrates (Yang et al. 2009). On the other hand, FTIR spectrum of synthesized Ag nanoparticles (Fig. 5b) shows the presence of major peaks at 3,435, 2,931, 1,649, 1,085, 1,045, and 680 cm^{-1} which are associated with –OH stretching vibrations, C–H of alkanes, amide 1 band of proteins, C–N stretching vibration of amines, C–N stretching vibration of aliphatic amines and α-glucopyranose rings deformation of carbohydrates, respectively. After bio-reduction, there is a shift in the absorption band of 1,630–1,649 cm^{-1}, 1,087–1,085 cm^{-1} and 1,032–1,045 cm^{-1}. The shifting of the band from 1,630 to 1,649 cm^{-1} may be due to the binding of (NH)C=O group with the nanoparticles. The (NH)C=O groups within the cage of cyclic peptides are involved in stabilizing the nanoparticles. Thus, the peptides may play an important role in the reduction of $AgNO_3$ into Ag nanoparticles. The above-mentioned shift was observed in *codium capitatum* (Kannan et al. 2013).

Particles morphology (AFM and HRTEM measurements)

The surface morphology and size of the Ag nanoparticles harvested after 120 h of incubation were studied by AFM. The two- and three-dimensional images of the nanoparticles are shown in Fig. 6a, b. From the 2D view, well-separated spherical particles are seen. The sizes of the particles are in the range of 5–22 nm. However, most of the particles are in the range of 18 nm. The 3D view revealed that the growth direction of all the particles was almost same.

To confirm, the morphology of the synthesized nanoparticles, HRTEM image was recorded. Figure 7a shows the HRTEM image of the sample synthesized at 120 h of incubation, revealing the formation of almost monodispersed nanoparticles with average diameter ~2 nm. The TEM image shown in Fig. 7b depicts the formation of well-defined spherical nanoparticles with sizes from 15 to 20 nm. The particles size measured from TEM images matches well with the values arrived from the corresponding XRD pattern and AFM measurements. Figure 7c shows the selected-area electron diffraction (SAED) pattern of the sample. The SAED pattern shows the well-defined electron diffraction spots, confirming the single crystalline nature of the cubic phase of Ag nanoparticles.

Antibacterial activity

The antibacterial activity of the bio-synthesized Ag nanoparticles against different human pathogens is shown

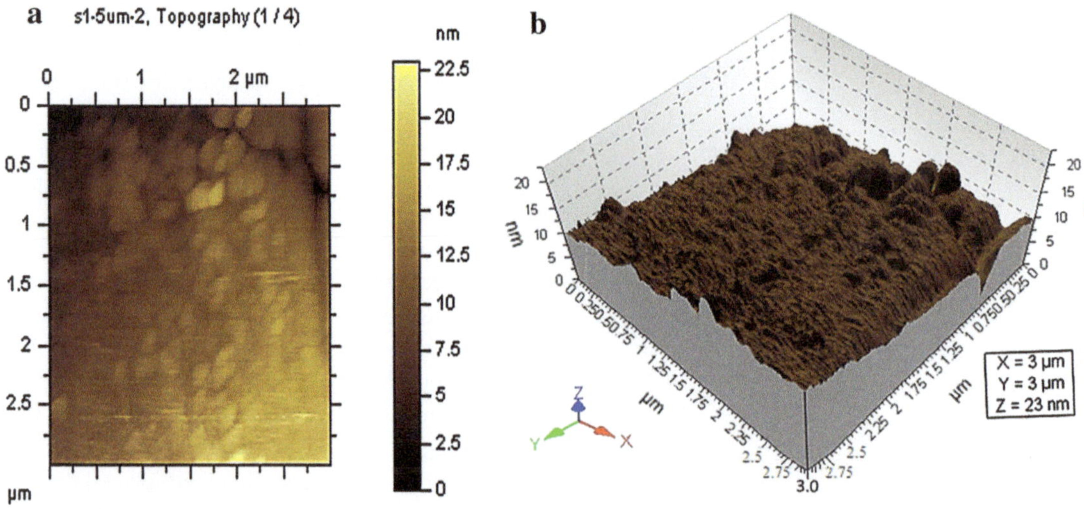

Fig. 6 **a** AFM images of synthesized silver nanoparticles using extract of *Sargassum wightii* and **b** corresponding 3D view

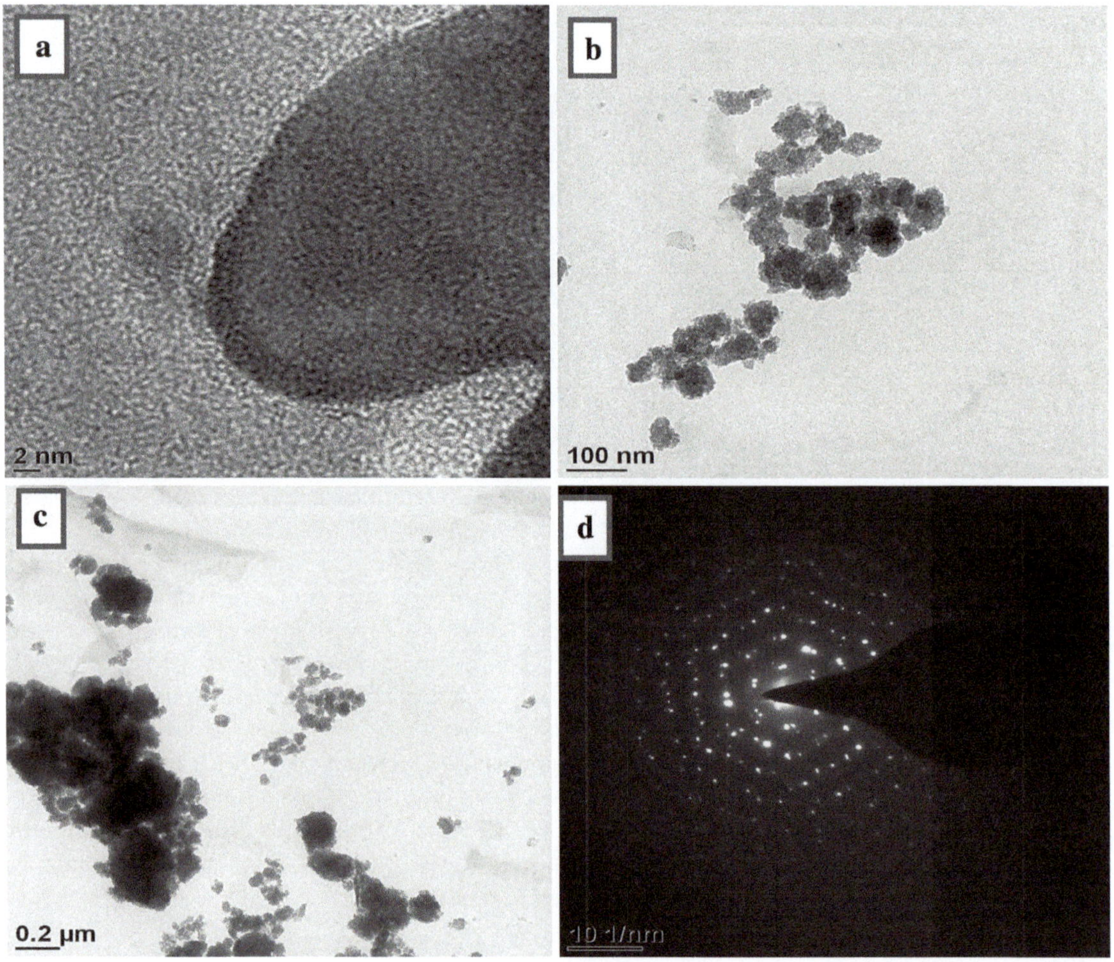

Fig. 7 TEM images of synthesized silver nanoparticles using extract of *Sargassum wightii*. **a** HRTEM, **b**, **c** TEM images and **d** corresponding SAED pattern

in Fig. 8. It is apparent that the Ag nanoparticles showed inhibition zone against almost all the test organisms (Table 1). Maximum zone of inhibition was found against gram positive *S. aureus* (12 mm) and it was higher than that of the positive control ampicillin (6 mm), and minimum zone of inhibition was obtained against

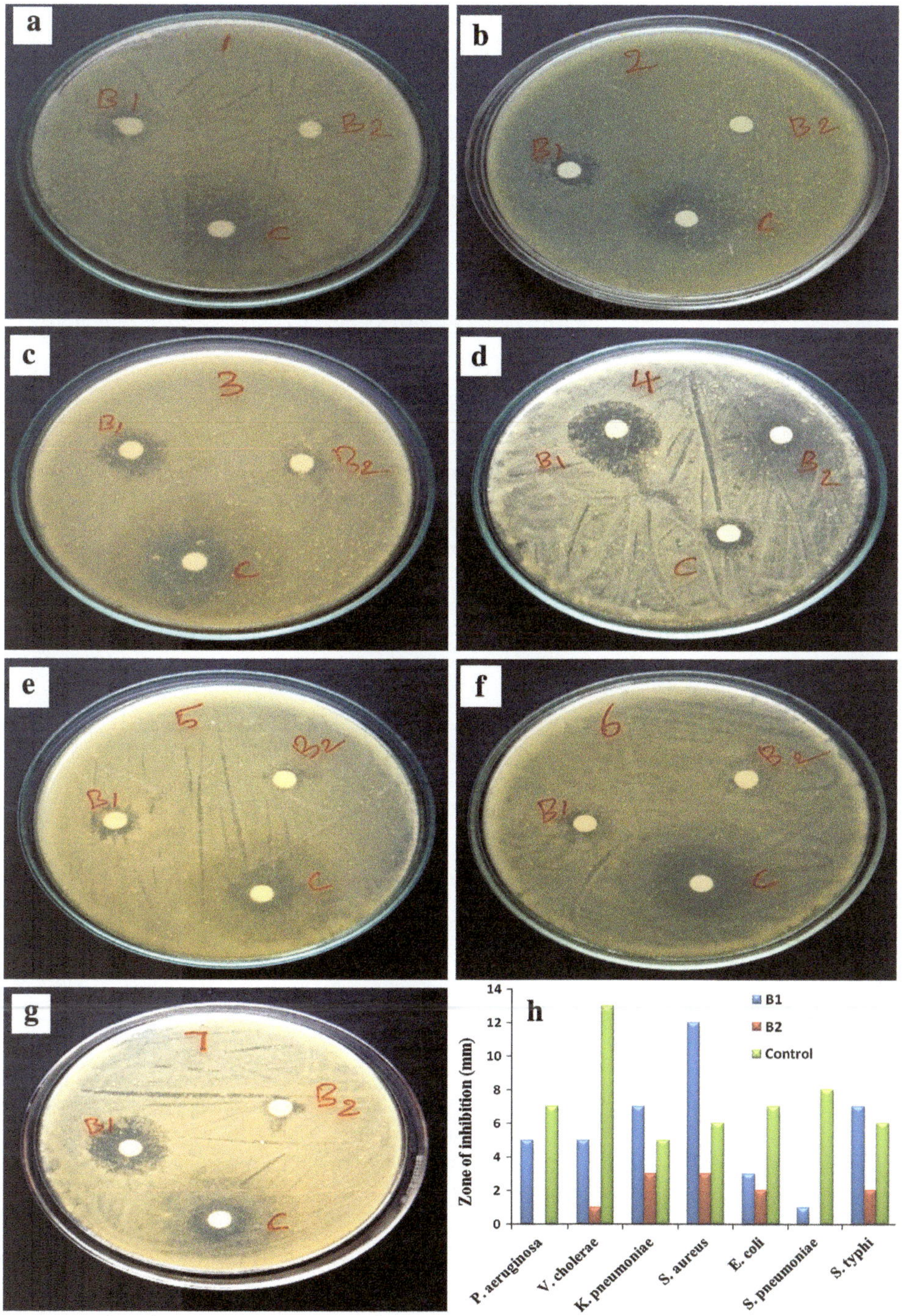

Fig. 8 Antibacterial activity of synthesized silver nanoparticles against **a** *P. aeruginosa*, **b** *V. cholerae*, **c** *K. pneumoniae*, **d** *S. aureus*, **e** *E. coli*, **f** *S. pneumoniae*, **g** *S. typhi* and **h** corresponding *bar* diagram

S. pneumoniae (1 mm). Similarly, the gram negative bacteria *S. typhi* also showed higher zone of inhibition (7 mm) than that of positive control (5 mm). The potential reason for the antibacterial activity of silver is that AgNPs may attach to the surface of the cell membrane disturbing permeability and respiration functions of the cell (Kvítek et al. 2005). Smaller AgNPs having the large surface area available for interaction would give more antibacterial effect than the larger AgNPs. It is also possible that AgNPs not only interact with the surface of membrane, but can also penetrate inside the bacteria (Morones et al. 2005).

Table 1 Antibacterial activity of synthesized silver nanoparticles against some human pathogens

Species	Zone of inhibition in mm for AgNPs (B_1)	Zone of inhibition in mm for $AgNO_3$ (B_2)	Zone of inhibition in mm for positive control
P. aeruginosa	5	0	7
V. cholera	5	1	13
K. pneumoniae	7	3	5
S. aureus	12	3	6
E. coli	3	2	7
S. pneumoniae	1	0	8
S. typhi	7	2	6

Conclusion

It has been predicted that the extract of marine seaweed *Sargassum wightii* is capable of producing Ag nanoparticles extracellularly and these nanoparticles are quite stable in solution due to capping likely by the proteins present in the extract. This is an efficient, eco-friendly and simple process. The Ag NPs showed potential antibacterial activity against some human pathogens. Therefore, nanoparticles of silver in combination with commercially available antibiotics could be used as an antimicrobial agent after further trials on experimental animals.

Acknowledgments The authors wish to thank Dr. S. Barathan, Professor and Head, Department of Physics, Annamalai University, for providing necessary facilities to carry out this work. We also thank Mr. D. Kannan, NFMTC lab, IIT Madras, for recording HR-TEM. Authors cordially thank Dr. N. Krishnakumar, Assistant Professor, Annamalai University, for his stimulating discussions during the performance of this work.

References

Alt V, Bechert T, Steinrucke P, Wagener M, Seidel P, Dingeldein E, Domann E, Schnettler R (2004) An in vitro assessment of the antibacterial properties and cytotoxicity of nanoparticulate silver bone cement. Biomaterials 25:4383–4391

Al-Warthan A, Kholoud MM, El-Nour A, Eftaiha A, Ammar RAA (2010) Synthesis and applications of silver nanoparticles. Arabian J Chem 3:135–140

Binupriya AR, Sathishkumar M, Vijayaraghavan K, Yun SI (2010) Bioreduction of trivalent aurum to nano-crystalline gold particles by active and inactive cells and cell-free extract of *Aspergillus oryzae* var. *viridis*. J Hazard Mater 177:539–545

Chen H, Hao F, He R, Cui DX (2007) Chemiluminescence of luminol catalyzed by silver nanoparticles. J Colloids Interface Sci 315:158–163

Chou WL, Yu DG, Yang MC (2005) The preparation and characterization of silver-loading cellulose acetate hollow fiber membrane for water treatment. Polym Adv Technol 16:600–607

Elechiguerra JL, Burt JL, Morones JR, Bragado AC, Gao X, Lara HH, Yocaman M (2005) Interaction of silver nanoparticles with HIV-1. J Nanobiotechnol 3:6

He ST, Yao JN, Jiang P, Shi DX, Zhang HX, Xie SS, Pang SJ, Gao HJ (2001) Formation of silver nanoparticles and self-assembled two-dimensional ordered superlattice. Langmuir 17(5):1571–1575

Jain P, Pradeep T (2005) Potential of silver nanoparticle-coated polyurethane foam as an antibacterial water filter. Biotechnol Bioeng 90:59–63

Jeong SH, Yeo SY, Yi SC (2005) The effect of filler particle size on the antibacterial properties of compounded polymer/silver fibers. J Mater Sci 40:5407–5411

Jin M, Zhang XT, Nishimoto S, Liu ZY, Tryk DA, Emeline AV, Murakami T, Fujishima A (2007) Light-stimulated composition conversion in TiO_2-based nanofibers. J Phys Chem C 8:658–665

Kannan RRR, Stirk WA, Van Staden J (2013) Synthesis of silver nanoparticles using the seaweed Codium capitatum P.C. Silva (Chlorophyceae). S Afr J Bot 86:1–4

Karting T, Still F, Reinhaler F (1991) Antimicrobial activity of the essential oil of young pine shoots. J Ethnopharmacol 35:155–157

Kathiresan K, Asmathunisha N (2013) A review on biosynthesis of nanoparticles by marine organisms. Colloids Surf B 103: 283–287

Krishnaraj C, Jagan EG, Ramachandran R, Abirami SM, Mohan N, Kalaichelvan PT (2012) Effect of biologically synthesized silver nanoparticles on *Bacopa monnieri* (Linn.) Wettst. plant growth metabolism. Process Biochem 47:651–658

Kvítek L, Prucek R, Panáček A, Novotný R, Hrbác J, Zbořil R (2005) The influence of complexing agent concentration on particle size in the process of SERS active silver colloid synthesis. J Mater Chem 15:1099–1105

Lee HY, Park HK, Lee YM, Kim K, Park SB (2007) A practical procedure for producing silver nanocoated fabric and its antibacterial evaluation for biomedical applications. Chem Commun 28:2959–2961

Li S, Shen Y, Xie A, Yu X, Qiu L, Zhang L, Zhang Q (2007) Green synthesis of silver nanoparticles using *Capsicum annuum* L. extract. Green Chem 9:852–858

Morones JR, Elechiguerra JL, Camacho A, Holt K, Kouri JB, Ramírez JT, Yacaman MJ (2005) The bactericidal effect of silver nanoparticles. Nanotechnology 16:2346–2353

Nino-Martinez N, Martinez-Castanon GA, Aragon-Pina A, Martinez-Gutierrez F, Martinez Mendoza JR, Ruiz F (2008) Characterization of silver nanoparticles synthesized on titanium dioxide fine particles. Nanotechnology 19(6):065711/1–065711/8

Rai G, Vyjayanti VN, Dorjsuren D, Simeonov A, Jadhav A, Wilson DM, Maloney DJ (2012) Synthesis, biological evaluation, and structure–activity relationships of a novel class of apurinic/apyrimidinic endonuclease 1 inhibitors. J Med Chem 55(7): 3101–3112

Russel AD, Hugo WB (1994) Antimicrobial activity and action of silver. Prog Med Chem 31:351–370

Song JY, Jang HK, Kim BS (2009) Biological synthesis of gold nanoparticles using *Magnolia kobus* and *Diopyros kaki* leaf extracts. Process Biochem 44:1133–1138

Susanto H, Feng Y, Ulbricht M (2009) Fouling behavior of aqueous solutions of polyphenolic compounds during ultrafiltration. J Food Eng 91:333–340

Venkatpurwar V, Pokharkar V (2011) Green synthesis of silver nanoparticles using marine polysaccharide: study of in vitro antibacterial activity. Mater Lett 65:999–1002

Williams D (2008) The relationship between biomaterials and nanotechnology. Biomaterials 29:1737

Xu J, Han X, Liu H, Hu Y (2006) Synthesis and optical properties of silver nanoparticles stabilized by gemini surfactant. Colloids Surf A 273:179–183

Yang W, Yang C, Sun M, Yang F, Ma Y, Zhang Z, Yang X (2009) Green synthesis of nanowire-like Pt nanostructures and their catalytic properties. Talanta 78:557–564

4

Investigation of the optical property and photocatalytic activity of mixed phase nanocrystalline titania

Susmita Paul · Amarjyoti Choudhury

Abstract Mixed phase nanocrystalline titania are prepared by simple sol–gel method. The physico-chemical characteristics of the prepared nanoparticles are studied with X-ray diffraction, high-resolution transmission electron microscopy, RAMAN, BET, UV–Vis, steady state and time resolved photoluminescence. X-ray diffraction and Raman spectra clearly demarcate the anatase and rutile phase as both the phases give different diffraction patterns and Raman peaks. A comparison in the band gap indicates that pure anatase and rutile phase have band gap in the UV region, whereas a mixture of these phases has lower band gap and corresponds to the visible region. Steady state and time resolved photoluminescence are employed to understand the emissivity and carrier lifetime. The photocatalytic activity is evaluated by monitoring the degradation of phenol under visible light illumination. Due to the synergistic effect of mixed anatase and rutile phases, mixed phase nanocrystalline titania exhibit superior photocatalytic activity.

Keywords Mixed phase · Band gap · Carrier lifetime · Photocatalysis

Introduction

Titanium oxide (TiO_2) is one of the most important semiconductor oxides. Since the initial report by Fujishima and Honda (1972) on the photolysis of water by TiO_2, intense interests in the properties of nanocrystalline TiO_2 have been generated due to their promising applications in photocatalysis (Paul et al. 2013) and photoelectrochemical (Shin et al. 2013) energy conversion. TiO_2 belongs to the family of transition metal oxides. Four commonly known polymorphs of TiO_2: anatase, brookite, rutile, and TiO_2 (B) monoclinic exist in nature. Besides these two additional high pressure forms, TiO_2 (II) with a PbO_2 structure and TiO_2 (H) with a hollandite polymorphs are also known. From the catalytic viewpoint, anatase and rutile are the most important phase. The brookite phase is the rarest of the natural TiO_2 polymorphs and is the most difficult phase to prepare in the laboratory (Gupta and Tripathi 2011).

Anatase TiO_2 has a band gap of 3.2 eV (385 nm), while the rutile phase has a smaller band gap of 3.0 eV (410 nm). It is generally accepted that anatase TiO_2 is a better photocatalytic candidate compared to rutile TiO_2, but recently the anatase–rutile mixture is found to have a magical effect on the charge transfer process (Knorr.et.al 2008). The presence of rutile phase extends the photoactive response to the visible region, harvesting more light, and stabilization of charge separation by electron transfer from rutile to anatase trapping sites slows down the recombination (Hurum et al. 2003), thereby improving the photocatalytic activity.

Although there are several methods to prepare nanocrystalline titania but the sol–gel method represents several advantages over other methods (Zhou et al. 2006). As the method is carried out in solution, tailoring of certain desired structural characteristics such as compositional homogeneity, grain size, particle morphology, and porosity is possible with this method. The process finds wide applications in the development of new materials for catalysis (Oye et al. 2006), photochromic applications (Volkan et al. 2005), chemical sensors (Qui et al. 2011), and also in a diverge range of scientific and engineering

S. Paul · A. Choudhury (✉)
Department of Physics, Tezpur University,
Napaam 784028, India
e-mail: ajc@tezu.ernet.in

fields such as ceramic industry (Marinel et al. 2013), electronic industry (Lin 2013).

In the present investigation, nanocrystalline titania are prepared by simple sol–gel method. The phase transformation rearranges the bonding arrangement of TiO_2 and introduces many defects in the mixed phase sample. X-ray diffraction, TEM, Raman spectroscopy have been investigated to clearly distinguish the basic anatase and rutile phase, their sizes and the defects present. A detailed analysis on the optical properties of the reduced anatase, anatase–rutile mixed phase and of rutile phase are reported. Due to the synergistic effect of mixed anatase and rutile phases, an enhanced photocatalytic performance is noticed in the mixed phase titania.

Experimental methods

Material synthesis

The reagents used for the synthesis were titanium isopropoxide (TTIP, purity 97 %) and isopropanol (99 %). The synthesis was started by adding 15 ml of isopropanol to 5 ml titanium isopropoxide solution. The solution was allowed to stir for about 15 min and then 1 ml of water was added to initiate the hydrolysis process. The solution became turbid and the stirring was continued until gel formation (Dutta et al. 2009). The transformation from initial sol to gel took about 6–7 h. The gel was left for aging for about 12 h. This was followed by centrifugation with ethanol and water for 5 times. The centrifuged product was dried in an oven at 80 °C. Finally, the dried product was annealed at 450, 650, 850 °C for 3 h and the samples were labeled as T45, T65, T85, respectively.

Characterization details

The structure of all the samples is determined using Rigaku Miniflex CD 10041 XRD unit with copper target and $\lambda = 1.54$ Å at a scanning rate of 1°/min and in the scanning range of 10°–80°. High-resolution transmission electron microscopic images for morphology and particle size determination are observed with JEM-2100, 200 kV JEOL. Raman spectra of the samples are acquired with Reinshaw inVia spectrometer. The 514.5-nm emission of argon-ion laser is used as the excitation source. Surface area and pore size distribution of all the samples are studied with the help of a quantachrome instrument. Diffuse reflectance spectra (DRS) of the samples are taken with Shimadzu-2450 UV–Vis spectrometer. The photoluminescence (PL) measurements at room temperature are recorded with PERKIN ELMER LS 55 fluorescence spectroscopy. Time resolved photoluminescence spectra (TRPL) are recorded on an IBH Horiba-Yvon TCSPC using 340 nm Nano-LED (FWHM = 750 ps) and 375 nm nano-LED (FWHM = 300 ps) as excitation source for the samples with a time scale of 0.007 ns/channel.

Photocatalytic activity

Photocatalytic activity of the prepared TiO_2 nanoparticles was studied by examining the degradation of aqueous solution of phenol. For visible light irradiation, a 25-W white lamp was used. In order to carry out the process, 50 mg of the photocatalyst was mixed with 50 ml water in a 100-ml beaker. To this mixture, 5 mg of phenol was added and the solution was stirred for about half an hour in dark to obtain the absorption–desorption equilibrium (Choudhury et al. 2011). The catalyst loaded phenol solution was then placed at a distance of 6 cm and irradiated for 20, 40, 60 and 80 min. After completion of the irradiation process, the samples were centrifuged at 10,000 rpm to make it free from any catalyst. 5 ml of the aliquot was taken to measure the absorbance. The degradation efficiency is calculated using the following equation:

$$\%D = \left(\frac{A_O - A_t}{A_O}\right) \times 100 \quad (1)$$

where A_O is the initial absorbance and A_t is the absorbance after time t.

Results and discussions

Structural and morphological study

The XRD patterns are depicted in Fig. 1. The XRD pattern of the as-prepared powders shows the presence of brookite phase at 36° (JCPDS-761934). This phase is usually formed at low temperatures and gets disappeared at high annealing temperatures (So et al. 2001). On careful study of the phase transition of the samples, it is found that T45 resulted in pure anatase phase. The phase transition began at 650 °C and complete transformation to rutile phase occurred at 850 °C, in agreement with earlier results (Song et al. 2010; Li et al. 2008). The peaks in the spectra are in agreement with the JCPDS file nos. 782486 and 894202 of tetragonal systems for pure anatase and rutile phase, respectively. The average crystallite size of each sample (Table 1) is obtained taking the (101) peak for anatase, (110) for rutile and both (101), (110) for mixed phase using Scherer's formula:

$$d = \frac{0.9\lambda}{\beta\cos\theta} \quad (2)$$

Here d is the crystallite size, λ the wavelength of the X-ray (Cu $K_\alpha = 1.54$ Å), β the full width at half maximum, and θ

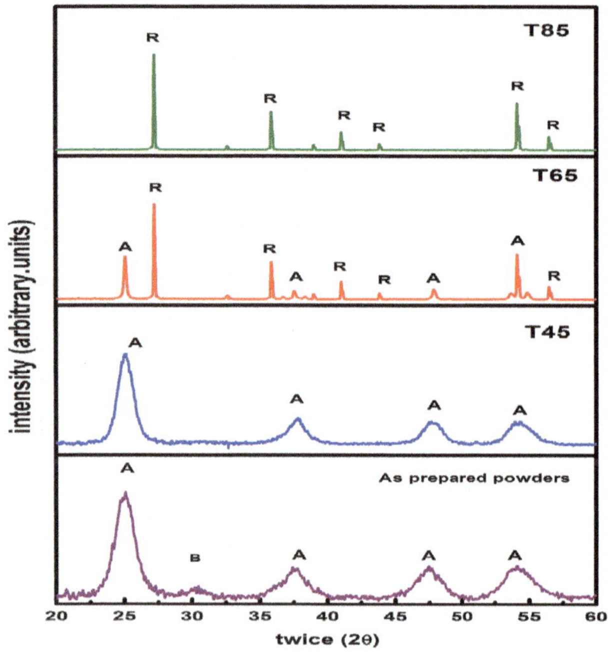

Fig. 1 X-ray diffraction pattern of all the samples

Table 1 Crystallite size, phase composition, band gap, Urbach energy details of the samples calcined at different temperatures

SI. no.	Calcination temp (°C)	Crystallite size (nm)		Fraction (%)		Band gap (eV)	Urbach energy (meV)
		Anatase	Rutile	Anatase	Rutile		
1.	450 (T45)	10	0	100	0	3.25	260
2.	650 (T65)	36	67	63.6	35.4	2.82	475
3.	850 (T85)	–	70	0	100	3.08	343

is the diffraction angle. The crystallite size is calculated to be in the nanoscale and a steady increase in size is noticed with the increase in annealing temperatures. Each of the anatase and rutile mass fractions of the samples is calculated using Spurr's formula (Spurr and Myers 1957):

$$f_a = \frac{1}{1 + 1.26\frac{I_a}{I_r}} \tag{3}$$

$$f_r = (1 - f_a) \tag{4}$$

where f_a and f_r are the anatase and rutile fractions, and I_a and I_r are the integrated intensities of the most intense anatase (101) and rutile (110) peak, respectively. The parameters calculated using Eqs. (3) and (4) are depicted in Table 1.

Figure 2a and b shows the TEM images of T45 and T65, respectively. The average particle size is small in case of pure anatase nanoparticles and increases with its transformation to mixed phase and then to pure rutile phase. For T45, the particle size distribution is within the range of ~16 nm, while for T65 it is 30–65 nm (Fig. 2b) and for T85 (Fig. 2c) the range is 126–230 nm. It is worth mentioning that we have obtained differences in the size calculated from XRD and TEM. The magnitude of the crystallite and particle size agrees well only for the case of stable, monodisperse, single crystals and hard spheres, and any deviation from this criteria leads to variation in measured sizes (Baalousha et al. 2012).

Raman spectroscopy is further employed in the characterization of crystalline structure of TiO_2 samples. Figure 3 shows the Raman spectra of all the samples. According to the group factor analysis, anatase has six active Raman modes at ($A_{1g} + 2B_{1g} + 3E_g$) (Zhang et al. 2000), while for rutile TiO_2 group factor analysis gives four Raman active modes ($A_{1g} + B_{1g} + B_{2g} + E_g$) (Aita 2007; Orendorz et al. 2007; Mazza et al. 2007). In our study in T45, a strong E_g peak appears at 144 cm^{-1} followed by the appearance of low intense peaks at 197 (E_g), 399 (B_{1g}), 513 ($A_{1g} + B_{1g}$) and 641 cm^{-1} (E_g). All these peaks correspond to the pure anatase phase. For T65 sample, E_g peak of anatase remains at 144 cm^{-1}, while additional peaks corresponding to the rutile phase appear at 236, 448, 613 cm^{-1} (Narksitipan and Thongtem 2012). In T85, all the Raman peaks identical to rutile TiO_2 predominate.

For a perfect crystal, the phonon contribution comes from the center of the Brilliouin zone satisfying the phonon momentum selection rule, $q \approx 0$ (Zhang et al. 2000). In semiconductor nanoparticles, the phonon selection rule breaks and the phonon momentum distribution increases. The phonon scattering thus involves phonon contribution from the entire Brillouin zone (Brojcin et al. 2005, Mo and Ching 1995). For nanoparticles of diameter (d), the intensity for first-order Raman scattering is given by:

$$I(\omega) = \frac{C0, k^2}{[\omega - \omega(k)]^2 + (\frac{\Gamma_o}{2})^2} d^3q \tag{5}$$

where $w(k)$ is the phonon dispersion curve, Γ_o the natural full line width and C (0, k) is the fourier co-efficient of the phonon confinement function and is often in use as

$$|C(0,k)|^2 = \left(-k^2d^2/16\Pi^2\right) \tag{6}$$

where d is the average size of the nanocrystal. The changes in the Raman peaks are a result of this integration in Eq. (5) due to changes in grain size (Sharma et al. 2011).

In our study compared to T45, the E_g mode of T65 is slightly red shifted and the FWHM decreases. The phonon confinement model predicts that with the increase in crystallite size, the E_g peak is shifted to lower wave numbers and its FWHM and band intensity decrease (Orazen 2007).

Presence of oxygen defects also affects the Raman line, shape and position. The Raman spectra involve vibration of the oxygen atoms only (Georgescu et al. 2012). As the

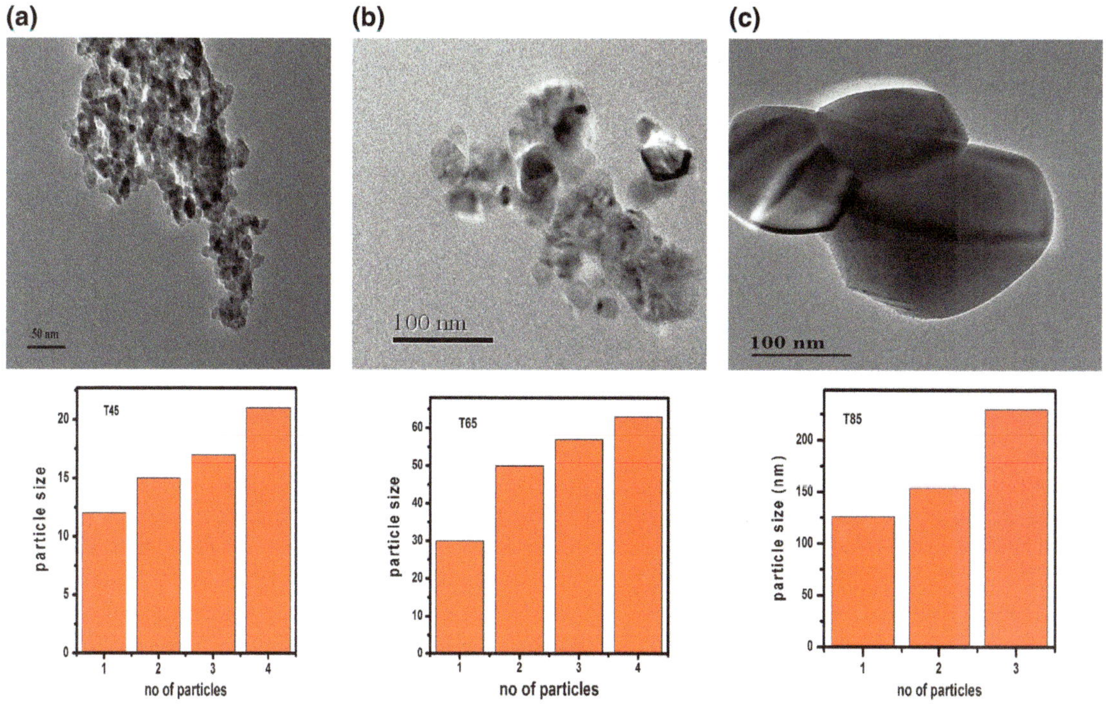

Fig. 2 TEM images of **a** T45, **b**T65, **c** T85

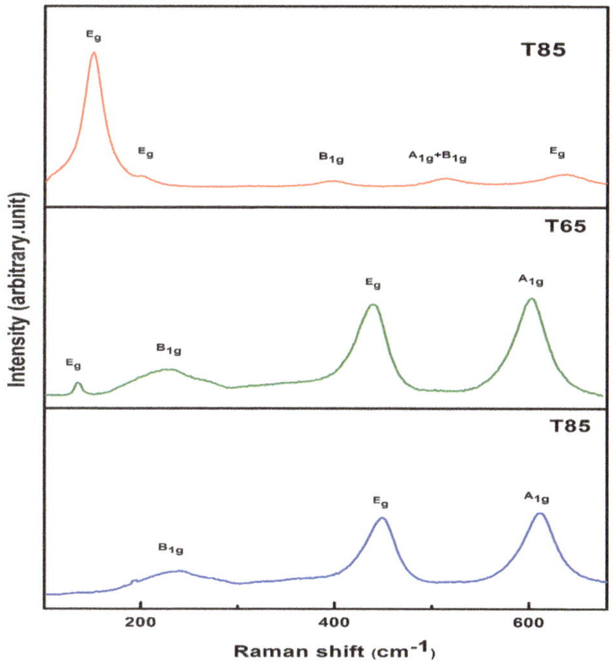

Fig. 3 Raman spectra of T45, T65, T85

temperature treatment changes the local co-ordination of oxygen atoms surrounding T_i^{4+}, switching from anatase to rutile phase involves the re-arrangement of TiO_6 octahedra that affects the Raman lines. We have calculated the phonon lifetime via the energy–time uncertainty relation $1/\tau = \Delta E/h = 1/2\ \pi c\Gamma$ (Mali et al. 2011), where ΔE is the uncertainty in energy, h is the Planck's constant and Γ is the FWHM of the Raman peaks in cm^{-1}. For T45 and T65, the phonon lifetime using the E_g mode is calculated to be 0.22 and 0.57 ps, respectively, while taking the A_{1g} mode for T65 and T85, the lifetime corresponds to 0.10 and 0.13 ps, respectively. The increased lifetime for T65 is attributed to the generation of increased defects due to the anatase–rutile contacts. In T85, an increase in crystal size adds to the increased phonon distribution further increasing the phonon lifetime (Wang et al. 2008).

Figure 4 shows the typical nitrogen adsorption isotherm. From the analysis, it is observed that the surface area and the pore size distribution determined from the BJH analysis exhibit a decrease with the increase in annealing temperatures (Table 2). The surface area of T45 is calculated to be 95 m^2/g, while for T65 and T85 samples, the surface area corresponds to 25 and 4 m^2/g, respectively, The lowering in the surface area may be due to the increase in the grain size and particle growth during high-temperature annealing.

Optical study

Diffuse reflectance spectroscopy was carried out to analyze the optical spectral response of all the samples. Figure 5 shows the absorbance curve. The absorption band around 370 nm can be assigned to the band gap excitation of anatase TiO_2 that corresponds to the band to band transition from Ti 3d level to O 2p levels (Devi et al. 2010).

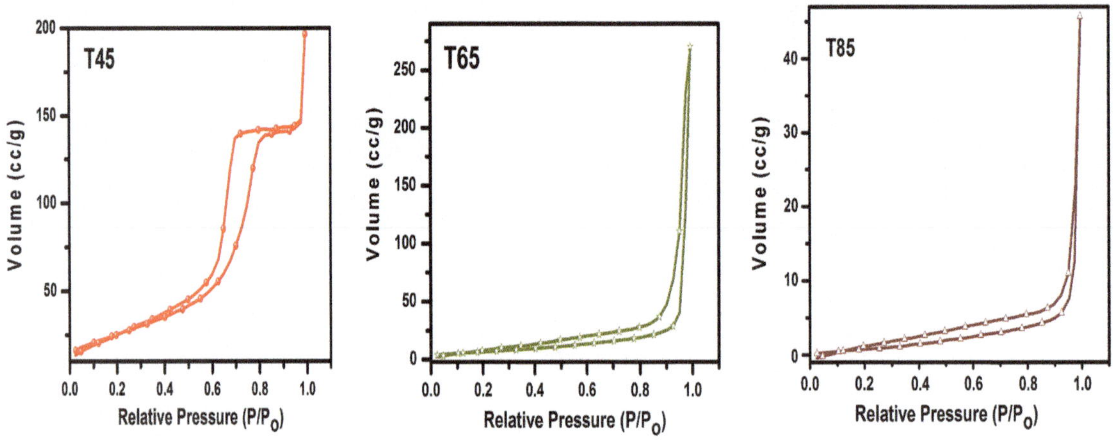

Fig. 4 Isotherm curve of all the samples

Table 2 Surface area and pore size distribution of all the samples

SI. no.	Sample	Surface area (m^2/g)	Pore size (nm)
1.	T45	95	31
2.	T65	25	5
3.	T85	4	3

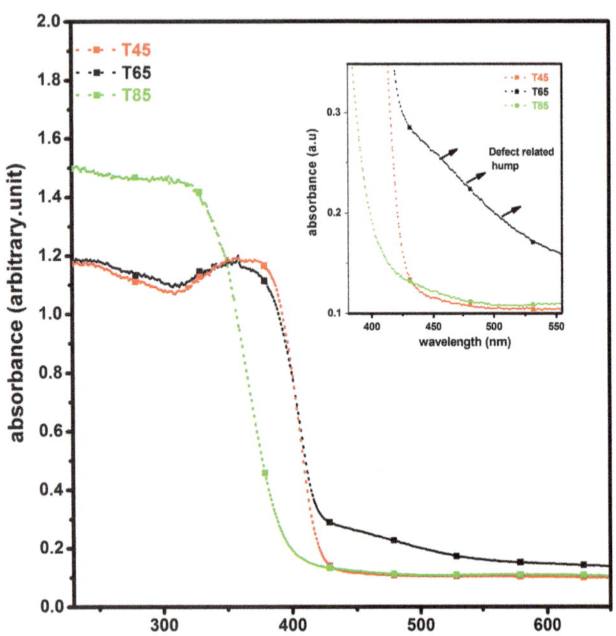

Fig. 5 UV–Vis absorption study annealed at different temperatures

Compared to T45, the absorption spectra of the mixed phase sample, T65 shows a hump in the visible region (420–540 nm) (inset, Fig. 5) due to the defects present at the anatase–rutile contacts. In T85 sample, an enhancement in the absorption is noticed and this is possibly due to the presence of rutile content that has smaller band gap in comparison to pristine anatase phase (Zhang et al. 2010).

TiO_2 is an indirect band gap material and the band gap is evaluated by drawing a line on the linear part of the curve, $[F(R)h\nu]^{1/2}$ vs $h\nu$ at $[F(R)h\nu]^{1/2} = 0$. Band gap values of each sample are shown in Fig. 6. The band gap of T45 is slightly higher than that of bulk anatase and this is due to quantum confinement effect. The band gap of T85 matches with the band gap of bulk rutile TiO_2, while T65 has a band gap lower than that of either anatase (3.2 eV) or rutile (3.0 eV) TiO_2. This lowering in the band gap in T65 is due to the presence of defect bands that perturb the band structure by forming additional states within the band gap in the form of tails. This tail is referred to as the Urbach tail and the associated energy as the Urbach energy, expressed by the formula (Lethy et al. 2008):

$$\alpha = \alpha_o \exp \frac{E}{E_u} \tag{7}$$

α is the absorption co-efficient, E the photon energy and E_u is the Urbach energy. For the calculation of Urbach energy, we have plotted ln A (or ln $F(R)$ in DRS mode) vs E (eV) (Fig. 7). The reciprocal of the slopes of the linear portion below the optical band gap gives the Urbach energy. Table 1 gives the band gap and the corresponding Urbach energy of all the samples. The Urbach energy is calculated to be maximum for T65. In the TiO_2 nanoparticles, Ti^{4+} is surrounded by six oxygen ions forming the TiO_6^{2-} octahedra. High-temperature annealing at 650 °C alters the arrangement of oxygen ions around Ti^{4+} and generates a considerable amount of oxygen vacancies due to breaking of the Ti–O bond. In T65 the anatase and rutile phase being in contact with each other the defect by and large localizes at the contact of these two phases. These increase the width of the defect band and in turn increase the Urbach energy. In T85, a decrease in oxygen defect density decays the Urbach tail and increases the band gap (Zhang et al. 2008).

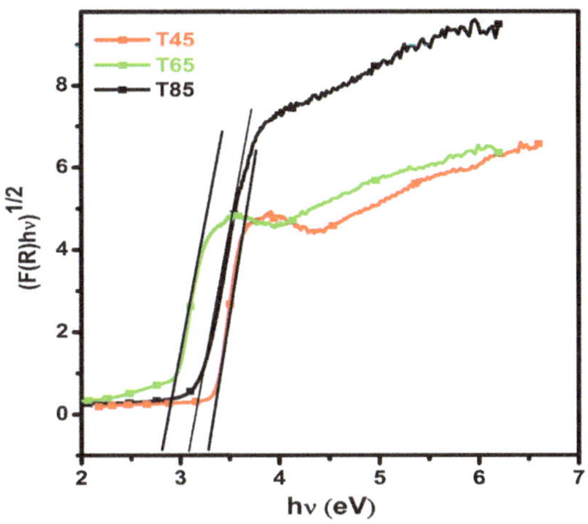

Fig. 6 Kubelka–Munk plot showing the variation in band gap

Photoluminescence is a very versatile and commonly used tool to disclose the efficiency of charge carrier trapping, immigration and transfer in semiconductors. Figure 8 exhibits the PL spectra of the samples at an excitation wavelength of 320 nm (3.87 eV). T45 has a UV emission at 392 nm and is the band edge emission of host TiO_2 arising from X_{1b} transition to Γ_3 (Serpone et al. 1995). In addition, there appear two blue emission peaks at 457 (2.7 eV) and 490 nm (2.5 eV) and one green emission peak at 537 nm (2.37 eV). The peak at 425 nm (2.9 eV) is probably due to the self-trapped exciton on the TiO_6 octahedra (Knorr Fritz et al. 2008; Wan et al. 2010). However, the same peak in rutile TiO_2 is due to recombination of free excitons (Yamada and Kanemitsu 2010). The blue emission peak at 490 nm is due to the charge transfer transition from Ti^{3+} to oxygen anion in a $TiO_6{}^{2-}$ complex associated with oxygen vacancies (Yu et al. 2002). Another blue emission peak at 457 nm and green emission peak at 537 nm are due to the color centers associated with oxygen vacancies (Nick 2006).

The PL emission intensity is contributed by both radiative and non-radiative centers. Radiative recombination increases the emission intensity, while non-radiative recombination decreases the intensity. In our study, the emission intensity of T65 decreases as compared to the T45 sample. The decrease in intensity is due to the presence of defects that act as non-radiative centers. Another reason may be due to the trap to trap relocation of electron until a recombination center is anticipated. The process restrains the electron–hole recombination and quenches the emission intensity. In T85 sample, high-temperature annealing removes some of the defects and so the emission intensity increases inferring increase in charge carrier recombination rate compared to T65 sample (Jung and Kim 2009).

To better understand the recombination process, we have studied the PL decay curve (Fig. 9). The curves have been fitted with tri-exponential decay response functions defined by the following equation:

$$Y = y_0 + A_1 \times \exp(-(x - x_0)/t_1) + A_2 \times \exp(-(x - x_0)/t_2) + A_3 \times \exp(-(x - x_0)/t_3) \quad (8)$$

where y_0 is the baseline correction, t_1, t_2, t_3 the decay constants and A_1, A_2, A_3 are the pre-exponential factors. The average decay $< \tau_{avg} >$ time is calculated using the following formula (Das et al. 2009):

$$< \tau_{avg} > = \frac{\sum_{i=1}^{i=3} A_i t_i^2}{\sum_{i=1}^{i=3} A_i t_i} \quad (9)$$

Using the above equation, the average lifetime for T45, T65, T85 are calculated to be 1.99, 2.13 and 1.65 ns, respectively. T65 has a longer lifetime compared to T45 and T95. The charge carrier lifetime is affected by several

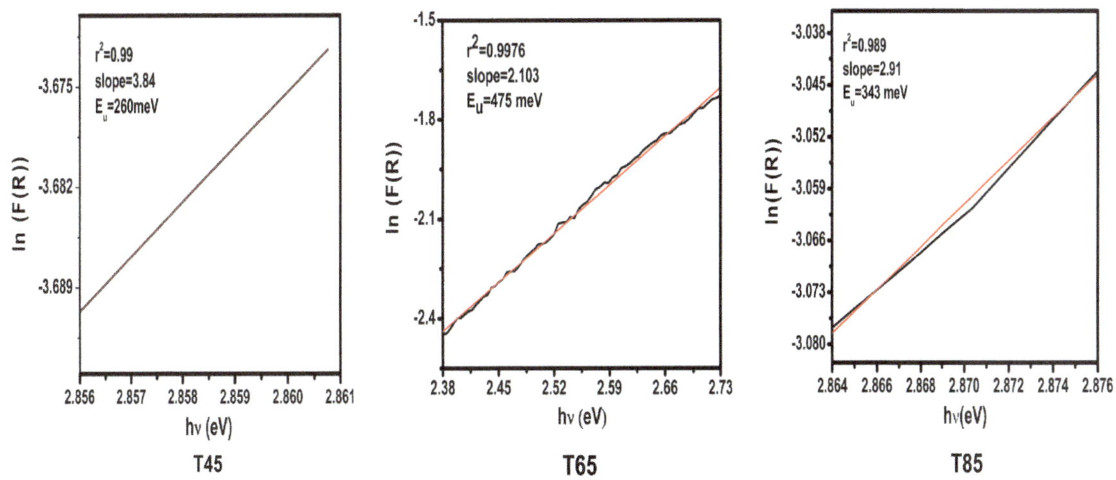

Fig. 7 Urbach energy plots for all the samples

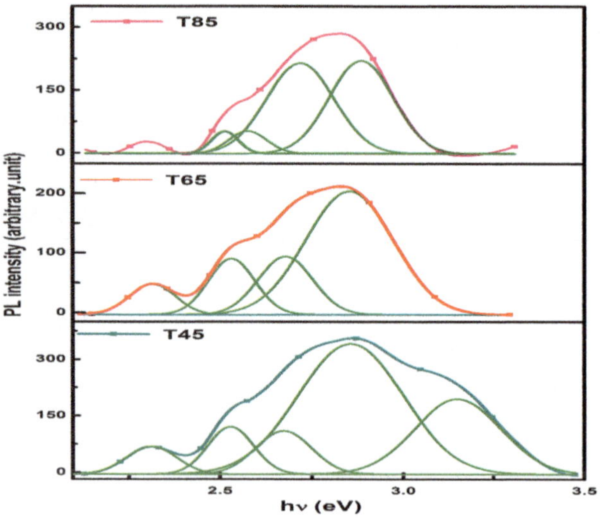

Fig. 8 Room-temperature photoluminescence after excitation at 320 nm

factors such as band structure, carrier trapping, mobility of the carriers, etc. Anatase TiO_2 being an indirect semiconductor, the conduction band electron and the valence band holes are in different positions in the Brillouin zone (Yamada and Kanemitsu 2012). The extent of overlapping of the wave function is therefore low. Inclusion of defects within the band gap further lowers the strength of overlapping and increases the lifetime of the charge carriers. Baiju et al. (2009) stated that the charge carriers have a longer lifetime when the anatase and rutile phases are in intimate contact. In case of T85, decrease in lifetime is due to increased recombination process (Jung and Kim 2009).

Thus, in the present investigation, the presence of peaks in the visible region, maximum quenching in emission intensity and increased lifetime for T65 samples throw light on the feasibility of the sample to act as a good photocatalytic candidate.

Photocatalytic activity

The photocatalytic activity of the samples is evaluated by analyzing the degradation of phenol under visible light. In the photocatalysis process, after light absorption by host TiO_2, electrons and holes so formed combine with water or OH group, and water-dissolved oxygen get adsorbed on different active sites of the photocatalyst to form the hydroxyl radical and the super oxide anion. These hydroxyl radical and super oxide anion act as the potential agents in the photocatalytic degradation process.

$$\mathrm{TiO_2} \longrightarrow hv\mathrm{e}^- + \mathrm{h}^+$$

$$OH^{\cdot} + h^+ \longrightarrow hvOH^{\cdot}$$

$$\mathrm{O_{2+}e^-} \longrightarrow hv\mathrm{O_2^-}$$

$$\text{Phenol} +^{\cdot} \mathrm{O_2^-} \rightarrow \text{Products}$$

$$\text{Phenol} + \mathrm{OH^{\cdot}} \rightarrow \text{Products}$$

Figure 10 shows the photocatalytic degradation curve. From the graph, T65 is found to exhibit better degradation compared to T45 and T85. Under Visible light irradiation, the photogenerated electrons rapidly transfer from rutile to lower energy anatase trapping site (Fig. 11). This leads to a more stable charge separation and in turn contributes to increased photocatalytic activity (Hurum et al. 2003).

It is worth mentioning that although the surface area and pore size in T65 are lower than T45, such high photocatalytic activity is attributed to the intimate anatase–rutile contacts that generates considerable amount of defects. These defects promote the charge separation and restrain the electron–hole recombination which in turn increases the lifetime of the photoinduced charge carriers as seen from the luminescence decay curve. The generated charge carriers interact with the phenol before they recombine and hence results in higher degradation activity.

In T85, high-temperature annealing removes most of the defects and thereby increases the carrier mobility. The

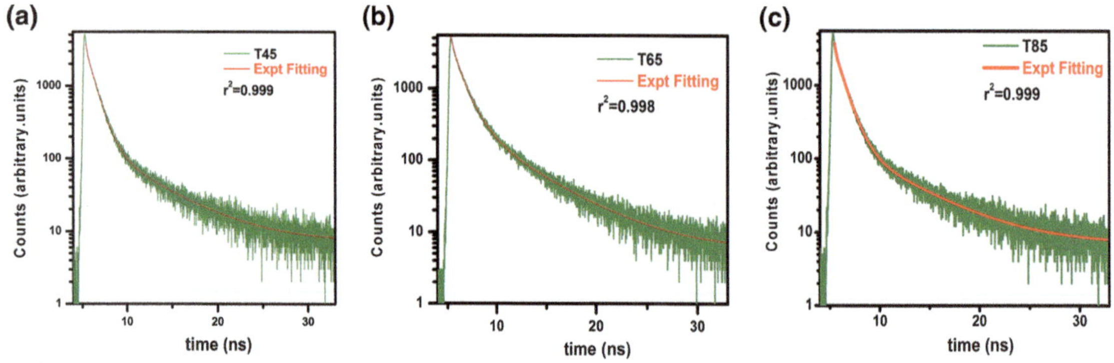

Fig. 9 Luminescence decay study

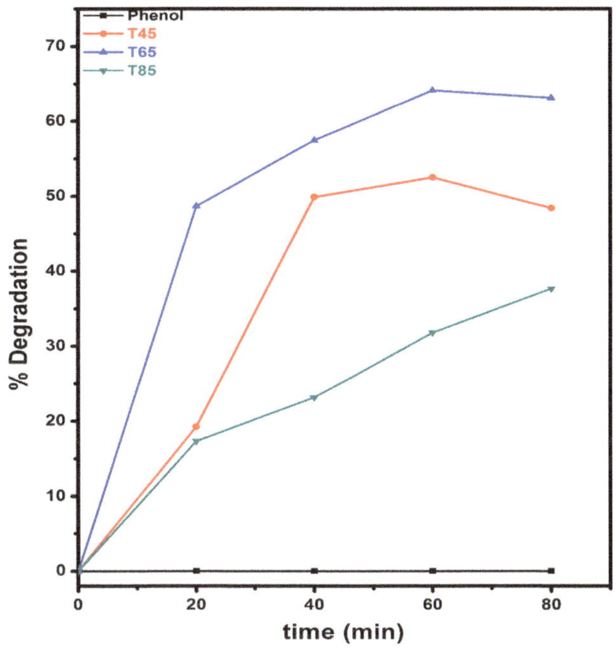

Fig. 10 Degradation curve of phenol under visible light illumination

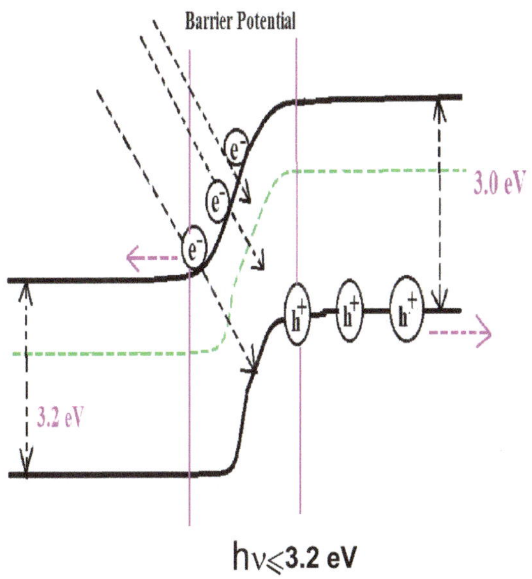

Fig. 11 Energy band profile with $h\nu < 3.2$ eV

increased carrier mobility induces the increased recombination process and hence lowers the photocatalytic activity (Jung and Kim 2009).

Conclusion

In summary, mixed phase nanocrystalline titania have been prepared employing a simple wet-chemical techniques. Thermal treatment of anatase-type titania precursor leads to transformation of anatase to rutile at 650 °C. In the photodegradation experiment, mixed phase titania (T65) are found to exhibit improved photocatalytic activity compared to pure anatase and pure rutile phase. Such enhancement in the photocatalytic activity is attributed to the formation of defects at the anatase–rutile interface. The presence of these defects lowers the band gap and increases the charge separation, thereby favoring two important criteria for an efficient photocatalytic process.

Acknowledgments We are thankful to Department Of Chemistry, IIT Mumbai for helping us in carrying out the TRPL measurement, Sophisticated Analytical Instrument Facility (SAIF), NEHU, Shillong for helping us in providing the HRTEM images. We are also thankful to University Grants Comission (UGC) and Department Of Science and Technology (DST) for financial support to the project F.No.42-785/2013 (SR) and SR/NM/NS-98/2010(G), respectively.

References

Aita CR (2007) Raman scattering by thin film nanomosaic rutile TiO_2 Appl. Phys Lett 90:213112–213115

Baalousha M, Nam YJ, Cole PA, Gaiser B, Fernandes TF, Hriljac JA, Jepson MA, Stone V, Tyler CR, Lead JR (2012) Characterisation of cerium oxide nanoparticles: Part 1 size measurements Environmental toxicology and chemistry. Environ Toxicol 31:983–993

Baiju KV, Zachariah A, Shukla S, Biju S, Reddy MLP, Warrier KGK (2009) Correlating photoluminescence and photocatalytic activity of mixed-phase nanocrystalline titania. Catal Lett 130: 130–136

Brojčin MG, Šćepanović MJ, Dohčević-Mitrović ZD, Hinić I, Matović B, Stanišić G, Popović ZV (2005) Infrared study of laser synthesized anatase TiO2 nanoparticles. J Phys D Appl Phys 38:1415–1420

Choudhury B, Borah B, Choudhury A (2011) Extending visible light photocatalytic activity of TiO_2 nanoparticles to visible region of illumination by doping of cerium. Photochem Photobiol 88:257–264

Das K, Sharma SN, Kumar M, De SK (2009) Morphology dependent luminescence properties of co doped TiO2 nanostructures. J Phys Chem C 113:14783–14792

Devi LG, Kottam N, Murthy BN, Kumar SG (2010) Enhanced photocatalytic activity of transition metal ions Mn^{2+}, Ni^{2+}, Zn^{2+} doped polycrystalline titania for the degradation of aniline blue under UV/solar light'. J Mol Catal A: Chem 328:44–52

Dutta N, Mohanta D, Sulania I, Choudhury A (2009) Studies of optical properties and SHI irradiation on PbS sensitized nanoporous TiO_2 network. J Opt 38:169–176

Fujishima A, Honda K (1972) Electrochemical photolysis of water at a semiconductor. Nature 238:37–38

Georgescu D, Baia L, Ersen O, Baia M, Simon S (2012) Experimental assessment of the phonon confinement in TiO2 anatase nanocrystallites by Raman spectroscopy. J Raman Spect 43:876–883

Gupta SM, Tripathi M (2011) A review on TiO_2 nanoparticles. Chin Sci Bull 56:1639–1657

Hurum DC, Agrios AG, Gray KA, Rajh T, Thurnauer C (2003) Explaining the enhanced photocatalytic activity of degussa P25 mixed-phase TiO_2 using EPR. J Phys Chem B 107:4545–4549

Jung HS, Kim H (2009) Origin of low photocatalytic activity of rutile TiO_2. Electron Mater Lett 5:73–76

Knorr Fritz J, Mercado Candy C, McHale Jeanne L (2008) Trap-state distributions and carrier transport in pure and mixed-TiO_2: influence of contacting solvent and interphasial electron transfer. J Phys Chem C 112:12786–12794

Lethy KJ, Beena D, Mahadevan Pillai VP, Ganesan V (2008) Bandgap renormalization in titania modified nanostructured tungsten oxide thin films prepared by pulsed laser deposition technique for solar cell applications. J Appl Phys 104:033515 (12 pp)

Li GH, Dimitrijevic NM, Chen L, Nichols JM, Rajh T, Gray KA (2008) The important role of tetrahedral Ti^{4+} sites in the phase transformation and photocatalytic activity of TiO_2 nanocomposites. J Am Chem Soc 130:5402–5403

Lin Y-C (2013) Applying Ag–TiO_2/functional filter for abating odor exhausted from semiconductor and opti-electronic industries. Clean Technol Environ Policy 15:359–366

Mali SM, Betty CA, Bhosale PN, Patil PS (2011) Hydrothermal synthesis of rutile TiO_2 with hierarchical microspheres and their characterization. Cryst Eng Commun 13:6349–6351

Marinel S, Choi DH, Heuguet R, Agarwal D, Lanagan M (2013) Broadband dielectric characterization of TiO_2 ceramics sintered through microwave and conventional processes. Ceram Int 39:299–306

Mazza T, Barborini E, Piseri P, Milani P, Cattaneo D, Bassi AL, Bottani CE, Ducati C (2007) Raman spectroscopy characterization of rutile TiO_2 nanocrystals. Phys Rev 75:045416 (5 pp)

Mo S, Ching WY (1995) Electronic and optical properties of three phases of titanium dioxide: rutile, anatase and brookite. Phys Rev B 51:13023–13032

Narksitipan S, Thongtem S (2012) Preparation and characterisation of rutile TiO2 films. J Ceram Process 13:35–37

Nick S (2006) Is the band gap of pristine TiO2 narrowed by anion- and cation-doping of titanium dioxide in second-generation photocatalysts. J. Phys Chem B 110:24287–24293

Orazen Erin L (2007) Raman spectroscopy on the nanoscale: an investigation of silver triangle arrays, polymer laminates and gold-loaded mesoporous titania fims. Dissertation, University Of California

Orendorz A, Brodyanski A, Losch J, Bai LH, Chen ZH, Le YK, Ziegler C, Gnaser H (2007) Phase transformation and particle growth in nanocrystalline anatase TiO2 films analyzed by X-ray diffraction and Raman spectroscopy. Surf Sci 601:4390–4394

Oye G, Glomn WR, Vralstad T, Volden S, Magnusson H, Stocker M, Sjoblom J (2006) Synthesis, functionalisation and characterisation of mesoporous materials and sol–gel glasses for applications in catalysis, adsorption and photonics. Adv Colloid Interface Sci 123–126:17–32

Paul S, Choudhury A, Bojja S (2013) Effect of nickel doping on the optical property and photocatalytic activity of titanium dioxide nanoparticles. Micro Nano Lett 8:184–187

Qui J, Zhang S, Zhao H (2011) Recent applications of TiO_2 nanomaterials in chemical sensing in aqueous media Sens. Actuat B: Chem 160:875–890

Serpone N, Lawless D, Khairutdinov R (1995) Size effects on the photophysical properties of colloidal anatase TiO2 particles: size quantization versus direct transitions in this indirect semiconductor. J Phys Chem 99:16646–16654

Sharma S, Chaudhary S, Kashyap SC, Malik VK (2011) DC magnetization investigations in $Ti_{1-x}Mn_xO_2$ nanocrystalline powder. J Alloy Compd 509:7434–7438

Shin K, J-B Yoo, Park JH (2013) Photoelectrochemical cell/dye-sensitized solar cell tandem water splitting systems with transparent and vertically aligned quantum dot sensitized TiO2 nanorod arrays. J Power Sour 225:263–268

So WW, Park SB, Kim KJ, Shin CH, Moon SJ (2001) The crystalline phase stability of titania particles prepared at room temperature by the sol–gel method. J Mater Sci 36:4299–4305

Song S, Liu Z, He Z, Zhang A (2010) Impacts of morphology and crystallite phases of titanium oxide on the catalytic oxidation of phenol. Environ Sci Technol 44:3913–3918

Spurr RA, Myers H (1957) Quantitative analysis of anatase-rutile mixtures with an X-ray diffractometer. Anal Chem 29:760–762

Volkan M, Stokes DL, Vo-Dinh T (2005) A sol–gel derived AgCl photochromic coating on glass for SERS chemical sensor application. Sens Actuat B: Chem. 106:660–667

Wan YW, Chang YM, Ting JM (2010) Room-temperature synthesis of single-crystalline anatase TiO_2 nanowires. Cryst Growth Des 10:1646–1651

Wang D, Zhao J, Chen B, Zhu C (2008) Lattice vibration fundamentals in nanocrystalline Anatase investigated with Raman scattering. J Phys: Condens Matter 20:085212 (7 pp)

Yamada Y, Kanemitsu Y (2010) Blue photoluminescence of highly photoexcited rutile TiO_2: nearly degenerate conduction-band effects. Phys Rev B 82:113103 (4 pp)

Yamada Y, Kanemitsu Y (2012) Determination of electron and hole lifetimes of rutile and anatase TiO_2 single crystals. Appl Phys Lett 101:133907

Yu JC, Yu J, Ho W, Jiang Z, Zhang L (2002) Effects of F^- doping on the photocatalytic activity and microstructures of nanocrystalline TiO_2 powders. Chem Mater 14:3808–3816

Zhang WF, He YL, Zhang MS, Yin Z, Chen Q (2000) Raman scattering on anatase TiO_2 nanocrystals. J Phys D: Appl Phys 33:912–916

Zhang H, Xie C, Zhang Y, Liu G, Li Z, Liu C, Ma X, Zhang W (2008) Effects of thermal treatment under different atmospheres on the spectroscopic properties of nanocrystalline TiO_2. J Appl 103:103107 (4 pp)

Zhang Y, Chen J, Li X (2010) Preparation and photocatalytic performance of anatase/rutile mixed-phase TiO_2 nanotubes. Catal Let 139:129–133

Zhou J, Zhang Y, Zhao XS, Ray AK (2006) Photodegradation of benzoic acid over metal-doped TiO_2. Ind Eng Chem Res 45:3503–3511

5

Efficient adsorption of congo red dye from aqueous solution using green synthesized coinage nanoparticles coated activated carbon beads

Jolly Pal · Manas Kanti Deb

Abstract In this paper, the removal of congo red (CR) dye by adsorption on coinage nanoparticles [silver nanoparticles (AgNPs) and gold nanoparticles (AuNPs)] coated activated carbon (AC) has been discussed. The effect of various process parameters has been investigated by following the adsorption technique. Equilibrium adsorption data of CR were carried out at room temperature. The equilibrium time is independent of the initial CR concentration and the percentage removal of CR increased with increase in contact time. The adsorption data were analyzed by using adsorption isotherm studies. The characteristic parameters for isotherm and related correlation coefficients were determined from graphs of their linear equations. Kinetic studies showed that the adsorption of CR followed pseudo-first-order kinetics. AgNPs- and AuNPs-coated AC is found to be suitable adsorbent for the adsorption of CR. CR was effectively removed 88.0 $\pm$ 0.8 % from aqueous solution using AuNPs beads as the adsorption process. Desorption studies were made to elucidate recovery of the adsorbate and adsorbent for the economic competitiveness of the removal system. The PVP-supported AgNPs and AuNPs-coated AC were successfully recycled for ten successive adsorption–desorption cycles indicating its high reusability.

Keywords Silver nanoparticles · Gold nanoparticles · Activated carbon · Microwave irradiation · Polyvinylpyrrolidone · Congo red

J. Pal · M. K. Deb (✉)
School of Studies in Chemistry, Pandit Ravishankar Shukla University, Raipur 492 010, Chhattisgarh, India
e-mail: debmanas@yahoo.com

Introduction

Water is the most essential requirement in daily life that has been contaminated by the disposal of domestic, municipal, and industrial wastes. Anything, which is not needed, finds its way to the nearest watercourse or land, which further pollutes the ground water. So, there is a need to utilize the available resources effectively without polluting the water (Prasad and Kumar 2010). Textile industries have shown a significant increase in the use of synthetic complex organic dyes as coloring materials (Talarposhti et al. 2001). A dye is carcinogenic, affects reproductive organs and develops toxicity and neurotoxicity (Lakshmi 1987). Therefore, the dyes are to be necessarily removed from water and wastewater. Congo red (CR) is an anionic dye widely used in textiles, paper, rubber, and plastic industries. Figure 1 shows the chemical structures of CR.

Different processes for color removal typically include physical, chemical, and biological schemes. Some processes such as electrochemical techniques and ion-pair extractions are relatively new for textile waste treatment, while others have been used in the industry for a long time. Adsorption has been found to be superior to other techniques for water reuse in terms of initial cost, simplicity of design, use of operation, and insensitivity to toxic substances (Meshko et al. 2001). The lower generation of residues, easy metal recovery, and the possibility to reuse adsorbent are the greatest advantages of this method (Gurnani et al. 2003). The removal of colored and colorless organic pollutants from industrial wastewater is considered as an important application of adsorption processes (Al-Qodah 2000). The treatment of industrial effluents is a challenging topic in environmental science, as control of water pollution has become of increasing importance in

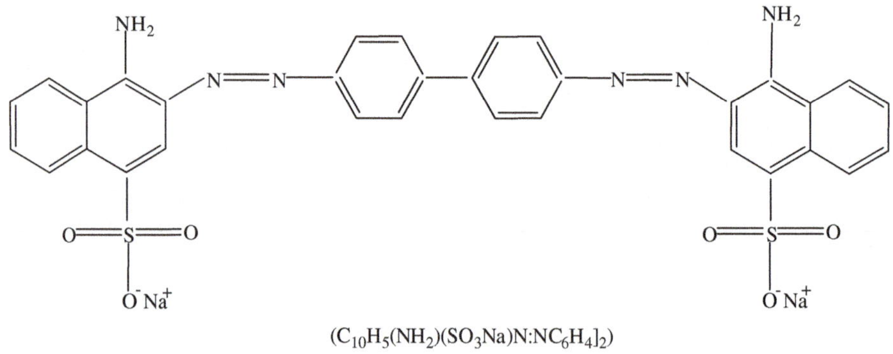

Fig. 1 Chemical structure of CR

recent years. Phenomenologically, adsorption is generally described by a graphic representation of the distribution ratios of adsorbate adsorbed per unit mass of the adsorbent and the concentration of the unadsorbed adsorbate at constant temperature. This graphic representation is known as the adsorption isotherm. Several types of adsorption isotherm have been reported in the literature, but the most widely used are the Freundlich and Langmuir isotherms. One of the most used and successful method for the removal of organic pollutants is their adsorption on activated carbon (AC) arranged in different arrays such as filters (Sabio and Zamora 2006). AC is characterized by a high porosity and a very large surface area, which enables it to efficiently adsorb many kinds of pollutants. In spite of its high adsorption capacity, the use of AC on a large scale is limited by process engineering difficulties such as the dispersion of the AC powder and the cost of its regeneration. AC is the generic term used to describe a family of carbonaceous adsorbents with a highly crystalline form and extensively developed internal pore structure.

Metal nanoparticles (NPs) with controlled size and shape are of great interest because of their morphology-dependent properties (Schmid 1992) and potential applications in a lot of fields (Andres et al. 1996). Metal NPs have attracted considerable interest because of their novel properties and their potential application (Du et al. 2008). Binding capacity of noble metal NPs with dyes are more when compared with other NPs. Novel metal crystallites such as silver and gold provide a more interesting research field due to their close lying conduction and valence bands in which electrons move freely. The free electrons give rise to a surface plasmon absorption band, which depends on both the particle size and chemical surrounding. Right now, shape-controlled synthesis of NPs has been achieved either by using geometric templates (Zande et al. 2000), or by using some additive, such as polymers (Ahmadi et al. 1996), or inorganic anions (Filankembo et al. 2003), to regulate the particle growth. Accurate controls of size, composition, morphology and stability, and the use of environment-friendly procedures are highly desirable for the synthesis of NPs. There have been a number of techniques for NPs synthesis developed over the years using a range of metals (Pal and Deb 2012a, b, c, d, 2013; Pal et al. 2009, 2012, 2013a, b; Shah et al. 2010, 2012; Nafees et al. 2013). The most widely used substances for the stabilization of metal NPs are ligands and polymers, specially natural or synthetic polymers with a certain affinity toward metals, which are soluble in suitable solvents (Hirai et al. 1985). Such substances can also control the reduction rate of the metal ions and the aggregation process of zerovalent metal atoms. The polymers also control the aggregation of metal atoms in solution. Polyvinylpyrrolidone (PVP) is soluble in water and other polar solvents. In solution, it has excellent wetting properties and readily forms films. This makes it good as a coating or an additive to coating. PVP is used in as a binder and complexation agent in agro applications such as crop protection, seed treatment, and coating. Studies have been carried out aiming at developing more effective and selective adsorbent materials, which are abundant in nature, requiring little processing in order to decrease cost (Reddad et al. 2002). In recent years, metal NPs have attracted much research attention due to their unique electric, catalytic, and optical properties originating from the quantum-scale dimensions (Duan et al. 2005). Metal NPs have been of increasing interest in applications to biological and chemical nanosensors. One aspect of the fantastic researches on metal NPs is focused on the phenomena of aggregation or flocculation of metal NPs in solution (Liu et al. 2004). Particles aggregation results in further color changes of metal NPs solutions due to mutually induced dipoles that depend on interparticle distance and aggregate size (Lazarides and Schatz 2000). Recently, metal NPs have been applied in biosensors because of their dimensional similarities with biomacromolecules and significant size-dependent optical and electronic properties (Bukasov and Shumaker-Parry 2007). A number of attempts have been made using different experimental conditions to prepare metal NPs of different sizes and shapes.

Microwave irradiation (Rao and Ramesh 1995) is one of the novel techniques developed during the last years for the synthesis of solid materials. Microwave dielectric heating results from dipolar polarization as a result of dipole–dipole interaction between polar molecules and electromagnetic field. It promotes the nucleation of NPs without interfering in the particle growth process. The main advantage of microwave irradiation is that it produces a uniform heating of the solution, so that a more homogeneous nucleation is obtained as well as a shorter crystallization time, when compared with conventional heating, and it is therefore very useful for the formation of monodisperse metal colloids. Further advantages are short thermal induction period, absence of convection processes, easy control, and low cost.

In this paper, green methods to synthesize silver nanoparticles (AgNPs) and gold nanoparticles (AuNPs) are described. The preparation of AgNPs and AuNPs was carried out by irradiating the solution under microwave. NPs were characterized by UV–visible spectroscopy, transmission electron microscopy (TEM), and X-ray diffraction (XRD). The adsorption behavior of CR by the PVP-supported AgNPs and AuNPs-coated AC are also given in detail in this paper. Adsorption studies were carried out to study the effects of various experimental parameters such as initial concentration, contact time, pH, and adsorbent dose were evaluated. In addition, the equilibrium isotherms, adsorption kinetics, and desorption studies of CR onto the AgNPs- and AuNPs-coated AC were also investigated.

Experimental

Materials and apparatus

$AgNO_3$ was obtained from Merck, $HAuCl_4$ was obtained from Aldrich, glucose was obtained from Molychem, PVP was obtained from Himedia and CR was obtained from LOBA. All aqueous solutions were prepared in triple-distilled water. 1×10^{-4} mol/L $AgNO_3$, 1×10^{-4} $HAuCl_4$, 0.5 mol/L glucose, 1 % PVP, and parts per million solutions of CR were used.

A Samsung CE2877 domestic microwave oven (850 W), Samsung India Electronics Ltd, New Delhi, India, was employed for irradiating solutions. The particle size and morphology of the AuNPs were characterized by Morgagni 268D transmission electron microscope operating at 80 KB (Mega view III Camera CCD) at the All India Institute of Medical Sciences (AIIMS), New Delhi. Varian Carry 50 UV–vis spectrophotometer was used for spectral studies. The XRD measurements were carried out using Bruker D8 Advance X-ray diffractometer at UGC-DAE Consortium for Scientific Research INDORE-CENTRE. The X-rays were produced using a sealed tube and the wavelength of X-ray was 0.154 nm (Cu K-alpha). The X-rays were detected using a fast counting detector based on silicon strip technology (Bruker-LynxEye detector).

Preparation of AgNPs and AuNPs

Silver nanoparticles were synthesized by using PVP as protective agent and using $AgNO_3$ as precursor and glucose as reducing agent. The reaction solution were prepared by dissolving 0.5 mol/L glucose, 1.0×10^{-4} mol/L $AgNO_3$, and 1 % PVP in triple distilled water in a 50-ml conical flask to obtain a homogeneous reaction mixture. Then the conical flask was placed on the turntable of the microwave oven. The mixture was irradiated at a power of 300 W for duration of 4 min. The reaction was carried out discontinuously to prevent an increase of pressure. After irradiation, the dilute colloidal solution with pale yellow color was cooled to room temperature and stored in a refrigerator at −4 °C for further characterization. In a similar approach, AuNPs was prepared using $HAuCl_4$ in place of $AgNO_3$. The colors of the AuNPs were purple (Pal and Deb 2012c; Pal et al. 2013b).

Fabrication of NPs

First, charcoals were locally collected, cut into small spherical shape (10 mm diameter). In order to remove the contaminants from its surface, charcoals were washed by keeping for hours the pieces into porcelain dish filled with distilled water; and after removing, dried (250 °C) on hot plate. In this process, charcoals convert to AC. Such freshly prepared AC has a clean surface. This AC produces a surface with high adsorptive capacity. Sufficient amounts of small pieces of AC were dipped into a solution of the prepared AgNPs capped with PVP, in a glass beaker (250 ml). The beaker was kept on a hot plate and the content was heated gently at 50 °C till dryness and complete removal of any moisture content. The PVP used here provides a polymeric support to fasten AgNPs over the surface of AC. The AC thus coated with AgNPs was taken for adsorption studies. In a similar approach, AuNPs-coated AC was prepared using $HAuCl_4$ in place of $AgNO_3$ (Pal and Deb 2012c; Pal et al. 2013b).

Adsorption studies

Adsorption isotherm describes the equilibrium relationship between bulk activity of adsorbate in solution and the moles adsorbed on the surface, at constant temperature. The efficiency of the removal of CR from aqueous solutions using PVP-supported AgNPs and AuNPs-coated AC was experimentally studied by recording adsorption isotherms. Adsorption experiment was conducted in which aliquots of CR solution with known concentrations were

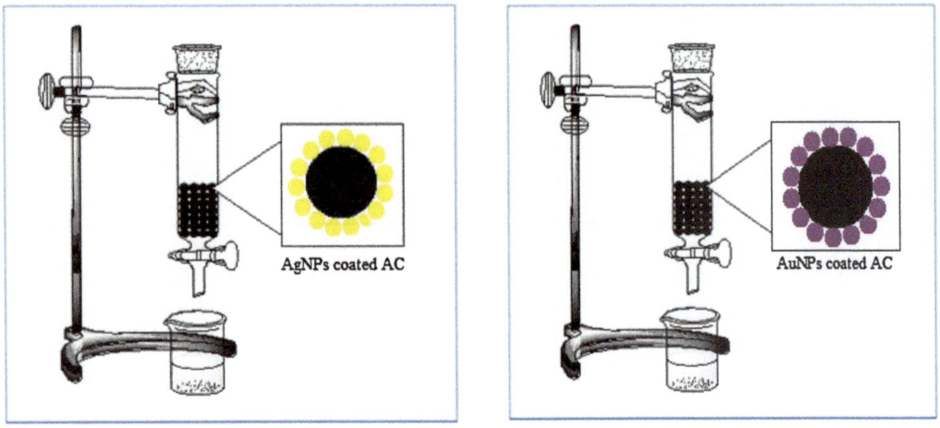

Fig. 2 Schematic diagram of column adsorption experiment

introduced into column filled with accurately weighed amount of PVP-supported AgNPs and AuNPs-coated AC. The adsorption of CR by PVP-supported AgNPs and AuNPs-coated AC was investigated in aqueous solutions at room temperature (27 °C). Figure 2 shows the schematic diagram of column adsorption experiments carried out in this work.

To determine the equilibrium adsorption capacity of the CR by PVP-supported AgNPs and AuNPs-coated AC, 100 ml of CR solution was placed separately in columns (10 cm height and 4 cm diameter) filled with accurately weighed amounts of PVP-supported AgNPs and AuNPs-coated AC. The columns, at room temperature, were sealed with stopper and left for equilibration. Aliquots of the CR solution were drained out after equilibrium time to determine the CR equilibrium concentration at a wavelength of maximum absorbance (λ_{max}), i.e., at 500 nm for CR using a UV–visible spectrophotometer.

Effects of various process parameters on the extent of removal of CR were studied. The data were analyzed statistically and interpreted. Percentage removal was calculated using the following equations:

$$\text{Percentage removal} = 100(C_i - C_e)/C_i$$

where C_i and C_e are initial and equilibrium (final) concentration of CR (mg/L), respectively. The data were analyzed statistically and interpreted. The data were modelled with Freundlich and Langmuir adsorption isotherms.

Desorption studies

Desorption of CR from PVP-supported AgNPs and AuNPs-coated AC surface were carried out as follows: after adsorption experiments with CR under optimum conditions, the PVP-supported AgNPs and AuNPs-coated AC was separated and washed with 0.05 N HNO_3 to remove adsorbed CR. The color intensity of the CR desorbed was measured UV–visible spectrophotometer. The %desorption

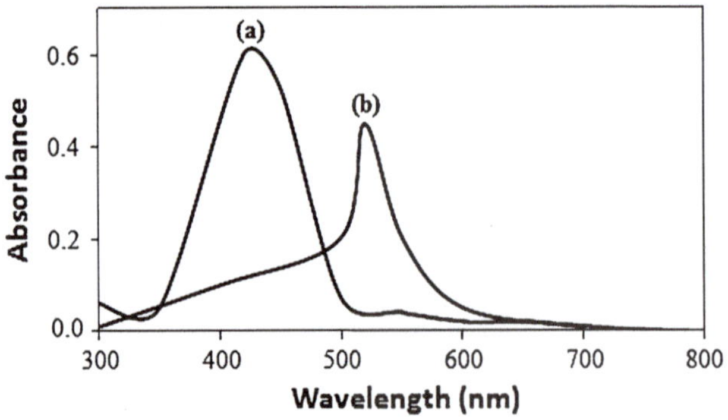

Fig. 3 UV–visible spectrum of the prepared **a** AgNPs and **b** AuNPs. Varian Carry 50 UV–visible spectrophotometer

of the CR were calculated with the help of following equation:

$$\%\text{Desorption} = \frac{\text{amount of CR liberated by acid}}{\text{amount of CR adsorbed on adsorbents}} \times 100$$

The above washed adsorbents were reused to study further the change in removal efficiency of used adsorbents.

Results and discussions

Formation of AgNPs and AuNPs

It is interesting to find that AgNPs and AuNPs can be synthesized with PVP, glucose and $AgNO_3$ for AgNPs, and $HAuCl_4$ for AuNPs promoted by microwave irradiation. Figure 3a shows the UV–visible spectrum of the prepared AgNPs in aqueous solution. The colloidal silver solution thus formed exhibits a single and strong absorption at 420 nm. The yellow color of the colloidal silver sample provides clear evidence for the formation of AgNPs. Figure 3b shows the UV–visible spectrum of the prepared

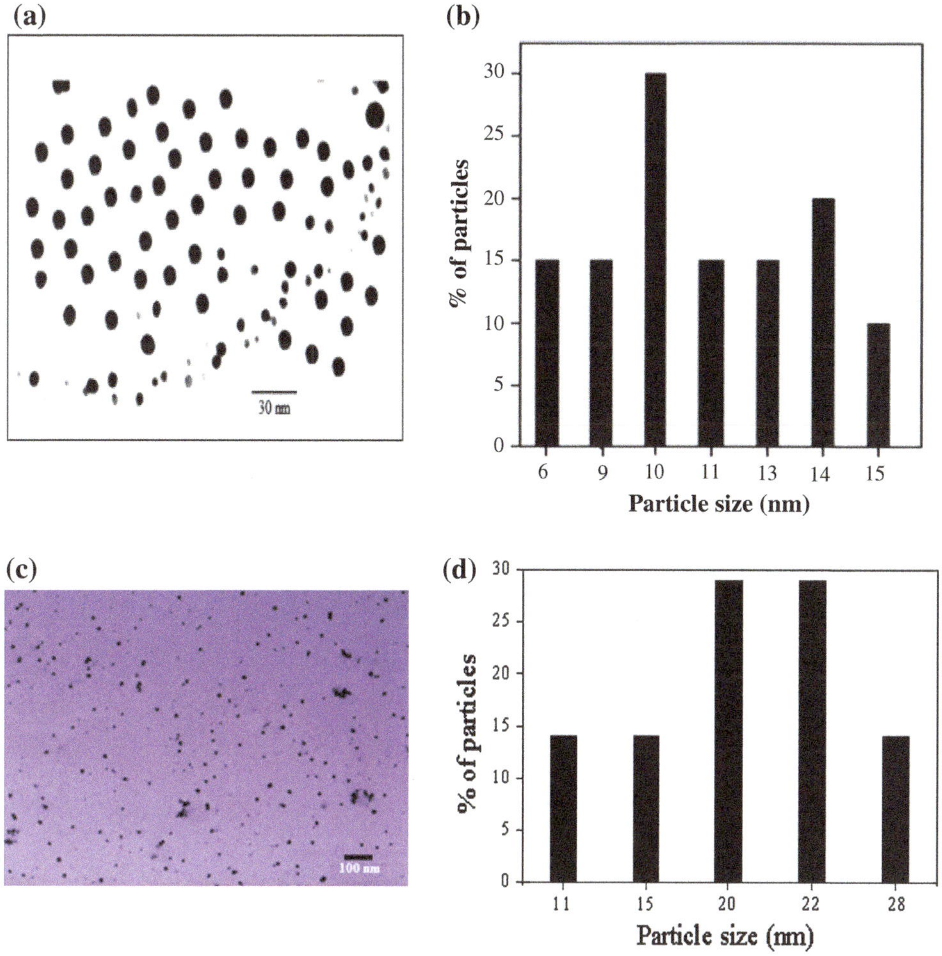

Fig. 4 **a** TEM image of the AgNPs prepared employing glucose as reducing agent and PVP as stabilizing agent (300 W; 4 min irradiation). Magnification: 13,000×; resolution: 1,376 × 1,032 × 16; image intensity: gray value; accelerating voltage: 70 kV; microscope: Morgagni 268; camera type: Keen View FW. **b** Size distributions of the AgNPs synthesized under optimum experimental conditions. **c** TEM image of the AuNPs prepared employing glucose as reducing agent and PVP as stabilizing agent (300 W; 4 min irradiation). Magnification: 25,000×; resolution: 1,376 × 1,032 × 16; image intensity: gray value; accelerating voltage: 70 kV; microscope: Morgagni 268; camera type: Keen View FW. **d** Size distributions of the AuNPs synthesized under optimum experimental conditions

AuNPs in aqueous solution. The colloidal gold solution thus formed exhibits a single peak with strong absorption at 530 nm. The purple color of the colloidal gold sample provides clear evidence for the formation of AuNPs (Pal and Deb 2012d; Pal et al. 2013b). The UV–visible spectra of the prepared AgNPs and AuNPs in aqueous solution should exhibit a single surface plasmon band, which means NPs was small and spherical (Mie 1908).

Transmission electron microscopy image was further used to characterize formed NPs. Figure 4a shows the TEM image and the corresponding particle size distribution histogram, Fig. 4b of the silver colloidal solution. The photograph shows that most of the particles are nearly spherical. The size distribution histogram reveals that such silver particles range from 6 to 15 nm in size. The 30 % of these NPs consist of particles with 10 nm diameter, 20 % particles were of 14 nm size remaining particles were in the range 6, 9, 11, 13, and 15 nm diameter sizes. The average size of AgNPs is about 11 nm via TEM images. It may be noted that all these particles are well separated from each other. AgNPs thus formed were free from flocculation or aggregation for several weeks, suggesting that the polymer serves as a very effective protective agent for AgNPs. Figure 4c shows a typical TEM image and Fig. 4d the corresponding particle size distribution histogram of AuNPs produced. The photograph shows that most of the particles are nearly spherical. The size distribution histogram reveals that such gold particles range from 11 to 28 nm in size. The 14 % of these NPs consist of particles with 11 and 15 nm diameter, 28 % particles were of 20 and 22 nm. The average size of AuNPs is about 20 nm via TEM images. It was noticed that all these particles were well separated from each other (Pal and Deb 2012d; Pal et al. 2013b).

The XRD patterns of AC- and PVP-supported AgNPs and AuNPs-coated AC, shown in Fig. 5, were also able to confirm the existence of AgNPs and AuNPs. AC shows peaks at $2\theta = 24.7°$, 29.5°, 39.5°, 43.3°, 47.4°, and 48.6°. After coating of PVP-supported AgNPs and AuNPs over AC, a significant change in XRD pattern was observed. In the PVP-supported AgNPs and AuNPs-coated AC sample new peaks appeared at $2\theta = 23.0°$, 26.6°, 29.4°, 36.0°, 39.4°, 43.2°, 47.4°, 48.4°, 57.4°, 60.6°, 64.6° and $2\theta = 23.2°$, 29.5°, 36.0°, 39.5°, 43.3°, 47.6°, 48.5°, and 57.4°, 60.6°, and 64.6°, respectively. The crystallinity of AC was apparently lower than that of the PVP-supported

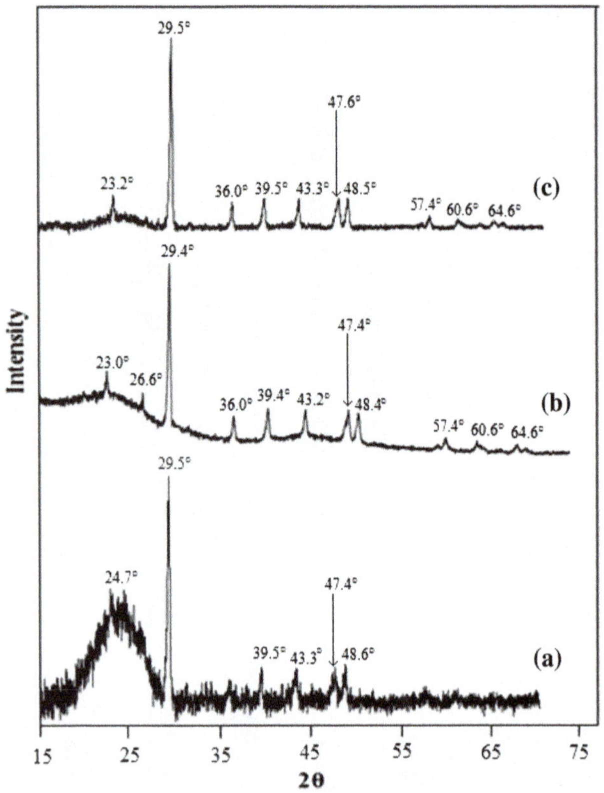

Fig. 5 XRD of **a** AC, **b** PVP-supported AgNPs-coated AC, and **c** PVP-supported AuNPs-coated AC. Bruker D8 advance X-ray diffractometer; Wavelength of S-ray was 0.154 nm (Cu K-alpha); detector based on silicon strip technology (Bruker-LynxEye detector)

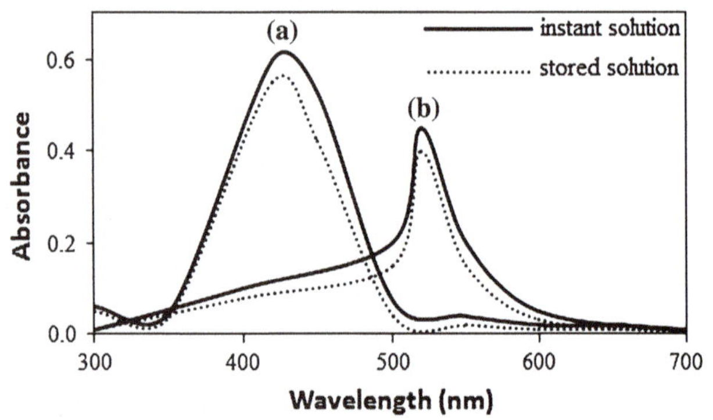

Fig. 6 UV–visible spectra of **a** AgNPs, **b** AuNPs after it is stored for 8 months in refrigerator (−4 °C)

AgNPs and AuNPs-coated AC. The results indicate that the crystallinity of PVP-supported AgNPs and AuNPs-coated AC were increased because of the formation of NPs (Pal and Deb 2012d; Pal et al. 2013b).

Effect of stabilization of AgNPs and AuNPs

As shown in Fig. 6a, b, the absorbance spectra of silver and gold colloidal solution were observed to have changed only slightly in their relative absorbance intensities, after the solution was stored for 8 months in refrigerator (−4 °C). The solution remained perfectly transparent with no obvious change in color while storage. The results implied that the AgNPs an AuNPs prepared by this method were very stable with negligible aggregation.

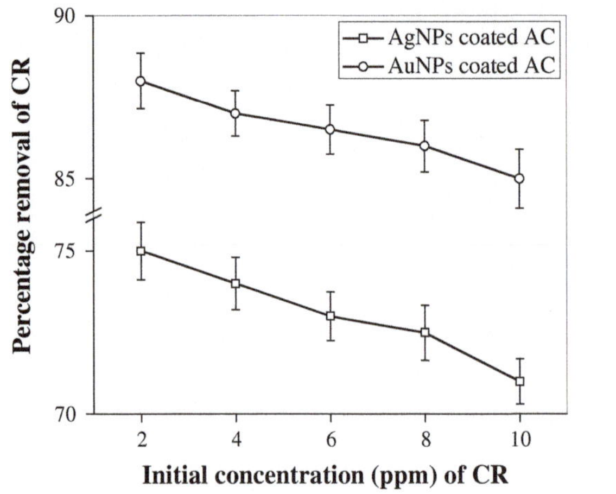

Fig. 7 Effect of initial concentration on the removal of CR by PVP-supported AgNPs and AuNPs-coated AC at room temperature ($n = 5$)

Effect of initial concentration

The effect of initial concentrations on the percentage removal of CR is shown in Fig. 7 with error bars showing the values of standard deviation ($n = 5$). The experiments were conducted at pH 6.5 ± 0.8 for CR, respectively, at room temperature. The concentrations of CR were varied from 2 to 10 mg/L. The results show that the value of percentage removal of CR decreases from 75.0 ± 0.8 to 71.0 ± 0.7 % by PVP-supported AgNPs-coated AC and 88.0 ± 0.8 to 85.0 ± 0.9 % by PVP-supported AuNPs-coated AC, respectively. Thus, the percentage removal is a direct function of initial concentrations of CR. This may probably be due to the limited number of available active sites on the surface of PVP-supported AgNPs and AuNPs-coated AC to accommodate higher concentration of CR (Li et al. 2010).

Effect of contact time

The percentage removal increased with increase in contact time and reached a constant value. This may be due to the attainment of equilibrium condition of contact time for PVP-supported AgNPs and AuNPs-coated AC. The effect of contact time on adsorption of CR with error bars showing the values of standard deviation ($n = 5$) are shown in Fig. 8. The effect of contact time was studied with initial CR concentration of 2 mg/L at pH 6.5 ± 0.8

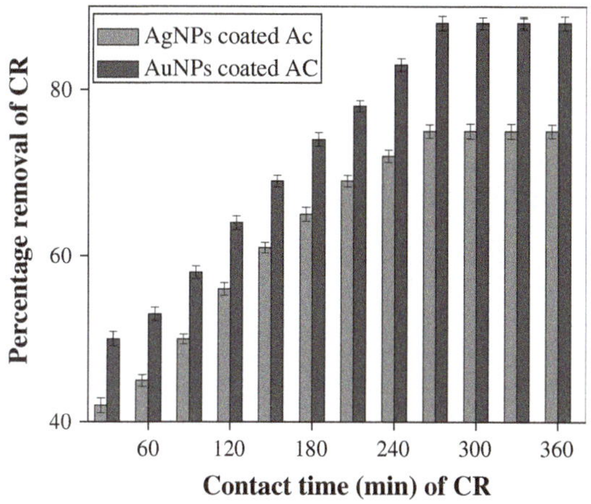

Fig. 8 Effect of contact time on the removal of CR by PVP-supported AgNPs and AuNPs-coated AC at room temperature ($n = 5$)

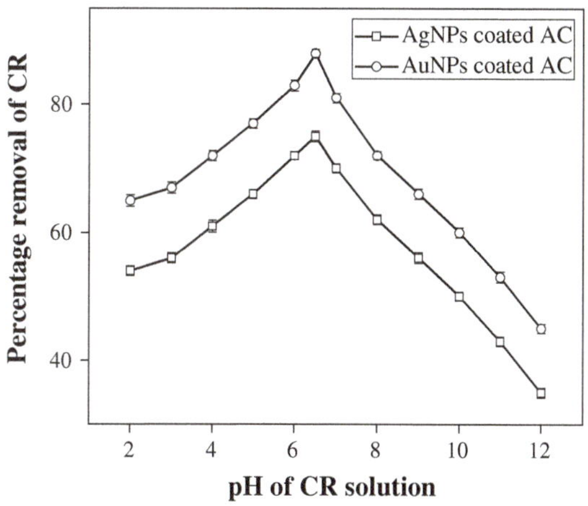

Fig. 9 Effect of pH on the removal of CR by PVP-supported AgNPs and AuNPs-coated AC at room temperature ($n = 5$)

for CR at room temperature. The percentage removal of CR for a contact time period between 30 and 270 min is 42.0 ± 0.8 to 75.0 ± 0.8 % and 50.0 ± 0.8 to 88.0 ± 0.8 % for PVP-supported AgNPs and AuNPs-coated AC, respectively.

The results show a significant improvement in percentage removal of CR by the AgNPs- and AuNPs-coated AC. At the initial stage, the rate of removal of CR was higher, due to the availability of more than required number of active sites on the surface of adsorbents and became lower at the later stages of contact time, due to the decreased or lesser number of available active sites (Kannan and Karupasamy 1998).

Influence of pH

To study the influence of pH on the percentage removal of CR by PVP-supported AgNPs and AuNPs-coated AC, experiments were carried out at different pH values varying

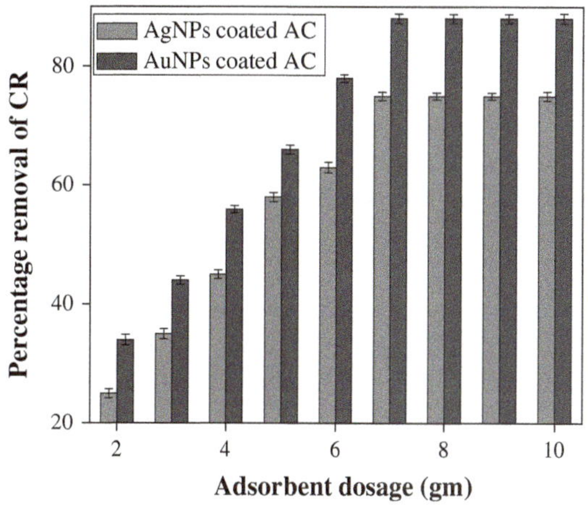

Fig. 10 Effect of adsorbent dose on the removal of CR by PVP-supported AgNPs and AuNPs-coated AC at room temperature ($n = 5$)

from 2 to 12. The observed data are presented in Fig. 9 for CR with error bars showing the values of standard deviation ($n = 5$).

It may be seen that as pH increases, the extent of removal increases, reaches to a maximum value and then decreases steadily. Therefore, the optimum pH of aqueous medium for removal of CR was fixed at 6.5 ± 0.8 for PVP-supported AgNPs and AuNPs-coated AC adsorbent systems.

Effect of adsorbent dosage

Adsorbent dose is an important parameter in the determination of adsorption capacity. The effect of the adsorbent dose was investigated by the addition of various amounts of adsorbent from 2 to 10 g in 100 mL aqueous solution of CR (2 mg/L) at room temperature for equilibrium time. The result is shown in Fig. 10. It is evident from the plots that the percentage removal of CR from the aqueous solution increases with increase in the adsorbent dosage.

It was observed that the removal efficiency increased from 25 ± 0.8 to 75 ± 0.8 % and 34.0 ± 0.8 to 88.0 ± 0.8 % for CR with the adsorbent dose varying from 2 to 7 g of PVP-supported AgNPs and AuNPs-coated AC, respectively, and there after reached a constant value. It is readily understood that the number of available adsorption sites increases by increasing the adsorbent dose and it therefore results in an increase in the percentage of CR adsorbed.

Adsorption isotherm

The adsorption characteristics for wide range of adsorbate concentrations are mostly described by adsorption isotherm (Freundlich and Langmuir), which relates the equilibrium adsorbate concentration in the bulk and the uptake of adsorbate on the adsorbent surface. Freundlich isotherm is presented by the following relation:

$$\ln Q_e = \ln K_F + \ln C_e$$

where K_F and n are Freundlich constant, a characteristic of the system indicating the adsorption capacity (mg/g), and adsorption intensity or surface heterogeneity, respectively.

The Langmuir isotherm is valid for monolayer adsorption onto a surface containing a finite number of identical sites.

$$\frac{C_e}{Q_e} = \frac{b}{Q_0} + \frac{C_e}{Q_0}$$

where C_e is the equilibrium dye concentration in the solution (mg/L), Q_e is the equilibrium dye concentration on the adsorbent (mg/g), Q_0 is the maximum adsorption capacity of the dye (forming a monolayer) per unit weight of adsorbent (mg/g), and b is a constant related to the affinity of the binding sites (L/mg).

The essential characteristics of the Langmuir isotherm can be expressed by a separation factor R_L (Mall et al. 2005), which is defined in the following equation:

$$R_L = \frac{1}{1 + bC_0}$$

The R_L value shows the nature of the adsorption process to be unfavorable ($R_L > 1$), linear ($R_L = 1$), favorable ($0 < R_L < 1$), or irreversible ($R_L = 0$).

The adsorption isotherm of CR was investigated at optimized condition pH 6.5 ± 0.8 at room temperature. As seen from the Fig. 11a, the Freundlich model described the equilibrium adsorption process. The values for CR on the Freundlich parameters n and K_F were 1.12, 24.82 ± 0.45 mg/g for PVP-supported AgNPs-coated AC; and 1.14, 28.40 ± 0.55 mg/g for PVP-supported AuNPs-coated AC, respectively. Figure 11b present the Langmuir plots for the adsorption of CR onto PVP-supported AgNPs and AuNPs-coated AC. The values for CR on the Langmuir parameters Q_0 and b were 0.47 ± 0.45 mg/g, 0.09 L/mg for PVP-supported AgNPs-coated AC; and 0.50 ± 0.55 mg/g, 0.22 L/mg for PVP-supported AuNPs-coated AC, respectively. The results obtained from adsorption isotherms, according to Freundlich and Langmuir models are shown in Table 1. Freundlich values indicated high adsorption capacity of CR on PVP-supported AgNPs and AuNPs-coated AC when compared to the Langmuir isotherm model, in which adsorption is based on heterogeneous surface of adsorbents and Freundlich isotherm is not restricted to the formation of the monolayer (Mohammadi et al. 2011). The adsorption capacity was much faster when PVP-supported AuNPs-coated AC was the adsorbent as compared with PVP-supported AgNPs-coated AC. It is estimated from the adsorption experiment that loading in the case of PVP-supported AuNPs-coated AC is much more compared to PVP-supported AgNPs-coated AC (Saha et al. 2009).

Adsorption kinetics

The percentage removal of CR at a fixed adsorbent dose (7 g) was monitored at different time intervals. The kinetics of CR removal by PVP-supported AgNPs and AuNPs-coated AC indicated rapid binding of CR to the PVP-supported AgNPs and AuNPs-coated AC during first few minutes, followed by a slow increase until a state of equilibrium time was reached. No change in the uptake capacity was observed with further increase in equilibrium time. The initial rapid phase may be due to increased number of vacant sites available at the initial stage. Generally, when adsorption involves a surface reaction process, the initial adsorption is rapid. Then, a slower adsorption would follow as the available adsorption site gradually decreases (Kannan and Karrupasamy 1998).

The adsorption kinetics can be described by the pseudo-first-order kinetic model and the pseudo-second-order kinetic model (Eftekhari et al. 2010; Dogan et al. 2007). The pseudo-first-order equation is generally expressed as following:

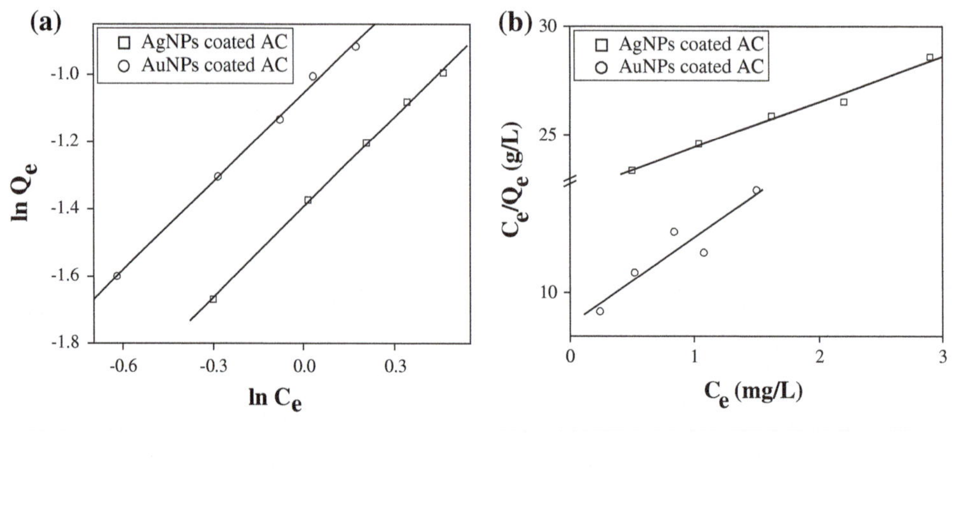

Fig. 11 **a** Freundlich isotherm. **b** Langmuir isotherm plot for adsorption of CR on PVP-supported AgNPs and AuNPs-coated AC at room temperature

$$\left(\frac{1}{q_t}\right) = \left(\frac{k_1}{q_e}\right)\left(\frac{1}{t}\right) + \left(\frac{1}{q_e}\right)$$

Kinetics of adsorption was modelled by the pseudo-second-order equation as shown below.

$$\frac{t}{q_t} = \frac{t}{q_e} + \frac{1}{q_e^2 k_2}$$

where, q_e and q_t are the equilibrium adsorption capacity (mg/g) and the adsorption capacity at time t, respectively; k_1 is the rate constant of pseudo-first-order adsorption (time^{-1}). The plot of q_t versus t should give a linear relationship from which k_1 and q_e can be determined from the slope and intercept of the plot, respectively. Here, k_2 is the pseudo-second-order rate constant of adsorption (g/mg/time). The plot is a linear relationship between t/q_t and t, q_e and k_2 can be determined from the slope and intercept of the plot of t/q_t versus t, respectively. Graphical representation of pseudo-first-order equation is shown in Fig. 12a for CR when adsorbed onto PVP-supported AgNPs and AuNPs-coated AC. Graphical representation of pseudo-second-order equation is shown in Fig. 12b for CR when adsorbed onto PVP-supported AgNPs and AuNPs-coated AC. The correlation coefficients (R^2), adsorption capacity (q_e), and the rate constant for adsorbents are summarized in Table 2. For all the systems studied, good correlation coefficients were obtained by fitting the experimental data to pseudo-first-order and pseudo-second-order kinetics. The adsorption kinetic data, however, fitted best in pseudo-first-order model, where linear plot of t versus q_t was obtained. Pseudo-first-order values indicated that the adsorption of CR on PVP-supported AgNPs and AuNPs-coated AC were high adsorption capacity and correlation coefficients than pseudo-second-order, which can be ascribed to the effective physical adsorption of CR onto adsorbents.

Table 1 Adsorption isotherm data for removal of CR by PVP-supported AgNPs and AuNPs-coated AC at room temperature

Adsorbent	Freundlich constants			Langmuir constants			
	K_F (mg/g)	n	R^2	Q_0 (mg/g)	b (L/mg)	R^2	R_L
AgNPs-coated AC	24.82 ± 0.45	1.12	0.99	0.47 ± 0.45	0.09	0.99	0.84
AuNPs-coated AC	28.40 ± 0.55	1.14	0.99	0.50 ± 0.55	0.22	0.88	0.70

Desorption studies

Desorption studies help to elucidate the nature of adsorption and to the recovery of the valuable adsorbate and adsorbent. To make the adsorbent economically competitive, the prepared composite materials should be reused for 'n' number of cycles.

Attempts were made to the effective desorption of CR from the spent PVP-supported AgNPs and AuNPs-coated AC using 0.05 N HNO_3. Percentage removal of CR by PVP-supported AgNPs and AuNPs-coated AC is shown in Fig. 13.

It was observed that the percentage removal of PVP-supported AgNPs and AuNPs-coated AC was almost same after ten adsorption–desorption cycles. This result shows that PVP-supported AgNPs and AuNPs-coated AC is good reusable adsorbent for the removal of CR from aqueous solution.

Comparison with other studies

Table 3 shows a list of adsorbents used in wastewater treatment for CR. Because of its great capacity to remove CR, AC is the most effective adsorbents. The loss of AC during recycling process by HNO_3 was prevented by the coating of NPs over it. Additionally, NPs provide more surface area for adsorption to occur on it and consequently this enhances the adsorption efficiency. Table 4 shows the extent of removal of the tested CR by different adsorbents at the optimized condition. It is clear from the observed data that the maximum degree of adsorption for different

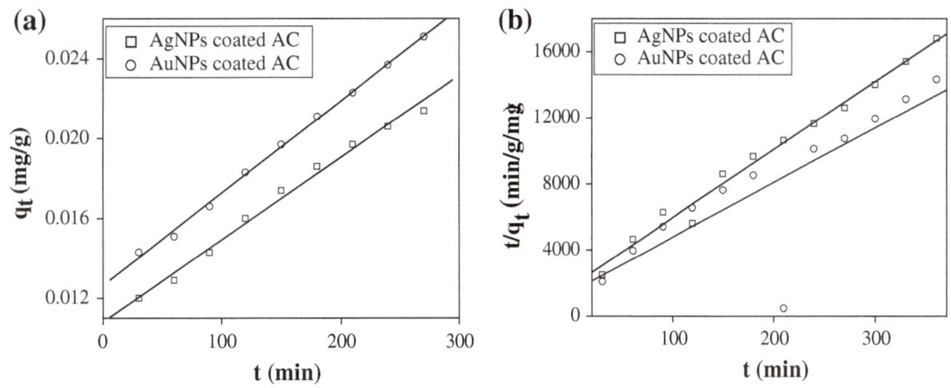

Fig. 12 **a** Pseudo-first-order and **b** pseudo-second-order kinetic plot for the removal of CR by PVP-supported AgNPs and AuNPs-coated AC

Table 2 Adsorbents value calculated using the pseudo-first-order and pseudo-second-order kinetic models for the CR at room temperature

S. no.	Adsorbent	Pseudo-first-order			Pseudo-second-order		
		q_e (mg/g)	k_1 (h^{-1})	R^2	q_e (mg/g)	k_2 (g/mg/h)	R^2
1.	AgNPs-coated AC	79.36 ± 0.45	0.039	0.998	0.025 ± 0.451	0.73	0.991
2.	AuNPs-coated AC	92.59 ± 0.55	0.037	0.990	0.029 ± 0.552	0.57	0.986

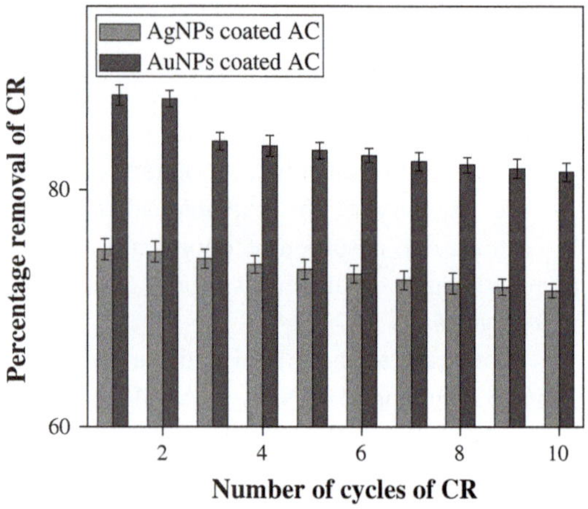

Fig. 13 Percentage removal of CR by PVP-supported AgNPs and AuNPs-coated AC after adsorption–desorption cycles ($n = 5$)

adsorbent systems, at a minimum concentration of CR tested, is in the order AC-PVP-AuNPs > AC-PVP-AgNPs > AC-PVP > AC.

Conclusion

Silver nanoparticles and AuNPs have been synthesized rapidly in chemically green condition. AgNPs and AuNPs were synthesized from $AgNO_3$ and $HAuCl_4$, respectively, using glucose as reducing agent and PVP as stabilizing agent. The particles were characterized by UV–visible spectroscopy, TEM, and XRD analysis and they were found to be spherical and crystalline and with average size of 11 nm AgNPs and 20 nm AuNPs. The investigation on the formation of AgNPs and AuNPs using UV–visible spectroscopy has shown that the AgNPs and AuNPs formed are nanosized, uniformly distributed and absorption peak at 420 for AgNPs and at 530 nm for AuNPs. In this work, the newly synthesized adsorbent, i.e., AC modified with NPs was applied successfully for separation of CR. The experiments were performed with CR initial concentrations varying from 2 to 10 mg/L. The percentage removal increases with increase in contact time. The adsorption equilibrium was reached at about 270 min for CR. The pH 6.5 ± 0.8 for CR were found most favorable and at this pH the percentage removal is high at room temperature (27 °C). The data are well represented by Freundlich isotherm and Langmuir isotherm indicating favorable adsorption of CR by the PVP-supported AgNPs and AuNPs-coated AC. The characteristic parameters for isotherm and related correlation coefficients were determined from graphs of their linear equations. The maximum adsorption capacity from Freundlich isotherm model were 24.82 ± 0.45 mg/g for PVP-supported AgNPs-coated AC and 28.40 ± 0.55 mg/g for PVP-supported AuNPs-coated AC of CR. This indicates the adsorption is based on heterogeneous surface of adsorbents and Freundlich isotherm is not restricted to the formation of the monolayer. The data on kinetic studies indicated that the adsorption kinetics of CR on PVP-supported AgNPs and AuNPs-coated AC followed the pseudo-first-order and pseudo-second-order model at room temperature. The data on kinetic studies indicated that the adsorption kinetics of CR on PVP-supported AgNPs and AuNPs-coated AC followed the pseudo-

Table 3 Recent reported adsorbent systems for CR

S. no.	Adsorbents	Adsorption capacities (mg/g)	Percentage removal (%)	Economic viability	Sources
1	Rice husk carbon activated by steam	–	99	–	Sharma and Janveja, 2008
2	Anilinepropylsilica xerogel	22.62	97	–	Pavan et al., 2008
3	Hydrogen peroxide treated tendu waste	134.4	73	–	Nagda and Ghole, 2009
4	AgNPs-coated AC	64.80	75	Percentage removal almost same after 10 cycles	Present work
5	AuNPs-coated AC	71.05	88	Percentage removal almost same after 10 cycles	Present work

Table 4 Removal of CR by different adsorbents at optimized condition (pH ~6.5, 27 °C, 270 min)

S. no.	Dye (mg/L)	AC (%)	PVP-supported AgNPs-coated AC (%)	PVP-supported AuNPs-coated AC (%)	PVP-supported AC (%)
1.	2.0	53.0	75.0	88.0	63.0
2.	4.0	50.3	74.0	87.0	60.0
3.	6.0	48.5	73.0	86.0	59.3
4.	8.0	47.0	72.5	86.5	57.6
5.	10	46.2	71.0	85.0	56.1

first-order kinetics indicating physicosorptions. The regeneration of PVP-supported AgNPs and AuNPs-coated AC is found almost same even after at least ten cycles. Adsorption capacity of PVP-supported AuNPs-coated AC seems to be superior than the PVP-supported AgNPs-coated AC for CR because the loading of PVP-supported AuNPs-coated AC is much high when compared with PVP-supported AgNPs-coated AC in adsorption experiments.

References

Ahmadi TS, Wang ZL, Green TC, Henglein A, El-Sayed MA (1996) Shape-controlled synthesis of colloidal platinum nanoparticles. Science 272:1924–1926

Al-Qodah Z (2000) Adsorption of dyes using shale oil ash. Water Res 34:4295–4303

Andres RP, Bielefeld JD, Henderson JI, Janes DB, Kolagunta VR, Kubiak CP, Mahoney WJ, Osifchin RG (1996) Self-assembly of two-dimensional superlattice of molecularly linked metal clusters. Science 273:1690–1693

Bukasov R, Shumaker-Parry JS (2007) Highly tunable infrared extinction properties of gold nanocrescents. Nano Lett 7:1113–1118

Dogan M, Ozdemir Y, Alkan M (2007) Adsorption kinetics and mechanism of cationic methyl violet and methylene blue dyes onto sepiolite. Dyes Pigm 75:701–713

Du BD, Phu DV, Duy NN, Lan NTK, Lang VTK, Thanh NVK, Phong NTP, Hien NQ (2008) Preparation of colloidal silver nanoparticles in ploy (*N*-vinylpyrrolidone) by γ-irradiation. J Exp Nanosci 3(3):207–213

Duan HY, Chen F, Ai XP, He ZK (2005) The interaction between propranolol and gold nanoparticles and its analytical application. Chin Chem Lett 16:947–950

Eftekhari A, Yangjeh H, Sohrabnezhad S (2010) Application of AlMCM-41 for competitive adsorption of methylene blue and rhodamine B: thermodynamic and kinetic studies. J Hazard Mater 178:349–355

Filankembo A, Giorgio S, Lisiecki I, Pileni MP (2003) Is the anion the major parameter in the shape control of nanocrystals. J Phys Chem B 107:7492–7500

Gurnani V, Singh AK, Venkataramani B (2003) Cellulose functionalized with 8-hydroxyquinoline: new method of synthesis and applications as a solid phase extractant in the determination of metal ions by flame atomic absorption spectrometry. Anal Chim Acta 485:221–232

Hirai H, Chawanya H, Toshima N (1985) Synthesis and properties of a chelating resin containing a macrocyclictetraaza system as active group. React Polym 3:127–135

Kannan N, Karrupasamy K (1998) Low cost adsorbents for the removal of phenyl acetic acid from aqueous solution. Indian J Environ Prot 18:683–690

Lakshmi D (1987) Influence of the dyestuff effluents of a textile processing unit, Cauvery river, erode on the activity and glycogen content of brain, liver, heart and muscle tissue. M. Phil. Dissertation. Bharathiar University, Coimbatore

Lazarides AA, Schatz GC (2000) DNA-linked metal nanosphere materials: structural basis for the optical properties. J Phys Chem B 104:460–467

Li G, Du Y, Tao Y, Deng H, Luo X, Yang J (2010) Iron (II) cross-linked chitin-based gel beads: preparation, magnetic property and adsorption of methyl orange. Carbohydr Polym 82:706–713

Liu XL, Yuan H, Pang DW (2004) Resonance light scattering spectroscopy study of interaction between gold colloid and thiamazole and its analytical application. Spectrochim Acta A 60:385–389

Mall ID, Srivastava VC, Agarwal NK, Mishra IM (2005) Removal of congo red from aqueous solution by bagasse fly ash and activated carbon: kinetic study and equilibrium isotherm analyses. Chemosphere 61:492–501

Meshko V, Markovska L, Minchev M, Rodrigues AE (2001) Adsorption of basic dyes on granular activated carbon and natural zeolite. Water Res 35:3357–3366

Mie G (1908) Beitrage zur Optik truber Medien, speziell kolloidaler Metallosungen. Ann Phys 25:377–445

Mohammadi N, Khani H, Gupta VK, Amereh E, Agarwal S (2011) Adsorption process of methyl orange dye onto mesoporous carbon material—kinetic and thermodynamic studies. J Colloid Interface Sci 362:457–462

Nafees M, Ali S, Idrees S, Rashid K, Shafique M (2013) A simple microwave assists aqueous route to synthesis CuS nanoparticles and further aggregation to spherical shape. Appl Nanosci 3:119–124

Nagda GK, Ghole VS (2009) Biosorption of congo red by hydrogen peroxide treated tendu waste. Iran J Environ Health Sci Eng 6:195–200

Pal J, Deb MK (2012a) Effective removal of brilliant green dye from aqueous solution by adsorption onto biopolymer supported silver nanoparticles beads. J Indian Chem Soc 89:1689–1695

Pal J, Deb MK (2012b) Green formation and catalytic activity of palladium nanoparticles on brilliant green in aqueous solution. Indian J Environ Prot 32:574–578

Pal J, Deb MK (2012c) Microwave green synthesis of PVP stabilized gold nanoparticles and their adsorption behaviour for methyl orange. J Exp Nanosci.

Pal J, Deb MK (2012d) Microwave synthesis of polymer coated silver nanoparticles by glucose as reducing agent. Indian J Chem Sec A 51:821–824

Pal J, Deb MK (2013) Efficient sorption of basic organic dyes from aqueous solution using green synthesized silver nanoparticles beads. J Dispers Sci Technol 34:1193–1201

Pal A, Shah S, Devi S (2009) Microwave-assisted synthesis of silver nanoparticles using ethanol as a reducing agent. Mater Chem Phys 114:530–532

Pal J, Deb MK, Deshmukh DK, Sen BK (2012) Microwave-assisted synthesis of platinum nanoparticles and their catalytic degradation of methyl violet in aqueous solution. Appl Nanosci.

Pal J, Deb MK, Deshmukh DK (2013a) Microwave-assisted synthesis of silver nanoparticles using benzo-18-crown-6 as reducing and stabilizing agent. Appl Nanosci.

Pal J, Deb MK, Deshmukh DK, Verma D (2013b) Removal of methyl orange by activated carbon modified by silver nanoparticles. Appl Water Sci 3:367–374

Pavan FA, Dias SLP, Lima EC, Benvenutti EV (2008) Removal of congo red from aqueous solution by anilinepropylsilica xerogel. Dyes and Pigm 76:64–69

Prasad A, Kumar P (2010) Biosortion of cadmium: equilibrium, kinetics and thermodynamics by *Erythrina variegata* orientalis leaf powder. Intern J Chem Eng Res 2:173–180

Rao KJ, Ramesh PD (1995) Use of microwaves for the synthesis and processing of materials. Bull Mater Sci 18:447–465

Reddad Z, Gerente C, Andres Y, Cloirec P (2002) Adsorption of several metal ions onto a low cost biosorbent: kinetic and equilibrium studies. Environ Sci Technol 36:2067–2073

Sabio E, Zamora F (2006) Adsorption of p-nitrophenol on activated carbon fixed-bed. Water Res 40:3053–3060

Saha S, Pal A, Pande S, Sarkar S, Panigrahi S, Pal T (2009) Alginate gel-mediated photochemical growth of mono- and bimetallic gold and silver nanoclusters and their application to surface-enhanced Raman scattering. J Phys Chem C 113(18):7553–7560

Schmid G (1992) Large clusters and colloids—metals in the embryonic state. Chem Rev 92:1709–1727

Shah S, Pal A, Gude R, Devi S (2010) Synthesis and characterization of thermo responsive copolymeric nanoparticles of poly (methyl methyacrylate-co-*N*-vinylcaprolactam. Eur Polym J 46(5):958–967

Shah B, Tailor R, Shah A (2012) Equilibrium, kinetics, and breakthrough curve of phenol sorption on zeolitic material derived from BFA. J Dispers Sci Technol 33:41–51

Sharma J, Janveja B (2008) A study on removal of congo red dye from the effluents of textile industry using rice husk carbon activated by steam. Rasayan J Chem 4:936–942

Talarposhti AM, Donnelly T, Anderson GK (2001) Colour removal from a simulated dye wastewater using a two-phase anaerobic packed bed reactor. Water Res 35:425–432

Zande BMI, Bohmer MR, Fokkink LGJ, Sehonenberger C (2000) Colloidal dispersions of gold rods: synthesis and optical properties. Langmuir 16:451–458

Explicit numerical study of unsteady hydromagnetic mixed convective nanofluid flow from an exponentially stretching sheet in porous media

O. Anwar Bég · M. S. Khan · Ifsana Karim · Md. M. Alam · M. Ferdows

Abstract A numerical investigation of unsteady magnetohydrodynamic mixed convective boundary layer flow of a nanofluid over an exponentially stretching sheet in porous media, is presented. The transformed, non-similar conservations equations are solved using a robust, explicit, finite difference method (EFDM). A detailed stability and convergence analysis is also conducted. The regime is shown to be controlled by a number of emerging thermophysical parameters i.e. combined porous and hydromagnetic parameter (R), thermal Grashof number (G_r), species Grashof number (G_m), viscosity ratio parameter (Λ), dimensionless porous media inertial parameter (∇), Eckert number (E_c), Lewis number (L_e), Brownian motion parameter (N_b) and thermophoresis parameter (N_t). The flow is found to be accelerated with increasing thermal and species Grashof numbers and also increasing Brownian motion and thermophoresis effects. However, flow is decelerated with increasing viscosity ratio and combined porous and hydromagnetic parameters. Temperatures are enhanced with increasing Brownian motion and thermophoresis as are concentration values. With progression in time the flow is accelerated and temperatures and concentrations are increased. EFDM solutions are validated with an optimized variational iteration method. The present study finds applications in magnetic nanomaterials processing.

Keywords Nanofluid · Exponentially stretching sheet · Mixed convective flow · Magnetic field · Porous media · Transient flow · Brownian motion · Explicit finite difference method (EFDM) · Stability analysis · Variational iteration method (VIM)

O. A. Bég
Gort Engovation Research (Propulsion/Biomechanics), 15 Southmere Avenue, Bradford BD73NU, UK

M. S. Khan · I. Karim · Md. M. Alam
Mathematics Discipline, Khulna University, Khulna 9208, Bangladesh

M. Ferdows
Department of Mathematics, Dhaka University, Dhaka, Bangladesh

Present Address:
M. Ferdows
Quantum Beam Science Directorate, Japan Atomic Energy Authority, Tokyo, Japan

O. A. Bég (✉)
Narvik, Norway
e-mail: gortoab@gmail.com

Nomenclature

B_0	Magnetic field strength
C	Nanoparticle concentration
C_0	Reference concentration
C_w	Nanoparticle concentration at stretching surface
C_∞	Ambient nanoparticle concentration as y tends to infinity
$\overline{C}$	Dimensionless concentration
c_p	Specific heat capacity
D_B	Brownian diffusion coefficient
D_T	Thermophoresis diffusion coefficient
g	Acceleration due to gravity
G_r	Thermal Grashof number
G_m	Species (mass) Grashof number
L_e	Lewis number
L	Reference length
N_b	Brownian motion parameter
N_t	Thermophoresis parameter

P	Fluid pressure
P_r	Prandtl number
R	Combined porous and magnetic parameter
R_e	Local Reynolds number
t	Time
K	Permeability of the porous regime
T	Fluid temperature
$\overline{T}$	Dimensionless temperature
T_0	Reference temperature
T_w	Temperature at the stretching surface
T_∞	Ambient temperature as y tends to infinity
u, v	Velocity components along x and y axes, respectively
U, V	Dimensionless velocity components
U_0	Reference velocity
x, y	Cartesian coordinates measured along stretching surface

Greek symbols

ν	Kinematic viscosity
$\tilde{\nu}$	Reference kinematic viscosity
$(\rho c)_p$	Effective heat capacity of the nanofluid
$(\rho c)_f$	Heat capacity of the fluid
α	Thermal diffusivity
β_T	Coefficient of thermal expansion
β_c^*	Coefficient of mass expansion
τ	Dimensionless time

Introduction

Magnetohydrodynamic (MHD) boundary layer flows with heat and mass transfer from a continuously stretching surface with a given temperature distribution moving in an otherwise quiescent fluid medium have stimulated considerable interest in recent years. Magnetic fields can be used to manipulate thermal or mechanical energy in flowing electrically conducting polymers and can yield significant cost savings in manufacturing processes (Garnier 1992). Many robust applications of MHD materials processing have been developed including micro-structural modification by heat transfer control (Asai 2012), induction heating of ceramic metal matrix composites (Garnier 1996), liquid metal stirring operations (Fautrelle et al. 2009) and boundary-layer separation control via Lorentzian forces (Ishikwa et al. 2007). Transport phenomena in *stretching sheet flows* are of particular relevance to boundary layer modeling (Abdou and Soliman 2012). After the pioneering mathematical studies of Sakiadis (1961) and Crane (1970), several researchers further investigated stretching sheet boundary layer flows for different types of stretching velocity. Magyari and Keller (2000) considered the steady boundary layer heat and mass transfer flow from an *exponentially* stretching continuous surface with an exponential temperature distribution, deriving solutions which exhibited an exponential dependence on the temperature distribution in the direction parallel to that of the stretching. Partha et al. (2005) studied the effect of viscous dissipation on the mixed convective boundary layer heat transfer from an exponentially stretching surface. Elbashbeshy (2001) considered wall transpiration (suction) effects on heat transfer in boundary layer flow from an exponentially continuous stretching surface. Khan (2006) studied non-Newtonian (viscoelastic) boundary layer fluid flow over an exponentially stretching sheet. Sanjayanand and Khan (2006) considered viscoelastic heat and mass transfer from an exponentially stretching sheet. A numerical simulation of boundary layer flow over an exponentially stretching sheet with thermal radiation was undertaken by Bidin and Nazar (2009). The exponential stretching sheet scenario is important both for practical reasons and also for mathematical treatment. In real polymer stretching systems, Stastna et al. (1991) have shown that relaxational processes in polymer systems can be characterized by a relaxation function which exhibits a stretched exponential behaviour. By stretching polymer sheets at exponential rates, the relaxation of elastic stresses is overcome and a homogenous material distribution is achieved. Analytically exponential sheet models have also been studied by Elbashbeshy (2001) who obtained similarity solutions based on the exponential stretching velocity distribution in the stretching direction. Therefore, exponential stretch rates are both physically relevant and also allow mathematical similarity analysis. Although the original study by Ishikwa et al. (2007) considers only linear stretching, quadratic stretching has been investigated by Kumaran and Ramanaiah (1996). A generalized form of stretching can also be achieved and indeed has been formulated by Weidman and Magyari (2010) for continuous surfaces stretching with arbitrary polynomial velocities, designated as "super-stretching". In the context of actual materials processing operations, the exponential model for stretching achieves the best efficiency (Bataller 2008).

Convective transport in porous media also has extensive applications in industrial systems including energy storage, filtration of liquids, drying processes, etc. Gorla and Zinolabedini (1987) investigated free convective heat transfer from a vertical surface to a saturated porous medium with an arbitrary varying surface temperature. Bég et al. (2009) obtained non-similar solutions for hydromagnetic transport in a porous medium from a stretching sheet with cross-diffusion effects.

In recent years, a fundamental new development in thermo-fluid mechanics has been the introduction of nanofluids. Choi (1995) described such fluids as being suspensions comprising nanometer-sized metallic particles strategically deployed in common working fluids (air, water) which achieve highly enhanced thermal properties. Kang et al. (2006) presented an experimental study of nanofluid thermal conductivities. Crainic et al. (2003, 2007) identified the significant *magnetohydrodynamic* properties of specific nanofluids which are exploitable in their manufacture. The performance-enhancing characteristics of magnetic nanofluids in industrial MHD pumps used in materials processing have also been emphasized by Shahidian et al. (2011). *Boundary layer flows* of nanofluids in either non-porous (purely fluid) and porous media, have also received significant attention following the study of Kuznestov and Nield (2010) in which buoyancy effects were considered. This study highlighted that *Brownian motion* and *thermophoresis* are significant mechanisms in nanofluid performance. Bég and Tripathi (2012) further showed the substantial role of Brownian motion and thermophoresis in augmenting heat transfer in peristaltic nanofluid transport. Khan and Pop (2010, 2011) studied, respectively, the laminar boundary layer flow of a nanofluid past a stretching sheet and also free convection boundary layer nanofluid flow in a porous medium. Hamad and Pop (2011) addressed the stagnation-point nanofluid boundary layer flow on a permeable stretching sheet in a porous medium in the presence of a heat sink or source. Bég et al. (2012) studied the free convection nanofluid boundary layer from a spherical body to a porous medium using a homotopy analysis method and a Darcian drag force model. Rana et al. (2012) employed a variational finite element method to study natural convection nanofluid boundary layer flow from a tilted surface in porous media. Magnetohydrodynamic nanofluid boundary layer flows have also recently garnered interest. Hamad et al. (2011) used a group theoretical approach and shooting quadrature to study magneto-nanofluid natural convection boundary layer flow from a vertical plate. Khan et al. (2011) analyzed numerically the thermal radiative flux effects on hydromagnetic nanofluid boundary layer flow from a stretching surface. Recently Rana et al. (2013) studied using a finite element technique, the transient magnetohydrodynamic boundary layer flow in an incompressible rotating nanofluid over a stretching continuous sheet, showing that both Brownian motion and thermophoresis enhance wall mass transfer rates (Sherwood number). Very recently Abbasbandy and Ghehsareh (2012) used the Hankel–Padé expansion method to study nanofluid hydromagnetic boundary-layer flows.

The vast majority of nanofluid boundary layer flow models have been steady-state in nature. In the present article we therefore simulate the transient MHD dissipative mixed convective boundary layer nanofluid flow over an *exponentially* stretching sheet adjacent to a non-Darcian porous medium. The governing equations are transformed into dimensionless, strongly coupled, non-linear partial differential equations featuring a number of thermophysical parameters. An explicit finite difference computational algorithm is employed to yield solutions. Elaboration of the stability and convergence characteristics is also included. The present study is relevant to the manufacturing of magnetic nanofluids (Baron et al. 2007) and chemical engineering operations involving electro-conductive nanofluid suspensions (Stephens et al. 2010) and has not been, to the authors' knowledge, thus far reported in the literature.

Mathematical model

Consider the time-dependent (unsteady) two-dimensional flow of an incompressible viscous and electrically conducting nanofluid induced by a stretching sheet in a porous medium saturated with quiescent ambient nanofluid. The sheet uniform temperature and species concentration are raised to $T_w(>T_\infty)$ and $C_w(>C_\infty)$ respectively, which are thereafter maintained constant, where T_w, C_w are temperature and species (nanoparticle) concentration at the wall and T_∞, C_∞ are temperature and species concentration far away from the sheet, respectively. The x-axis is orientated along the exponentially stretching sheet in the direction of the motion and y-axis is perpendicular to it. A variable strength magnetic field, $B(x)$ is applied normal to the sheet and induced magnetic field is neglected, which is justified for MHD flow at small magnetic Reynolds number. The physical configuration and coordinate system are shown in Fig. 1. Under the above assumptions and usual boundary layer approximation, the transient MHD mixed convective nanofluid transport is described by the following equations, which extend the earlier formulation of Kuznetsov and Nield (2010) to consider magnetic field and inertial porous medium drag effects:

Mass conservation

$$\frac{\partial u}{\partial x}+\frac{\partial v}{\partial y}=0 \tag{1}$$

Momentum conservation

$$\frac{\partial u}{\partial t}+u\frac{\partial u}{\partial x}+v\frac{\partial u}{\partial y}=\tilde{v}\frac{\partial^2 u}{\partial y^2}-\frac{v}{K}u-c^*\varepsilon^2u^2-\frac{\sigma B^2}{\rho}u+g\beta_T(T-T_\infty)+g\beta_c^*(C-C_\infty) \tag{2}$$

Energy conservation

$$\frac{\partial T}{\partial t}+u\frac{\partial T}{\partial x}+v\frac{\partial T}{\partial y}=\alpha\frac{\partial^2 T}{\partial y^2}+\frac{\tilde{v}}{c_p}\left(\frac{\partial u}{\partial y}\right)^2+\frac{(\rho c)_p}{(\rho c)_f}\left\{D_B\left(\frac{\partial T}{\partial y}\cdot\frac{\partial C}{\partial y}\right)+\frac{D_T}{T_\infty}\left(\frac{\partial T}{\partial y}\right)^2\right\} \tag{3}$$

Fig. 1 Physical model and coordinate system

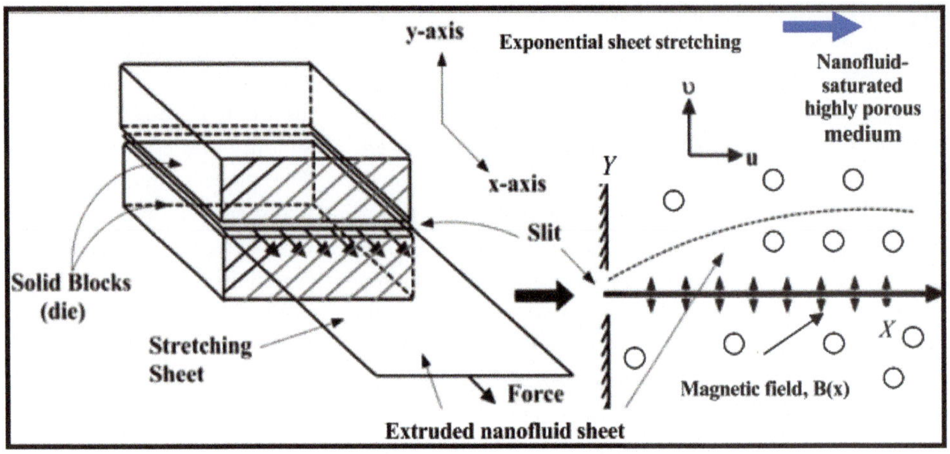

Species (nano-particle concentration) conservation

$$\frac{\partial C}{\partial t}+u\frac{\partial C}{\partial x}+v\frac{\partial C}{\partial y}=D_{\mathrm{B}}\frac{\partial^2 C}{\partial y^2}+\frac{D_{\mathrm{T}}}{T_\infty}\frac{\partial^2 T}{\partial y^2} \tag{4}$$

where u and v are the velocities in the x- and y-directions, respectively, t is time, ρ is the fluid density, v the kinematic viscosity, $\tilde{v}$ the reference kinematic viscosity, K the permeability of the porous regime, c_p the specific heat at constant pressure, T and C the fluid temperature and concentration in the boundary layer, $c^*\varepsilon^2$ is the inertia parameter, α is the thermal diffusivity, $(\rho c)_p$ is effective heat capacity of the nanofluid, $(\rho c)_{\mathrm{f}}$ is heat capacity of the fluid, D_{B} is the species diffusivity and D_{T} is the thermophoresis diffusion coefficient. In Eq. (2) (Newton's second law), the first term on the left hand side is the temporal velocity gradient, the second and third terms are the convective acceleration terms. The first term on the right hand side denotes viscous shear, the second represents the Darcian porous media drag (linear), the third designates second order Forchheimer porous media drag, the fourth is the magnetohydrodynamic Lorentz body force, the fifth is the thermal buoyancy term and the last term on the right hand side of Eq. (2) is the species buoyancy force. In Eq. (3) which is a statement of *Fourier's law of heat conservation*, the first term on the left hand side denotes the transient temperature gradient and the second and third terms on the left hand side represent the convective heat transfer terms. The first term on the right hand side signifies thermal diffusion, the second term is viscous heating, and the last dual component term denotes Brownian motion and thermophoresis contributions. In Eq. (4), which is a statement of *Fick's law of mass (species) diffusion*, the first term on the left hand side is the transient concentration gradient, and the second and third terms are the convective mass transfer terms. The first term on the right hand side denotes the species diffusion and the last term is the relative contribution of thermophoresis to Brownian motion.

The relevant initial and boundary conditions are:

$$\begin{aligned}
&t=0,\ u=0,\ v=0,\ T=T_\infty,\ C=C_\infty \quad \text{everywhere}\\
&t\geq 0,\ u=0,\ v=0,\ T=T_\infty,\ C=C_\infty \quad \text{at}\, x=0\\
&u=u_w=U_0 e^{x/L},\ v=0,\ T=T_w,\ T_\infty+T_0 e^{2x/L},\ C=C_w=C_\infty+C_0 e^{2x/L} \quad \text{at}\, y=0\\
&u=0,\ v=0,\ T\to T_\infty,\ C\to C_\infty \quad \text{at}\, y\to\infty
\end{aligned} \tag{5}$$

where U_0 is the reference velocity, T_0, C_0 the reference temperature and concentration, respectively, and L is the reference length. To obtain similarity solutions, it is assumed that the magnetic field $B(x)$ and the variable thermal conductivity K^* are of the form:

$$B=B_0 e^{\frac{x}{2L}} \tag{6}$$

$$K=K_0 e^{\frac{x}{L}} \tag{7}$$

where B_0 denotes constant magnetic field and K_0 is the constant thermal conductivity. This exponential formulation has also been adopted by several other researchers in magnetohydrodynamics, as elaborated by Yakovlev et al. (2013) who have also adopted exponential magnetic field relations. Sadooghi and Taghinavaz (2012) have also explained the validity of employing exponentially decaying functions of magnetic field. Exponential decay in thermal conductivity has also been related to energy transport in nanomaterials, as elaborated by Wang (2012). Furthermore the formulations adopted in Eqs. (6), (7) in addition to being physically valid, also provide an elegant simplification in the mathematical complexity of the non-dimensional transport equations, as will be demonstrated now. Introducing the following non-dimensional variables;

$$X = \frac{xU_0}{\nu} e^{\frac{x}{L}},\ Y = \frac{yU_0}{\nu} e^{\frac{x}{L}},\ U = \frac{u}{U_0} e^{-\frac{x}{L}},\ V = \frac{v}{U_0} e^{-\frac{x}{L}},$$
$$\tau = \frac{tU_0^2}{\nu} e^{\frac{2x}{L}},\ \bar{T} = \frac{T - T_\infty}{T_w - T_\infty},\ \bar{C} = \frac{C - C_\infty}{C - C_\infty}. \tag{8}$$

From the above transformations the, non-linear, coupled partial differential Eqs. (1)–(4) become non-dimensional as follows:

$$\frac{\partial U}{\partial X} + \frac{\partial V}{\partial y} = 0 \tag{9}$$

$$\frac{\partial U}{\partial \tau} + U\frac{\partial U}{\partial X} + V\frac{\partial U}{\partial Y} = \wedge \frac{\partial^2 U}{\partial Y^2} - \frac{1}{R_e}\left(RU + \Delta U^2 - G_r\bar{T} - G_M\bar{C}\right) \tag{10}$$

$$\frac{\partial \bar{T}}{\partial \tau} + U\frac{\partial \bar{T}}{\partial X} + V\frac{\partial \bar{T}}{\partial Y} = \frac{1}{P_r}\left(\frac{\partial^2 \bar{T}}{\partial Y^2}\right) + \wedge E_c\left(\frac{\partial U}{\partial Y}\right)^2 + N_b\left(\frac{\partial \bar{T}}{\partial Y}\cdot\frac{\partial \bar{C}}{\partial Y}\right) + N_t\left(\frac{\partial \bar{T}}{\partial Y}\right)^2 \tag{11}$$

$$\frac{\partial \bar{C}}{\partial \tau} + U\frac{\partial \bar{C}}{\partial X} + V\frac{\partial \bar{C}}{\partial Y} = \frac{1}{L_e}\left[\frac{\partial^2 \bar{C}}{\partial Y^2} + \left(\frac{N_t}{N_b}\right)\frac{\partial^2 \bar{T}}{\partial Y^2}\right]. \tag{12}$$

The non-dimensional boundary and initial conditions transform to:

$$\begin{aligned} &\tau \le 0,\ U = 0,\ V = 0,\ \bar{T} = 0,\ \bar{C} = 0 \quad \text{everywhere} \\ &\tau > 0,\ U = 0,\ V = 0,\ \bar{T} = 0,\ \bar{C} = 0 \quad \text{at X = 0} \end{aligned} \tag{13}$$

$$\begin{aligned} &U = 1,\ V = 0,\ \bar{T} = 1,\ \bar{C} = 1 \quad \text{at } Y = 0 \\ &U = 0,\ V = 0,\ \bar{T} = 0,\ \bar{C} = 0 \quad \text{at } Y \to \infty \end{aligned} \tag{14}$$

where the notation primes denote differentiation with respect to η and the parameters are defined as follows: $R = \left(\frac{\nu L}{K_0 U_0} + \frac{\sigma B_0^2 L}{\rho U_0}\right)$ is the combined porous and magnetic parameter, $\wedge = \frac{\nu}{\nu}$ is viscosity ratio parameter, $\nabla = c^* \varepsilon^2 L$ is dimensionless porous media inertia parameter, $G_r = \frac{g\beta T_0 L}{U_0^2}$ is thermal Grashof number, $G_m = \frac{g\beta C_0 L}{U_0^2}$ is species Grashof number, $P_r = \frac{\upsilon}{\alpha}$ is Prandtl number, $E_c = \frac{U_0^2}{c_p T_0}$ is Eckert number, $N_B = \frac{(\rho c)_p D_B (C_w - C_\infty)}{\nu(\rho c)_f}$ is Brownian motion parameter, $N_t = \frac{(\rho c)_p D_T (T_w - T_\infty)}{\nu T_\infty (\rho c)_f}$ is thermophoresis parameter, $L_e = \frac{\nu}{D_B}$ is the Lewis number and $R_{e_x} = \frac{LU_0 e^{\frac{x}{L}}}{\nu}$ is the local Reynolds number.

Explicit numerical solutions

In order to solve the non-similar unsteady coupled non-linear partial differential equations (9)–(12), under boundary conditions (13, 14), an *explicit* finite difference method (EFDM) algorithm, as described by Carnahan et al. (1969) has been developed. Finite difference algorithms are still employed widely in *unsteady* multi-physical materials processing flows and generally yield very accurate and stable solutions and can easily accommodate time variables. Finite difference simulations have been conducted recently by, for example, Mohiddin et al. (2010) who studied transient double-diffusive non-Newtonian convection form a cone. Further studies using difference methods include Prasad et al. (2011) who analyzed unsteady viscoelastic natural convection boundary layer flow from a vertical surface and Prasad et al. (2011) who studied radiative-convective non-Newtonian flow from a cone. An extensive review of applications of finite difference (and other numerical) methods in hydromagnetic materials processing flows has recently been conducted by Bég (2012). In the explicit approach, a rectangular region of the flow field is chosen and the region is divided into a grid of lines parallel to the X and Y axes, where X-axis is taken along the sheet and the Y-axis is normal to the sheet, as depicted in Fig. 2.

Here the sheet of height $X_{max}(=100)$ is considered i.e. X varies from 0 to 100 and assumed $Y_{max}(=25)$ as corresponding to $Y \to \infty$ i.e. Y varies from 0 to 25. There are $m_{max}(=125)$ and $n_{max}(=125)$ grid spaces in the X and Y directions, respectively (Fig. 2). It is assumed that ΔX, ΔY are constant mesh sizes along the X and Y directions, respectively, and are prescribed as follows: $\Delta X = 0.8(0 \le X \le 100)$ and $\Delta Y = 0.2(0 \le Y \le 25)$ with the smaller time-step, $\Delta\tau = 0.005$. Let U', V', $\bar{T}'$ and $\bar{C}'$ denote the values of U, V, $\bar{T}$ and $\bar{C}$ at the end of a time-step, respectively. Using the explicit finite difference approximation, the following system of finite difference equations is obtained:

$$\frac{U'_{i,j} - U'_{i-1,j}}{\Delta X} + \frac{V_{i,j} - V_{i,j-1}}{\Delta Y} = 0 \tag{15}$$

Fig. 2 Finite difference space grid

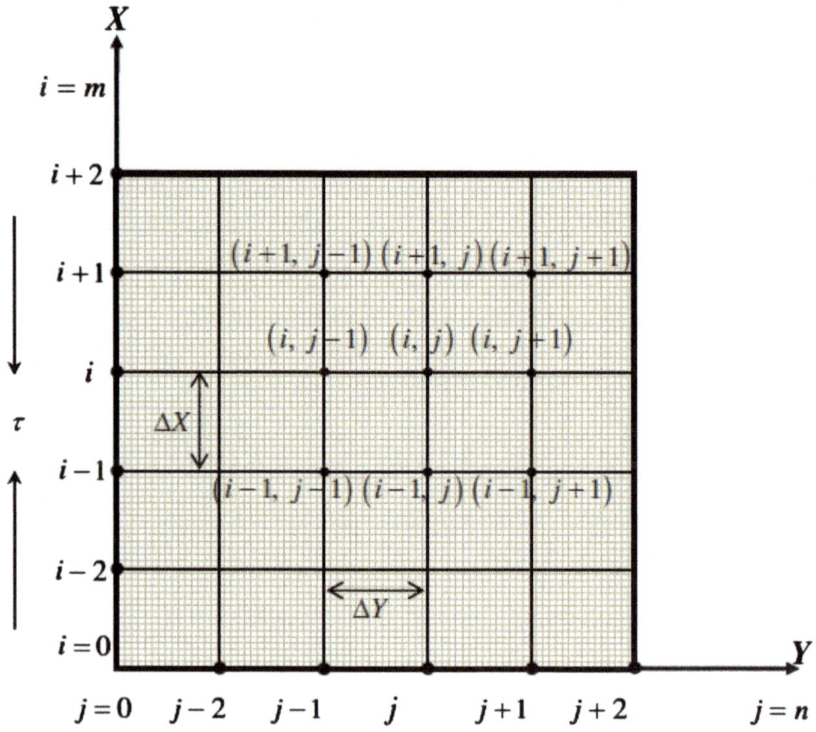

$$\frac{U'_{i,j}-U_{i,j}}{\Delta\tau}+U_{i,j}\frac{U_{i,j}-U_{i-1,j}}{\Delta X}+V_{i,j}\frac{U_{i,j+1}-U_{i,j}}{\Delta Y} = \wedge\frac{U_{i,j+1}-2U_{i,j}+U_{i,j-1}}{(\Delta Y)^2} - \frac{1}{R_e}\left(RU_{i,j}+\nabla U^2_{i,j}-G_r\bar{T}_{i,j}-G_m\bar{C}_{i,j}\right) \qquad (16)$$

$$\frac{\bar{T}'_{i,j}-\bar{T}_{i,j}}{\Delta\tau}+U_{i,j}\frac{\bar{T}_{i,j}-\bar{T}_{i-1,j}}{\Delta X}+V_{i,j}\frac{\bar{T}_{i,j+1}-\bar{T}_{i,j}}{\Delta Y} = \frac{1}{P_r}\left(\frac{\bar{T}_{i,j+1}-2\bar{T}_{i,j}+\bar{T}_{i,j-1}}{(\Delta Y)^2}\right)+\wedge E_c\left(\frac{U_{i,j+1}-2U_{i,j}+U_{i,j-1}}{(\Delta Y)^2}\right) + N_b\left(\frac{\bar{T}_{i,j+1}-\bar{T}_{i,j}}{\Delta Y}.\frac{\bar{C}_{i,j+1}-\bar{C}_{i,j}}{\Delta Y}\right)+N_t\left(\frac{\bar{T}_{i,j+1}-\bar{T}_{i,j}}{\Delta Y}\right)^2 \qquad (17)$$

$$\frac{\bar{C}'_{i,j}-\bar{C}_{i,j}}{\Delta\tau}+U_{i,j}\frac{\bar{C}_{i,j}-\bar{C}_{i-1,j}}{\Delta X}+V_{i,j}\frac{\bar{C}_{i,j+1}-\bar{C}_{i,j}}{\Delta Y} = \frac{1}{L_e}\left[\left(\frac{\bar{C}_{i,j+1}-2\bar{C}_{i,j}+\bar{C}_{i,j-1}}{(\Delta Y)^2}\right)+\frac{N_t}{N_b}\left(\frac{\bar{T}_{i,j+1}-2\bar{T}_{i,j}+\bar{T}_{i,j-1}}{(\Delta Y)^2}\right)\right] \qquad (18)$$

with initial and boundary conditions:

$$U^0_{i,j}=0,\ V^0_{i,j}=0,\ \bar{T}^0_{i,j}=0,\ \bar{C}^0_{i,j}=0$$
$$U^n_{0,j}=0,\ V^n_{0,j}=0,\ \bar{T}^n_{0,j}=0,\ \bar{C}^n_{0,j}=0 \qquad (19)$$

$$U^n_{i,0}=1,\ V^n_{i,0}=0,\ \bar{T}^n_{i,0}=1,\ \bar{C}^n_{i,0}=1$$
$$U^n_{i,L}=0,\ V^n_{i,L}=0,\ T^n_{i,L}=0,\ C^n_{i,L}=0,\ \text{where } L\to\infty. \qquad (20)$$

Here the subscripts i and j designate the grid points with X and Y coordinates, respectively, and the superscript n represents a value of time, $\tau = n\cdot\Delta\tau$ where $n = 0, 1, 2, \ldots$

Stability and convergence analysis

Since an explicit procedure is being employed, a discussion of the stability and convergence aspects of the finite difference scheme is warranted. For *constant* mesh sizes the stability criteria of the scheme may be established as follows. Equation (15) will be ignored since $\Delta\tau$ does not feature in it. The general terms of the Fourier expansion for U, $\bar{T}$ and $\bar{C}$ at a time arbitrarily called $\tau = 0$ are all $e^{i\alpha X}e^{i\beta X}$, apart from a constant, where $i=\sqrt{-1}$. At a time τ, these terms become:

$$U : \psi(\tau)e^{i\alpha X}e^{i\beta Y}$$
$$\bar{T} : \theta(\tau)e^{i\alpha X}e^{i\beta Y} \qquad (21)$$
$$\bar{C} : \phi(\tau)e^{i\alpha X}e^{i\beta Y}$$

After the time-step these terms will become:

$$U : \psi'(\tau)e^{i\alpha X}e^{i\beta Y}$$
$$\bar{T} : \theta'(\tau)e^{i\alpha X}e^{i\beta Y} \qquad (22)$$
$$\bar{C} : \phi'(\tau)e^{i\alpha X}e^{i\beta Y}$$

Substituting (21) and (22) into Eqs. (16)–(18), regarding the coefficients U and V as constants over any one time-step, we obtain the following equations upon simplification,

$$\frac{\psi'(\tau)-\psi(\tau)}{\Delta\tau}+U\frac{\psi(\tau)(1-e^{i\alpha\Delta X})}{\Delta X}+V\frac{\psi(\tau)(e^{i\beta\Delta Y}-1)}{\Delta Y} = \wedge\left\{\frac{2\psi(\tau)(\cos\beta\Delta Y-1)}{(\Delta Y)^2}\right\} - \frac{R}{R_e}\psi(\tau)-\frac{\nabla U}{R_e}\psi(\tau)+\frac{G_r}{R_e}\theta'+\frac{G_m}{R_e}\varphi' \tag{23}$$

$$\frac{\theta'(\tau)-\theta(\tau)}{\Delta\tau}+U\frac{\theta(\tau)(1-e^{-i\alpha\Delta X})}{\Delta X}+V\frac{\theta(\tau)(e^{i\beta\Delta Y}-1)}{\Delta Y} = \frac{1}{P_r}\frac{2\theta(\tau)(\cos\beta\Delta Y-1)}{(\Delta Y)^2} + \wedge E_c U\psi(\tau)\left(\frac{e^{i\beta\Delta Y}-1}{\Delta Y}\right)^2 + N_b\bar{C}\theta(\tau)\left\{\frac{(e^{i\beta\Delta Y}-1)}{\Delta Y}\right\}^2 + N_t\bar{T}\theta(\tau)\left\{\frac{(e^{i\beta\Delta Y}-1)}{\Delta Y}\right\}^2 \tag{24}$$

$$\frac{\phi'(\tau)-\phi(\tau)}{\Delta\tau}+U\frac{\phi(\tau)(1-e^{-i\alpha\Delta X})}{\Delta X}+V\frac{\phi(\tau)(e^{i\beta\Delta Y}-1)}{\Delta Y} = \frac{1}{L_e}\left[\left\{\frac{2\phi(\tau)(\cos\beta\Delta Y-1)}{(\Delta Y)^2}\right\} + \left(\frac{N_t}{N_b}\right)\cdot\left\{\frac{2\theta(\tau)(\cos\beta\Delta Y-1)}{(\Delta Y)^2}\right\}\right] \tag{25}$$

The Eqs. (23)–(25) can be written in the following form:

$$\psi' = A\psi + B\theta' + C\phi' \tag{26}$$

$$\theta' = D\psi + E\theta \tag{27}$$

$$\phi' = F\theta + G\phi \tag{28}$$

where

$$A = 1 - U\frac{\Delta\tau}{\Delta X}(1-e^{-ia\Delta X}) - V\frac{\Delta\tau}{\Delta Y}(e^{i\beta\Delta Y}-1)+\frac{2\Delta\tau}{(\Delta Y)^2} \wedge(\cos\beta\Delta Y-1) - \frac{R}{R_e}\Delta\tau - \frac{\nabla U}{R_e}\Delta\tau \tag{29}$$

$$B = \frac{G_r}{R_e}\Delta\tau \tag{30}$$

$$C = \frac{G_m}{R_e}\Delta\tau \tag{31}$$

$$D = \frac{\Delta\tau}{(\Delta Y)^2}\wedge E_c U(e^{i\beta\Delta Y}-1)^2 \tag{32}$$

$$E = 1 - U\frac{\Delta\tau}{\Delta X}(1-e^{-i\alpha\Delta X}) - V\frac{\Delta\tau}{\Delta Y}(e^{i\beta\Delta Y}-1) + \frac{1}{P_r}\frac{2(\cos\beta\Delta Y-1)}{(\Delta Y)^2}\Delta\tau + N_b\bar{C}\left\{\frac{(e^{i\beta\Delta Y}-1)}{\Delta Y}\right\}^2\Delta\tau + N_t\bar{T}\left\{\frac{e^{i\beta\Delta Y}-1}{\Delta Y}\right\}^2\Delta\tau, \tag{33}$$

$$F = \frac{1}{L_e}\left(\frac{N_t}{N_b}\right)\frac{2\Delta\tau}{(\Delta Y)^2}(\cos\beta\Delta Y-1) \tag{34}$$

and

$$G = 1 - U\frac{\Delta\tau}{\Delta X}(1-e^{-i\alpha\Delta X}) - V\frac{\Delta\tau}{\Delta X}(e^{i\beta\Delta Y}-1) + \frac{1}{L_e}\frac{2\Delta\tau}{(\Delta Y)^2}(\cos\beta\Delta Y-1). \tag{35}$$

Again using Eqs. (27) and (28) in (26):

$$\psi' = H\psi + I\theta + J\phi \tag{36}$$

where,

$$H = A + BD,\quad I = BE + CF \text{ and}.\quad J = CG \tag{37}$$

Therefore Eqs. (26)–(28) can be expressed as:

$$\psi' = H\psi + I\theta + J\phi \tag{38}$$

$$\theta = D\psi + E\theta \tag{39}$$

$$\phi' = F\theta + G\phi \tag{40}$$

Furthermore the Eqs. (38)–(40) can be expressed in matrix–vector form as follows:

$$\begin{bmatrix}\psi'\\ \theta'\\ \phi'\end{bmatrix} = \begin{bmatrix}H & I & J\\ D & E & 0\\ 0 & F & G\end{bmatrix}\begin{bmatrix}\psi\\ \theta\\ \phi\end{bmatrix} \tag{41}$$

that is, $\eta' = T\eta$ where

$$\eta' = \begin{bmatrix}\psi'\\ \theta'\\ \phi'\end{bmatrix}, T = \begin{bmatrix}H & I & J\\ D & E & 0\\ 0 & F & G\end{bmatrix} \text{ and } \eta = \begin{bmatrix}\psi\\ \theta\\ \phi\end{bmatrix} \tag{42}$$

To determine the stability condition, it is necessary to evaluate the eigenvalues of the amplification matrix T but this task is very difficult since all the elements of T are different. Hence the problem requires that the Eckert Number E_c is assumed to be very small and tends to zero. With this consideration $D = 0$ and the amplification matrix becomes:

$$T = \begin{bmatrix}H & I & J\\ 0 & E & 0\\ 0 & F & G\end{bmatrix} \tag{43}$$

Hence the problem requires that after simplification of the matrix T, we get the following eigenvalues, $\lambda_1 = H$, $\lambda_2 = E$ and $\lambda_3 = G$. For stability, each of the eigenvalues λ_1, λ_2 and λ_3 must not exceed unity in modulus. Hence the stability condition is $|H| \leq 1$, $|E| \leq 1$ and $G \leq 1$, for all α, β Now we assume that U is everywhere *non-negative* and V is everywhere *non-positive*. Details for this assumption are given in Carnahan et al. (1969) and are required for well-posedness of the boundary value problem and to ensure the elimination of spurious modes of oscillation. The assumption is founded on numerical analysis, and is more a computational aspect than a physical one. It follows that, $H = A + BD = A$; since $D = 0$.

$$\therefore H = 1 - U\frac{\Delta\tau}{\Delta X}(1 - e^{-i\alpha\Delta X}) - V\frac{\Delta\tau}{\Delta Y}(e^{i\beta\Delta Y} - 1) + \frac{2\Delta\tau}{(\Delta Y)^2}\wedge(\cos\beta\Delta Y - 1) - \frac{R}{R_e}\Delta\tau - \frac{\nabla U}{R_e}\Delta\tau$$

$$\Rightarrow H = 1 - 2\left[a + b + 2\wedge c - \frac{R}{2R_e}\Delta\tau - \frac{\nabla U}{2R_e}\Delta\tau\right] \quad (44)$$

where $a = U\frac{\Delta\tau}{\Delta X}$, $b = |V|\frac{\Delta\tau}{\Delta Y}$ and $c = \frac{\Delta\tau}{(\Delta Y)^2}$. The coefficients a, b and c are all *real* and non-negative. We can demonstrate that the maximum modulus of H occurs when $\alpha\Delta X = m\pi$ and $\beta\Delta Y = n\pi$, where m and n are integers and hence H is real. The value of $|H|$ is greater when both m and n are odd integers. To satisfy the $|H| \leq 1$, the most negative allowable value is $H = -1$. Therefore, the *first stability condition* is:

$$2\left[a + b + 2\wedge c - \frac{R}{2R_e}\Delta\tau - \frac{\nabla U}{2R_e}\Delta\tau\right] \leq 2 \quad (45)$$

$$\therefore \left[U\frac{\Delta\tau}{\Delta X} + |V|\frac{\Delta\tau}{\Delta Y} + 2\wedge\frac{\Delta\tau}{(\Delta Y)^2} - \frac{R}{2R_e}\Delta\tau - \frac{\nabla U}{2R_e}\Delta\tau\right] \leq 1 \quad (46)$$

Likewise, the *second stability condition* $|E| \leq 1$ requires that;

$$\therefore \left[U\frac{\Delta\tau}{\Delta X} + |V|\frac{\Delta\tau}{\Delta Y} + 2\frac{\Delta\tau}{(\Delta Y)^2}\left(\frac{1}{P_r} + N_b\bar{C} + N_t\bar{T}\right)\right] \leq 1 \quad (47)$$

Similarly the *third stability condition* $|G| \leq 1$ requires that;

$$\left[U\frac{\Delta\tau}{\Delta X} + |V|\frac{\Delta\tau}{\Delta Y} + \frac{2}{L_e}\frac{\Delta\tau}{(\Delta Y)^2}\right] \leq 1 \quad (48)$$

Therefore, the *stability conditions* of the method are;

$$U\frac{\Delta\tau}{\Delta X} + |V|\frac{\Delta\tau}{\Delta Y} + 2\wedge\frac{\Delta\tau}{(\Delta Y)^2} - \frac{R}{2R_e}\Delta\tau - \frac{\nabla U}{2R_e}\Delta\tau \leq 1 \quad (49)$$

$$U\frac{\Delta\tau}{\Delta X} + |V|\frac{\Delta\tau}{\Delta Y} + 2\frac{\Delta\tau}{(\Delta Y)^2}\left(\frac{1}{P_r} + N_b\bar{C} + N_t\bar{T}\right) \leq 1 \quad (50)$$

$$U\frac{\Delta\tau}{\Delta X} + |V|\frac{\Delta\tau}{\Delta Y} + \frac{2}{L_e}\frac{\Delta\tau}{(\Delta Y)^2} \leq 1. \quad (51)$$

From the initial condition, $U = V = \bar{T} = \bar{C} = 0$ at $\tau = 0$ and the consideration due to stability and convergence analysis is $R \geq 0.5$ and $R_e \geq 0.5$. Hence the *convergence criteria* of the method are $\wedge \leq 4.01$, $P_r \geq 0.25$ and $L_e \geq 0.25$.

Validation with variational iteration method

To verify the EFDM solutions, the present non-linear boundary value problem has also been solved with an optimized He variational iteration method (VIM). VIM is also a very powerful semi-analytical/numerical technique developed by He (1999). This procedure was originally developed for environmental chemical engineering pollution problems and approximate solutions for the problem of seepage flow in a porous medium with fractional derivatives were obtained (Anwar Bég 2013). VIM has been subsequently successfully applied to variety of other fluid dynamic phenomena including shallow water hydrodynamics (Tari et al. 2007), bio-thermal tissue treatment simulations (Elsayed 2013), electro-thermal thruster simulation for spacecraft (Anwar Bég 2013), brain tissue chemo-mechanics (Anwar Bég 2013), Von Kármán swirling flows in porous regimes (Shahmohamadi et al. 2012). VIM is a robust method therefore in obtaining exact and approximate solutions of linear and non-linear differential equations. In this method, *general Lagrange multipliers* are introduced to construct *correction functionals* for the problem. The multipliers can be identified optimally via the variational theory. There is no need for linearization or discretization, and excessive computational work and round-off errors are thereby avoided. Time is easily accommodated as a third dimension to the two spatial dimensions (X, Y). VIM demonstrates exceptional stability, *exponential* convergence and accuracy and is ideal for non-linear transport phenomena encountered in chemical engineering problems. For readers not familiar with this procedure, we provide here a brief overview. Consider the following non-linear differential equation:

$$Lu + Nu = g(t), \quad (52)$$

where L, N and $g(t)$ are the linear operator, the non-linear operator and a heterogeneous term, respectively. VIM uses a correction functional (He 1999) and for Eq. (52) this can be written as:

$$u_{n+1}(t) = u_n(t) + \int_0^t \lambda \left[L u_n(\tau) + N \tilde{u}_n(\tau) - g(\tau) \right] d\tau \quad n \geq 0, \tag{53}$$

The successive approximations, u_j, $j \geq 0$ can be established by determining λ, a general Lagrangian multiplier, which can be identified *optimally* via the *variational* theory. The function $\tilde{u}_n$ is a restricted variation which means $\delta \tilde{u}_n = 0$. Therefore, we first determine the Lagrange multiplier λ that will be identified optimally via integration by parts. The successive approximations $u_{n+1}(t)$, $n \geq 0$ of the solution $u(t)$ will be readily obtained upon using the obtained Lagrange multiplier and by using any selective function u_0. When λ has been determined, then several approximations $u_j(t)$, $j \geq 0$, follow immediately. Consequently, the exact solution may be obtained by using:

$$u = \lim_{n \to \infty} u_n. \tag{54}$$

We therefore construct correction functionals and thereafter apply the variational iteration formula. Very lengthy algebraic expressions result from the functionals and are omitted here for brevity. The series expansions are evaluated in a purpose-built Matlab-based code, TRANSNANOVIM (Anwar Bég 2013), developed for transient nanofluid dynamic flows. Computations on a dual-processor Unix workstation are achieved in tens of seconds. Comparison of the EFDM and VIM solutions are documented in Tables 1, 2, 3, for the velocity, temperature and nanoparticle concentration fields, for various combinations of the governing thermofluid parameters and *different* time steps. In all cases excellent agreement is obtained, testifying to the accuracy of the EFDM computations, the latter being used to present all graphical solutions in the next section. Confidence in the EFDM results is therefore justifiably very high. In fact both methods demonstrate exceptional accuracy, stability and fast convergence characteristics and show excellent promise in simulating non-linear problems in nanophysical flows. It is also evident from the current simulations that these methods are an elegant alternative to other popular but computationally intensive methods for nonlinear boundary value problems e.g. Chebyschev spectral collocation methods (Anwar Bég et al. 2013) which are also popular for electrical transport phenomena simulations.

Table 1 EFDM and VIM computations compared for velocity field (U) with $= 5$, $R = 4.0$, $G_r = 3.0$, $G_m = 2.0$, $\Lambda = 1.5$, $\nabla = 1.0$, $E_c = 0.01$, $L_e = 5.0$, $N_b = N_t = 0.1$, $P_r = 1.0$, $R_e = 0.5$, $X = 10.0$

Y	U (EFDM)	U (VIM)
0	1.0000	1.0000
2.5	0.6710	0.6709
5.0	0.0200	0.0201
10.0	0.0000	0.0001
15	0.0000	0.0000
25	0.0000	0.0000

Table 2 EFDM and VIM computations compared for temperature field ($\bar{T}$) with $= 20$, $R = 4.0$, $G_r = 4.0$, $G_m = 2.0$, $\Lambda = 1.5$, $\nabla = 1.0$, $E_c = 0.01$, $L_e = 5.0$, $N_b = N_t = 0.3$, $P_r = 1.0$, $R_e = 0.5$, $X = 10.0$

Y	$\bar{T}$ (EFDM)	$\bar{T}$ (VIM)
0	1.0000	1.0000
5.0	0.5610	0.5614
10.0	0.1642	0.1647
15.0	0.0325	0.0329
20.0	0.0010	0.0013
25.0	0.0000	0.0000

Table 3 EFDM and VIM computations compared for nanoparticle concentration field ($\bar{C}$) with $\tau = 60$, $R = 4.0$, $G_r = 4.0$, $G_m = 2.0$, $\Lambda = 1.5$, $\nabla = 1.0$, $E_c = 0.01$, $L_e = 5.0$, $N_b = N_t = 0.1$, $P_r = 1.0$, $R_e = 0.5$, $X = 10.0$

Y	$\bar{C}$ (EFDM)	$\bar{C}$ (VIM)
0	1.0000	1.0000
5.0	0.3125	0.3128
10.0	0.0985	0.0987
15.0	0.0253	0.0258
20.0	0.0000	0.0000

Results and discussion

Extensive numerical solutions have been obtained for the system governing Eqs. (15)–(18) under boundary conditions (19) and (20), with the EFDM algorithm. The values of the governing parameters are chosen to be physically representative of actual nanofluids (Kuznetsov and Nield 2010; Khan et al. 2011; Rana et al. 2013; Abbasbandy and Ghehsareh 2012). Non-dimensional velocity, temperature and species concentration are computed for different values of combined porous and hydromagnetic parameter (R), thermal Grashof number(G_r), species Grashof number (G_m), viscosity ratio parameter (Λ), dimensionless porous media inertial parameter (∇), Eckert number (E_c), Lewis number (L_e), Brownian motion parameter (N_b),

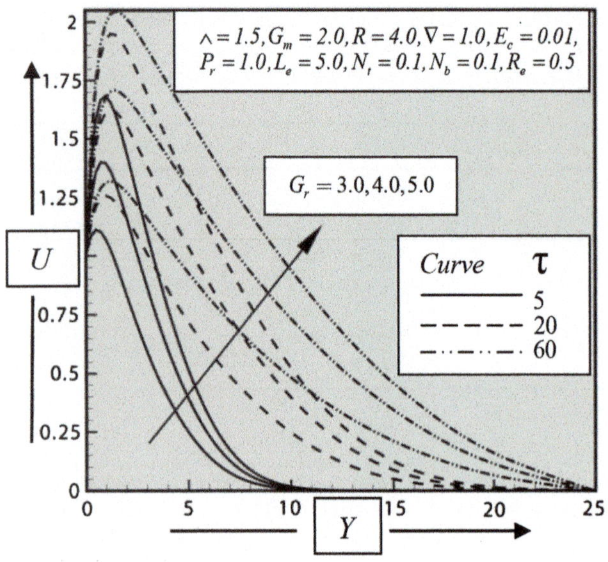

Fig. 3 Thermal Grashof number (G_r) effect on velocity profiles

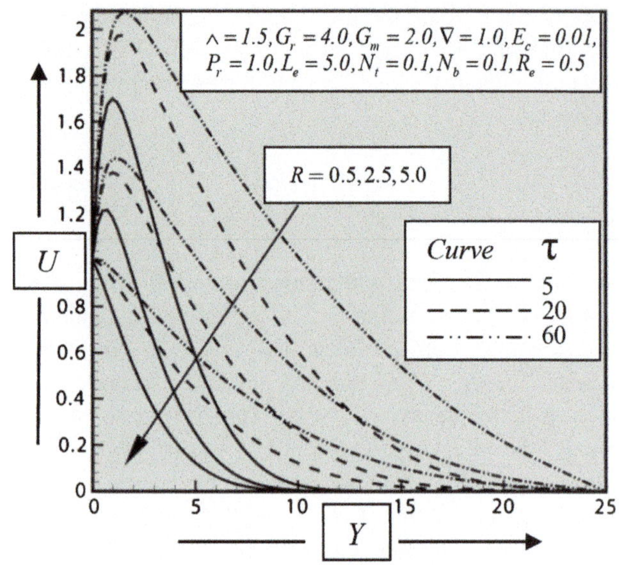

Fig. 5 Combined porous and magnetic parameter effect on velocity profiles

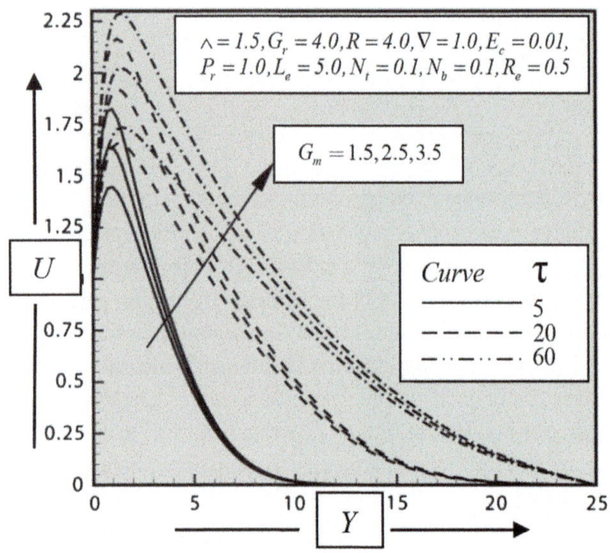

Fig. 4 Species Grashof number (G_m) effect on velocity profiles

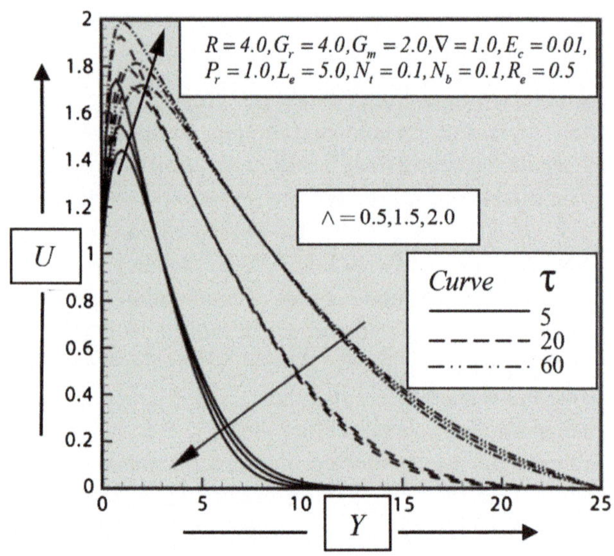

Fig. 6 Viscosity ratio parameter $(\wedge)$ effect on velocity profiles

thermophoresis parameter (N_t) and Prandtl number (P_r) and local Reynolds number (R_e). To obtain the *steady-state* solutions, the calculations are executed for a range of non-dimensional times, $\tau = 5$ to 80. Velocity, temperature and concentration profiles do not exhibit any subsequent variation after $\tau = 60$. Therefore, the solution for $\tau \geq 60$ is taken as the steady-state solution. The distributions of the flow variables are illustrated in Figs. 3, 4, 5, 6, 7, 8, 9, 10, 11, 12, 13. In all cases, X is prescribed a value of 10.

Figure 3 represents the evolution of dimensionless velocity (U) with Y for different values of G_r and non-dimensional times, τ. An increase in thermal Grashof number clearly enhances velocity i.e. induces a strong acceleration in the flow. A distinct velocity shoot arises for all profiles near the sheet surface $(Y = 0)$ and this is accentuated with increasing thermal Grashof number. With increasing thermal Grashof number the thermal buoyancy force is increased which aids in momentum development in the boundary layer. Velocity boundary layer thickness is therefore increased with increasing G_r. With greater elapse of time, τ, the velocity is also found to be enhanced substantially.

Figure 4 illustrates the distribution of dimensionless velocity distribution (U) with transverse coordinate, Y for various species (mass) Grashof numbers, G_m and non-dimensional times, τ. Again a significant elevation in

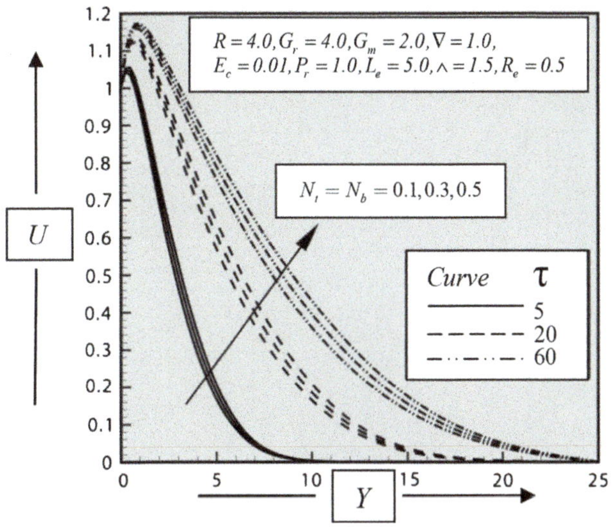

Fig. 7 Thermophoresis and Brownian motion parameter (N_t and N_b) effect on velocity profiles

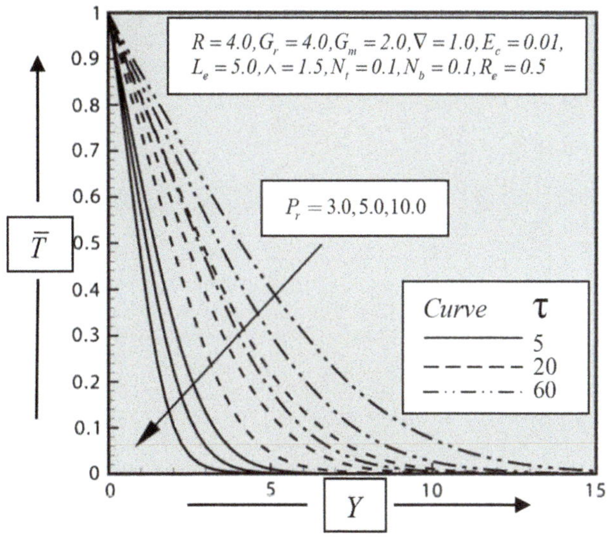

Fig. 9 Prandtl number (P_r) effect on temperature profiles

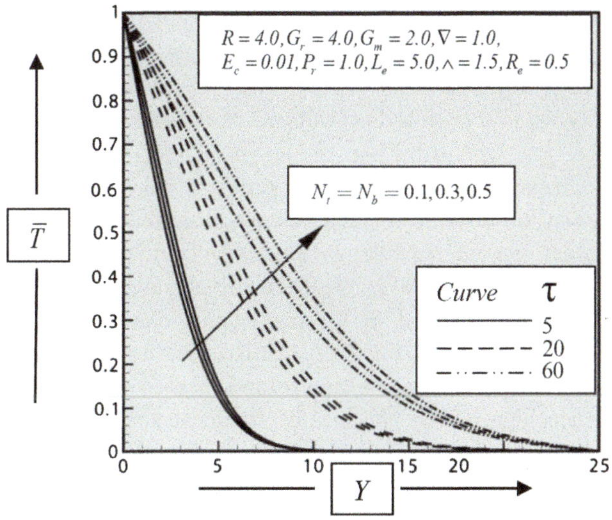

Fig. 8 Thermophoresis and Brownian motion parameter (N_t and N_b) effect on temperature profiles

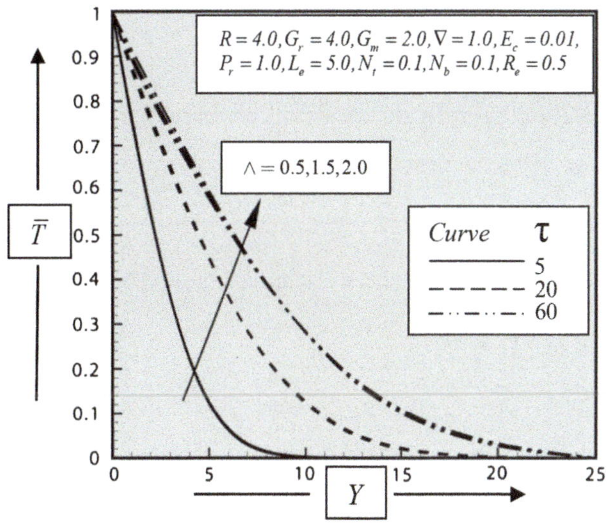

Fig. 10 Viscosity ratio parameter ($\wedge$) effect on temperature profiles

velocity is observed with progressively greater velocity shoots near the stretching sheet, as species Grashof number is increased. Greater values of G_m imply a greater species buoyancy force associated with the mass diffusion of nanoparticles in the regime. With greater values of τ the nanofluid flow is also strongly accelerated.

Figure 5 displays the dimensionless velocity profiles (U) versus Y for different values of combined porous media drag and hydromagnetic drag parameter, R, and non-dimensional time, τ. A marked *deceleration* is witnessed in the vicinity of exponentially stretching sheet with increasing R. In the dimensionless momentum equation (10), the components of $R = \left(\frac{\upsilon L}{K_\circ U_\circ} + \frac{\sigma B_\circ^2 L}{\rho U_\circ}\right)$ both serve to inhibit the momentum development in the boundary layer. The first term i.e. *Darcian linear drag* causes greater impedance to the flow of the nanofluid. The second term i.e. *Lorentzian hydromagnetic drag*, acts transverse to the magnetic field i.e. in the negative X-direction and also retards the flow considerably. For $R = 0.5$, 2.5 a velocity shoot is still present near the sheet surface; however, this vanishes for $R = 5.0$. With increasing time, τ, velocity is again found to be enhanced. Progression in time therefore once again accelerates the flow in the porous regime.

Figure 6 presents the dimensionless velocity distributions (U) versus Y for different values of the viscosity ratio parameter, Λ and also non-dimensional time, τ. The parameter Λ features both in the momentum conservation

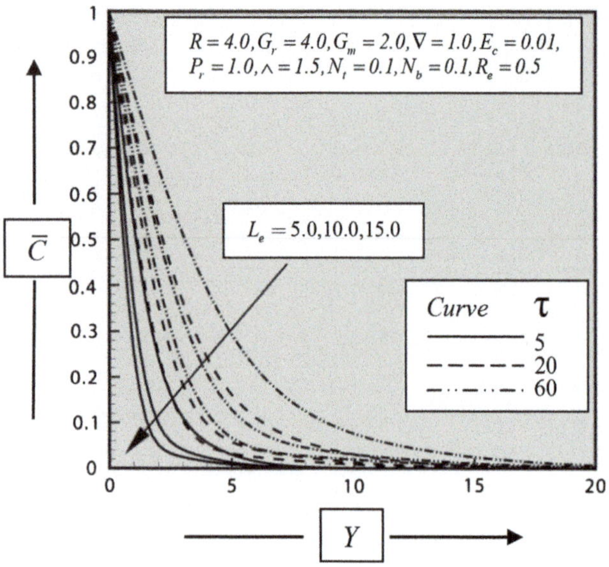

Fig. 11 Lewis number effect on nanoparticle concentration profiles

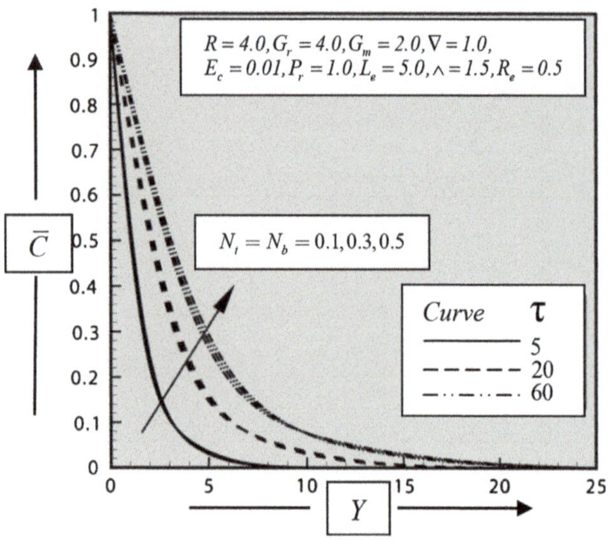

Fig. 12 Thermophoresis and Brownian motion parameter effect on nanoparticle concentrations

Eq. (10) and also the energy conservation Eq. (11), where it is associated with the momentum diffusion term and viscous heating term, respectively. Close to the stretching sheet, an increase in Λ, causes a notable rise in velocity magnitudes, however, further from the sheet surface this trend is reversed and the flow is slightly decelerated. The contribution of viscosity is progressively reduced from the sheet surface (wall) towards the free stream. Momentum boundary layer thickness is therefore found to be increased closer to the sheet surface. In all cases a velocity shoot is computed in close proximity to the wall. With progression in time, τ, the velocity is strongly accelerated. However,

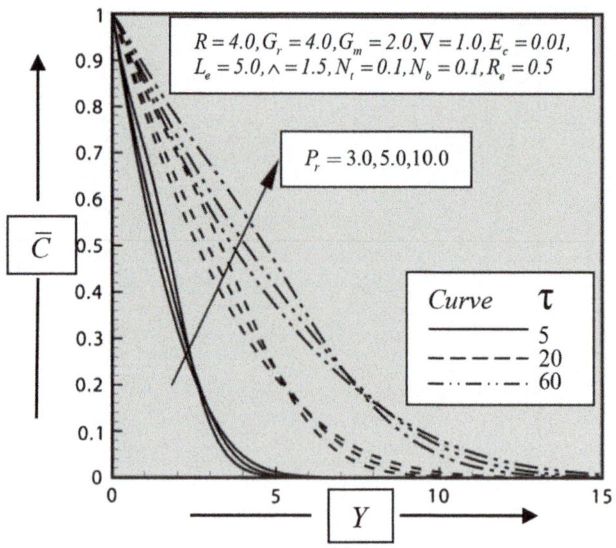

Fig. 13 Prandtl number effect on nanoparticle concentration profiles

with greater time values, there is a more gradual decay in velocity from the near-wall regime to the free stream.

Figure 7 depicts the response in dimensionless velocity profiles (U) through the boundary layer, transverse to the sheet surface for different values of thermophoresis parameter, N_t, and Brownian motion parameter, N_b, and also with various non-dimensional times, τ. Thermophoresis serves to warm the boundary layer and simultaneously exacerbates nanoparticle deposition away from the fluid regime (on to the surface). This effectively accentuates momentum development in the regime and elevates the velocity, as observed in Fig. 6. A similar effect is induced with increasing Brownian motion parameter values i.e. the flow is accelerated. The presence of smaller nanoparticles, which corresponds to higher N_b values, implies a stronger contribution from Brownian motion. This assists the boundary layer flow and increases velocity boundary layer thickness. Similar trends have been observed by, among others, Kuznetsov and Nield (2010) and also Khan and Pop (2010, 2011). As in the other velocity distributions, an increase in time is again shown to significantly accelerate the flow throughout the boundary layer.

Figure 8 displays the evolution of dimensionless temperatures $(\bar{T})$ with transverse coordinate, Y for different values of thermophoresis parameter, N_t, Brownian motion parameter, N_b, and various non-dimensional times, τ. As expected, the boundary layer profiles exhibit similar patterns to those for regular heat transfer fluids i.e. base fluids. The temperature in the boundary layer increases with an increase in both the thermophoresis and Brownian motion parameters. In numerous studies reported in the literature e.g. (Choi 1995; Kang et al. 2006), various mechanisms have been proposed to account for the thermal conduction enhancement in nanofluids. These explanations include interfacial ordering of liquid molecules on the surface of nanoparticles, ballistic transport of energy carriers within individual nanoparticles and between nanoparticles that are in contact, as furthermore the geometrical nanoparticle networking. A popular explanation is the direct contribution of nanoparticles which transport thermal energy since nanoparticles are often in the form of agglomerates and/or aggregates. As elucidated earlier, for larger nanoparticles, Brownian motion is weak and the parameter N_b will have small values, and vice versa for smaller nanoparticles. A greater concentration of smaller nanoparticles is therefore expected to enhance thermal conduction more effectively than a smaller concentration of larger nanoparticles. The influence of Brownian motion on thermal fields is generally very strong and tends to thicken thermal boundary layers as does the thermophoresis effect. The temperatures are also generally enhanced with an increase in time for all values of Y. Thermal boundary layer thickness is therefore increased with elapse in time.

Figure 9 illustrates dimensionless temperature distribution ($\bar{T}$) versus Y for different values of Prandtl number, P_r and non-dimensional time, τ. Prandtl number signifies the relative contribution of momentum diffusion to thermal diffusion in the boundary layer regime. For $P_r > 1$, momentum diffusion rate exceeds thermal diffusion rate. As a result the temperatures in the nanofluid regime will be decreased with a rise in P_r from 3, through 5 to 10. Thermal boundary layer thickness will also be markedly decreased. With increasing time, temperatures are again observed to be strongly enhanced throughout the regime i.e. for all values of Y from the sheet surface through to the free stream.

Figure 10 shows the dimensionless temperature profiles ($\bar{T}$) versus Y for different values of viscosity ratio parameter, Λ and non-dimensional time, τ A strong increase in temperature accompanies a rise in viscosity ratio, and this pattern is sustained for all distances into the boundary layer. Thermal boundary layer thickness is therefore enhanced with increasing viscosity ratio. With greater values of time, τ, nanofluid temperatures are also elevated.

Figures 11, 12, 13 present the response of nanoparticle concentration profiles for various Lewis number, Brownian motion and thermophoresis numbers, and Prandtl numbers, respectively. In all plots the effect of non-dimensional time, τ, is also studied. An increase in Lewis number significantly reduces the nanoparticle concentration values (Fig. 11) in the regime. Lewis number defines the ratio of thermal diffusivity to mass (nanoparticle species) diffusivity. It is used to characterize fluid flows where there is simultaneous heat and mass transfer by convection. Effectively it is also the ratio of Schmidt number and the Prandtl number. For $L_e > 1$, thermal diffusion rate exceeds species diffusion rate and nanoparticle concentrations are therefore suppressed. Concentration boundary layer thickness is also reduced with increasing L_e values. Conversely an increase in Brownian motion and thermophoresis parameters (Fig. 12) acts to enhance nanoparticle concentration values. These two mechanisms therefore assist the diffusion of nanoparticles in the boundary layer and elevate the concentration boundary layer thickness. Similarly increasing Prandtl number (Fig. 13) is also found to initially boost the nanoparticle concentration values (closer to the sheet surface); however, further from the sheet surface (wall) this behaviour is reversed and an increase in Prandtl number is observed to marginally decrease concentration values. In all cases (Figs. 11, 12, 13) an increase in time, τ, generally enhances concentration values.

Conclusions

A mathematical model for transient hydromagnetic mixed convection boundary layer flow of an electrically conducting, nanofluid over an exponentially stretching sheet embedded in an isotropic, homogenous porous medium has been developed. The non-dimensionalized conservation equations have been solved with a robust, EFDM, with details of the stability and convergence characteristics included. Validation has been obtained with an optimized transient VIM algorithm. A detailed study of the effects of several key thermophysical parameters controlling the flow characteristics has been conducted. The computations have shown that:

1. Velocity and momentum boundary layer thickness are enhanced with increasing thermal Grashof Number, species Grashof number, Brownian motion parameter and thermophoresis parameter, whereas they are decreased with increasing Darcian porous media drag parameter, hydromagnetic parameter and viscosity ratio parameter.
2. Nanofluid temperature and thermal boundary layer thickness are elevated with increasing thermophoresis, Brownian parameter and viscosity ratio parameter, whereas they are suppressed with increasing Prandtl number.
3. Nanoparticle concentration and concentration boundary layer thickness are both increased with increasing thermophoresis, Brownian parameter and Prandtl number, whereas they are reduced with increasing Lewis number.
4. Increase in time generally accelerates the flow i.e. increases momentum boundary layer thickness, increases temperatures and enhances nanoparticle concentration values in the boundary layer regime.

The resent study has demonstrated the excellent accuracy and stability of the explicit finite difference numerical approach in non-linear unsteady two-dimensional magnetohydrodynamic nanofluid transport simulation in porous media. Future studies will consider the application of this method to *bio-convection* nanofluid flows, of interest in microbial fuel cell technology (Anwar Bég et al. 2013) and will be communicated in due course.

References

Abbasbandy S, Ghehsareh HR (2012) Solutions of the magnetohydrodynamic flow over a nonlinear stretching sheet and nano boundary layers over stretching surfaces. Int J Numer Methods Fluids 70:1324–1340

Abdou MA, Soliman AA (2012) New explicit approximate solution of MHD viscoelastic boundary layer flow over stretching sheet. Math Methods Appl Sci 35:1117–1125

Anwar Bég O (2012) Numerical methods for multi-physical magnetohydrodynamics. In: New developments in hydrodynamics research, Chap 1. Nova Science, New York, pp 1–110

Anwar Bég O (2013) Benchmarking geofluid dynamics computations generated with a finite element code (GEOFEM) with He's variational iteration method groundwater solutions. Technical Report-GEO-45-97-F-K, Gort Engovation-Aerospace Engineering Sciences, Bradford, UK

Anwar Bég O (2013) ELECTROVIM—a new variational iteration method code for simulating electrostatic and electrodynamic thruster flows with the Jahn formulation. Technical report—ELEC-J61, Gort Engovation-Aerospace Engineering Sci., Bradford, UK

Anwar Bég O (2013) NEUROVIM—optimized variational iteration method code for modeling brain swelling from automotive and aerospace crash incidents with a chemo-mechanical deformation constitutive model. Technical report-NEURO-J-61, Gort Engovation-Aerospace Engineering Sciences, Bradford, UK

Anwar Bég O (2013) TRANSNANOVIM—a variational iteration method program in MATLAB for transient nonlinear nanofluid dynamics simulation. Technical report-NANO-H-61, Gort Engovation-Aerospace Engineering Sciences, Bradford, UK

Anwar Bég O, Hameed M, Bég TA (2013a) Chebyshev spectral collocation simulation of nonlinear boundary value problems in electrohydrodynamics (EHD). Int J Comput Methods Eng Sci Mech 14(2):104–115

Anwar Bég O, Prasad VR, Vasu B (2013) Numerical study of mixed bioconvection in porous media saturated with nanofluid containing oxytactic microorganisms. J Mech Med Biol 13(4):1350067.1–1350067.25

Anwar BO, Bakier AY, Prasad VR (2009) Numerical study of free convection magnetohydrodynamic heat and mass transfer from a stretching surface to a saturated porous medium with Soret and Dufour effects. Comput Mater Sci 46(1):57–65

Asai S (2012) Magnetohydrodynamics in materials processing. Electromagn Process Mater Fluid Mech Appl 99:49–86

Baron A, Szewieczek D, Nowosielski R (2007) Selected manufacturing techniques of nano-materials. J Achiev Mater Manuf Eng 20:83–86

Bataller RC (2008) Similarity solutions for flow and heat transfer of a quiescent fluid over a nonlinearly stretching surface. J Mater Process Technol 203(1–3):176–183

Bég OA, Tripathi D (2012) Mathematica simulation of peristaltic pumping with double-diffusive convection in nanofluids: a bio-nano-engineering model. Proc IMechE-Part N J Nanoeng Nanosyst 225:99–114

Bég O, Anwar Bég TA, Rashidi MM, Asadi M (2012) Homotopy semi-numerical modelling of nanofluid convection boundary layers from an isothermal spherical body in a permeable regime. Int J Microscale Nanoscale Thermal Fluid Transp Phenom 3(4):367–396

Bidin B, Nazar R (2009) Numerical solution of the boundary layer flow over an exponentially stretching sheet with thermal radiation. Eur J Sci Res 33:710–717

Carnahan B, Luther HA, Wilkes JO (1969) Applied numerical methods. Wiley, New York

Choi SUS (1995) Enhancing thermal conductivity of fluids with nanoparticles. In: Siginer DA, Wang HP (eds) Development and applications of non-newtonian flows. ASME MD, vol. 231 and FED, vol. 66. USDOE, Washington, DC, pp 99–1

Crainic N, Marques AT, Bica D et al (2003) The usage of the nanomagnetic fluids and the magnetic field to enhance the production of composite made by RTM-MNF. In: 7th international conference on frontiers of polymers and advanced materials, Bucharest, June 10–15

Crainic N, Bica D, Torres Marques A et al (2007) Magnetic nanocomposites obtained using high evaporation rate magnetic nanofluids. Int J Nanomanuf 1:784–798

Crane LJ (1970) Flow past a stretching plate. J Appl Math Phys (ZAMP) 21:590–595

Elbashbeshy EMA (2001) Heat transfer over an exponentially stretching continuous surface with suction. Arch Mech 53:643–651

Elsayed AF (2013) Comparison between variational iteration method and homotopy perturbation method for thermal diffusion and diffusion thermo effects of thixotropic fluid through biological tissues with laser radiation existence. Appl Math Model 37:3660–3673

Fautrelle Y, Ernst R, Moreau R (2009) Magnetohydrodynamics applied to materials processing. Int J Mater Res 100:1389–1398

Garnier M (1992) Magnetohydrodynamics in materials processing. Philos Trans R Soc Phys Sci Eng 344:249–263

Garnier M (1996) Present and future prospect in electromagnetic processing of materials. Magnetohydrodynamics 32(2):109–115

Gorla RSR, Zinolabedini A (1987) Free convection from a vertical plate with non-uniform surface temperature embedded in a porous medium. ASME J Energy Res Technol 109:26–30

Hamad MAA, Pop I (2011) Scaling transformations for boundary layer stagnation-point flow towards a heated permeable stretching sheet in a porous medium saturated with a nanofluid and heat absorption/generation effects. Transp Porous Media 87:25–39

Hamad MAA, Pop I, Ismail AI (2011) Magnetic field effects on free convection flow of a nanofluid past a semi-infinite vertical flat plate. Nonlinear Anal Real World Appl 12:1338–1346

He JH (1999) Variational iteration method—a kind of non-linear analytical technique: some examples. Int J Non-Linear Mech 34:699–708

Ishikwa M, Yuhara M, Fujino T (2007) Three-dimensional computation of magnetohydrodynamics in a weakly ionized plasma with strong MHD interaction. J Mater Process Technol 181:254–259

Kang HU, Kim SH, Oh JM (2006) Estimation of thermal conductivity of nanofluid using experimental effective particle volume. Exp Heat Transf 19:181–191

Khan SK (2006) Boundary layer viscoelastic fluid flow over an exponentially stretching sheet. Int J Appl Mech Eng 11:321–335

Khan WA, Pop I (2010) Boundary-layer flow of a nanofluid past a stretching sheet. Int J Heat Mass Trans 53:2477–2483

Khan WA, Pop I (2011) Free convection boundary layer flow past a horizontal flat plate embedded in a porous medium filled with a nanofluid. ASME J Heat Trans 133:9

Khan MS, Alam MM, Ferdows M (2011) Finite difference solution of MHD radiative boundary layer flow of a nanofluid past a stretching sheet. In: Proceedings of the international conference on mechanical engineering (ICME 11), FL-011, BUET, Dhaka, Bangladesh

Kumaran V, Ramanaiah G (1996) A note on the flow over a stretching sheet. Acta Mech 116:229–233

Kuznetsov AV, Nield DA (2010) Natural convective boundary-layer flow of a nanofluid past a vertical plate. Int J Thermal Sci 49:243–247

Magyari E, Keller B (2000) Heat and mass transfer in the boundary layers on an exponentially stretching continuous surface. J Phys D Appl Phys 32:577–585

Mohiddin SG, Prasad VR, Anwar Bég O (2010) Numerical study of unsteady free convective heat and mass transfer in a Walters-B viscoelastic flow along a vertical cone. Int J Appl Math Mech 6:88–114

Partha MK, Murthy PVSN, Rajasekhar GP (2005) Effect of viscous dissipation on the mixed convection heat transfer from an exponentially stretching surface. Heat Mass Transf 41:360–366

Prasad VR, Vasu B, Anwar Bég O, Parshad R (2011a) Unsteady free convection heat and mass transfer in a Walters-B viscoelastic flow past a semi-infinite vertical plate: a numerical study. Thermal Sci-Int Sci J 15(2):S291–S305

Prasad VR, Vasu B, Anwar Bég O (2011) Numerical modeling of transient dissipative radiation free convection heat and mass transfer from a non-isothermal cone with variable surface conditions. Elixir-Appl Math 41:5592–5603

Rana P, Bhargava R, Anwar Bég O (2012) Numerical solution for mixed convection boundary layer flow of a nanofluid along an inclined plate embedded in a porous medium. Comput Math Appl 64(9):2816–2832

Rana P, Bhargava R, Anwar Bég O (2013) Finite element simulation of unsteady MHD transport phenomena on a stretching sheet in a rotating nanofluid. Proc IMechE-Part N J Nanoeng Nanosyst 227:77–99

Sadooghi N, Taghinavaz F (2012) Local electric current correlation function in an exponentially decaying magnetic field. Phys Rev D 85:125035

Sakiadis BC (1961) Boundary-layer behavior on continuous solid surfaces. AIChE J 7:26–28

Sanjayanand E, Khan SK (2006) On heat and mass transfer in a viscoelastic boundary layer flow over an exponentially stretching sheet. Int J Thermal Sci 45:819–828

Shahidian A et al (2011) Effect of nanofluid properties on magnetohydrodynamic pump (MHD). Adv Mater Res 403:663–669

Shahmohamadi H, Rashidi MM, Anwar Bég O (2012) A new technique for solving steady flow and heat transfer from a rotating disk in high permeability media. Int J Appl Math Mech 8(7):1–17

Stastna J, De Kee D, Harrison B (1991) Non-Markovian diffusion process in polymers and stretched exponential relaxation. Rheol Acta 30:263–269

Stephens JR, Beveridge JS, Latham AH, Williams ME (2010) Diffusive flux and magnetic manipulation of nanoparticles through porous membranes. Anal Chem 82:3155–3160

Tari H, Ganji DD, Rostamian M (2007) Approximate solutions of K(2,2), KdV and modified KdV equations by variational iteration method, homotopy perturbation method and homotopy analysis method. Int J Nonlinear Sci Numer Simul 8:203–210

Wang X (2012) Exp Micro/Nanoscale Thermal Transp. Wiley, New York

Weidman PD, Magyari E (2010) Generalized Crane flow induced by continuous surfaces stretching with arbitrary velocities. Acta Mech 209:353–362

Yakovlev NL, Tay YY, Tay ZJ, Chen HV (2013) Distribution of switching fields in thin films with uniaxial magnetic anisotropy. J Magn Magn Mater 329:170–177

7

Group analysis and numerical computation of magneto-convective non-Newtonian nanofluid slip flow from a permeable stretching sheet

M. J. Uddin · M. Ferdows · O. Anwar Bég

Abstract Two-dimensional magnetohydrodynamic boundary layer flow of non-Newtonian power-law nanofluids past a linearly stretching sheet with a linear hydrodynamic slip boundary condition is investigated numerically. The non-Newtonian nanofluid model incorporates the effects of Brownian motion and thermophoresis. Similarity transformations and corresponding similarity equations of the transport equations are derived via a linear group of transformations. The transformed equations are solved numerically using Runge–Kutta–Fehlberg fourth-fifth order numerical method available in the Maple 14 software for the influence of power-law (rheological) index, Lewis number, Prandtl number, thermophoresis parameter, Brownian motion parameter, magnetic field parameter and linear momentum slip parameter. Validation is achieved with an optimized Nakamura implicit finite difference algorithm (NANO-NAK). Representative results for the dimensionless axial velocity, temperature and concentration profiles have been presented graphically. The present results of skin friction factor and reduced heat transfer rate are also compared with the published results for several special cases of the model and found to be in close agreement. The study has applications in electromagnetic nanomaterials processing.

M. J. Uddin
American International University-Bangladesh,
Dhaka 1206, Bangladesh

M. Ferdows
Department of Mathematics, University of Dhaka,
Dhaka 1000, Bangladesh

M. Ferdows
Nuclear Science Research Institute, Japan Atomic
Energy Agency, 2-4, Ibaraki 319-1184, Japan

O. A. Bég (✉)
Gort Engovation-Propulsion, Nanomechanics and Biophysics,
Southmere Avenue, Bradford, UK
e-mail: gortoab@gmail.com

Keywords Non-Newtonian nanofluids · MHD flow · Hydrodynamic slip · Linear group transformations · Maple · Nakamura method

List of symbols

Roman

A	Real number
a	Generalized momentum slip parameter
c_p	Specific heat at constant pressure of the fluid
C	Nanoparticle volume fraction
D_B	Brownian diffusion
D_T	Thermophoretic diffusion coefficient
$f(n)$	Dimensionless stream function
fw	Suction/injection velocity
g	Acceleration due to gravity
k	Thermal conductivity
K	Consistency coefficient of the fluid
L	Characteristic length
Le	Generalized Lewis number
n	Power-law rheological index
N_1	Slip factor
Nb	Generalized Brownian motion parameter
Nt	Generalized thermophoresis parameter
$Nu_{\bar{x}}$	Local Nusselt number
Pr_x	Generalized Prandtl number
Re	Reynolds number
$Re_{\bar{x}}$	Local Reynolds number
$Sh_{\bar{x}}$	Local Sherwood number

T Fluid temperature
U_r Reference velocity
$\bar{u}$ Dimensional velocity in $\bar{x}$-direction
$\bar{v}$ Dimensional velocity in $\bar{y}$-direction
$\bar{v}_w$ Lateral mass flux velocity
$\bar{x}$ Coordinate along the surface
y Coordinate normal to the surface

Greek

α Thermal diffusivity
α_i Real numbers
ν Kinematic viscosity of the fluid
Γ Group
ρ_f Fluid density
ρ_p Nanoparticle mass density
$(\rho c)_f$ Heat capacity of the fluid
$(\rho c)_p$ Heat capacity of the nanoparticle material
τ Ratio between the effective heat capacity of the nanoparticle material and heat capacity of the fluid
η Similarity independent variable
θ Dimensionless temperature
ρ Density of the fluid
ψ Dimensionless stream function

Subscripts

w Conditions at the wall
∞ Ambient condition

Superscripts

$'$ Prime denotes the derivative with respect to η

Introduction

Nanofluid transport phenomena have received extensive attention in the past decade or so, following seminal studies by researchers at the Argonne Energy Laboratory in Illinois, USA, in the 1990s (Eastman et al. 2004). Although initially interests in nanofluids were directed at thermal enhancement in transport engineering (automotive, aerospace etc.), applications have significantly diversified in the past decade. New areas in which nanofluids have been deployed include bioconvection in microbial fuel cells (Bég et al. 2013a), electronics cooling (Townsend and Christianson 2009), building physics and contamination control (Kulkarni et al. 2009), pharmacological administration mechanisms (Mody et al. 2013), peristaltic pumps for diabetic treatments (Bég and Tripathi 2012) and solar collectors (Rana et al. 2013a). Experimental works have been complemented in recent years by numerous theoretical and computational simulations. The latter have deployed an extensive range of sophisticated algorithms. Rashidi et al. (2012) used the homotopy analysis method (HAM) and Mathematica symbolic software to analyze coating flows of a vertical cylinder with nanofluids. Rashidi et al. (2013) studied the Von Karman swirl flow of a nanofluid with entropy generation using differential transform methods. Bég et al. (2013b) studied both single and two-phase nanofluid dynamics in circular conduits using finite volume numerical codes. Although nanofluids have largely been simulated as Newtonian fluids, their rheological properties have been established for some time. Important experimental confirmation of nanofluid rheology has been made by Chen et al. (2007) for ethylene glycol-based titania nanofluids. Kedzierski et al. (2010) described the results of a significant series of colloidal dispersions collected as part of the International Nanofluid Property Benchmark Exercise (INPBE). They considered the rheology of seven different fluids that include dispersions of metal-oxide nanoparticles in water, and in synthetic oil, elaborating on the effects of particle shape and concentration on the viscosity of these same nanofluids and benchmarking computations with classical theories on suspension rheology. Further significant studies exploring the non-Newtonian characteristics of nanofluids include Anoop et al. (2009) who considered agglomeration. Kim et al. (2011) showed that rod-like alumina nanoparticles have a sol- or weakly flocculated gel-structure depending on particle concentration and that thermal conductivity rate is enhanced with concentration and is markedly faster in the sol state than in the weakly flocculated gel state. They also observed that when the nanofluid becomes a strongly flocculated gel thermal conductivity remains almost the same as the pure liquid value and that Brownian motion is the key mechanism contributing to enhancing thermal conductivity.

As with Newtonian nanofluid dynamics, non-Newtonian modeling of nanofluid transport phenomena has also attracted significant attention very recently. A diverse range of rheological models have been implemented to simulate different flow characteristics of nanofluid suspensions including Oswald–DeWaele power-law models, viscoelastic models and micro-morphic models. These studies have revealed many interesting features of non-Newtonian nanofluids. Sheu (2011) employed an Oldroyd-B viscoelastic model to study nanofluid convection in a horizontal layer of a saturated porous medium. Goyal and Bhargava (2013) used the Reiner-Rivlin second order rheological model to simulate the effect of velocity slip boundary condition on the flow and heat transfer of non-Newtonian nanofluid over a stretching sheet. They employed a finite element algorithm and examined Brownian motion and thermophoresis effects also. Gorla and Kumari (2012) utilized the Eringen micromorphic model in their analysis of mixed convection non-Newtonian nanofluid flow from a nonlinearly stretching sheet. Hojjat et al. (2012) conducted both analytical and experimental studies of frictional pressure drop of non-Newtonian nanofluids flowing through a horizontal circular tube in both the laminar and the turbulent regimes, for three different types of oxide nanoparticles, γ-Al_2O_3, CuO, and TiO_2, were used for the preparation of nanofluids. They observed that improved heat transfer coefficients are achieved with nanofluids without requiring an increase in the pressure drop. Uddin et al. (2013a) utilized Maple quadrature and a power-law rheological model to investigate nanofluid boundary layer flow in porous media with heat source effects. He et al. (2012) considered laminar forced convective heat transfer of dilute liquid suspensions of nanoparticles (nanofluids) in a straight pipe. Using high shear mixing and ultrasonication methods, they observed that the non-Newtonian character of nanofluids influences the overall thermal enhancement. Estellé et al. (2013) performed experiments on the steady-state rheological behaviour of carbon nanotube (CNT) water-based nanofluid and considered the influence of a controlled pre-shear history on the viscosity. They examined both the effect of the influence of stress rate during pre-shear and the effect of resting time before viscosity measurement, and demonstrated that CNT water-based nanofluid behaves as a viscoelastic media at low shear rate and as a shear-thinning fluid at higher shear rates, with the observed behaviour being a function of shear history due to the breakdown in the structural network of nanofluid agglomerates. It is also observed that the nanofluid can reform at rest. Ellahi et al. (2012) studied analytically using a homotopy method, the fully-developed flow of an incompressible, thermo-dynamically-compatible non-Newtonian third-grade nanofluid in coaxial cylinders, employing the Reynolds and Vogel variable viscosity models. Akbar and Nadeem (2013) implemented the Eyring-Prandtl rheological fluid model to study peristaltic nanofluid transport in a diverging tube. They used a homotopy perturbation technique to compute the effects of Brownian motion and thermophoresis on velocity, temperature, nanoparticle, and pressure gradient under five different peristaltic waves. Very recently Bég (2013a) used a Lattice Boltzmann method to simulate nanofluid transport in elliptical tubes with the Eringen microstretch model, examining deformability of suspending nanoparticles and gyratory motions of micro-elements. Further studies of nano-rheological flows include Susrutha et al. (2012) who examined experimentally the behaviour of gold-nanofluids with poly (vinylidene fluoride) molecules and Hojjat et al. (2011) who studied power-law nanofluid thermal convection in tubes.

Recently great progress has also been made in a new generation of *magnetic nanofluids* which provide very desirable features in materials processing, energy applications and also medical engineering. Nkurikiyimfura et al. (2011) have described the excellent thermal and flow control characteristics of magnetic nanofluids which respond to magnetic fields. Such fluids offer many rich problems for mathematical modeling since they amalgamate nanofluid dynamics with the established field of magnetohydrodynamics (MHD). Finite element simulations of the heat and mass transfer characteristics of transient magneto-nanofluid flows in rotating manufacturing processes were reported by Rana et al. (2013b). Ferdows et al. (2013) used Maple software to study the effects of radiative heat transfer on unsteady MHD nanofluid transport from a stretching surface in high-temperature materials processing. Boundary layer flow of magnetic nanofluid with buoyancy forces was examined by Hamad et al. (2011). The most popular mathematical model for boundary layer flows of nanofluids is the Kuznetsov and Nield (2010) theory which incorporates Brownian motion and thermophoresis effects. This model has been implemented successfully in increasingly more complex multi-physical nanofluid problems. For example, Bég et al. (2012) investigated nanofluid transport phenomena in porous media from a sphere using homotopy methods and the Kuznetsov–Nield model.

The above studies have all assumed the no-slip condition at the wall. However in various industrial processes, slip effects can arise at the boundary of pipes, walls, curved surfaces etc. A frequently adopted approach in simulating slip phenomena is the Navier velocity slip condition. Thermal slip may also arise. Recent studies on hydrodynamic slip include the article by Wang (2007) who developed analytical solutions for stagnating Newtonian flows on a cylindrical body. Uddin et al. (2012a) used Lie algebra to derive similarity solutions for double-diffusive convective boundary layer flows with slip. Prasad et al. (2013) used the Keller box finite difference method to simulate numerically the heat transfer in boundary layer slip flow of Casson non-Newtonian fluid from a cylinder. Bég et al. (2011a) employed the network electro-thermal simulation code (PSPICE) to study magneto-convective Von Karman slip flow with radiative heat transfer and slip effects. Hamad et al. (2013) analyzed the composite heat and mass transfer in boundary layer flow with variable diffusivity and hydrodynamic slip and thermal convective boundary conditions. Tripathi et al. (2012) studied the rhythmic flow of Oldroyd-B fractional viscoelastic fluids with wall slip effects. Several studies have also addressed nanofluid slip flows. Uddin et al. (2012b) and Bég (2013b) have studied computationally the magneto-convective Sakiadis nanofluid slip flow with heat source effects. Yang et al. (2012) have considered wall mass flux and slip effect on nanofluid convection from a porous two-dimensional wedge configuration.

In the present work, we extend the work of Andersson and Bech (1992) to analyze the transport of non-Newtonian nanofluids from a linear stretching surface with a hydrodynamic slip boundary condition. A linear group of transformation has been used to develop similarity variables for the two-point boundary value problem. Numerical solutions are obtained with both Maple quadrature and also the Nakamura implicit finite difference technique, showing excellent correlation. Further validation is achieved with simpler models from the published literature. The effect of the emerging thermophysical parameters dictating transport characteristics, on the dimensionless axial velocity, temperature and nanoparticle volume fraction are elaborated in detail. To the authors' best knowledge, the present problem has not been communicated thus far in the literature and has important applications in nano-technological materials processing.

Governing equations

Consider the two-dimensional $(\bar{x}, \bar{y})$ steady-state MHD boundary layer slip flow of a an electrically-conducting non-Newtonian power law nanofluid from a heated porous stretching sheet. The nanofluid is assumed to be single phase, in thermal equilibrium and there is a slip velocity between base fluid and particles. The nanoparticles are assumed to have uniform shape and size. The coordinate system and flow regime are illustrated in Fig. 1. Thermophysical properties are assumed to be invariant. A transverse magnetic field of constant strength is imposed in the $\bar{y}$-direction. Hall current and Alfven wave effects are neglected as are magnetic induction and Ohmic dissipation. The sheet surface is also assumed to be *non-conducting*. The governing flow field equations in dimensional form are obtained by combining the models of Andersson and Bech (1992) and Makinde and Aziz (2011):

$$\frac{\partial \bar{u}}{\partial \bar{x}} + \frac{\partial \bar{v}}{\partial \bar{y}} = 0, \tag{1}$$

$$\bar{u}\frac{\partial \bar{u}}{\partial \bar{x}} + \bar{v}\frac{\partial \bar{u}}{\partial \bar{y}} = -\frac{K}{\rho}\frac{\partial}{\partial \bar{y}}\left(-\frac{\partial \bar{u}}{\partial \bar{y}}\right)^{n} - \frac{\sigma B_0^2}{\rho}\bar{u}, \tag{2}$$

$$\bar{u}\frac{\partial T}{\partial \bar{x}} + \bar{v}\frac{\partial T}{\partial \bar{y}} = \alpha\frac{\partial^2 T}{\partial \bar{y}^2} + \tau D_B \frac{\partial T}{\partial \bar{y}}\frac{\partial C}{\partial \bar{y}} + \tau\left(\frac{D_T}{T_\infty}\right)\left(\frac{\partial T}{\partial \bar{y}}\right)^2, \tag{3}$$

$$\bar{u}\frac{\partial C}{\partial \bar{x}} + \bar{v}\frac{\partial C}{\partial \bar{y}} = D_B\frac{\partial^2 C}{\partial \bar{y}^2} + \left(\frac{D_T}{T_\infty}\right)\frac{\partial^2 T}{\partial \bar{y}^2}, \tag{4}$$

where $\alpha = \frac{K}{(\rho c_p)_f}$ is the thermal diffusivity of the fluid and $\tau = \frac{(\rho c)_p}{(\rho c)_f}$ is the ratio of heat capacities, and all other terms are defined in the notation section. The boundary conditions at the wall (sheet) and in the free stream are prescribed thus:

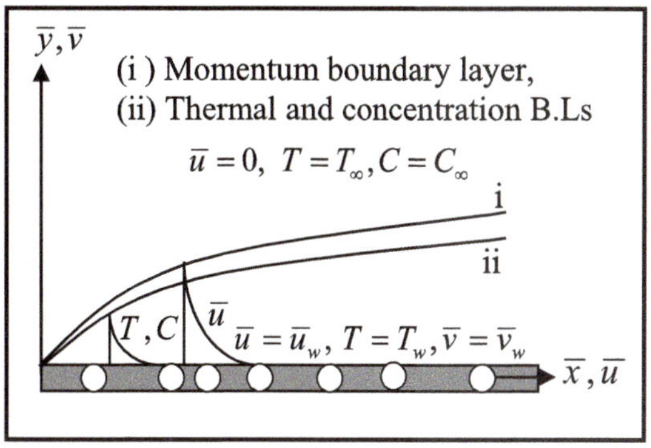

Fig. 1 Coordinate system and flow model

$$\begin{aligned} &\bar{u} = \bar{u}_w + \bar{u}_{slip} = U_r\left(\frac{\bar{x}}{L}\right) + N_1 v\left(-\frac{\partial \bar{u}}{\partial y}\right)^{n-1}\left(\frac{\partial \bar{u}}{\partial y}\right), \\ &v = \bar{v}(\bar{x}), T = T_w, C = C_w \quad \text{at } y = 0, \\ &\bar{u} \to 0, T \to T_\infty, C \to C_\infty \quad \text{as } y \to \infty. \end{aligned} \tag{5}$$

Here U_r is a reference velocity, L is the characteristic of length, and N_1 is the slip factor having the dimension (seconds/metres). It needs to be mentioned that for the non-Newtonian power law model, the case of $N < 1$ is associated with shear-thinning fluids (pseudoplastic fluids), $n = 1$ corresponds to Newtonian fluids and $n > 1$ applies to the case of shear-thickening (dilatant).

Non-dimensionalization

Express the governing equations into dimensionless using following dimensionless boundary layer variables:

$$\bar{x} = \frac{x}{L}, \bar{y} = \frac{x}{L}Re_L^{\frac{1}{n+1}}, u = \frac{\bar{u}}{U_r}, v = \frac{\bar{v}}{L}Re_L^{\frac{1}{n+1}}, \theta = \frac{T - T_\infty}{T_w - T_\infty}, \\ \phi = \frac{C - C_\infty}{C_w - C_\infty}. \tag{6}$$

We introduce also a stream functions ψ is defined by:

$$u = \frac{\partial \psi}{\partial y}, \quad -v = -\frac{\partial \psi}{\partial x}. \tag{7}$$

The normalized transformed partial differential conservation equations are thereby reduced to:

$$\frac{\partial \psi}{\partial y}\frac{\partial^2 \psi}{\partial x \partial y} - \frac{\partial \psi}{\partial x}\frac{\partial^2 \psi}{\partial y^2} - n\left(-\frac{\partial^2 \psi}{\partial y^2}\right)^{n-1}\frac{\partial^2 \psi}{\partial y^3} + M\frac{\partial \psi}{\partial y} = 0, \tag{8}$$

$$\frac{\partial \psi}{\partial y}\frac{\partial \theta}{\partial x} - \frac{\partial \psi}{\partial x}\frac{\partial \theta}{\partial y} - \frac{1}{Pr_L}\frac{\partial^2 \theta}{\partial y^2} - Nb_L\frac{\partial \theta}{\partial y}\frac{\partial \phi}{\partial y} \\ - Nt_L\left(\frac{\partial \theta}{\partial y}\right)^2 = 0, \tag{9}$$

$$\frac{\partial \psi}{\partial y}\frac{\partial \phi}{\partial x} - \frac{\partial \psi}{\partial x}\frac{\partial \phi}{\partial y} - \frac{1}{Le_L}\frac{\partial^2 \phi}{\partial y^2} - \frac{1}{Le_L}\frac{Nt_L}{Nb_L}\frac{\partial^2 \theta}{\partial y^2} = 0, \tag{10}$$

The transformed boundary conditions assume the form:

$$\frac{\partial\psi}{\partial y}=x+\frac{N_1 v}{L^n}U_{\mathrm{r}}^{n-1}Re_L^{\frac{n}{n+1}}\left(-\frac{\partial^2\psi}{\partial y^2}\right)^{n-1}\left(\frac{\partial^2\psi}{\partial y^2}\right),\frac{\partial\psi}{\partial x}=\mathrm{fw}_L,$$
$$\theta=\psi=1 \quad \text{at } y=0,$$
$$\frac{\partial\psi}{\partial y}=\theta=\psi=0 \quad \text{as } y\to\infty. \tag{11}$$

In Eqs. (8)–(11), the parameters M, Re_L, Pr_L, Nb_L, Nt_L, Le_L, fw_L denote the magnetic body force parameter, uniform Reynolds number, uniform Prandtl number, uniform Brownian motion parameter, uniform thermophoresis parameter, uniform Lewis number and uniform suction, respectively, and are defined as

$$M=\frac{\sigma B_0^2 L}{\rho c U_{\mathrm{r}}},\ Re_L=\frac{\rho L^n U_{\mathrm{r}}^{2-n}}{K},\ Pr_L=\frac{U_{\mathrm{r}}L}{\alpha}\left(\frac{\rho L^n U_{\mathrm{r}}^{2-n}}{k}\right)^{-\frac{2}{n+1}},$$
$$Nb_L=\tau D_{\mathrm{B}}\Delta C\left(\frac{U_{\mathrm{r}}^{\frac{3-3n}{2}}L^{\frac{n-1}{2}}\rho}{K}\right)^{\frac{2}{n+1}},$$
$$Nt_L=\frac{\tau D_{\mathrm{T}}\Delta T}{T_\infty}\left(\frac{\rho U_{\mathrm{r}}^{\frac{3-3n}{2}}}{K}\right)^{\frac{2}{n+1}},\ Le_L=\frac{LU_{\mathrm{r}}}{D_{\mathrm{B}}}\left(\frac{\rho L^n U_{\mathrm{r}}^{2-n}}{K}\right)^{-\frac{2}{n+1}},$$
$$\mathrm{fw}_L=\frac{v_{\mathrm{w}}Re_L^{1/(n+1)}}{U_{\mathrm{r}}} \tag{12}$$

where K is the consistency coefficient of the nano-rheological fluid.

Applications of linear group of transformations

The governing partial differential Eqs. (8)–(10) with the boundary conditions (11) comprise a strongly coupled nonlinear boundary value problem. Totally analytical solutions for such a system are intractable. Additionally numerical solutions of these equations are complicated and computationally expensive. Similarity solutions have been demonstrated to provide an excellent mechanism for further transforming boundary value problems and reducing computational costs, while retaining essentially the physical accuracy of mathematical models. We therefore now implement a group theoretical method which combines the independent two variables (x, y) into a single independent variable (η) and reduce the governing partial differential equations into ordinary differential equations. For this purpose we scale all independent and dependent variables as:

$$\Gamma: x^*=xA^{\alpha_1},\ y^*=yA^{\alpha_2},\ \psi^*=\psi A^{\alpha_3},\ \theta^*=\theta A^{\alpha_4},$$
$$\phi^*=\psi A^{\alpha_5}, \tag{13}$$

where $A(\alpha_i, i=1,2,3\ldots,5)$ are all arbitrary *real* constants. We seek the values of α_i such that the form of Eqs. (8)–(11) is *invariant under the transformations* in (13). Substituting new variables in Eq. (13) into Eqs. (8)–(11), equating powers of A (to confirm the invariance of the Eqs. (8)–(11) under this group of transformations), we have, following Hansen (1964) and Shang (2010):

$$2\alpha_3-2\alpha_2-\alpha_1=(n-1)\alpha_3-2(n-1)\alpha_2+\alpha_3-3\alpha_2$$
$$=\alpha_3-\alpha_2,$$
$$\alpha_4+\alpha_3-\alpha_2-\alpha_1=\alpha_4-2\alpha_2=\alpha_4+\alpha_5-2\alpha_2=2\alpha_4-2\alpha_2,$$
$$\alpha_4+\alpha_3-\alpha_2-\alpha_1=\alpha_5-2\alpha_2=\alpha_4-2\alpha_2, \tag{14}$$

solving the *linear* system defined by Eq. (14), we arrive in due course at:

$$\alpha_1=\frac{n+1}{2n}\alpha_3,\alpha_2=\frac{n-1}{2n}\alpha_3,\alpha_4=\alpha_5=0. \tag{15}$$

Here at least one of $\alpha\neq 0$ ($i=1,2,3,4,5$)-if all $\alpha_i=0$, this group method is invalidated, as further elaborated by Hansen (1964). Next, we seek "absolute invariants" under this elected group of transformations. Absolute invariants are the functions possessing the same form before and after the transformation. It is apparent from Eqs. (13) and (15) that:

$$\frac{y}{x^{\frac{n-1}{2n}}}=\frac{y^*}{x^{\frac{n-1}{2n}}}. \tag{16}$$

This combination of variables is therefore *invariant* under the present group of transformations and consequently, it constitutes an *absolute invariant*. We denote this functional form by:

$$\eta=\frac{y}{x^{\frac{n-1}{2n}}}=\frac{y^*}{x^{\frac{n-1}{2n}}}. \tag{17}$$

By the same argument, other absolute invariants are readily yielded:

$$\psi=x^{\frac{2n}{n+1}}f(\eta),\theta=\theta(\eta),\phi=\phi(\eta). \tag{18}$$

Here η is the similarity independent variable, $f(\eta)$, $\theta(\eta)$, $\phi(\eta)$ are the dimensionless stream function, temperature and nanoparticle concentration functions, respectively. Using Eqs. (17) and (18), it emerges that Eqs. (8)–(11) may be written as the following "local similarity" ordinary differential equations:

Momentum conservation

$$n(-f'')^{n-1}f'''+\frac{2n}{n+1}ff''-f'^2-Mf'=0, \tag{19}$$

Energy conservation

$$\frac{1}{Pr_x}\theta''+Nb_x\theta'\phi'+Nt_x\theta'^2+\frac{2n}{n+1}f\theta'=0, \tag{20}$$

Species (Nanoparticle concentration) conservation

$$\frac{1}{Le_x}\phi''+\frac{1}{Le_x}\frac{Nt_x}{Nb_x}\theta''+\frac{2n}{n+1}f\phi'=0. \tag{21}$$

The final forms of the wall and edge boundary layer conditions associated with the Eqs. (19)–(21) now become

$$f'(0) = 1 + a_x[-f''(0)]^{n-1}f''(0),\ f(0) = fw_x,$$
$$\theta(0) = \phi(0) = 1,\ f'(\infty) = \theta(\infty) = \phi(\infty) = 0. \tag{22}$$

Here $Pr_x = Pr_L x^{\frac{2n-2}{n+1}}$ is the local Prandtl number, $Nb_x = Nb_L x^{\frac{2-2n}{n+1}}$ is the local Brownian motion parameter, $Nt_x = Nt_L x^{\frac{2-2n}{n+1}}$ is the local thermophoresis parameter, $Le_x = Le_L x^{\frac{2-2n}{n+1}}$ is the local Lewis number, $a_x = \frac{N_1 \nu}{L^n} U_{\mathrm{r}}^{n-1} Re_L^{\frac{n}{n+1}} x^{\frac{n-1}{n+1}}$ is the local momentum slip parameter and $\mathrm{fw}_x = \mathrm{fw}_L \frac{n+1}{2n} x^{\frac{1-n}{1+n}}$ is the local lateral mass flux (wall suction/injection) parameter. In adopting local similarity solutions, we have deployed a subscript x. This subscript $(\cdot)_x$ is now dropped for brevity when referring to the local dimensionless numbers hereafter.

It interesting to note that the transformed local similarity model developed reduces exactly to the Newtonian nanofluid transport model of Makinde and Aziz (2011) and also to that of Noghrebadi et al. (2012) for $n = 1$ (Newtonian fluids), $M = 0$ (vanishing magnetic field), $\mathrm{fw} = 0$ (solid wall, i.e., suction/injection absent) and no wall momentum slip. ($a = 0$) Furthermore the current model retracts to that of Andersson and Bech (1992) for the impermeable wall case ($\mathrm{fw} = 0$) with nanofluid characteristics neglected (i.e., Eq. (21) dropped and $Nb = Nt = 0$) and with no wall slip ($a = 0$) Finally the model presented by Xu and Liao (2009) is retrieved in absence of magnetohydrodynamic, species diffusion (nanoparticle) and momentum slip effects. These cases therefore provide excellent benchmarks for validating the numerical solutions developed for the present more general boundary value problem.

Physical quantities

Since our motivation is nano-technological materials processing, a number of engineering design quantities are of great interest. These are the local friction factor, $C_{\mathrm{f}_{\bar{x}}}$, the local Nusselt number, $Nu_{\bar{x}}$ and the local Sherwood number, $Sh_{\bar{x}}$ respectively. Physically, $C_{\mathrm{f}_{\bar{x}}}$ represents the wall shear stress, $Nu_{\bar{x}}$ defines the heat transfer rate and $Sh_{\bar{x}}$ designates nanoparticle volume fraction gradient (wall species diffusion rate). These quantities can be calculated from the following relations:

$$C_{\mathrm{f}_{\bar{x}}} = \frac{2K}{\rho \bar{u}_{\mathrm{w}}^2}\left(-\frac{\partial \bar{u}}{\partial \bar{y}}\right)^n_{\bar{y}=0},\ Nu_{\bar{x}} = \frac{-\bar{x}}{T_{\mathrm{w}} - T_\infty}\left(\frac{\partial T}{\partial \bar{y}}\right)_{\bar{y}=0},\ Sh_{\bar{x}} = \frac{-\bar{x}}{C_{\mathrm{w}} - C_\infty}\left(\frac{\partial C}{\partial \bar{y}}\right)_{\bar{y}=0}. \tag{23}$$

Using Eqs. (6), (17) and (18), we obtain:

$$C_{\mathrm{f}_{\bar{x}}} Re_{\bar{x}}^{\frac{1}{n+1}} = 2[f''(0)]^n,\ Nu_{\bar{x}} Re_{\bar{x}}^{-\frac{1}{n+1}} = -\theta'(0),$$
$$Sh_{\bar{x}} Re_{\bar{x}}^{-\frac{1}{n+1}} = -\phi'(0), \tag{24}$$

where $Re_{\bar{x}} = \frac{\bar{u}_{\mathrm{w}} \bar{x}}{\nu}$ is the local Reynolds number. It is pertinent to highlight that the local skin friction factor, local Nusselt number and local Sherwood number are directly proportional to the numerical values of $[-2f''(0)]^n$, $-\theta'(0)$ and $-\phi\prime(0)$ respectively. In the parlance of Kuznetsov and Nield (2010), $Nu_{\bar{x}} Re_{\bar{x}}^{-\frac{1}{n+1}}$ and $Sh_{\bar{x}} Re_{\bar{x}}^{-\frac{1}{n+1}}$ may be referred as the *reduced local Nusselt number, Nur, and reduced local Sherwood number, Shr*, for non-Newtonian nanofluids, and are represented by $-\theta'(0)$ and $-\phi\prime(0)$, respectively.

Numerical solutions with maple quadrature

Equations (19)–(21) subject to the boundary conditions (22) have been solved using Runge–Kutta–Fehlberg fourth-fifth order numerical method provided in the symbolic computer software Maple 14. Thermophysical problem solving with Maple has been addressed succinctly in the monograph of Aziz (2010). Maple has become increasingly popular in the 21st century and has been deployed by the authors to resolve many highly coupled, nonlinear thermomechanics problems. These include bioconvection flows as elaborated by Uddin et al. (2013b, c), viscoelastic channel flows reader as examined by Bég and Makinde (2011), hydromagnetic entropy problems in hybrid aerospace propulsion as studied by Makinde and Bég (2010) and thermoelasticity of functional graded materials as described by Tounsi et al. (2013). In the current problem, the asymptotic boundary conditions given in Eq. (22) are replaced by a finite value in the range 15–20 depending on the parameters values. The choice of $\eta_{\max}$ must be selected judiciously to ensure that all numerical solutions approached to the asymptotic values correctly. The selection of sufficiently large value for $\eta_{\max}$ is imperative for maintaining desired accuracy in boundary layer flows, and is a common pitfall encountered in numerous studies.

Validation with Nakamura difference scheme

The seventh order nonlinear boundary value problem defined by Eqs. (19)–(21), with boundary conditions (22), has also been solved with the exceptionally efficient implicit Nakamura Tridiagonal finite difference Method (NTM) (Nakamura 1994). The NANONAK code has been developed to implement the Nakamura method for nonlinear nano-rheological flow problems. As with other difference schemes, a reduction in the higher order differential equations, is also fundamental to this method. It is also particularly effective at simulating highly nonlinear flows as characterized by nanofluids and also non-Newtonian fluids. The application of Nakamura's method (and numerous other computational algorithms) to nonlinear magnetofluid dynamics problems has been presented very recently by Bég (2012). NTM works well for both one-dimensional (ordinary differential) similar and two-dimensional (partial differential) non-similar flows. NTM entails a combination of the following aspects.

- The flow domain for the convection field is discretized using an *equi-spaced* finite difference mesh in the η direction.
- The partial derivatives for f, θ, ϕ, with respect to η are evaluated by central difference approximations.
- A single iteration loop based on the method of successive substitution is utilized due to the high nonlinearity of the momentum, energy and species (nanoparticle) conservation equations.
- The finite difference discretized equations are solved as a linear second order boundary value problem of the ordinary differential equation type on the η domain.

For the energy and species conservation Eqs. (20), (21) which are *second order* equations, only a *direct* substitution is needed. However a reduction is required for the momentum Eq. (19) which is of third order. Setting:

$$P = f' \tag{25a}$$

$$Q = \theta \tag{25b}$$

$$R = \phi \tag{25c}$$

The Eqs. (19)–(21) then assume the form:Nakamura momentum equation

$$A_1P'' + B_1P' + C_1P = S_1 \tag{26}$$

Nakamura energy equation

$$A_2Q'' + B_2Q' + C_2Q = S_2 \tag{27}$$

Nakamura species (nanoparticle diffusion) equation

$$A_3R'' + B_3R' + C_3R = S_3 \tag{28}$$

where $A_{i=1\ldots3}$, $B_{i=1\ldots3}$, $C_{i=1\ldots3}$ are the *Nakamura matrix coefficients*, $S_{i=1\ldots3}$ are the *Nakamura source terms* containing a mixture of variables and derivatives associated with the lead variable. The Nakamura Eqs. (26)–(28) are transformed to finite difference equations and these are formulated as a "tridiagonal" system which is solved iteratively.

Results and discussion

Prior to describing full numerical solutions to the present problem, we validate the Maple solutions via selected comparisons with previous studies and furthermore with the NANONAK code. We compare the value of $-f''(0)$ with asymptotic solutions of Andersson and Bech (1992) and homotopy solutions of Xu and Laio (2009) for non-Newtonian fluid in Table 1. Table 2 shows comparison of $-f''(0)$ for various magnetic parameters (M) and rheological power-law index (n) again with Andersson and Bech (1992) and the NANONAK solutions. The numerical results for $-\theta'(0)$ computed by Maple 14 are additionally benchmarked with those reported by Grubka and Bobba (1985), Chen (1998) and Ishak (2010) for Newtonian fluids, as presented in Table 3. In all cases we have found very close agreement is achieved and thus great confidence is assured in the accuracy of the Maple numerical results to be reported subsequently. Tables 4 and 5 also document the direct comparison of Maple 14 solutions with NANONAK and once again very close correlation is achieved, further testifying to the validity of the Maple computations.

Figures 2a–c, 3, 4, 5, 6a, b document the graphical computations for non-Newtonian power-law nanofluid flow characteristics, i.e., dimensionless axial velocity, temperature and nanoparticle concentration for the effects of the dictating thermophysical parameters. Figure 2a–c depict the influence of linear momentum slip (a), mass transfer velocity (fw) and magnetic field (M) on velocity profiles with transverse coordinate (η) for the pseudoplastic cases ($n = 0.5$) and also $n = 1$ (Newtonian fluids).

Inspection of Fig. 2a–c, indicates that the velocity profiles are larger for a *pseudoplastic* fluid than for *Newtonian* fluid. Evidently with lower rheological index (n), the viscosity of the nanofluid is depressed and the boundary layer flow is accelerated. Since dilatant fluids are relatively rarely encountered in materials processing, the case of $n > 1$ is not examined here. Velocity is also found to decrease with increasing parameter momentum slip (Fig. 2a), wall mass flux (Fig. 2b) and magnetic parameter (Fig. 2c). Increased momentum slip (Fig. 2a) imparts a retarding effect to the flow at the wall and this influences the entire flow through the boundary layer, transverse to the wall. With increased mass flux into the sheet regime via the

Table 1 Comparison of $-f''(0)$ for various n when $M = 0$, $a = 0$, fw $= 0$

n	Maple 14	Xu and Liao (2009)	Andersson and Bech (1992)	NANONAK
0.4	1.27306	–	1.273	1.27304
0.6	1.09644	–	1.096	1.09650
0.8	1.02883	1.02853	1.029	1.02879
1.0	1.00000	1.00000	1.000	1.00021
1.5	1.19281	0.98237	0.981	0.98342

Table 2 Comparison of $-f''(0)$ for various M and n when $a = 0$, fw = 0

n	Maple 14				Andersson and Bech (1992)			
	$M = 0.5$	$M = 1.0$	$M = 1.5$	$M = 2.0$	$M = 0.5$	$M = 1.0$	$M = 1.5$	$M = 2.0$
0.4	1.812	1.777	2.719	3.127	1.811	2.284	2.719	3.127
0.6	1.463	1.544	2.060	2.321	1.463	1.777	2.060	2.321
0.8	1.308	1.414	1.754	1.945	1.309	1.544	1.754	1.945
1.0	1.225	1.384	1.581	1.732	1.225	1.414	1.581	1.732
1.2	1.252	1.280	1.507	1.621	1.175	1.333	1.472	1.596
1.5	1.168	2.284	1.381	1.475	1.131	1.257	1.367	1.466

Table 3 Comparison of $-\theta'(0)$ with published results for various Pr

Pr	Maple 14	Ishak (2010)	Chen (1998)	Grubka and Bobba (1985)
0.01	0.128936	0.0099	0.0099	0.0099
0.72	0.465014	0.4631	0.46315	0.46315
1	0.582228	0.5820	0.58199	0.58199
3	1.165217	1.1652	1.16523	1.16523
10	2.307984	2.3080	2.30796	2.30796
100	7.765637	7.7657	7.76536	7.76536

wall (fw < 0, i.e., injection), the boundary layer flow is significantly accelerated, as observed in Fig. 2b. The converse behaviour accompanies increasing mass flux our of the boundary layer (fw > 0, i.e., suction). Evidently the suction effect induces greater adherence to the wall of the fluid regime and this also increases momentum boundary layer thickness. With greater injection, the momentum boost results in a reduction in momentum boundary layer thickness. These computations concur with many previous studies in transpiring boundary layer flows of nanofluids (Rashidi et al. 2013; Goyal and Bhargava 2013; Rana et al. 2013b; Ferdows et al. 2013; Hamad et al. 2011). The case of the impervious sheet, i.e., solid wall (fw = 0) naturally falls between the injection and suction cases. With an increase in magnetic parameter, there is an escalation in magnitude of the Lorentzian hydromagnetic drag force. This very clearly acts to depress velocity magnitudes (Fig. 2c) and even though the magnetic body force is only a linear term [$-Mf'$ in Eq. (18)] it exerts a very dramatic effect on the flow. Hydrodynamic (momentum) boundary layer thickness is therefore sizeably decreased with increasing values of M. In all three Fig. 2a–c, asymptotically smooth convergence of computations is achieved into the freestream.

Figure 3a–c present the response of temperature profiles, again for both pseudoplastic ($n = 0.5$) and Newtonian ($n = 1$) fluids, for a variation in momentum slip, wall mass flux and magnetic parameter. Temperatures are markedly elevated with pseudoplastic fluids compared with Newtonian fluids. Increasing momentum slip significantly increases the temperature magnitudes throughout the boundary layer regime and therefore enhances thermal boundary layer thickness (Fig. 3a). The presence of injection (fw < 0) as elaborated earlier, aids momentum development, i.e., viscous diffusion. This will also serve to encourage thermal diffusion leading to manifestly higher temperatures than for the suction case (fw > 0) which will depress temperatures (Fig. 3b). Injection therefore elevates thermal boundary layer thickness whereas suction decreases it. A similar observation has been reported by Makinde and Aziz (2011) and Rana et al. (2013a), who have also noted, as in the present case, that despite the strong influence of suction in decelerating the flow (and the corresponding cooling of the boundary layer), there is never any *flow reversal* (velocities are sustained as positive). The Lorentz force magnetohydrodynamic drag generated by the imposition of transverse magnetic field, has the tendency to not only slow down the flow but achieves this at the expense of increasing temperature (Fig. 3c). The supplementary kinetic energy necessary for dragging the magnetic nanofluid against the action of the magnetic field, even at relatively low values of M, is dissipated as thermal energy in the boundary layer. As elucidated by Bég et al. (2011b) this is characteristic of boundary layer hydromagnetics and reproduces the famous "Rossow results". The overall effect is to boost the temperature. Clearly for the *non-conducting* case, temperatures will be minimized as magnetic field will vanish. Moreover for the pseudoplastic nanofluid case, temperatures are the highest achieved with the infliction of a magnetic field, and evidently this is beneficial in the synthesis of electro-conductive nanopolymers, as highlighted in the recent theoretical studies by Rana et al. (2013b), Ferdows et al. (2013), and further emphasized in the experimental study of Crainic et al. (2003).

Figure 4a–c depict the evolution of temperature fields with a variation in thermophoresis number (Nt), Brownian motion parameter (Nb) and Prandtl number (Pr), respectively, in all cases for both Newtonian and pseudoplastic nanofluids. A consistent response is not

Table 4 Comparison of $-f''(0)$ between Maple and NANONAK for various M and n when $a = 0$, fw = 0

n	Maple 14				NANONAK			
	$M = 0.5$	$M = 1.0$	$M = 1.5$	$M = 2.0$	$M = 0.5$	$M = 1.0$	$M = 1.5$	$M = 2.0$
0.4	1.812	1.777	2.719	3.127	1.813	1.776	2.718	3.127
0.6	1.463	1.544	2.060	2.321	1.464	1.543	2.061	2.322
0.8	1.308	1.414	1.754	1.945	1.308	1.413	1.753	1.944
1.0	1.225	1.384	1.581	1.732	1.223	1.383	1.582	1.733
1.2	1.252	1.280	1.507	1.621	1.254	1.281	1.507	1.622
1.5	1.168	2.284	1.381	1.475	1.167	2.283	1.383	1.474

Table 5 Comparison of $-\theta'(0)$ between Maple and NANONAK for various Pr

Pr	Maple 14	NANONAK
0.01	0.128936	0.128940
0.72	0.465014	0.465011
1	0.582228	0.582230
3	1.165217	1.165220
10	2.307984	2.307982
100	7.765637	7.765640

observed in all the temperature profiles with rheological index, as computed in earlier graphs. In Fig. 4a, b, temperatures are maximized for the pseudoplastic case ($n = 0.5$), whereas in Fig. 4b they are greatest for the Newtonian fluid. Clearly in Fig. 4b, since Brownian motion effect is being varied, this parameter has a complex interaction with the thermal field, which alters the familiar response computed in other figures. Both an increase in Nt and Nb result in a substantial accentuation in temperatures (Fig. 4a, b respectively). This will therefore be accompanied with a significant increase in thermal boundary layer thickness. Thermophoresis relates to increasing nanoparticle deposition on the sheet surface in the direction of a decreasing temperature gradient. As such, the nanofluid temperature is elevated by this enhancement in particle migration, a trend also observed by numerous experimental studies including Eastman et al. (2004) and many theoretical simulations including Kuznetsov and Nield (2010). Within the framework of the Kuznetsov–Nield theory, a nanofluid behaves significantly more as a viscous fluid rather than the conventional solid–fluid mixtures in which relatively larger particles with micrometer or millimeter orders are suspended. Nevertheless, since the nanofluid is a two-phase fluid in nature and has some common features with solid–fluid mixtures, it does exhibit similar shear characteristics. The enhanced heat transfer by the nanofluid may result from either the fact that the suspended particles increase the thermal conductivity of the two-phase mixture or owing to chaotic movement of ultrafine particles accelerating energy exchange process in the fluid. However much needs to be resolved in elucidating more explicitly the exact mechanisms for the contribution of suspended particles to thermal enhancement, and in this regard the micro-structure of the nanofluid (whether Newtonian or non-Newtonian) may provide a more robust fluid dynamical framework. Indeed this has been explored by Gorla and Kumari (2012) with the Eringen micropolar model, and most recently by Bég (2013a) using the Eringen micro-stretch model with greater degrees of freedom of micro-elements. These studies have revealed that the gyratory motions of micro-elements containing the nanoparticles exert a significant influence on thermal transport, and capture features which are basically not realizable even in rheological nanofluid models which neglect microstructural properties. Although the present power-law model would indicate that thermal conductivity is boosted in the nanofluid, it is not possible to analyze whether the spin of nanoparticles is also a contributory factor. It is envisaged therefore that the "microstructural" rheological models of Eringen (2001) which have been reviewed comprehensively in *computational micropolar transport modeling* by Bég et al. (2011c) will also be deployed for stretching sheet magneto-nanofluid dynamics as explored in the near future. Figure 4c demonstrates that a rise in Prandtl number (Pr) effectively suppresses the nanofluid temperatures. Temperature is however for both Newtonian and non-Newtonian nanofluids, reduced markedly with an increase in Pr. Pr defines the ratio of momentum diffusivity to thermal diffusivity for a given nanofluid. With lower Pr nanofluids, heat diffuses faster than momentum (the energy diffusion rate exceeds the viscous diffusion rate) and vice versa for higher Pr fluids. With an increase in Pr, temperatures will therefore fall, i.e., the nanofluid boundary layer regime will be cooled as observed in Fig. 4c. Larger Pr values correspond to a thinner thermal boundary layer thickness and more uniform temperature

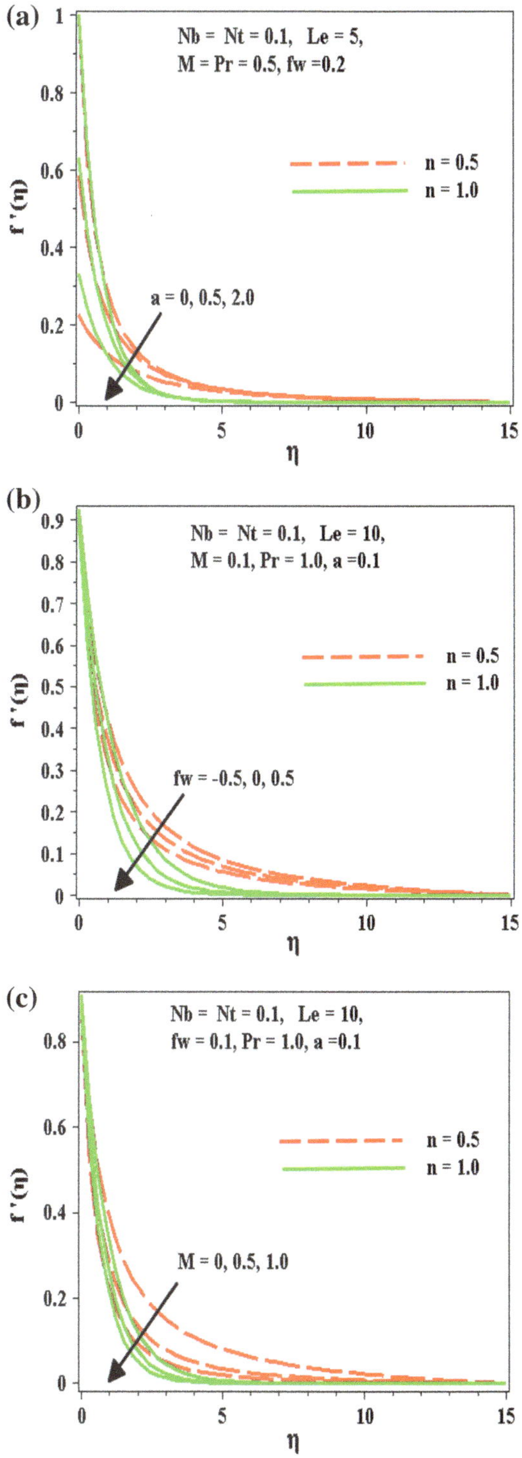

Fig. 2 Effects of **a** momentum slip, **b** lateral mass flux and **c** magnetic field on the dimensionless velocity profiles, for pseudoplastic and Newtonian nanofluids

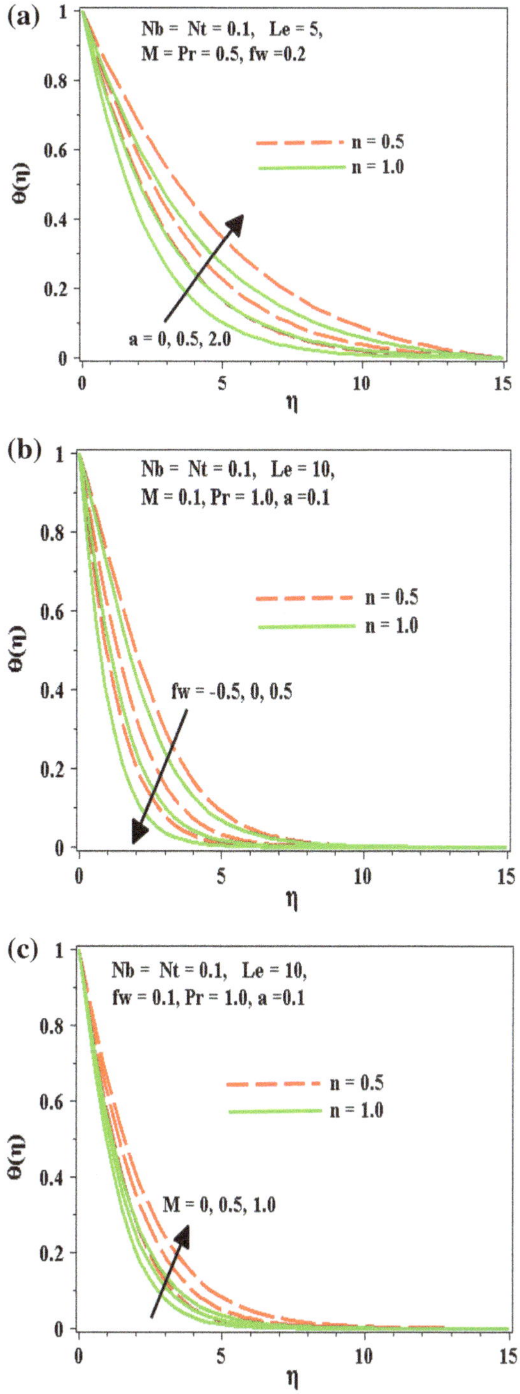

Fig. 3 Effects of **a** momentum slip, **b** lateral mass flux and **c** magnetic field on the dimensionless temperature profiles, for pseudoplastic and Newtonian nanofluids

distributions across the boundary layer. Smaller *Pr* nanofluids possess higher thermal conductivities so that heat can diffuse away from the vertical plate faster than for higher *Pr* fluids (low *Pr* fluids correspond to thicker boundary layers). We finally note that for the Newtonian case with maximum *Pr* (=6.8) temperature plummets very sharply from the sheet surface and assumes very low values very quickly. At the other extreme, with the lowest value of *Pr* (=0.72) a very gentle decay

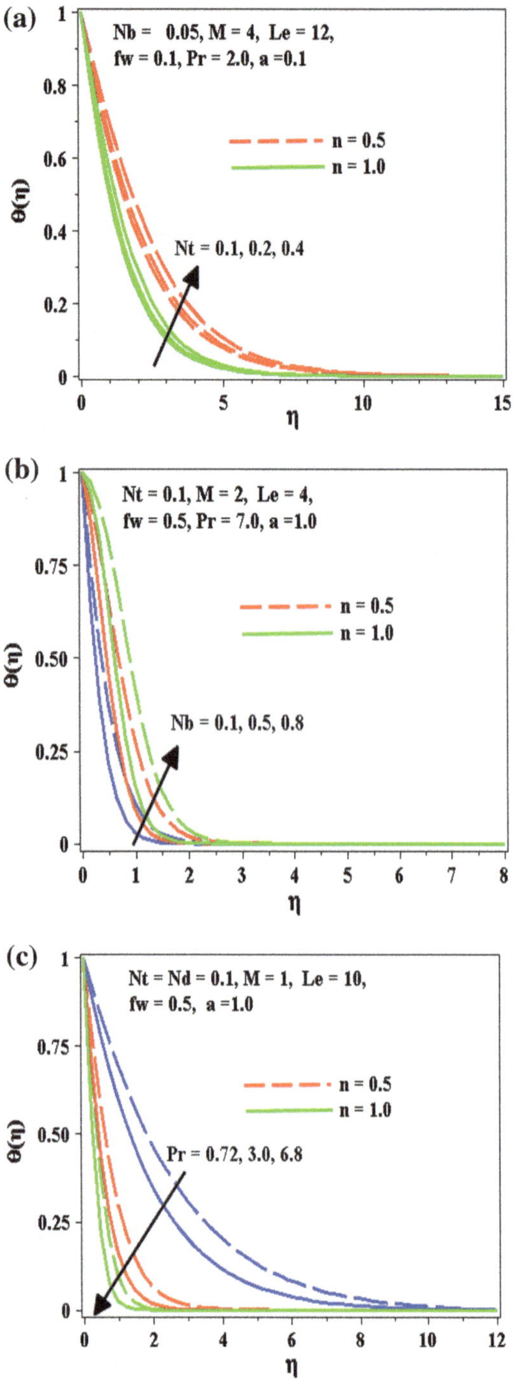

Fig. 4 Effects of **a** thermophoresis, **b** Brownian motion and **c** Prandtl number on the dimensionless temperature profiles, for pseudoplastic and Newtonian nanofluids

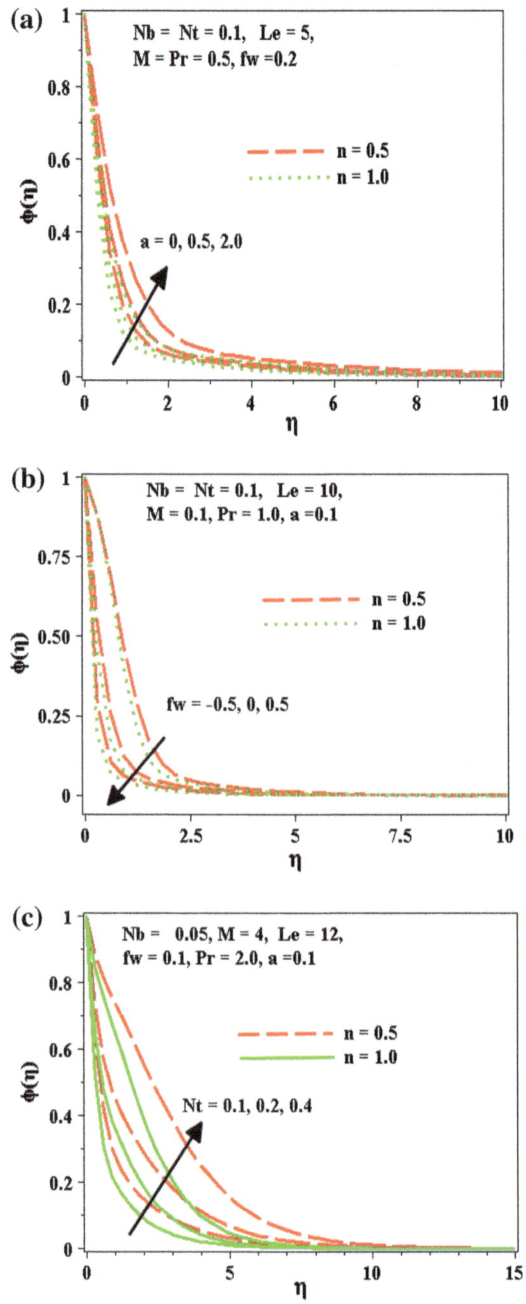

Fig. 5 Effects of **a** momentum slip, **b** lateral mass flux and **c** thermophoresis on the dimensionless concentration profiles, for pseudoplastic and Newtonian nanofluids

(monotonic descent) is observed from the wall to the free stream.

Figure 5a–c depicts the variation of nanoparticle concentration profiles (ϕ) for Newtonian and pseudoplastic nanofluids, with different values of the momentum slip (a), lateral mass flux (fw) and thermophoresis (Nt) parameters. In all the figures greater concentration magnitudes are attained with pseudoplastic fluids than Newtonian fluids. The concentration profiles increases with increasing wall slip and thermophoresis and furthermore are also enhanced with greater injection at the sheet surface. Enhanced migration of suspended nanoparticles via the mechanisms of thermophoresis, *increases* energy exchange rates in the fluid and simultaneously assists in nanoparticle diffusion, leading to greater values of (ϕ). The dispersion of nanoparticles will not only successfully heat the boundary layer

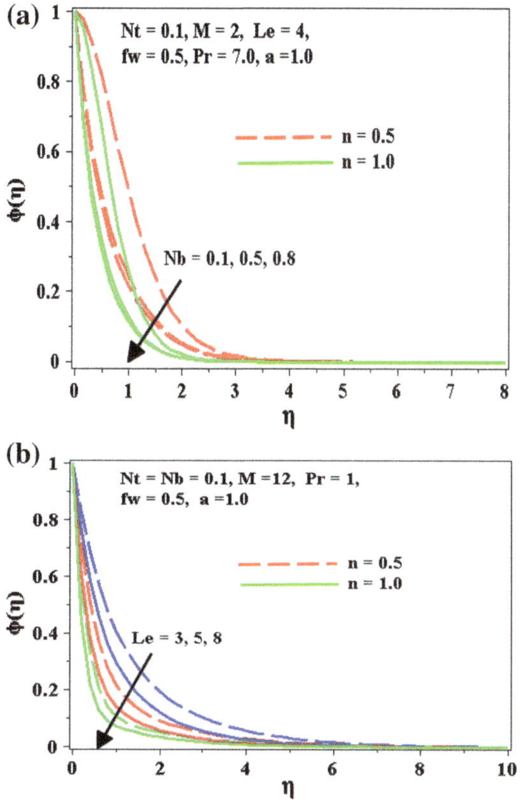

Fig. 6 Effects of **a** Brownian motion and **b** Lewis number on the dimensionless concentration profiles, for pseudoplastic and Newtonian nanofluids

and augment heat transfer rate *from* the nanofluid *to* the wall, but will also enhance concentration (species) boundary layer thicknesses.

Finally Fig. 6a, b, illustrates the nanoparticle concentration distributions with a variation in Brownian motion parameter (*Nb*) and Lewis number (*Le*), again for both Newtonian and pseudoplastic nanofluids. The general pattern observed in most of the computations described in earlier figures is confirmed in Fig. 6a, b, namely, that pseudoplastic fluids achieve greater concentration magnitudes relative to Newtonian nanofluids. As indicated in many studies including Kuznetsov and Nield (2010) for *Newtonian* nanofluids, Sheu (2011) for *viscoelastic* nanofluids and Uddin et al. (2013a) for *power-law* nanofluids, an increase in Brownian motion parameter effectively depresses nanoparticle concentrations in the nanofluid regime. This is attributable to the enhanced mass transfer to the sheet wall from the main body of the nanofluid (boundary layer) with increasing Brownian motion. Among others, Rana et al. (2013a, b) have demonstrated that the Brownian motion of nanoparticles achieves thermal conductivity enhancement either by way of a direct effect owing to nanoparticles that transport heat or alternatively via an indirect contribution due to micro-convection of fluid surrounding individual nanoparticles. For larger nanoparticles (and in the present study all nanoparticles in the nanofluid possess consistent sizes) Brownian motion is weakened so that *Nb* has lower values—this aids in species diffusion from the wall to the nanofluid and elevates nanoparticle concentration values. The exact opposite is manifested with higher *Nb* values. As anticipated the Brownian motion parameter causes a substantial modification in nanoparticle species distributions throughout the boundary layer regime. An even more dramatic effect is induced on the nanoparticle concentrations, via an increase in the Lewis number (Fig. 6b). This dimensionless number signifies the *ratio of thermal diffusivity to mass diffusivity* in the nanofluid. It is used to characterize fluid flows where there is simultaneous heat and mass transfer by convection. *Le also* expresses the ratio of the Schmidt number to the Prandtl number. For $Le = 1$, both heat and nanoparticle species will diffuse at the same rate. For $Le < 1$, heat will diffuse more slowly than nanoparticle species and vice versa for $Le > 1$. For the cases examined here, $Le \geq 1$, i.e., *thermal diffusion rate* substantially exceeds the *mass fraction (species) diffusion rate*. Inspection of Fig. 6b shows that nanoparticle concentration is clearly also suppressed with an increase in *Le*. As in all the other graphs, asymptotically smooth results are achieved in the free stream, testifying to the prescription of an adequately large value for infinity in the Maple code.

Conclusions

A laminar transport model has been developed for hydromagnetic nanofluid rheological flow from a stretching permeable sheet subjected to a transverse magnetic field, as a model of nano-magnetic materials processing. Slip effects have been incorporated in the model at the wall (sheet surface) via a Navier momentum slip parameter. Via a suitable group of Lie algebra transformations, the nonlinear two-point boundary value problem has been normalized and formulated as a system of coupled, multi-degree, multi-order nonlinear ordinary differential equations with physically justifiable boundary conditions. Both Maple 14 shooting quadrature and an implicit finite difference algorithm (Nakamura's method) have been implemented to obtain comprehensive numerical solutions. Additionally the Maple 14 and Nakamura solutions (NANONAK code) have been validated against each other, showing excellent correlation and furthermore have been benchmarked against published works, again achieving exceptionally close agreement. The present solutions obtained have indicated that:

1. Increasing momentum slip parameter decelerates the boundary layer flow (reduces velocities) whereas it

generally strongly increases temperatures and nanoparticle concentration values.

2. With increasing wall injection, velocity, temperature and concentration are generally enhanced with the converse behaviour observed with increasing wall suction.
3. Increasing magnetic field parameter markedly decelerates the flow (due to the Lorentzian retarding effect), increases momentum boundary layer thickness and moreover induces a noteworthy elevation in temperatures and thermal boundary layer thickness.
4. Increasing thermophoresis parameter boosts both the temperatures and nanoparticle concentration magnitudes throughout the boundary layer regime.
5. Increasing Brownian motion parameter enhances temperatures strongly whereas it markedly depresses nanoparticle concentration values.
6. Increasing Prandtl number substantially reduces temperatures.
7. Increasing Lewis numbers significantly stifles nanoparticle concentration values.
8. Generally the flow is accelerated, temperatures increased and nanoparticle concentrations boosted for pseudoplastic nanofluids with the converse for Newtonian nanofluids.

The present simulations have been confined to *steady-state flows* and *neglected micro-structural* rheological characteristics in nanofluids. Future studies will address both transient effects and utilize a variety of thermo-micromorphic constitutive models, and will be communicated imminently. Furthermore to generalize the study of electro-conductive nanopolymers, future analyses will also address *electrical field* effects (Bég et al. 2013c).

References

Akbar NS, Nadeem S (2013) Biomathematical study of non-Newtonian nanofluid in a diverging tube. Heat Transf-Asian Res 42:389–402

Andersson HI, Bech KH (1992) Magnetohydrodynamic flow of a power law fluid over a stretching sheet. Int J Non-Linear Mech 27:929–936

Anoop KB, Kabelac S, Sundararajan T, Das SK (2009) Rheological and flow characteristics of nanofluids: influence of electroviscous effects and particle agglomeration. J Appl Phys 106:034909

Aziz A (2010) Heat conduction with maple. Taylor and Francis, Philadelphia

Bég OA (2012) Numerical methods for multi-physical magnetohydrodynamics. In: New developments in hydrodynamics, Chap 1. Nova Science, New York, pp 1–110

Bég OA (2013) Lattice Boltzmann thermal solvers for micromorphic and microstretch Eringen nanofluids in spacecraft propulsion systems, Technical Report, AERO-NANO-2013-4, 127 pages, Gort Engovation, Bradford

Bég OA (2013) Magnetohydrodynamic flows of nanofluids in oscillating tubes with NMR applications for the Scandinavian medical industry, Technical Report, BIO-NANO-2013-5, 94 pages, Gort Engovation, Bradford

Bég OA, Makinde OD (2011) Viscoelastic flow and species transfer in a Darcian high-permeability channel. Petroleum Sci Eng 76:93–99

Bég OA, Tripathi D (2012) Mathematica simulation of peristaltic pumping in double-diffusive convection in nanofluids: a nano-bio-engineering model. Proc IMechE Part N: J Nanoeng Nanosyst 225:99–114

Bég OA, Zueco J, Lopez-Ochoa LM (2011a) Network numerical analysis of optically thick hydromagnetic slip flow from a porous spinning disk with radiation flux, variable thermophysical properties and surface injection effects. Chem Eng Commun 198:360–384

Bég OA, Ghosh SK, Bég Tasveer A (2011b) Applied magnetofluid dynamics: modelling and computation. Lambert Academic Publishing, Saarbrücken

Bég OA, Bhargava R, Rashidi MM (2011c) Numerical simulation in micropolar fluid dynamics. Lambert Academic Press, Saarbrucken

Bég OA, Tasveer-Bég A, Rashidi MM, Asadi M (2012) Homotopy semi-numerical modelling of nanofluid convection flow from an isothermal spherical body in a permeable regime. Int J Microscale Nanoscale Therm Fluid Transp Phenom 3:237–266

Bég OA, Prasad VR, Vasu B (2013a) Numerical study of mixed bioconvection in porous media saturated with nanofluid containing oxytactic micro-organisms. J Mech Med Biol.

Bég OA, Rashidi MM, Akbari M, Hosseini A (2013b) Comparative numerical study of single-phase and two-phase models for bio-nanofluid transport phenomena. J Mech Med Biol 13:31

Bég OA, Hameed M, Bég Tasveer A (2013c) Chebyschev spectral collocation simulation of nonlinear boundary value problems in electrohydrodynamics. Int J Comp Meth Eng Sci Mech 14:104–115

Chen CH (1998) Laminar mixed convection adjacent to vertical, continuously stretching sheets. Heat Mass Transf 33:471–476

Chen H, Ding Y, He Y, Tan C (2007) Rheological behaviour of ethylene glycol based titania nanofluids. Chem Phys Lett 444:333

Crainic N, Marques AT, Bica D (2003) The usage of nanomagnetic fluids and magnetic field to enhance the production of composite made by RTM-MNF. In: 7th International Conference. Frontiers-Polymers/Advanced Materials. Romania, pp 10–15

Eastman JA, Phillpot SR, Choi SUS, Keblinski P (2004) Nanofluids for thermal transport. Ann Rev Mat Res 34:219

Ellahi R, Raza M, Vafai K (2012) Series solutions of non-Newtonian nanofluids with Reynolds' model and Vogel's model by means of the homotopy analysis method. Math Comput Model 55:1876–1891

Eringen AC (2001) Microcontinuum field theories-II: fluent media. Springer, USA

Estellé P, Halelfadl S, Doner N, Maré T (2013) Shear history effect on the viscosity of carbon nanotubes water-based nanofluid. Curr Nanosci 9:225–230

Ferdows M, Khan MS, Bég OA, Azad MK, Alam MM (2013) Numerical study of transient magnetohydrodynamic radiative free convection nanofluid flow from a stretching permeable surface. Proc IMechE Part E: J Process Mech Eng.

Gorla RSR, Kumari M (2012) Mixed convection flow of a non-Newtonian nanofluid over a non-linearly stretching sheet. J Nanofluids 1:186–195

Goyal M, Bhargava R (2013) Boundary layer flow and heat transfer of viscoelastic nanofluids past a stretching sheet with partial slip conditions. Appl Nanosci (In press)

Grubka LJ, Bobba KM (1985) Heat transfer characteristics of a continuous, stretching surface with variable temperature. ASME J Heat Transf 107:248–250

Hamad MAA, Pop I, Ismail AI (2011) Magnetic field effects on free convection flow of a nanofluid past a semi-infinite vertical flat plate. Nonlinear Anal: Real World Appl 12:1338–1346

Hamad MAA, Uddin MJ, Ismail AI (2013) Investigation of combined heat and mass transfer by Lie group analysis with variable diffusivity taking into account hydrodynamic slip and thermal convective boundary conditions. Int J Heat Mass Transf 55:1355–1362

Hansen AG (1964) Similarity analysis of boundary layer problems in engineering. Prentice Hall Englewood Cliffs, New Jersey

He Y, Men Y, Liu X, Lu H, Chen H, Ding Y (2012) Study on forced convective heat transfer of non-Newtonian nanofluids. ASME J Therm Sci 18:20–26

Hojjat M, Etemad SG, Thibault J (2011) Convective heat transfer of non-Newtonian nanofluids through a uniformly heated circular tube. Int J Therm Sci 50:525–531

Hojjat M, Etemad SGh, Bagheri R, Thibault J (2012) Pressure drop of non-Newtonian nanofluids flowing through a horizontal circular tube. J Dispers Sci Technol 33:1066–1070

Ishak A (2010) Thermal boundary layer flow over a stretching sheet in a micropolar fluid with radiation effect. Meccanica 45:367–373

Kedzierski MA, Venerus D, Buongiorno J et al (2010) Viscosity measurements on colloidal dispersions (nanofluids) for heat transfer applications. Appl Rheol J 43:7

Kim S, Kim C, Lee W-H, Park S-R (2011) Rheological properties of alumina nanofluids and their implication to the heat transfer enhancement mechanism. J Appl Phys 110:034316

Kulkarni DP, Das DK, Vajjha RS (2009) Application of nanofluids in heating buildings and reducing pollution. Appl Energy 12:2566–2573

Kuznetsov AV, Nield DA (2010) Natural convective boundary-layer flow of a nanofluid past a vertical plate. Int J Therm Sci 49:243–247

Makinde OD, Aziz A (2011) Boundary layer flow of a nanofluid past a stretching sheet with a convective boundary condition. Int J Therm Sci 50:1326–1332

Makinde OD, Bég OA (2010) On inherent irreversibility in a reactive hydromagnetic channel flow. J Therm Sci 19:72–79

Mody VV, Cox A, Shah S, Singh A, Bevins W, Parihar H (2013) Magnetic nanoparticle drug delivery systems for targeting tumors. Appl Nanosci.

Nakamura S (1994) Iterative finite difference schemes for similar and non-similar boundary layer equations. Adv Eng Softw 21:123–130

Nkurikiyimfura I, Wang Y, Pan Z et al. (2011) Enhancement of thermal conductivity of magnetic nanofluids in magnetic field. In: International conference. Materials, renewable energy and environment (ICMREE), Shanghai

Noghrehabadi A, Pourrajab R, Ghalambaz M (2012) Effect of partial slip boundary condition on the flow and heat transfer of nanofluids past stretching sheet prescribed constant wall temperature. Int J Therm Sci 54:253–261

Prasad VR, Subba Rao A, Bhaskar Reddy N, Vasu B, Bég OA (2013) Modelling laminar transport phenomena in a Casson rheological fluid from a horizontal circular cylinder with partial slip. Proc IMechE Part E: J Process Mech Eng.

Rana P, Bhargava R, Bég OA (2013a) Finite element modeling of conjugate mixed convection flow of Al_2O_3–water nanofluid from an inclined slender hollow cylinder. Physica Scripta—Proc Royal Swedish Acad Sci 87:1–16

Rana P, Bhargava R, Bég O Anwar (2013b) Finite element simulation of unsteady MHD transport phenomena on a stretching sheet in a rotating nanofluid. Proc IMechE Part N: J Nanoeng Nanosyst.

Rashidi MM, Bég OA, Freidooni-Mehr N, Hosseini A, Gorla RSR (2012) Homotopy simulation of axisymmetric laminar mixed convection nanofluid boundary layer flow over a vertical cylinder. Theoret Appl Mech 39:365–390

Rashidi MM, Abelman S, Freidooni Mehr N (2013) Entropy generation in steady MHD flow due to a rotating porous disk in a nanofluid. Int J Heat Mass Transfer 62:515–525

Shang D (2010) Theory of heat transfer with forced convection film flows (Heat and Mass Transfer). Springer, Berlin

Sheu LJ (2011) Thermal instability in a porous medium layer saturated with a viscoelastic nanofluid. Transp Porous Media 88:461–477

Susrutha B, Ram S, Tyagi AK (2012) Effects of gold nanoparticles on rheology of nanofluids containing poly(vinylidene fluoride) molecules. J Nanofluids 1:120–127

Tounsi A, Houari MSA, Bég OA (2013) Thermoelastic bending analysis of functionally graded sandwich aero-structure plates using a new higher order shear and normal deformation theory. Int J Mech Sci (Accepted)

Townsend J, Christianson RJ (2009) Nanofluid properties and their effects on convective heat transfer in an electronics cooling application. ASME J Therm Sci Eng Appl 1:031006

Tripathi D, Bég OA, Curiel-Sosa JL (2012) Homotopy semi-numerical simulation of peristaltic flow of generalised Oldroyd-B fluids with slip effects. Comput Meth Biomech Biomed Eng.

Uddin MJ, Hamad MAA, Md AI (2012a) Ismail, Investigation of heat mass transfer for combined convective slip flow. A Lie group analysis. Sains Malaysiana 41:1139–1148

Uddin MJ, Khan WA, Md AI (2012b) Ismail, Scaling group transformation for MHD boundary layer slip of a nanofluid over a convectively heated stretching sheet with heat generation. Math Probl Eng 2012:1–20

Uddin MJ, Yusoff NHM, Bég OA, Ismail AI (2013a) Lie group analysis and numerical solutions for non-Newtonian nanofluid flow in a porous medium with internal heat generation. Phys Scr 87:1–14

Uddin MJ, Khan WA, Ismail AIM (2013b) Free convective flow of non-Newtonian nanofluids in porous media with gyrotactic microorganisms. AIAA J Thermophys Heat Transfer 27:326–333

Uddin MJ, Khan WA, Md AI (2013c) Ismail, effect of dissipation on free convective flow of a non-Newtonian nanofluid in a porous medium with gyrotactic microorganisms. Proc IMechE Part N—J Nanoeng Nanosyst 227:11–18

Wang CY (2007) Stagnation flow on a cylinder with partial slip- an exact solution of the Navier–Stokes equations. IMA J Appl Math 72:271–277

Xu H, Liao S-J (2009) Laminar flow and heat transfer in the boundary layer of non-Newtonian fluids over a stretching flat sheet. Comp Maths Applicns 57:1425–1431

Yang H, Fu Y, Tang Z (2012) Analytical solution of slip flow of nanofluids over a permeable wedge in the presence of magnetic field. Adv Mater Res 354:45–48

8

Size induced tuning of dielectric behavior in nanostructured $Y_{0.95}Ca_{0.05}MnO_3$ compounds

N. A. Shah

Abstract Size is the key factor of nanostructured materials, since all the structural, transport, electrical, magnetic and other physical properties can be tuned by this factor of materials. Only the condition is to choose appropriate inexpensive scale-processing method for material synthesis which offers good control over the stoichiometry, morphology and particle size distribution. Present communication deals with the studies on the sol–gel grown $Y_{0.95}Ca_{0.05}MnO_3$ (YCMO) nanostructured compounds for their size-induced tuning of dielectric behavior. Structural studies reveal the single phasic nature with improved crystallite size with sintering temperature. Dielectric constant (real and imaginary) is found to increase with temperature and crystallite size/sintering temperature. High dielectric loss has been observed in the present system. Size dependent activation energy (E_a), obtained from modulus measurement, showing the increase in E_a with crystallite size. The variation in various dielectric parameters and E_a has been discussed in the light of crystallite size, crystallite boundaries, oxygen vacancies and charge carrier hopping.

Keywords Size effect · Dielectric behavior · Sol–Gel · Nanostructure

Introduction

In periodic classification of the elements, there is a group of metals referred to as transition metals which use their valence electron to form compounds with other elements. During the development and use of various technologies, these transition metals play an important role. Without the contribution of transition metals in developing these technologies, such as semiconducting, superconducting, spintronic, etc., is not possible. Spintronic covers varieties of oxide/non-oxide compounds having different forms, their properties and applications, such as manganese oxide based polycrystalline bulk (Doshi et al. 2011), nanostructures (Solanki et al. 2010), thin films (Solanki et al. 2011), heterostructures (Uma et al. 2012), multilayers (Vachhani et al. 2011), devices (Ashish et al. 2012) and composites (Cheng and Wang 2007), multiferroic based polycrystalline bulk (Ravalia et al. 2011), thin films (Ashish et al. 2013), devices (Megha et al. 2013), heterostructures (Gajek et al. 2006) and composites (Cheng and Wang 2007), high temperature superconductor (HTSC) based polycrystalline bulk (Rayaprol et al. 2004), thin films (Mavani et al. 2007), devices (Cassinese et al. 2004) and composites (Dong et al. 1998), diluted magnetic semiconductors (DMS) based polycrystalline bulk (Megha et al. 2011) and nanostructures (Megha et al. 2011), etc.

Manganese oxide based multiferroics, such as $BiMnO_3$, $YMnO_3$, $HoMnO_3$, $TbMnO_3$, etc., have been recently receiving an attention due to their fundamental physics (Lorenz et al. 2004; Aikawa et al. 2005; Singh et al. 2010) and potential for various spintronic applications (Gajek et al. 2007; Li et al. 2009), such as read and write heads, field and temperature sensors, spin valves, spin filters, tunnel junctions, p-n junction diodes, non-volatile random access memory devices, etc. In 1990s, $YMnO_3$ (YMO) films were investigated for non-volatile ferroelectric memory devices, especially for metal–ferroelectric–semiconductor (MFS) devices (Fujimura et al. 1996a, b). YMO possesses hexagonal structure in its bulk form having electrical

N. A. Shah (✉)
Department of Electronics, Saurashtra University,
Rajkot 360 005, India
e-mail: snikesh@yahoo.com

transition from ferroelectric (FE) to paraelectric (PE) at temperature T_C is as high as ~1,000 K and magnetic transition from antiferromagnetic (AFM) to paramagnetic (PM) at temperature T_N ~ 120 K. In hexagonal YMO compounds, the magnetoelectric (ME) coupling and AFM order are found to be very weak which in turn results in the negligibly small net magnetization at low temperature (Fiebig et al. 2002; Bertaur et al. 1963). By engineering the preparation method, one can synthesize the orthorhombic YMO which shows essentially different FE order and magnetic properties from hexagonal YMO compound, in addition to exhibition of significantly strong ME coupling (Kalashnikova and Pisarev 2003; Lorenz et al. 2007).

Many reports exist on the studies on Mn-site substitution of transition metals in YMO (Gutierrez et al. 2002; Moure et al. 2002; Veres et al. 2006; Aikawa et al. 2005). Gutierrez have observed the structural phase transformation from hexagonal to orthorhombic upon Cu^{2+} substitution which has been ascribed to the reduction in Jahn–Teller Mn^{3+} ionic density (Gutierrez et al. 2002). At the same time, same group has reported the effect of Cu^{2+}, Ni^{2+} and Co^{2+} substitution at Mn^{3+} site on the structural behavior of doped YMO compounds and found that, the observed phase transition is governed by the structural parameters, such as size mismatch and tolerance factor (Moure et al. 2002). Veres et al. have reported the substitutional effect of Fe^{3+} at Mn-site on structural, transport and magnetic properties of $YMn_{1-x}Fe_xO_3$ (Veres et al. 2006). Studies from the view point of multiferroicity, Aikawa et al. have reported the substitutional effect of Ti^{4+} at Mn-site in YMO and found the structural phase transition and hence strong magnetodielectric behavior in the series of samples (Aikawa et al. 2005) Multiferroicity in doped YMO compounds has also been studied using the substitution of non-transition metal/non-magnetic Ga^{3+} at Mn-site (Zhou et al. 2005). Also, few reports are available on the studies on Y-site substitutional effect in YMO compounds (Lliev et al. 2005). At one hand, YMO oxides have been studied from the view point of substitution with transition metals and non-magnetic ions, on the other hand, very few reports are available on the multiferroic point of view studies on doped YMO oxides.

By keeping in mind what has been discussed above, in the present communication, the results of dielectric behavior of sol–gel grown nanostructured $Y_{0.95}Ca_{0.05}MnO_3$ (YCMO) oxides, sintered at various temperatures, have been discussed in the light of size effect.

Experimental details

Nanophasic $Y_{0.95}Ca_{0.05}MnO_3$ (YCMO) were grown using simple and low cost sol–gel technique (Solanki et al. 2009; Kuberkar et al. 2012). High purity Yttrium Acetate [Y $(CH_3CO_2)_3 \times 6H_2O$], Calcium Acetate [Ca $(CH_3CO_2)_2$] and Manganese Acetate [Mn $(CH_3CO_2)_2 \times 4H_2O$] were taken as starting materials in appropriate stoichiometric ratio. The precursor solution was prepared by dissolving the constituents (precursors/starting materials) in double distilled water (DDW) and acetic acid (AA) with desired stoichiometric composition. The optimum ratio of DDW and AA of 1:1 was maintained in proper volume to yield 0.3 M of solution. The solution was stirred at 90 °C using magnetic stirrer until a clean and transparent solution was obtained followed by the calcination of resultant brownish powder at 500 °C for 6 h in furnace. Final material product (in pellet form) was sintered at 700, 800, 900, 1,000 and 1,100 °C for 12 h in air. Hereafter, YCMO compounds sintered at 700, 800, 900, 1,000 and 1,100 °C are referred as YC7, YC8, YC9, YC10 and YC11, respectively. X-ray diffraction (XRD) patterns of YCMO compounds were recorded on Philips diffractometer (PW 3040/60, X'pert PRO) using CuKα radiation at room temperature (RT). Frequency dependent dielectric constant at various temperatures (in the range 80–300 K) was measured using Solatron impedance analyzer.

Results and discussion

XRD measurements were carried out on all the sol–gel grown nanostructured YCMO compounds sintered at various temperatures. Figure 1 depicts XRD patterns of YCMO compounds sintered at 700 °C (YC7) and 1,100 °C (YC11) which reveals that, all the compounds sintered at various temperatures are single phase in nature without any

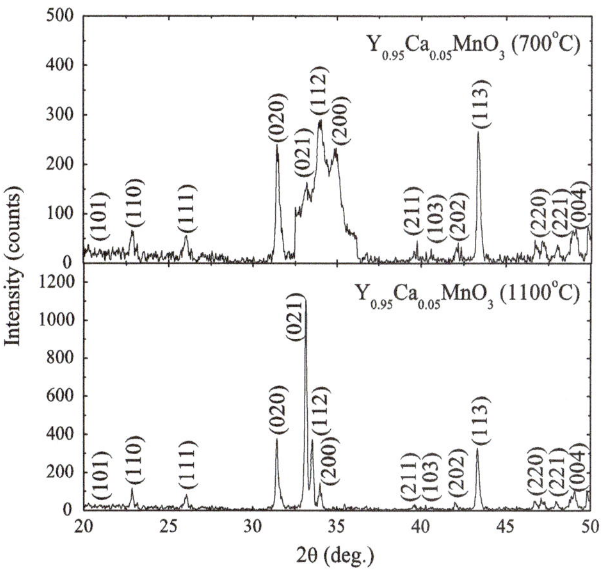

Fig. 1 XRD patterns of YC7 and YC11 compounds

detectable impurities within the measurement range (Lorenz et al. 2004). All the compounds possess orthorhombic structure with the *Pnma* space group (no. 62). All the XRD peaks are identified for their respective *(hkl)* parameters. It is clear that, large amorphous background exists in YC7 while a clear background is present in YC11. Figure 2 depicts an enlarged view of *(020)* peaks of all the nanostructured YCMO compounds. The peak intensity increases while the peak broadening (full width at half maximum—FWHM) decreases with sintering temperature. Peak shifts towards lower 2θ, from 33.531° (YC7) to 33.412° (YC11), indicating an increase in cell parameters and cell volume. Also *(020)* peak is splitting in all the compounds which reveal the possibility of orbital ordering in the compounds (Doshi et al. 2009), which requires further investigations. Further, crystallite size (CS) can be calculated using the scherer's formula: CS $= 0.9\lambda/B\cos\theta$, where λ is the X-ray wavelength, B is the FWHM and θ is the angle. It is observed that, CS increases with sintering temperature. The calculated values of CS are 7.92 nm (YC7), 22.67 nm (YC8), 25.57 nm (YC9), 26.82 nm (YC10) and 70.43 nm (YC11). Figure 3 shows the variation in CS with sintering temperature. Observed increased peak intensity, reduced background, decreased FWHM and increased CS indicate the improved crystallinity with sintering temperature.

To understand the effect of CS and sintering temperature on the dielectric behavior of presently studied sol–gel grown nanostructured YCMO compounds, temperature and frequency dependent complex impedance data were collected for all the YCMO compounds. The real and imaginary parts of complex permittivity or dielectric constant were calculated by inserting the obtained impedance data and sample dimensions into the standard equations given below:

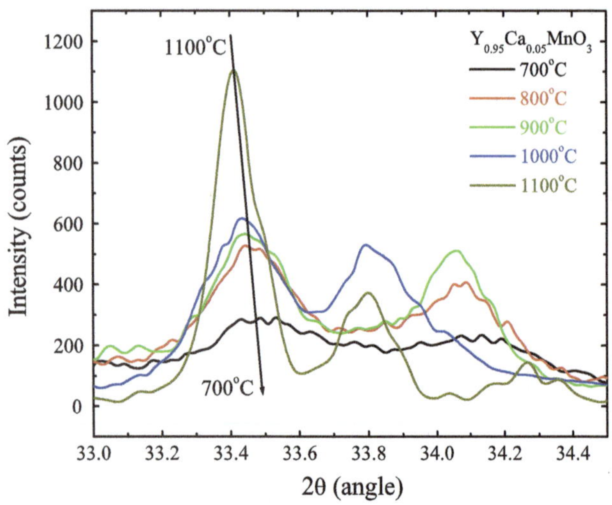

Fig. 2 Enlarged view of *(020)* peaks of all the YCMO compounds

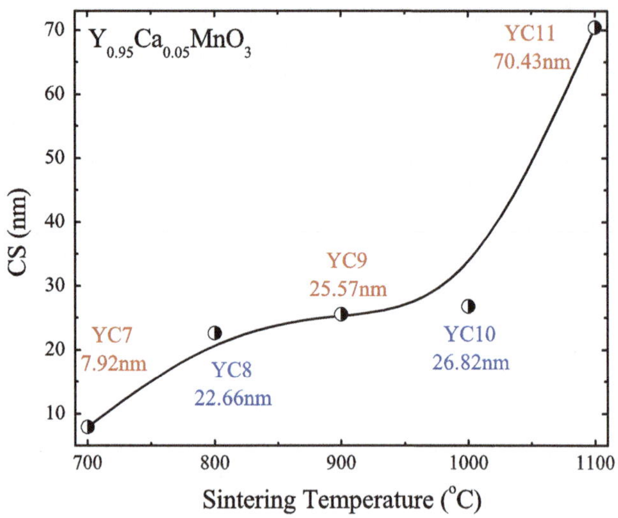

Fig. 3 Variation in crystallite size (CS) with sintering temperature for nanostructured YCMO compounds

$$\varepsilon^* = \varepsilon' + i\varepsilon''$$

$$\varepsilon' = -\frac{d}{\omega\varepsilon_0 A}\frac{Z''}{(Z'^2 + Z''^2)}$$

$$\varepsilon'' = -\frac{d}{\omega\varepsilon_0 A}\frac{Z'}{(Z'^2 + Z''^2)}$$

where, d is the thickness of pellet, A is area of electrode and ε_0 is permittivity in vacuum. The frequency response of real part of dielectric constant (ε') at various temperatures obtained from above equations for all the YCMO compounds are shown in Fig. 4. The value of ε' decreases with increasing frequency, which demonstrates a typical characteristic of FE materials (Ravalia et al. 2011). The observation can be understood by the phenomena of dipole relaxation where at low frequency the dipoles are able to follow the frequency of the applied field (Palkar et al. 2004). It is also clearly seen that, ε' increases with temperature indicating the FE nature of YCMO compounds studied. At low temperatures, may be due to freezing effect on the atomic/ionic movement in the structure, dipoles cannot easily follow the field direction while at higher temperatures, most of the electrical dipoles get enough exciting thermal energy to be able to follow the changes in the external field direction, which enhances the contribution of the dipoles to the polarization leading to an increase in the value of ε'.

It is also evident from the Fig. 4 that, dielectric constant increases with sintering temperature, which can be understood as—with increase in sintering temperature, the CS increases and hence crystallites become more compact in higher sintered compounds as compared to lower one. The agglomeration between the crystallites and the removal of crystallite boundaries result in the interfacial electric/dipole

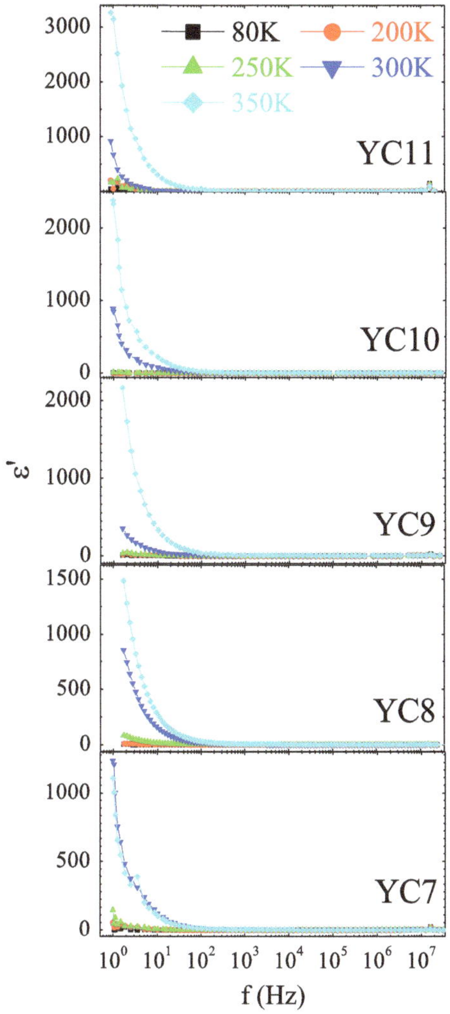

Fig. 4 Temperature dependent ε′ vs. f plots of nanostructured YCMO compounds

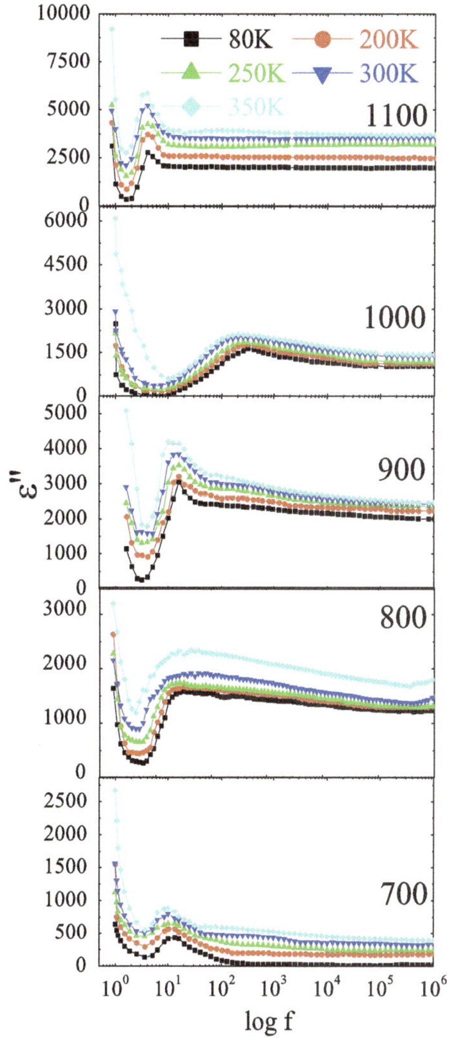

Fig. 5 Temperature dependent ε″ vs. f plots of nanostructured YCMO compounds

polarization (Manoj and Yadav 2007) and hence the resultant dielectric constant increases with increase in CS/sintering temperature. Furthermore, it is well established fact that, with increase in sintering temperature, the oxygen deficiency gets increased in oxide materials. In the present case, YC7 possesses less oxygen vacancies compared to that in YC11. Oxygen vacancies convert the Mn^{4+} (generated due to 5 % Ca^{2+} substitution) to Mn^{3+} in YCMO in order to maintain the valance in the system. Thus, with increase in sintering temperature, the hopping conduction decreases (due to reduced Mn^{4+} ionic density) which results in the enhancement in dielectric permittivity.

The variation in imaginary part of dielectric constant (ε″) with frequency at various temperatures for all the YCMO nanostructured compounds is shown in the Fig. 5. It can be seen that, with increase in temperature, ε″ increases. It is believed that, at higher temperatures, due to an increase in the mobility, enhanced conductivity causes an increase in ε″, associated with the conduction losses. A dielectric anomaly (ε″ peak) at low frequency is observed for all the compounds, which shifts towards lower frequency with temperature. The ε″ is almost doubled than ε′ (the ratio is ~2) revealing high dielectric loss present in the compounds which may be due to the substitution of divalent Ca^{2+} in multiferroic YMO compound.

To understand the relaxation mechanism in the nanostructured YCMO compounds and how CS affects the relaxation in the YCMO, the variation in dielectric modulus M'' with frequency in the temperature range of 80–350 K has been measured (Fig. 6). It is found that, with increase in temperature (from 80 to 350 K) and sintering temperature (700–1100 K), anomaly peak in M'' shifts towards higher frequency. From the peak shifting behavior of YCMO compounds, observed in M'' vs. f plots, one can find the activation energy for all the samples using the formula:

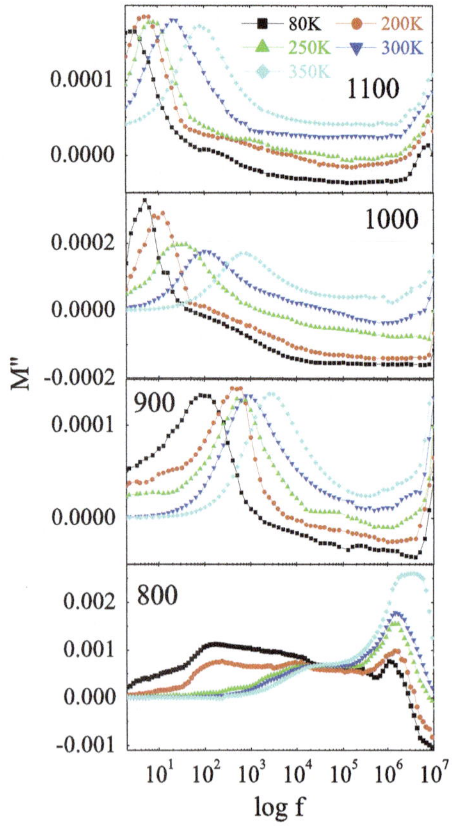

Fig. 6 Temperature dependent M'' vs. f plots of nanostructured YCMO compounds

$$f_m = f_0 exp\left(-\frac{E_a}{K_B T}\right)$$

where, f_m is the frequency corresponding to the loss peak, f_0 is the pre-exponential factor, E_a is the activation energy for relaxation, K_B is the Boltzmann constant and T is absolute temperature. Figure 7 shows the plots of ln f_m vs. 1,000/T (T 200–350 K) for YCMO nanostructured compounds. Values of activation energy, E_a, can be calculated using the slope obtained from the straight line fitting of the plots as shown in Fig. 7. The calculated values of E_a are 41.49, 70.89, 83.37 and 104.5 meV for YCMO samples sintered at 800, 900, 1,000 and 1,100 °C, respectively. It can be seen that, with increase in sintering temperature, E_a increases, which can be attributed to the sintering temperature induced enhancement in oxygen deficiency/vacancies and hence decrease in carrier hopping resulting in the enhancement in activation energy.

Effect of sintering temperature and crystallite size on the dielectric behavior of nanostructured YCMO compounds has been discussed in detail. Temperature dependent physical state of the materials can be understood as: mixed valent manganese oxides are intrinsically disordered systems having small scale (nanoscale) phases, such as

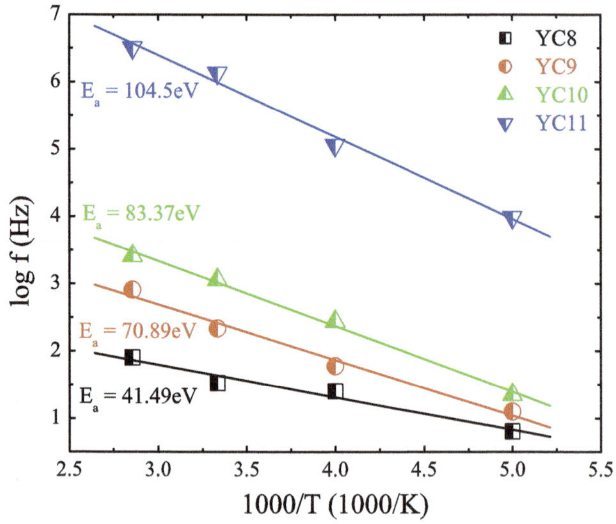

Fig. 7 Plots of ln f_m vs. 1000/T using M'' vs. f data for nanostructured YCMO compounds

ferromagnetic metallic (FMM), paramagnetic insulating (PMI), antiferromagnetic insulating (AFMI), charge-ordered insulating (COI), etc., depending upon the temperature, coexisting in single phase compound resulting in the electrical inhomogeneities (Kuberkar et al. 2012). The effect of sintering temperature on these temperature dependent nanoscale phases can be ascribed to the agglomeration of smaller crystallites forming the larger crystallites at higher sintering temperatures and related boundary modifications (Solanki et al. 2009). In the present case, the effect of sintering temperature on the inhomogeneous electronic phases can be understood, in the context of non-compositional oxygen content, as: in lower sintered samples, the boundaries are less oxygen deficient while in higher sintered samples the boundaries are more oxygen deficient which in turn results in the non-stoichiometric inhomogeneous phases in the samples. Across the boundaries, the oxygen deficiency results in the conversion of Mn^{4+} (generated due to the substitution of Ca^{2+}) to Mn^{3+} which controls the conductivity (leakage current) in the compounds. The larger the oxygen deficiency the smaller the conductivity (leakage) resulting in the stronger dielectric nature in the samples sintered at high temperatures.

Conclusions

In summary, $Y_{0.95}Ca_{0.05}MnO_3$ (YCMO) nanostructured compounds have been successfully synthesized using low cost and easy sol–gel route employing acetate precursor route. Increase in ε' and ε'' with temperature and CS has been ascribed to the thermal activation to the dipoles and interfacial polarization across the crystallite boundaries of the compounds sintered at higher temperatures, respectively. It is observed that, with increase in sintering temperature, the oxygen vacancies play an important role in governing the dielectric behavior, i.e. increase in E_a with sintering temperature is attributed to the enhancement in oxygen deficiency/vacancies and hence decrease in carrier hopping and increase in E_a. In conclusion, sintering temperature and hence CS is the key factor to maintain/govern the activation energy and electric dipole behavior in presently studied YCMO nanostructured compounds.

Acknowledgments Author is thankful to Government of Gujarat and Saurashtra University, Rajkot for providing financial support in the form of world class university grant for Nano Science Research. Author is also thankful to Mr. Davit Dhruv and Ms. Zalak Joshi for their help.

References

Aikawa Y, Katsufuji T, Arima T, Kato K (2005) Effect of Mn trimerization on the magnetic and dielectric properties of hexagonal $YMnO_3$. Phys Rev B 71(184418):1–5

Bertaur EF, Pauthenet R, Mercier M (1963) Proprietes magnetiques et structures du manganite d'yttrium. Phys Lett 7:110–111

Cassinese A, De Luca GM, Prigiobbo A, Salluzzo M, Vaglio R (2004) Field-effect tuning of carrier density in $Nd_{1.2}Ba_{1.8}Cu_3O_y$ thin films. Appl Phys Lett 84:3933–3935

Cheng Z, Wang X (2007) Room temperature magnetic field manipulation of electrical polarization in multiferroic thin film composite $BiFeO_3/La_{2/3}Ca_{1/3}MnO_3$. Phys Rev B 75(172406): 1–4

Dong ZW, Pai SP, Ramesh R, Venkatesan T, Johnson M, Chen ZY, Cavanaugh A, Zhao YG, Jiang XL, Sharma RP, Ogale S, Greene RL (1998) Novel high-T_C transistors with manganite oxides. J Appl Phys 83:6780–6782

Doshi RR, Solanki PS, Krishna PSR, Das A, Kuberkar DG (2009) Magnetic phase coexistence in Tb^{+3}- and Sr^{+2}-doped $La_{0.7}Ca_{0.3}MnO_3$ manganite: a temperature-dependent neutron diffraction study. J Magn Magn Mater 321:3285–3289

Doshi RR, Solanki PS, Khachar U, Kuberkar DG, Krishna PSR, Banerjee A, Chaddah P (2011) First order paramagnetic–ferromagnetic phase transition inTb3 p doped La0.5Ca0.5MnO3 manganite. Phys B 406:4031–4034

Fiebig M, Lottermoser Th, Frohlich D, Goltsev AV, Pisarev RV (2002) Observation of coupled magnetic and electric domains. Nature 419:818–820

Fujimura N, Azuma S, Aoki N, Yoshimura T, Ito T (1996a) Growth mechanism of $YMnO_3$ film as a new candidate for nonvolatile memory devices. J Appl Phys 80:7084–7088

Fujimura N, Ishida T, Yoshimura T, Ito T (1996b) Epitaxially grown $YMnO_3$ film: new candidate for nonvolatile memory devices. Appl Phys Lett 69:1011–1013

Gajek M, Bibes M, Varela M, Fontcuberta J, Herranz G, Fusil S, Bouzehouane K, Barthelemy A, Fert A (2006) "$La_{2/3}Sr_{1/3}MnO_3$–$La_{0.1}Bi_{0.9}MnO_3$ heterostructures for spin filtering". J Appl Phys 99:08 E504:1–3

Gajek M, Bibes M, Fusil S, Bouzehouane K, Fontcuberta J, Barthelemy A, Fert A (2007) Tunnel junctions with multiferroic barriers. Nature Mater 6(296):1–7

Gutierrez D, Pena O, Duran P, Moure C (2002) Crystalline structure and electrical properties of $YCu_xMn_{1-x}O_3$ solid solutions. J Eur Ceram Soc 22:2939–2944

Kalashnikova AM, Pisarev RV (2003) Electronic structure of hexagonal rare-earth manganites $RMnO_3$. JEPT Lett 78:143–147

Khachar U, Solanki PS, Choudhary RJ, Phase DM, Ganesan V, Kuberkar DG (2012) Current–voltage characteristics of PLD grown manganite based $ZnO/La_{0.5}Pr_{0.2}Sr_{0.3}MnO_3/SrNb_{0.002}Ti_{0.998}O_3$ thin film heterostructure. Solid State Commun 152:34–37

Kuberkar DG, Doshi RR, Solanki PS, Khachar U, Vagadia M, Ravalia A, Ganesan V (2012) Grain morphology and size disorder effect on the transport and magneto transport in sol–gel grown nanostructured manganites. Appl Sur Sci 258:9041–9046

Li SZ, Yan ZB, Wei T, Luo SJ, Liu B, Wang KF, Liu JM (2009) Preparation of epitaxial orthorhombic $YMnO_3$ thin films and the current–voltage rectifying effect. Appl Phys A 94:975–980

Lliev MN, Lorenz B, Litvinchuk AP, Wang YQ, Sun YY, Chu CW (2005) Structural, transport, magnetic properties and Raman spectroscopy of orthorhombic $Y_{1-x}Ca_xMnO_3$ $(0 \le x \le 0.5)$. J Phys Condens Matter 17:3333–3342

Lorenz B, Wang YQ, Sun YY, Chu CW (2004) Large magneto dielectric effects in orthorhombic $HoMnO_3$ and $YMnO_3$. Phys Rev B 70(212412):1–4

Lorenz B, Wang YQ, Chu CW (2007) Ferroelectricity in perovskite $HoMnO_3$ and $YMnO_3$. Phys Rev B 76(104405):1–5

Kumar M, Yadav KL (2007) Rapid liquid phase sintered Mn doped $BiFeO_3$ ceramics with enhanced polarization and weak magnetization. Appl Phys Lett 91(242901):1–3

Mavani KR, Rana DS, Rayaprol S, Parmar RN, Kuberkar DG, Kumar R, Tonouchi M, John J, Nagarajan R (2007) 200 MeV Ag^{+15} ion irradiation created columnar defects and enhanced critical current density of La-2125 type superconducting thin films. Solid State Commun 142:462–465

Moure C, Gutierrez D, Pena O, Duran P (2002) Structural characterization of $YMe_xMn_{1-x}O_3$ (Me = Cu, Ni, Co) perovskites. J Solid State Chem 163:377–384

Palkar VR, Kundaliya DC, Malik SK, Bhattacharya S (2004) "Magnetoelectricity at room temperature in the $Bi_{0.9-x}Tb_xLa_{0.1}FeO_3$ system". Phys Rev B 69(212102):1–3

Ravalia A, Vagadia M, Vachhani PS, Choudhary RJ, Phase DM, Asokan K, Kuberkar DG (2012) 200 MeV Ag^{15+} ion induced surface modification and transport behaviour in manganite based thin film devices. Appl Sur Sci 258:4203–4206

Ravalia A, Vagadia M, Trivedi P, Solanki PS, Asokan K, Ojha S, Thakur OP, Choudhary RJ, Phase DM, Kuberkar DG (2013) Role of oxygen in multiferroic behavior of $BiFeO_3$ films grown on 0.2% Nb doped $SrTiO_3$. Solid State Commun 169:10–13

Ravalia AB, Vagadia MV, Khachar UD, Doshi RR, Solanki PS, Savalia BT, Shah NA, Kuberkar DG (2011) Dielectric and Magnetic Behavior of Sol–Gel Grown $BiFeO_3$ Multiferroic. AIP Conf Proc 1349:1143–1144

Rayaprol S, Thaker CM, Chakraborthy Keka R, Krishna KSR, Ramanadham M, Kuberkar DG (2004) Structural and magnetic studies on $La_{2-x}Dy_xCa_{2x}Ba_2Cu_{4+2x}O_z$ type superconducting oxides. J Phys Condens Matter 16:6551–6560

Singh AK, Patnaik S, Kaushik SD, Siruguri V (2010) Dominance of magnetoelastic coupling in multiferroic hexagonal $YMnO_3$. Phys Rev B 81(184406):1–6

Solanki PS, Doshi RR, Thaker CM, Pandya S, Ganesan V, Kuberkar DG (2009) Transport and magnetotransport studies on sol–gel

grown nanostructured $La_{0.7}Pb_{0.3}MnO_3$ manganites. J Nanosci Nanotechnol 9:5681–5686
Solanki PS, Doshi RR, Khachar UD, Vagadia MV, Ravalia AB, Kuberkar DG, Shah NA (2010) Structural, microstructural, transport, and magnetotransport properties of nanostructured $La_{0.7}Sr_{0.3}MnO_3$ manganites synthesized by coprecipitation. J Mater Res 25:1799–1802
Solanki PS, Doshi RR, Khachar UD, Choudhary RJ, Kuberkar DG (2011) Thickness dependent transport and magnetotransport in CSD grown $La_{0.7}Pb_{0.3}MnO_3$ manganite films. Mater Res Bull 46:1118–1123
Vagadia M, Ravalia A, Khachar U, Solanki PS, Doshi RR, Rayaprol S, Kuberkar DG (2011) Size and grain morphology dependent magnetic behaviour of Co-doped ZnO. Mater Res Bull 46: 1933–1937
Vagadia M, Ravalia A, Solanki PS, Choudhary RJ, Phase DM, Kuberkar DG (2013) Improvement in resistive switching of Ba-doped $BiFeO_3$ films. Appl Phys Lett 103(033504):1–5
Vachhani PS, Solanki PS, Doshi RR, Shah NA, Rayaprol S, Kuberkar DG (2011) Substrate dependent transport and magnetotransport in manganite multilayer. Phys B 406:2270–2272
Veres A, Noudem JG, Fourrez S, Bailleul G (2006) The influence of iron substitution to manganese on the physical properties of $YMnO_3$. Solid State Sci 8:137–141
Zhou HD, Denyszyn JC, Goodenough JB (2005) Effect of Ga doping on the multiferroic properties of $RMn_{1-x}Ga_xO_3$ (R = Ho, Y). Phys Rev B 72(224401):1–5

9

Synthesis, characterization and adsorption behavior of Mo(VI) and W(VI) ions on titanium dioxide nanoparticles containing anatase modification

Polina A. Demina · Alexey N. Kuz'michev · Andrei M. Tsybinsky · Lubov N. Obolenskaya · Galina M. Kuz'micheva · Elena N. Domoroshchina · Elena V. Savinkina

Abstract Adsorption of Mo(VI) and W(VI) ions from water with nanosized anatase has been studied. This process depends on sorption conditions (temperature, contact duration, etc.) and sample characteristics (composition, specific surface area, coherent scattering regions, etc.). Maximal recovery efficiency for Mo(VI) and W(VI) was achieved with the use of the samples of commercial Hombifine N ($R_{Mo} = 99.90$ %) and Degussa P25 ($R_{Mo} = 99.99$ %, $R_W = 99.75$ %) and nanosized anatase modified with H_2O_2 ($R_{Mo,W} = 99.75$ %) under optimal conditions of sorption.

Keywords Nanosized titania · Anatase · Adsorption · Water purification

Introduction

The removal of heavy metals from wastewaters is a matter of paramount importance due to the fact that their high toxicity causes major environmental pollution problems. One of the most efficient, applicable and low cost methods for the removal of toxic metals from aqueous solutions is their adsorption on an inorganic adsorbent. Nanosized titanium dioxide attracts interest due to high speed of As(III) and As(V) (Dadachov 2006a, b; Meng et al. 2003), As(V) (Valencia-Trejo et al. 2010; Deedar et al. 2009; Xiao et al. 2007), U(VI) (Dadachov 2006a, b), Bi(V) (Kuzmicheva et al. 2012), (Cu(II) (Georgaka and Spanos 2010), V(V) (Tella et al. 2010), Cu(II) and Mn(II) (Bleam and McBride 1986), Cd(II) (Gao et al. 2004) extracting from aqueous solutions. To achieve high efficiency of adsorption, it is important to understand the effect of the solution parameters and characteristics of adsorbents. The purpose of the project is to study adsorption properties of samples with nano-anatase prepared via a sulfate technique for extraction of Mo(VI) and W(VI) from model water systems. Adsorption of Mo(VI) and W(VI) with titanium dioxide nanoparticles has not been previously reported.

P. A. Demina · A. M. Tsybinsky · L. N. Obolenskaya · G. M. Kuz'micheva · E. N. Domoroshchina (✉) · E. V. Savinkina
Lomonosov University of Fine Chemical Technology, 119571 Moscow, Russia
e-mail: elena7820@gmail.com

A. N. Kuz'michev
Bonifatius College, Burg. Fockema Andreaelaan 7, 3582 Utrecht, The Netherlands

Experimental

Synthesis

Several types of titania nanoparticles were used in this study. Samples with anatase were prepared using the well-known sulfate method (Kuzmicheva et al. 2010; Savinkina et al. 2011a, b), the main advantages of which is a low temperature of synthesis (not higher than 98 °C) and the possibility of the formation of compounds with the required composition and morphology. The hydrolysis of solvated titanyl sulfate (Aldrich) was performed under mild conditions ($t < 98$ °C, $\tau = 5$–30 min). Various conditions of samples preparation were used (with or without solution stirring, presence or absence of hydrochloric acid, and, in certain instances, treatment with NaOH). The samples were washed with distilled water and acetone, dried and isolated from transparent solutions.

Fig. 1 Scheme of the synthesis

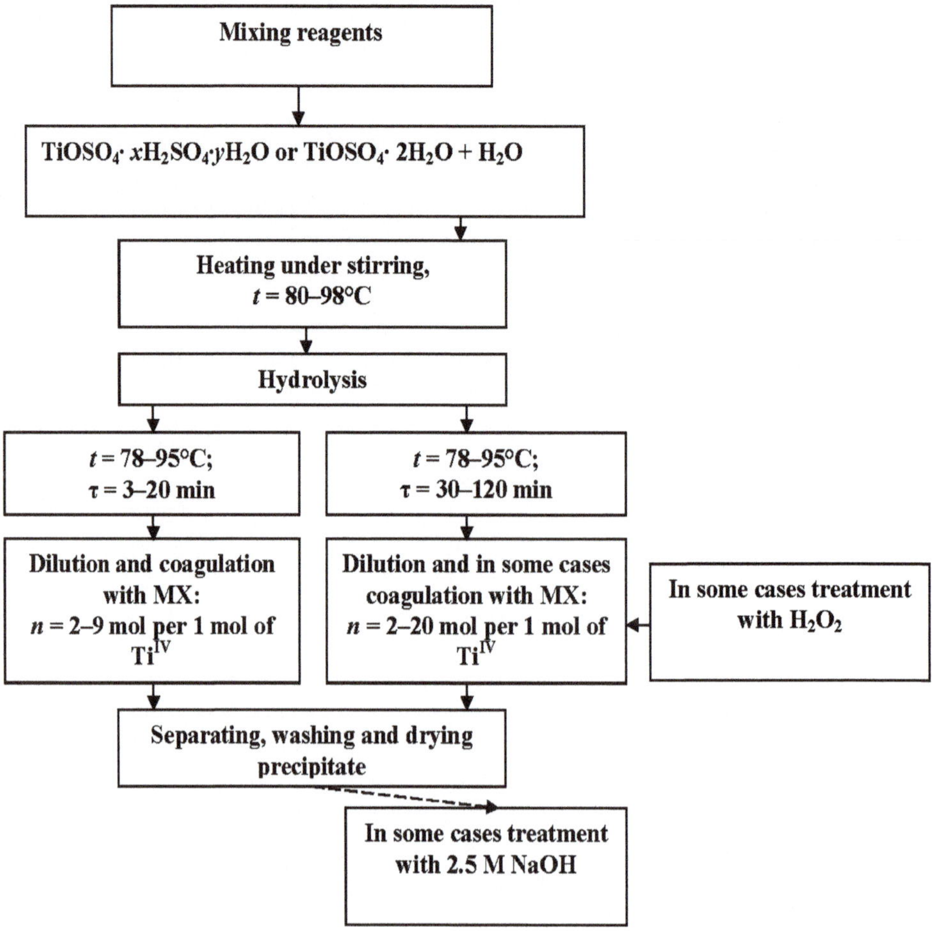

Anatase was modified with H_2O_2 by treatment of the reaction mixture with hydrogen peroxide while synthesizing titanium dioxide from titanyl sulfate by the earlier reported technique for preparing nanosize anatase (Savinkina et al. 2011a, b). The resulting yellow powders were >20 times washed with water, then with aqueous solution of Na_2SO_3 (to remove uncoordinated H_2O_2 molecules) and finally with acetone (to dry the samples) (Fig. 1).

Commercial Hombifine N and Hombikat UV100 with nanometer-scale anatase and Aeroxide Degussa (Evonik) P25 with the phase composition of ~85 % anatase and ~15 % rutile were also used in this study.

For modifying with H_2O_2, Hombikat UV100 powder (0.25 g) was dispersed in 50 ml of 10 M solution of H_2O_2 (TiO_2:H_2O_2 = 1:160), stirred for 40 min with a magnetic stirrer (30 mm rod) at room temperature, filtered off, washed with water (3–6 times) and acetone (once), dried at room temperature and ground.

X-ray study

All X-ray diffraction (XRD) patterns were recorded with sample rotation on an HZG-4 (Ni filter) diffractometer: CuKα radiation in the diffracted beam, stepwise mode (counting time 10 s, a step of 0.02°, 2θ 2°–50° or 80°). Qualitative XRD phase analyses were carried out using JCPDS PDF-2 and literature data. The coherent scattering region size L (Å) was evaluated using the Scherrer formula:

$$L = 0.9\lambda/\beta\cos\theta, \qquad (1)$$

where β is the measured diffraction peak width at half height ($2\theta \sim 25°$).

The sizes of nanoparticles N (nm) were calculated from the results of the small angle X-ray scattering (SAXSess diffractometer, CuK$_\alpha$) (Svergun 1992).

Scanning electron microscopy (SEM)

The microstructures of samples were examined by scanning electron microscopy (SEM) on a JEOL JSM-7500F field-emission scanning electron microscope. Micro- and nanostructures were examined using low-energy backscattered electron imaging, which insured a resolution of 1.4 nm at 1 kV and 1.0 nm at 20 kV. We also used gentle beam mode, intended to ensure high resolution at a low accelerating voltage.

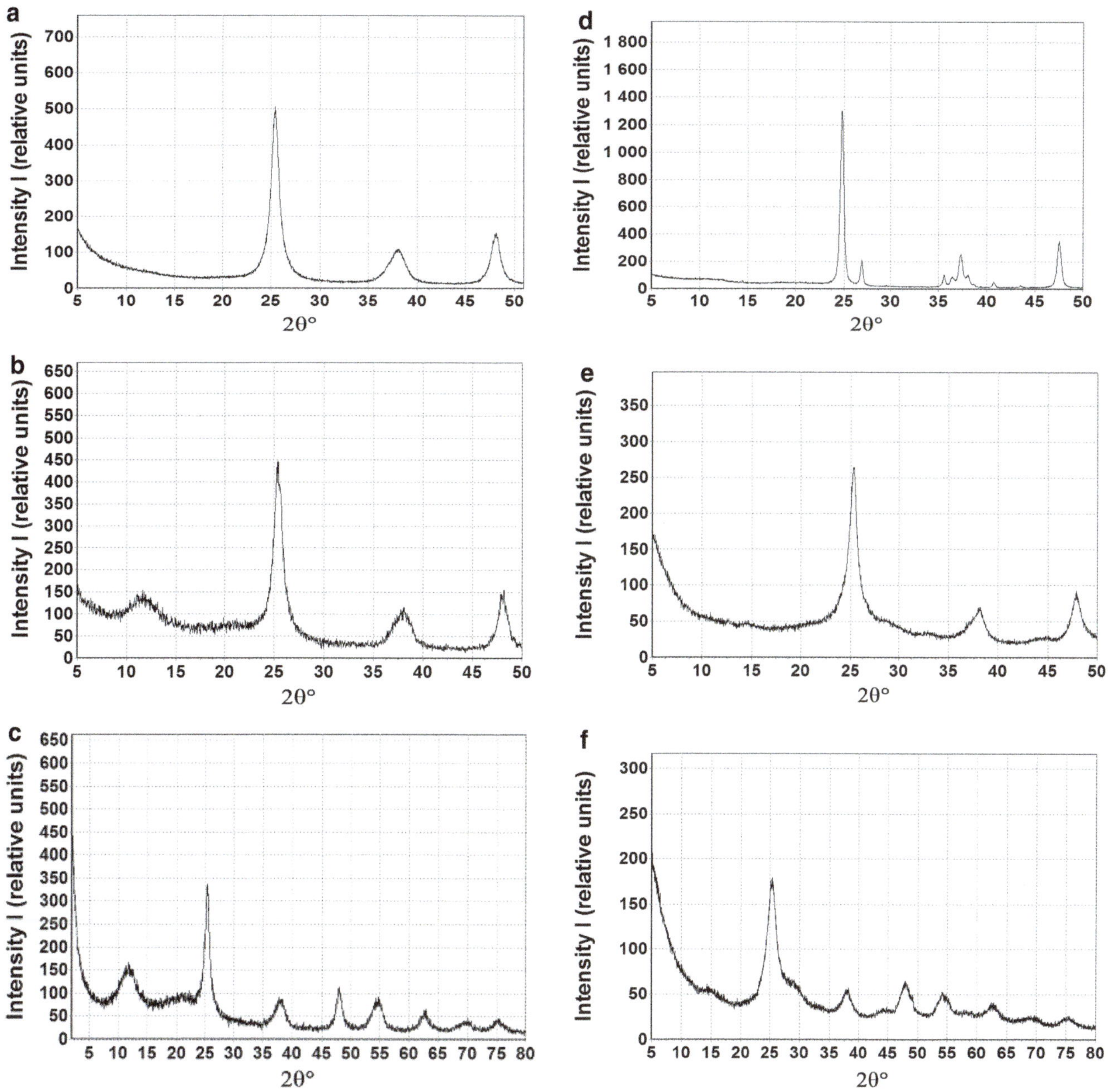

Fig. 2 Plot of the XRD patterns of nanoparticle-sized samples: Hombifine N (**a**), Hombikat UV100 (**b**), Hombikat UV100 modified with H_2O_2 (**c**), Degussa P25 (**d**), anatase (**e**), anatase modified with H_2O_2 (**f**)

Brunauer–Emmett–Teller (BET) method

The appearance and availability of high quality gas analyze sorption data onto the anatase powders using a Micromeritics ASAP 2020 apparatus allowed us to determine the texture characteristics (S m^2/g, the free specific surface area; V_N and V_U cm^3/g, nano- and ultrananopore volumes, correspondingly; N nm, the nanoparticles sizes) of materials and, as a result, to estimate the sizes of particles on whose surfaces adsorption occurs, using the formula (Gregg and Sing 1982; Dzis'ko et al. 1978):

$$N = \left[K_p/(\rho S)\right] \times 10^9, \quad (2)$$

where N is the average size of the particles; K_p is the shape factor of the particles ($K_p = 6$ for cubes and spheres). The shape of particles could be revealed from the form of the obtained adsorption isotherms (Obolenskaya et al. 2012; Zhilkina et al. 2013), ρ is the density of the material ($\rho = 3.84$ g/cm^3); and S is the specific surface area of the material. The specific surface area of the investigated samples was determined by the Brunauer–Emmett–Teller (S_{BET}, m^2/g) (Brunauer et al.

1938) and comparative (S_C, m^2/g) (Zagrafskaya et al. 1975) methods.

Adsorption experiments

Adsorption ability of the samples was studied at pH 0.35 with the use of the standard mixtures ICP-MS-68A-A (High-Purity Standards, USA) containing 1 ppm of analytes (C_i). The Mo(VI) and W(VI) concentration in the solutions after the sorption (C_f) was determined by inductively coupled plasma mass spectrometry (Agilent 7500c, Perkin Elmer ELAN DRC-E, USA). The degree of sorption was evaluated as

$$R = (1 - C_f/C_i) \times 100\ \%, \quad (3)$$

where C_f and C_i (the standard deviation was ±2 %) are final and initial concentrations of analytes, respectively.

Results and discussions

XRD results (Fig. 2) showed that Hombifine N (Fig. 2a), Hombikat UV100 (Fig. 2b) and the most representative samples (Fig. 2e, f) consist of anatase. Degussa P25 (Fig. 2d) contained a significant amount of rutile in addition to the anatase phase.

This is consistent with the literature data that the latter sample with anatase (Fig. 1e) contained, in addition to anatase, small amounts of β-TiO_2 ($2\theta \sim 15°$, $\sim 28°$, $\sim 45°$; JCPDS No. 46-1238). Hombikat UV100 (Fig. 1b) differs from Hombifine N (Fig. 1a) by the presence of hydrous amorphous TiO_2 with the general formula $TiO_{2-x}(OH)_{2x} \times yH_2O$ ($2\theta \sim 12°$).

The characteristics of samples are given in Table 1. Particle diameters for all the samples (N, nm) were took from SEM photos (N_{SEM}) and calculated from the surface areas determined by the comparative method (S_{CM}). These values for Hombifine N, Hombikat UV100 and Degussa P25 were similar. For very small nanoparticles (anatase and anatase modified with H_2O_2), the particle sizes determined from the BET surface area were somewhat larger than the primary grain sizes and similar to secondary ones (N_{SAXS}) estimated from the small angle X-ray scattering. This difference may be partly attributable to aggregation of individual crystallites, which would reduce the surface area available for N_2 adsorption and, in turn, yield surface area measurements corresponding to larger grain sizes.

SEM images (Kuzmicheva et al. 2010; Savinkina et al. 2011a, b) (in particular, of anatase) (Fig. 3) show agglomeration of primary nanoparticles into the secondary microparticles; the phenomenon is generally observed in nanoparticles synthesis.

Table 1 Characteristics of samples

Symbol	Sample	Value
L (Å)	Hombifine N	83 (4)
	Hombikat UV100	80 (4)
	Degussa P25[a]	250 (10)
	Anatase	84 (4)
	Anatase modified with H_2O_2[b]	40 (2)
	Hombikat UV100 modified with H_2O_2[b]	97 (5)
$N_{SEM,CM}/N_{SAXS}$ (nm)	Hombifine N	12/∼3 ÷ 20[a]
	Hombikat UV100	12/∼3 ÷ 20[c]
	Degussa P25	28/24
	Anatase	30/10 ÷ 20 (primary), 30 ÷ 45 (secondary)
S_{BET}/S_C (m^2/g)	Hombifine N	313 (3)/116
	Hombikat UV100	314/123
	Degussa P25	53.1/54.3
	Anatase	36.5 (4)/16.5
V_U/V_N (cm^3/g)	Hombifine N	0.083/0.343
	Hombikat UV100	0.082/0.344
	Degussa P25	0.016/0.163
	Anatase	0.013/0.084

[a] Data for anatase

[b] Characteristics of modified and non-modified samples differ by the coherent scattering region size L only

[c] Polydispersity system

The difference between the particle sizes N (nm) and the coherent scattering region size L (Å), that is assumed to be the crystallite size of all the samples (in particular for anatase and anatase modified with H_2O_2) except Degussa P25, is associated with amorphous shells on crystallites.

We have examined adsorption ability of the samples. It was found that the sorption degree (R, %) depends on the sorption conditions. The R value increases with the increase in the sorption duration (T_S) (Fig. 4), the sample weight (m) (Fig. 5) and the use of the membrane filter with 0.45 μm pores (Fig. 6).

It can be seen that the amount of Mo(VI) and W(VI) ions adsorbed by the TiO_2 nanoparticles from aqueous solutions first decreases and then increases with increasing the temperature from 4–10 to 10–85 °C, respectively, using both filtration through a membrane filter and centrifugation; the latter method allows to decrease duration of residue removal (Fig. 7).

This behavior suggests exothermic nature of the process (in range of 4–10 °C) which can be caused by physical modification of the adsorbents. Surfaces of metal oxides usually have positive charge at low pH (our case: pH 0.35, $i = 0.43(1)$, i-ionic strength). Mo(VI) and W(VI) ions in hydrated associates of different compositions probably have

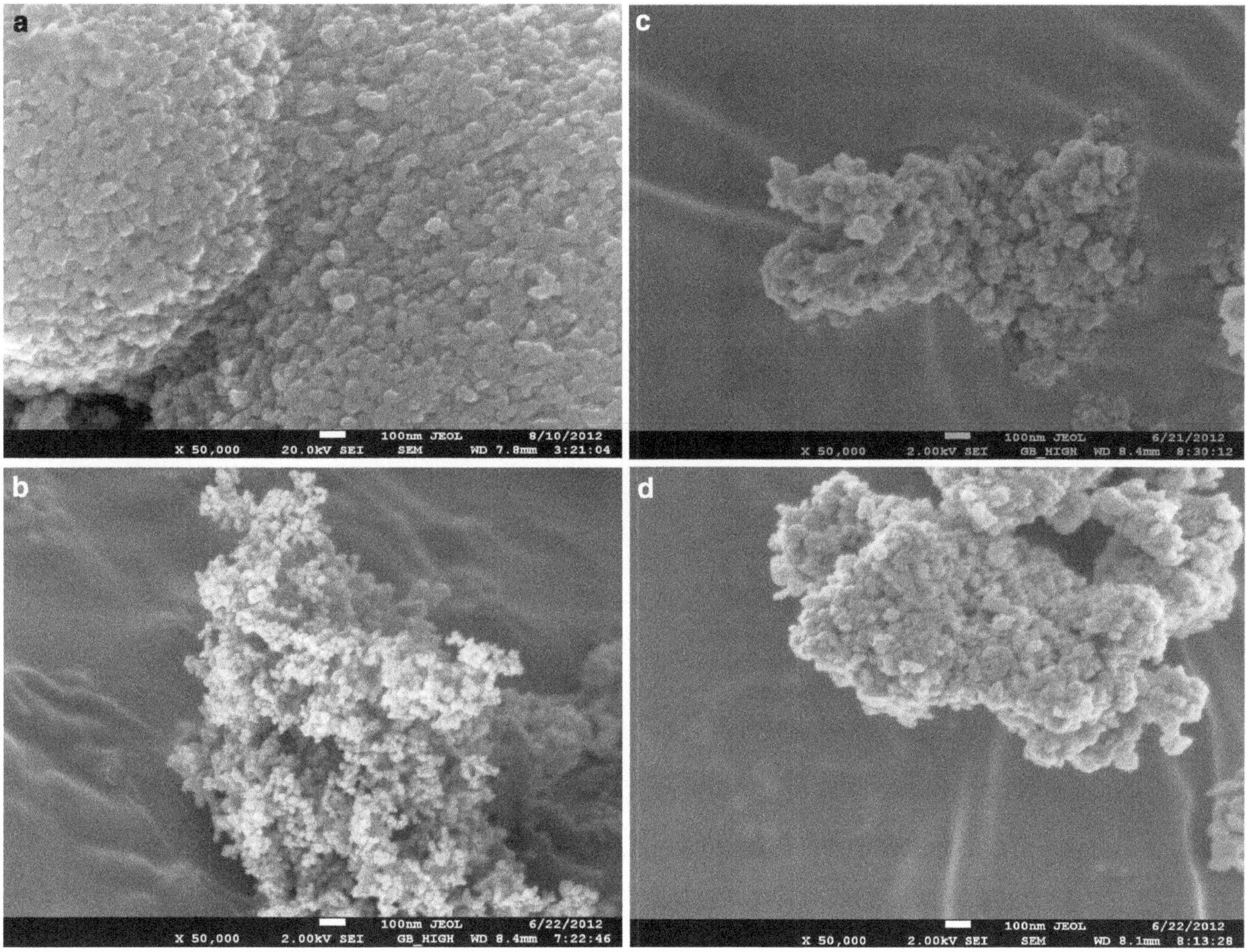

Fig. 3 SEM image of anatase (Fig. 2c) (**a**), Degussa P25 (Fig. 2b) (**b**), Hombifine N (Fig. 2a) (**c**), Hombikat UV100 (Fig. 2b) (**d**)

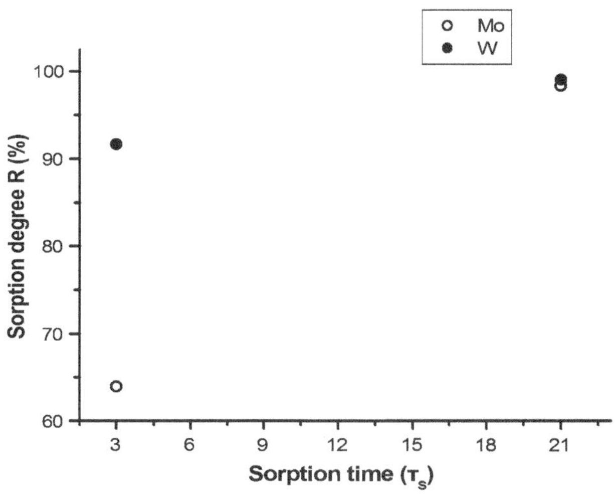

Fig. 4 The sorption degree (*R*, %) of anatase vs. duration of sorption time

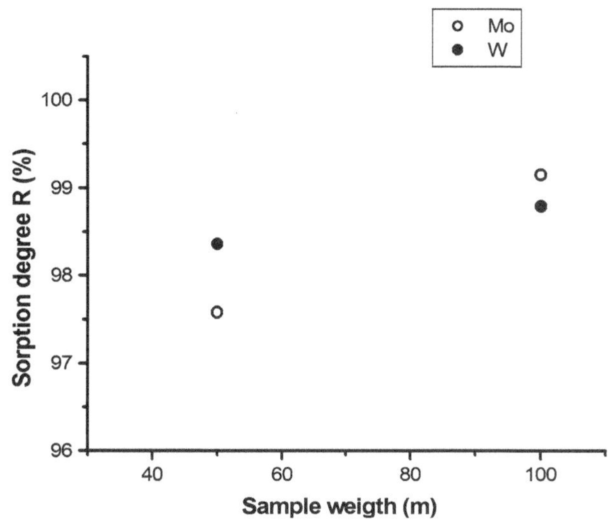

Fig. 5 The sorption degree (*R*, %) of Hombikat UV100 vs. sample weight

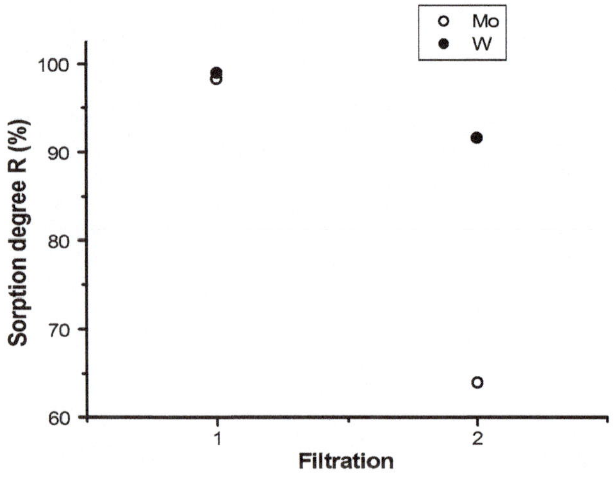

Fig. 6 The sorption degree (R, %) of anatase with *1* and without *2* filtration through a membrane filter

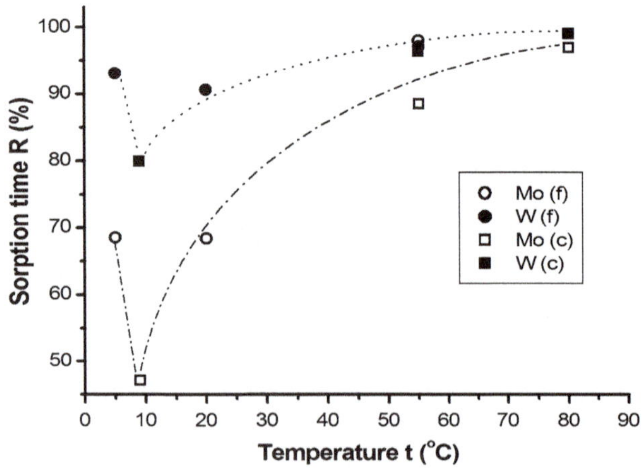

Fig. 7 The sorption degree (R, %) of anatase vs. sorption temperature: *f* filtration through a membrane filter, *c* centrifugation during 1 min

a negative charge (a similar to As(V) and Mo(VI) ions (Valencia-Trejo et al. 2010; Deedar et al. 2009; Xiao et al. 2007; Von Tytko et al. 1983). At low temperatures, they are adsorbed at the expense of forces of electrostatic character.

With temperature increasing from 4 to 10 °C, the attractive forces between TiO_2 and analytes are weakened and the adsorption decreases. Besides, at increasing temperature, the thickness of the boundary layer decreases due to the increasing tendency of the adsorbate to escape from the surface of the adsorbent to the solution phase, which results in a decrease in adsorption as previously reported by Igwe and Abia (2007).

It was found that in the temperature range from 10 to 85 °C the process is endothermic, since the adsorption increases with increasing temperature (Fig. 7). Therefore, the rate of physical adsorption and the amount of ions adsorbed by this way decrease, meanwhile the degree of chemical adsorption increases.

The increase in the sample dispersity under their grinding can result in either increase [for Mo(VI)] or decrease [for W(VI)] of the sorption degree.

Correlations between the sorption degree and sample characteristics are less obvious, since it is difficult to choose samples differing by only one parameter. However, in some cases (for our standard sorption conditions, $t = 20$ °C, $T_S = 21$ h, $m = 50$ mg, pH 0.35), the sorption degree was found to correlate with the samples characteristics (N, S, V_U, V_N). Particularly, these experimental relationships for Hombifine N and Degussa P25 and our samples with nanosized anatase (Table) can be described by equations (close to R_{max}):

$$R(N) = -23.44 \times 10^{-4}(N - 1.3346)^2 + 0.0875(N - 1.3346) + 99.1852\left(R_{max} \text{ for } N_{SEM,CM} \sim 20\,\text{nm}\right);$$

Fig. 8 The sorption degree (R, %) of studied samples with anatase vs. the coherent scattering region size (L value)

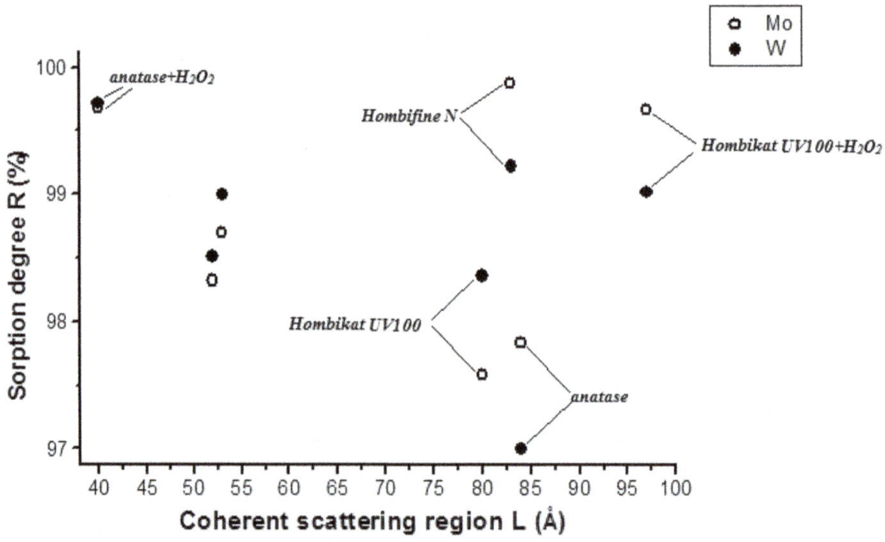

$$R(S) = -1.62 \times 10^{-4}(S + 7.2302)^2 + 0.0292(S + 7.2302) + 98.6778\left(R_{max} \text{ for } S_{CM} \sim 83\,\text{m}^2/\text{g}\right);$$

$$R(V_U) = -574.9033(V_U)^2 + 52.4378(V_U) + 98.7976\left(R_{max} \text{ for } V_U = 0.046\,\text{cm}^3/\text{g}\right);$$

$$R(V_N) = -79.1667(V_N)^2 + 38.391(V_N) + 95.3455\left(R_{max} \text{ for } V_N = 0.242\,\text{cm}^3/\text{g}\right) \text{ (for } W\text{)};$$

$$R(N) = -20.10 \times 10^{-4}(N + 2.4932)^2 + 0.0666(N + 2.4932) + 99.4433\left(R_{max} \text{ for } N_{SEM,CM} \sim 14\,\text{nm}\right);$$

$$R(S) = -15.9 \times 10^{-4}(S + 0.1481)^2 + 0.3421(S + 0.1481) + 81.5713(R_{max} \text{ for } N_{SEM,CM} \sim 14\,\text{nm})$$

$$R(V_U) = -129.845(V_U)^2 + 14.347(V_U) + 99.604\left(R_{max} \text{ for } V_U = 0.055\,\text{cm}^3/\text{g}\right);$$

$$R(V_N) = -19.217(V_N)^2 + 10.280(V_N) + 98.625\left(R_{max} \text{ for } V_N = 0.267\,\text{cm}^3/\text{g}\right) \text{ (for Mo)}.$$

Anatase modified with H_2O_2 has the smallest value of the coherent scattering region size ($L = 40$ Å) and the highest (comparing to other samples, excluding Degussa P25) sorption degree of W(VI) (Fig. 8; Table 1). Similar results were obtained earlier for As(III), As(V), Bi(V) and V(V) (Dadachov 2006a, b; Meng et al. 2003; Valencia-Trejo et al. 2010; Deedar et al. 2009).

However, there is no correlation between the coherent scattering region size L (Å) and the sorption degree (R, %). The maximal recovery efficiency from aqueous solutions of Mo(VI) ions has been obtained for Hombifine N with $L = 83$ Å and Degussa P25 with $L = 250$ Å (Fig. 10; Table 1). Hombikat UV100 has characteristics similar to Hombifine N (Table 1) except the specific surface area and composition of the sample. Nevertheless, Hombikat UV100 showed the worse adsorption ability as compared to Hombifine N (Fig. 10). The results led to the conclusion that the maximal recovery efficiency depends on concentration of adsorption centers—OH groups—on the adsorbent surface (Dadachov 2006a, b). Indeed, although the specific surface area of Hombikat UV100 is higher than for Hombifine N because of the presence of hydrous amorphous titanium dioxide (Fig. 2b), the concentration of OH groups on the surface of Hombikat UV100 is lower. Only the sample containing pure hydrous amorphous titanium dioxide has the characteristics ($S_{BET} = 317.5$ m^2/g and $S_C = 46.0$ m^2/g, $V_U = 0.099$ cm^3/g and $V_N = 0.229$ cm^3/g) and sorption degree ($R_{Mo} = 91.70$ % and $R_W = 97.72$ %) that are lower than for Hombikat UV100 with $R_{Mo} = 97.50$ % and $R_W = 98.40$ %. However, Hombikat UV100 modified with H_2O_2 by the cold impregnation (Fig. 2c) shows the sorption degree ($R_{Mo} = 99.70$ % and $R_W = 99.05$ %) second to Degussa P25 and the highest (among another samples, excluding Degussa P25) L value (Figs. 8, 10; Table 1). So, the treatment of the reaction mixture with hydrogen peroxide led decrease of the L value for anatase in contrast to the cold impregnation that increases this value (Table 1). Our experience suggests that the additional treatment of samples with hydrogen peroxide is of particular value for the

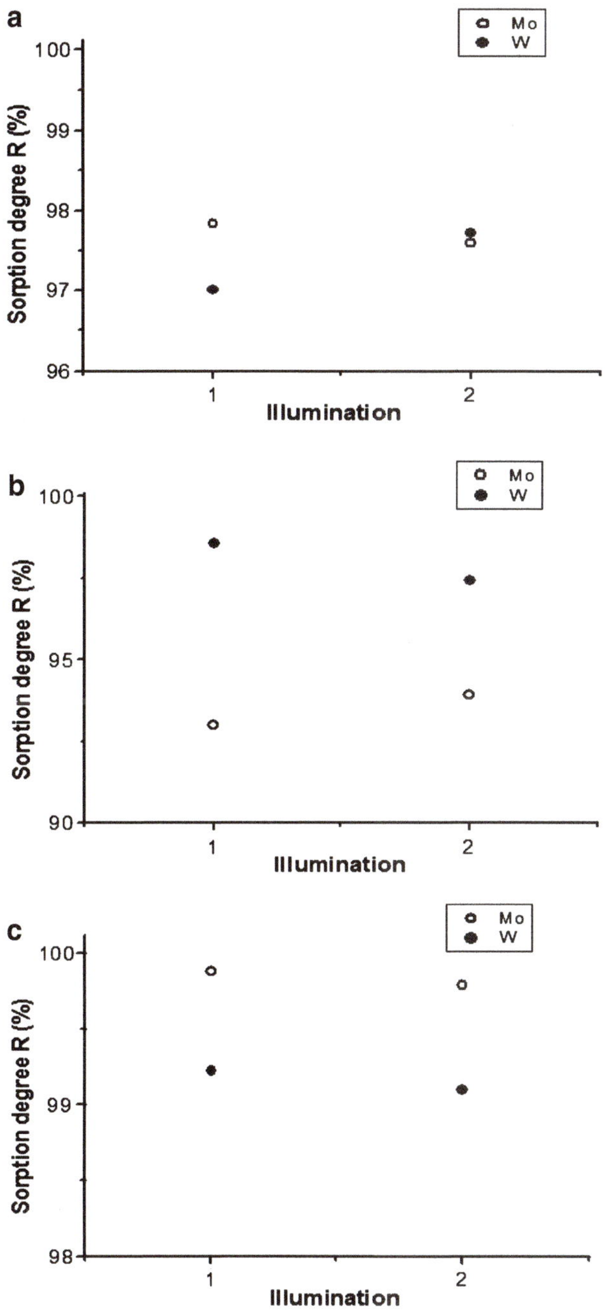

Fig. 9 The sorption degree (R, %) with *1* and without *2* illumination for different sorbents: anatase (**a**), Degussa P25 (**b**), Hombifine N (**c**)

Fig. 10 The sorption degree (*R*) vs. kinds of adsorbents and analytes

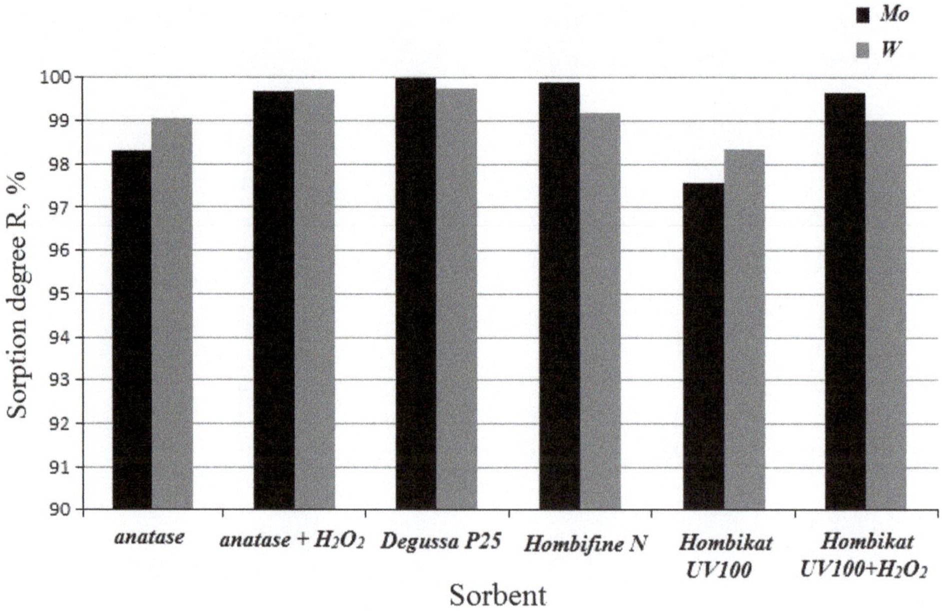

increase in the sorption degree associated with appearance of the O–O–H groups on the surface of particles.

On the other hand, nature of analytes affects the sorption degree. The nanoparticle surface of the studied samples contains cracks and pores of different scales which more or less deeply penetrate into the adsorbent. Their existence, especially ultrananopores, which size is comparable with a size of ions ($R_{Mo} = 0.69$ Å, $R_W = 0.60$ Å; ionic radius), considerably complicates the adsorption process. Large ions cannot penetrate into ultrananoporous of small sizes due to steric reasons; therefore, the ultraporous adsorbent possesses lower specific surface for such ions, as compared to smaller adsorbed ions.

The observed correlations between the *R* value and the sample illumination depend on the kind of analytes and the characteristics of adsorbents (specific surface area, volume of nano- and ultrananopores, nanoparticles and coherent scattering region sizes) (Fig. 9).

We have reached the maximal recovery efficiency of Mo(VI) ions sorption from aqueous solutions for the sample of commercial Hombifine N ($R = 99.90$ %; sorption conditions with the use of a membrane filter with 0.45 μm pores: $t = 20$ °C, $T_S = 21$ h, $m = 50$ mg), of Mo(VI) and W(VI) ions sorption for the sample of commercial Degussa P25 ($R_{Mo} = 99.99$ % and $R_W = 99.75$ %; sorption conditions with the use of centrifugation during 1 min: $t = 80$ °C, $T_S = 3$ h, $m = 50$ mg) and for nanosized anatase modified with H_2O_2 ($R = 99.75$ %; standard sorption conditions with the use of a membrane filter with 0.45 μm pores) (Fig. 10).

In general, the obtained experimental data show the role of sorption conditions, sorbent characteristics and analyte nature in the process of Mo(VI) and W(VI) extraction from aqueous solutions.

Conclusions

The adsorption of Mo(VI) and W(VI) ions in the form of associates of different nature on Degussa P25, Hombifine N, Hombikat UV100 and the synthesized samples with the anatase structure significantly depends on temperature, decrease in the 4–10 °C range and increase in the 10–85 °C range. This behavior indicates an exothermic and endothermic nature of the adsorption process in the given ranges, respectively. In addition, the sorption degree of Mo(VI) and W(VI) ions adsorption depends on the contact duration, sample weight, and kind treatment. The higher sorption degree correlates to the formation of the O–O–H groups on treatment of the samples with hydrogen peroxide.

As a result, we have found that all the samples with nano-sized anatase show high adsorption ability for Mo(VI) and W(VI) ions with $R \geq 99.75$ %. The similar properties were earlier obtained for As(III,V), Bi(V), V(V) (Kuzmicheva et al. 2012) and can be found for other ions. Due to its unique adsorption properties, nanosized anatase is a promising material for extraction of various ions from aqueous solutions.

Acknowledgments Authors would like to acknowledge financial support from Federal Targeted Programme (FTP) "Scientific and Scientific-Pedagogical Personnel of the Innovative Russia" (2012–2013) for supporting this work through the Grant No. 14.B37.21.1621.

References

Bleam WF, McBride MB (1986) The chemistry of adsorbed Cu(II) and Mn(II) in aqueous titanium dioxide. J Colloid Interface Sci 110:336–346

Brunauer S, Emmett PH, Teller E (1938) Adsorption of gases in multimolecular layers. J Am Chem Soc 60:309–319

Dadachov M (2006) Novel titanium dioxide, process of making and method of using same. US Pat. Appl. 0171877, 3 Aug 2006

Dadachov M (2006) Novel adsorbents and process of making and using same. US Pat. Appl. 0144793, 6 July 2006

Deedar N, Irfan A, Ishtiaq Q (2009) Evaluation of the adsorption potential of titanium dioxide nanoparticles for arsenic removal. J Environ Sci 21:402–408

Dzis'ko VA, Karnaukhov AP, Tarasova DV (1978) Physicochemical principles of oxide catalyst synthesis. Nauka, Novosibirsk (in Russian)

Gao Y, Wahi R, Kan AT, Falkner JC, Colvin VL, Tomson MB (2004) Adsorption of cadmium on anatase nanoparticles effect of crystal size and pH. Langmuir 20:9585–9593

Georgaka A, Spanos N (2010) Study of the Cu(II) removal from aqueous solutions. Glob NEST J 12:239–247

Gregg S, Sing K (1982) Adsorption, surface area, and porosity. Academic Press, New York

Igwe JC, Abia AA (2007) Studies on the effects of temperature and particle size on bioremediation of As(III) from aqueous solution using modified and unmodified coconut fiber. Glob J Environ Res 1:22–26

Kuzmicheva GM, Savinkina EV, Obolenskaya LN, Belogorokhova LI, Mavrin BN, Chernobrovkin MG, Belogorokhov AI (2010) Preparation, characterization and properties of nano-sized titanium dioxide modifications with the structures of anatase and η-TiO_2. Cryst Rep 55:866–871

Kuzmicheva GM, Natykan AA, Savinkina EV, Obolenskaya LN, Domoroshchina EN, Yakovenko AG (2012) A way of preparing adsorbent on the basis of nanosized titanium dioxide with anatase structure. RU Patent 2,463,252, 10 Oct 2012

Meng X, Dadachov M, Korfiatis GP, Christodoulatos C (2003) Methods of preparing a surface-activated and using same in water treatment processes. US Pat. Appl. 0155302, 23 August 2003

Obolenskaya LN, Kuzmicheva GM, Savinkina EV, Zhilkina AV, Sadovskaya NV, Prokudina NA, Chernyshev VV (2012) Influence of conditions of a sulphatic method on characteristics of samples with nanosized modification of anatase. Her RAS Chem 11:2032–2038

Savinkina EV, Kuz'micheva GM, Tabachkova NYu, Obolenskaya LN, Demina PA, Yakovenko AG (2011a) Synthesis and morphology of anatase and η-TiO2 nanoparticles. Inorg Mater 47:489–494

Savinkina EV, Kuzmicheva GM, Obolenskaya LN (2011b) Efficient synthesis and properties of η-titania. 2nd European Conference of Chemical Engineering, Puerto De La Cruz

Svergun DI (1992) Determination of the regularization parameter in indirect-transform methods using perceptual criteria. J Appl Crystallogr 25:495–503

Tella T, Panagiotou GD, Petsi T, Bouriras K, Kordulis C, Lycourghiotis A (2010) The mechanism of retention of vanadium oxo-species at the "titanium oxide/aqueous solution" interface. Glob NEST J 12:231–238

Valencia-Trejo E, Villicaña-Méndez M, Alfaro-Cuevas-Villanueva R, Garnica-Romo MG, Cortés-Martínez R (2010) Effect of Temperature on the Removal of Arsenate from Aqueous Solutions by Titanium Dioxide Nanoparticles. J Appl Sci Environ Sanit 5:171–184

Von Tytko KH, Baethe G, Hirschfeld ER, Mehmke K, Stellhorn D (1983) Über die Gleichgewichte in wäβrigen Polymolybdatlösungen Neuauswertung der potentiometrischen Meβdaten von Sasaki und Sillen. Z Anorg Allg Chem 5080:43–66

Xiao Y, Ling J, Qian Sh, Lin A, Zheng W, Xu W, Luo Yu, Zhang M (2007) Preconcentration of trace arsenite and arsenate with titanium dioxide nanoparticles and subsequent determination. Water Environ Res 79:1015–1022

Zagrafskaya RV, Karnaukhov AP, Fenelonov VB (1975) The globular model of porous bodies with corpuscular structure. III Study of random and partially ordered packed spheres. Kinet Catal 16:1583–1590

Zhilkina AV, Gordienko AA, Prokudina NA, Trusov LI, Kuz'micheva GM, Dulina NA, Savinkina EV (2013) Determination of the size of particles of highly dispersed materials by low temperature nitrogen adsorption. Russ J Phys Chem A 87:674–679

10

Preparation and characterization of nano structured Ba_2SnO_4 as a novel photocatalyst material for the chromate reduction

S. Omeiri · N. Allalou · G. Rekhila ·
Y. Bessekhouad · M. Trari

Abstract The photo-electrochemical properties of the layered perovskite Ba_2SnO_4, prepared by nitrate route are reported for the first time. The oxide exhibits a direct optical transition at 2.98 eV, attributed to the charge transfer O^{2-}: $2p \rightarrow Sn^{4+}$: $5s$ and possesses a chemical stability over the whole pH range. The Mott Schottky plot in acidic medium is characteristic of n type conductivity with a flat band potential E_{fb} of $-0.84\ V_{SCE}$ and an electron density N_D of 1.14×10^{17} cm^{-3}. The electrochemical impedance spectroscopy reveals the bulk and grain boundaries contributions. Ba_2SnO_4 is of interest for the environmental protection. At pH 2.5, its conduction band (CB, $-0.86\ V_{SCE}$) is below the $HCrO_4^-$ level, yielding a spontaneous reduction to Cr^{3+} under UV light. The reduction is optimized with respect to the Cr(VI) concentration. The photo reduction follows a pseudo-first-order kinetic and the data fit well the Langmuir–Hinshelwood (L–H) model with an apparent rate constant of 0.18 h^{-1} for an initial concentration of 50 ppm. The chromate reduction under solar light is studied under optimized conditions. In presence of oxalic acid, 73 % of initial concentration disappears after ~2 h of irradiation.

Keywords Perovskite · Ba_2SnO_4 · Photo-electrochemical · Chromate · Solar light

S. Omeiri · N. Allalou · G. Rekhila ·
Y. Bessekhouad · M. Trari (✉)
Laboratory of Storage and Valorization of Renewable Energies, Faculty of Chemistry, University of Science and Technology (USTHB), B P 32, El Alia, 16111 Algiers, Algeria
e-mail: solarchemistry@gmail.com

Introduction

Environment protection has become a field of high priority because of the large scale aquatic pollution (Karunakaran et al. 2009; Gherbi et al. 2011). Unlike organic molecules, heavy metals are non-biodegradable and various methods exist for their removal, including reverse osmosis, ion exchange, electrochemical-reduction and electro-dialysis (Paul et al. 2012). However, such methods are expensive and become inefficient at low concentrations and can consequently be used as post techniques. So, the photocatalysis is a clean and cheap process and is attractive to take the relay for the water decontamination (Kaur and Pal 2012); photocatalysis has emerged as a promising solution (Hernández-Gordillo et al. 2012; Puangpetch et al. 2008). In this regard, many semiconductor (SC) oxides have been used to convert inorganic pollutants into less harmful forms, the $CuFeO_2$-CB ($\sim -1\ V_{SCE}$) is more negative than the potential of many $M^{2+/0}$ couples which can be reduced spontaneously (Nasrallah et al. 2011; Omeiri et al. 2008). The CrO_4^{2-}/Cr^{3+} potential is too anodic with respect to $CuAlO_2$-BC level to be involved in a direct electron transfer. So, CdS is used as mediator and the new heterojunction $CuAlO_2$/CdS increases the photocatalytic efficiency. Moreover, salicylic acid as hole scavenger gives the high performance among hole scavengers, and $CuAlO_2$ approaches 100 % photostability at pH 7.5 (Brahimi et al. 2012). The applications of the photocatalysis are numerous and it is the environmental aspect that we develop through the reduction of chromate, recognized as a worldwide problem (World Health Organization International Agency for Research on Cancer 1990).

Solar energy is attractive because of its abundance, inexhaustibility and consistency. With an incident power of more than 1 kW m^{-2} (AM 1) at the sea level, the solar energy exceeds by far the energy the mankind needs. Chromate comes from industrial activities like tannery, batteries and microelectronic and is discharged in water at concentrations up to 500 ppm. It moves through the food chain and bio-accumulates in living organisms. The WHO has drastically limited the Cr(VI) concentration in water at 0.5 ppm and the effluents must be treated at the source.

On the other hand, a great interest has been paid to the elaboration of optically active materials, the key element in photocatalysis (Köferstein and Yakuphanoglu 2010;Noori et al. 2012). Oxides are privileged for photo-electrochemical (PEC) applications because of their chemical stability. In this respect, the stannates are expected to be good photocatalysts since they have CB of Sn-5*s* orbital ($\sim$4 eV below vacuum) capable to reduce any kind of inorganic ions (Omeiri et al. 2010). Comparatively, it has been reported by Fang et al. (2013) that the light-induced vapor generation at liquid-immersed metallic nanoparticles significantly enhances the photo activity. Interestingly, by increasing the incident light flux or lowering the inter-particle separation, the authors noticed the micrometer-sized bubbles resulting from the coalescence of nanoparticle-"bound" vapor envelopes.

To our knowledge, there are yet no papers on the photocatalytic activity of layered perovskites A_2SnO_4 (A = alkaline earth) and the PEC study has not been investigated before now. Ba_2SnO_4 has environmentally friendly characteristics, in addition to be low cost and chemically stable over the whole pH range (Hinatsu and Tezuka 1998). The conduction band must be positioned at energy as high as possible and barium contributes to decrease the electro affinity of the stannate. We are interesting by Ba_2SnO_4 because barium permits to increase the electro affinity leading to a large band bending, an attractive property for *n* type specimen. However, Ba_2SnO_4 is a low polaron semiconductor and the crystallite size plays a crucial role due to the increasing number of photocatalytic sites. Because of inherent defects in polycrystalline materials and low carrier mobility (Zidi et al. 2010), a large part of electron/hole (e^-/h^+) pairs are lost in the inter-grain regions, thus resulting in a poor photo activity. Therefore, we have prepared the oxide by the chemical method which has succeeded in the synthesis of homogenous powder (Wu et al. 2004) and the crystallite size becomes comparable with the diffusion length of the minority carriers. The present work deals with the chromate reduction upon UV light onto Ba_2SnO_4. The oxygen over-voltage is high on the stannates; the level O_2/H_2O is located far above the valence band and oxygen does not evolve on Ba_2SnO_4. So, the reduction is achieved in presence of oxalic acid as hole scavenger which promotes the charges separation and protects the oxide against photo corrosion (Lahmar et al. 2012; Iliev et al. 2010).

Experimental

Ba_2SnO_4 is synthesized by nitrate route. Sn (extra pure) is chemically polished with HCl (2 N), rinsed with water and dissolved in HNO_3 (9 N); $BaCO_3$ (Merck 99.9 %, dried at 250 °C) is added in stoichiometric amount and the solution is evaporated on the sand bath. The powder is ground in an agate mortar, pressed into circular discs (ϕ = 13 mm, thickness 0.5 mm) under 5 tons/cm^2 and heated at 1,100 °C; the density approximates 80 %. The completion of the reaction is checked by X-ray diffraction (XRD) using Cu Kα anticathode (λ = 0.154178 nm). The diffuse reflectance spectrum is plotted with a UV–Visible–NIR spectrophotometer (Specord plus).

The point of zero charge (pzc) is determined by measuring the equilibrium pH of an aqueous solution containing a suspension of finely powdered oxide. The sample has to be *n* type in order to be used as photoanode and this is achieved by heating the Ba_2SnO_4 pellet at 600 °C under dynamic vacuum (<1 mbar). A copper wire is fixed on the back pellet with silver cement. The pellet is isolated with epoxy resin so that the outside surface (0.70 cm^2) is exposed to the solution. Electrochemical measurements are conducted in the working solution ($HCrO_4^-$, pH 2.5) using a standard cell. A platinum foil serves as auxiliary electrode and all potentials are given with respect to a saturated calomel electrode (SCE) and controlled with a PGZ301 potentiostat (Radiometer). The frequency is set up at 10 kHz for the capacitance measurement. The electrochemical impedance spectroscopy (EIS) is measured at the open circuit potential (OCP = +0.236 V) by using small amplitude sine wave with a frequency analyzer in the range (10^{-2}–10^5 Hz).

Glassware flasks are immersed in sulfo-chromic acid overnight and rinsed with water. The stock solution (300 ppm) is prepared from $K_2Cr_2O_7$ (Merck, extra pure) in water. The concentrations down to 25 ppm are made up by dilution. The photocatalytic tests are done in a double walled Pyrex reactor containing 0.1 g of the catalyst powder in 100 cm^3 of $HCrO_4^-$ solution at variable concentrations, acidified by H_2SO_4 (pH 2.5). The chronopotentiometric profile is recorded in the $Ba_2SnO_4/HCrO_4^-$ suspension, the potential between the Pt electrode (Tacussel) and SCE is measured with a high impedance voltammeter (Tacussel 9A). The powder is maintained in the dark (4 h) to reach the adsorption equilibrium. The temperature is regulated at 25 °C with a thermo stated bath and the experiments are performed in air-equilibrated suspension. Then, the light is switched on, UV lamp (Phywe) serves as illumination source. The precipitate is allowed to deposit in the bottom of the reactor; the solution is centrifuged (3,000 rpm, 5 min) and the residual Cr(IV) concentration is measured with the same spectrophotometer using a 1 cm quartz cell. The efficiency of Cr(VI) photo reduction is given by:

$$\% = (A_o - A) \times 100/A_o$$

A_o and A are, respectively, the absorbance of $HCrO_4^-$ solution before and after illumination at saturation (~2 h). The solar experiments are performed on a sunny day in unstirred Pyrex reactor containing the chromate solution. The reactor is exposed to sunlight and the temperature reaches 33 °C at the end of the test. Distilled water (~0.8 MΩ cm) is used in all experiments.

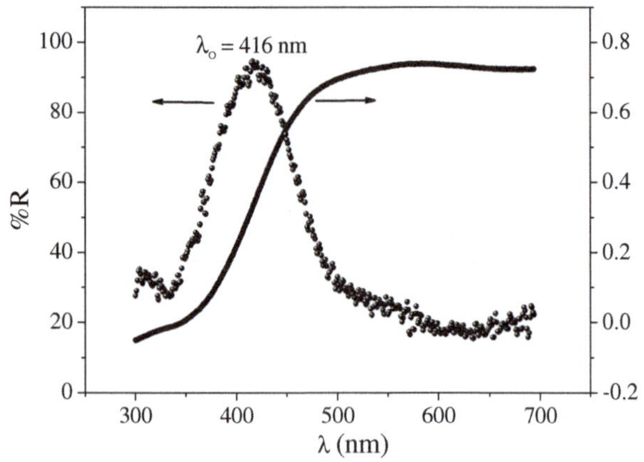

Fig. 2 The diffuse reflectance spectrum of Ba_2SnO_4 and its derivative curve

Results and discussion

The synthesis of oxides by solid state reactions requires long time of heat treatment and high temperatures leading to irregular crystallite sizes with low active surface. By contrast, the chemical route yields homogenous powder with narrow size distribution (Čerňanský et al. 2002). The XRD pattern (Fig. 1) is characteristic of single phase; all peaks are indexed in the perovskite structure with a quadratic symmetry (SG: I4/mmm, JCPDS) in agreement with the JCPDS card N° 40–0010. The crystallite size (D ~45 nm) is evaluated from the full width at half maximum (= $K\lambda/\beta$ cos), where $K = 0.94$ is the shape factor and β (rd) the width of the intense XRD peak (110). Assuming spherical and non-porous crystallites, a specific surface area of ~20 $m^2\ g^{-1}$ (= $6/\rho D$) is obtained, the experimental density ($\rho = 6.40$ g cm^{-3}) determined by hydrostatic method in toluene.

The optical properties are important in photo catalysis and the diffuse reflectance is used to get the gap (E_g); an accurate E_g value of 2.98 eV ($\lambda_o = 416$ nm) is obtained from the derivative curve (Fig. 2), which corresponds to the charge transfer O^{2-}: $2p \rightarrow Sn^{4+}$: $5s$. The transport properties of the stannates are altered by oxygen deficiency (Trari et al. 1994). The system is electrically neutral and the formation of oxygen vacancies, generated by heating under vacuum, is achieved concomitantly with reduction of Sn^{4+}:

$$Ba_2SnO_4 \rightarrow Ba_2SnO_{4-\delta} + 2\delta[e^-(CB)] + \delta/2\ O_2 \quad (1)$$

The conduction occurs by electron hopping between mixed valences $Sn^{4+/2+}$ accommodated in edge sharing octahedra. Ba_2SnO_4 is chemically stable over the whole pH range even in strong acid solutions. The intensity–potential J (V) curve in acidic medium exhibits a plateau region with a high oxygen over-voltage. The dark current is correlated to the space charge region and results from the weak electrons density (see below). The current becomes quickly large at potentials negative of ~0.4 V due to hydrogen

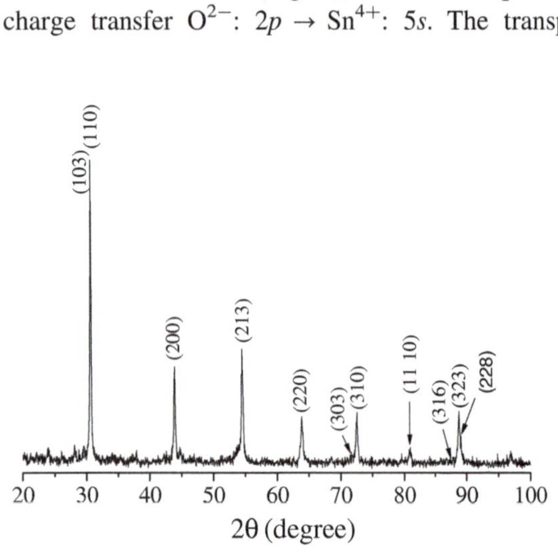

Fig. 1 XRD pattern of Ba_2SnO_4 synthesized by nitrate route at 1100 °C

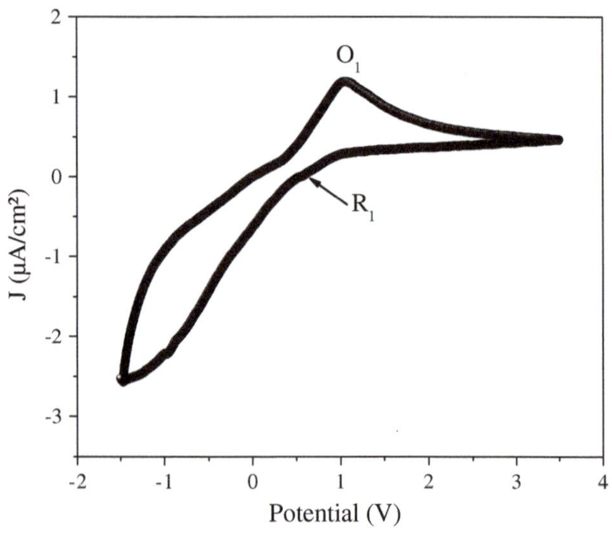

Fig. 3 Cyclic voltammogram of Ba_2SnO_4 in $HCrO_4^-$ solution (100 ppm, pH 2.5), scan rate 5 mV/s

evolution. The cathodic peak (*R*) and the corresponding anodic peak (O), separated by ~1.3 V are due to the internal redox process $Sn^{4+/2+}$ and not to $HCrO_4^-/Cr^{3+}$ couple since the peak persists in a chromate free solution (Fig. 3). The position of the electronic bands is determined from the variation of the interfacial capacitance with the applied potential (*V*):

$$\frac{1}{C^2} = \left(\frac{2}{e\varepsilon\varepsilon_0 N_D}\right)(V - V_{fb}) \quad (2)$$

where ε_o is the permittivity of vacuum and *e* the electronic charge. The intercept of the straight line with the potential axis (Fig. 4) gives a flat band potential (V_{fb}) of −0.84 V while the positive slope lends a further support of *n* type conductivity. The permittivity of the material (ε–5) is determined from the dielectric measurement at 10 kHz. The linear part (C^{-2}–*V*) indicates a constant density (N_D) whose value (1.14×10^{17} cm^{-3}) characterizes a lightly doped semiconductor with a wide depletion width (50 nm):

$$W = \left(\frac{2\varepsilon\varepsilon_0 \Delta V}{eN_D}\right)^{0.5} \quad (3)$$

($V - V_{fb}$ ~ 0.35 V) represents the optimal band bending at the interface, required for an efficient charges separation. The length *W* extends over many crystallographic units cells and this is advantageous in photocatalysis, since an increasing number of (e^-/h^+) pairs are separated by the junction electric field which contribute to the photoactivity. The potential V_{fb} gives the energetic position of Ba_2SnO_4-CB with respect to vacuum:

$$P = 4.75 + e\,V_{fb} - E_a \quad (4)$$

The *P* value (−0.86 V/−3.89 eV) indicates that CB is primarily made of Sn-5*s* orbital conferring to Ba_2SnO_4 a high reducing power under UV light. Ba_2SnO_4 is an ionic compound with a white colour and VB is made up of O^{2-}: 2*p* orbital, located at −6.87 eV ($P - E_g$).

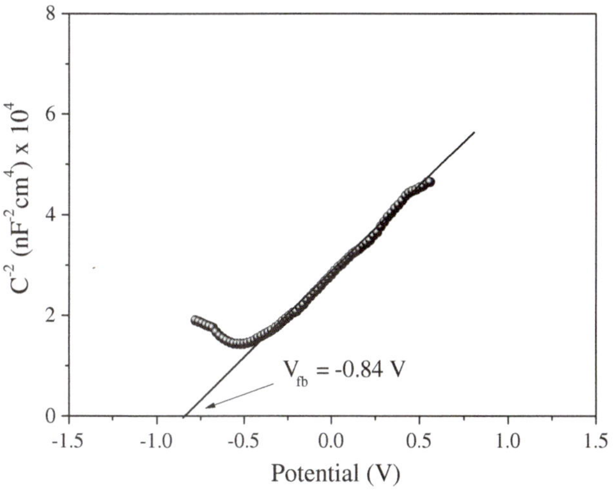

Fig. 4 The Mott–Schottky plot of Ba_2SnO_4 in the working solution ($HCrO_4^-$, 100 ppm, pH 2.5)

Photocatalysis

The solar energy continues to meet a growing demand and the photocatalysis is increasingly deployed to manage the environmental protection (Ketir et al. 2008). In view of the expensive techniques for the aquatic remediation, the exploitation of sunlight becomes highly recommended (Khataee and Kasiri 2010). Ba_2SnO_4 fills the photocatalytic requirements since it is non-toxic and chemically stable. However, it is responsive to UV light but the advantage resides in the high energy of the conduction band, able to reduce inorganic pollutant and the chromate reduction is used as test reaction. It is worthwhile to outline that the annual sunshine duration averages 12 h/day (Sahara) and more than 50 W m^{-2} of UV light is of interest for Ba_2SnO_4.

The stannates are small polaron oxides with low electron mobility (5.2×10^{-2} cm^2 V^{-1} s^{-1}).[1] Hence, the photocatalytic performance should be improved for small crystallite size (L), which must be comparable with the electron diffusion length and this can be achieved by preparing the oxide by nitrate route. The charge transfer is governed by the difference between Ba_2SnO_4-CB and $HCrO_4^-$ level. The further advantage is that Ba_2SnO_4-CB varies with pH (−0.06 V pH^{-1}) but less than $HCrO_4^-$ (0.14 V pH^{-1}) and this property has been exploited. The difference increases with decreasing pH and approaches the optimal band bending at pH 2.5, leading to an efficient separation of (e^-/h^+) pairs. With an acidity constant (pK_a) of 1.69, H_2CrO_4 predominates at low pH_s and converts to $HCrO_4^-$ above pH 1.7. The heterogeneous transfer requires a detailed knowledge of the interfacial chemistry through the junction as well as a judicious selection of the reducing agent to prevent the holes accumulation. Oxalic acid (OA) being favourable and its mineralization occurs via holes valence band process:

$$HCrO_4^- + 1.5\ C_2H_2O_4 + 4\ H_3O^+ \leftrightarrow Cr^{3+} + 3\ CO_2 + 8\ H_2O \quad (5)$$

The holes are consumed by OA, resulting in prolonged lifetime of carriers. The reactions occurring at the interface are the following:

$$Ba_2SnO_4 + h\nu \rightarrow h^+_{VB} + e^-_{CB} \quad (6)$$

$$HCrO_4^- + 7\ H_3O^+ + 3\ e^- \rightarrow Cr^{3+} + 11\ H_2O \quad (7)$$

$$3\ C_2H_2O_4 + 0.5\ O_2 + 4\ h^+ \rightarrow 6\ CO_2 + H_2O + 4\ H^+ \quad (8)$$

[1] Calculated from the $\sigma = e\,\mu_e\,N_D$.

$$2\,H_2O + 2e^- \rightarrow H_2 + 2\,OH^- \qquad (9)$$

The surface adsorption has a direct effect on the photoactivity, via the increasing number of the photocatalytic sites. The attachment of chromate on the catalyst powder makes easier the electron transfer. Therefore, the photocatalysis involves a dark adsorption below pzzp (= 12.7) and the enhanced activity is due simply to adsorbed $HCrO_4^-$ on the catalyst powder by electrostatic forces. Since the powder cannot be polarized, the additional criterion for the photocatalysis is that the free potential ($U_f = 0.041$ V) must be above the potential V_{fb} of n-Ba_2SnO_4 (−0.84 V). In the dark, the potential goes to cathodic values showing that the chromate adsorption takes place onto Ba_2SnO_4 and saturates at ~230 mn (Fig. 5). Then, the light is switched on and the potential decreases again indicating the reduction of $HCrO_4^-$ into Cr^{3+}. It is worth mentioning that the photolysis $HCrO_4^-$ (without catalyst) is weak; the concentration decreases by 2 % after 4 h irradiation and no redox reaction occurs between $HCrO_4^-$ and OA Hence, the performance of Ba_2SnO_4 is a light-induced process. However, the moderate photoactivity is attributed to the small exchange current density ($J_o \sim 200$ nA/cm^2) which is best though as the rate constant of electron transfer at zero potential.

The concentrations of chromate in industrial effluents can reach 300 ppm or more. Therefore, it is of interest to investigate the Cr(VI) reduction as a function of the initial concentration (C_o) in the range (25–300 ppm). The results show that the reduction decreases with increasing C_o (Fig. 6). At low concentrations the number of photocatalytic sites is more than the Cr(VI) ions and the photoactivity increases with raising C_o until saturation of all catalytic sites by $HCrO_4^-$ ion. The photons flux being constant, the chromate concentration increases and becomes greater than the number of photoelectrons, thus lowering the reduction process. The coloration of the solution also accounts for the regression in the photoactivity. Indeed, the absorption of $HCrO_4^-$ solution (50 ppm) is high over the visible region; ~30 % of light is absorbed for 1 cm path at 350 nm. It is well established that the photocatalysis is efficient at low concentrations. Therefore, we have selected a concentration of 50 ppm for the further experiments (Mekatel et al. 2012).

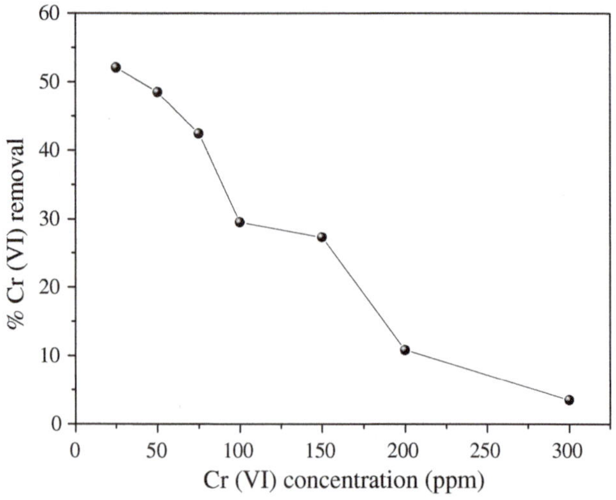

Fig. 6 The variation of the photo reduction with respect to the $HCrO_4^-$ concentration

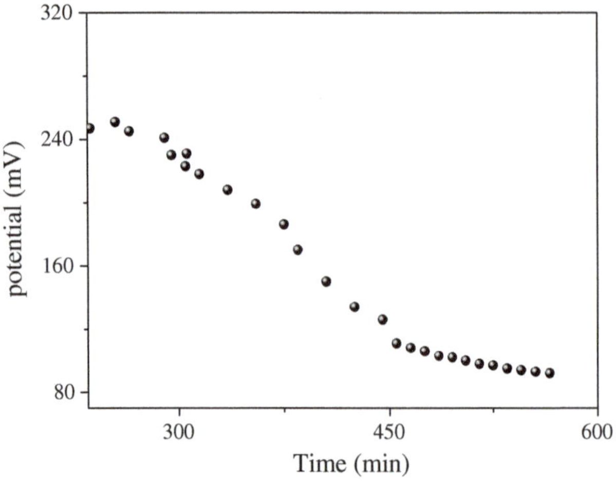

Fig. 5 The chrono-potentiometric profile of $HCrO_4^-$ solution (100 ppm) in Ba_2SnO_4 powder suspension both in the dark and under UV light

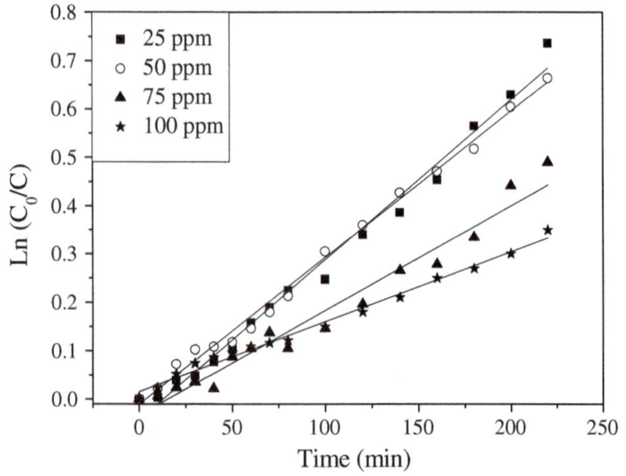

Fig. 7 Kinetics of the chromate photo reduction in presence of oxalic acid linear transform ln (C_o/C) as a function of time at different chromate concentrations (pH 2.5 $T = 25$ °C, and Ba_2SnO_4 dose = 1 mg/mL)

Table 1 The kinetic parameters of the chromate reduction according to the Langmuir–Hinshelwood

C_o (ppm)	25	50	75	100
K_{app} (min^{-1})	0.0033	0.0030	0.0021	0.0014
R^2	0.987	0.993	0.963	0.989
$t_{1/2}$	208.46	228.47	319.47	475.86

Kinetic analysis

The kinetic of the $HCrO_4^-$ reduction under illumination is given in Fig. 7. The reduction rate depends on the concentration C_o and is well described by a pseudo-first-order kinetic:

$$r = -\frac{dC}{dt} = k_{app}C \tag{10}$$

the integration gives:

$$\ln\left(\frac{C_0}{C}\right) = k_{app}t \tag{11}$$

k_{app} (mn^{-1}) is the apparent rate constant. The plot of ln (C_o/C) against the illumination time (t) for different concentrations C_o fits well the experimental data. The constants k_{app} and the correlation coefficients (R^2) are gathered in Table 1; k_{app}decrease with increasing C_o and the photo-reduction kinetic obeys to the Langmuir–Hinshelwood model (Poulios et al. 1998):

$$\frac{1}{k_{app}} = \frac{1}{k_r k_s} + \frac{C_o}{k_r} \tag{12}$$

The slope and the intercept of the linear plot of k_{app}^{-1} versus C_o (Fig. 8) give respectively the constants k_r (0.193 mg L^{-1} min^{-1}) and k_s (23.726 L mg^{-1}). The regression of the photoactivity may be due to the blocking of the catalytic sites by $Cr(OH)_3$ owing to the low solubility product ($K_s = 5.4 \times 10^{-31}$). Indeed, the color of Ba_2SnO_4 powder turns to green after the photocatalytic process, due to adsorbed hydroxide.

EIS is a powerful technique to differentiate the contributions of the grain boundaries, bulk and diffusion of the whole process. The spectrum of the junction Ba_2SnO_4/Cr(VI) solution is measured at the open circuit potential (+0.236 V). The occurrence of two semicircles in the Nyquist plot (Fig. 9) is due to the closeness of the time constants of the bulk and grain boundaries. The centers are located below the real axis due to constant phase element (CPE). CPE represents a deviation from the ideal capacitor, defined as $Z_{CPE} = [C(j\omega)^n]^{-1}$ where $-1 \le n \le 1$ and ω the angular frequency. The n value may be due to the non-uniform current distribution, the roughness of the electrode and defect states within the gap region. The bulk resistance ($R_b = 4.46$ kΩ cm^2) and grain boundaries resistance ($R_{gb} = 6.58$ kΩ cm^2) are due to the lack of the reversibility of the redox couple ($HCrO_4^-/Cr^{3+}$) over Ba_2SnO_4 i.e. a slow charge transfer of $HCrO_4^-$, thus corroborating the small exchange current density (J_o). The offset near the origin (1.42 kΩ cm^2) is due to the electrolyte resistance (R_{el}). The equivalent electrical circuit is established by the Zview Software (Fig. 9, inset), the absence of the Warburg diffusion implies that the overall process is kinetically governed by the electrons flow within the space charge region of Ba_2SnO_4. Indeed, the electrons move with low mobility in narrow Sn-5*s* orbital, not exceeding 2 eV (see above).

There is a great desire to extend the chromate reduction under sunlight. The experiments are conducted under direct

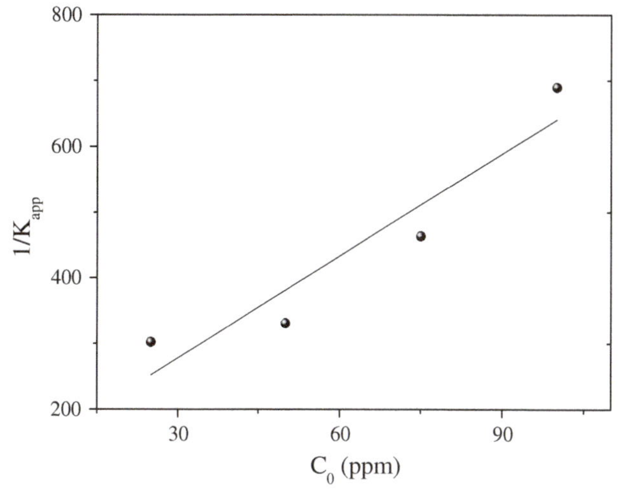

Fig. 8 The variation of $1/k_{app}$ as a function of the initial $HCrO_4^-$ concentration according to the Langmuir-Hinshelwood model. (pH 2.5, $T = 25$ °C and Ba_2SnO_4 dose = 1 mg/mL)

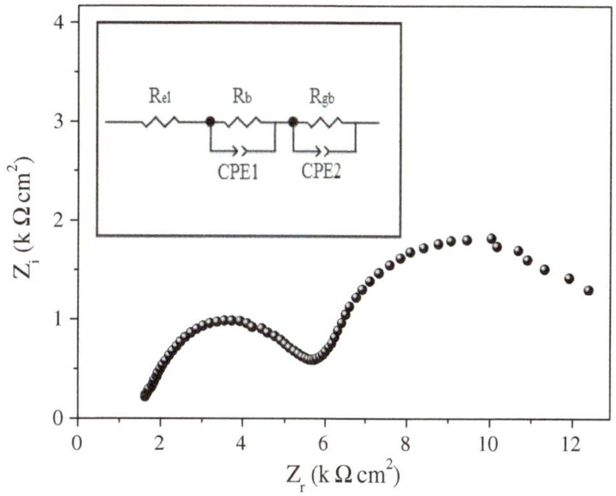

Fig. 9 The complex impedance plot of Ba_2SnO_4 in $HCrO_4^-$ solution (100 ppm) at pH 2.5. *Inset*: the equivalent electrical circuit

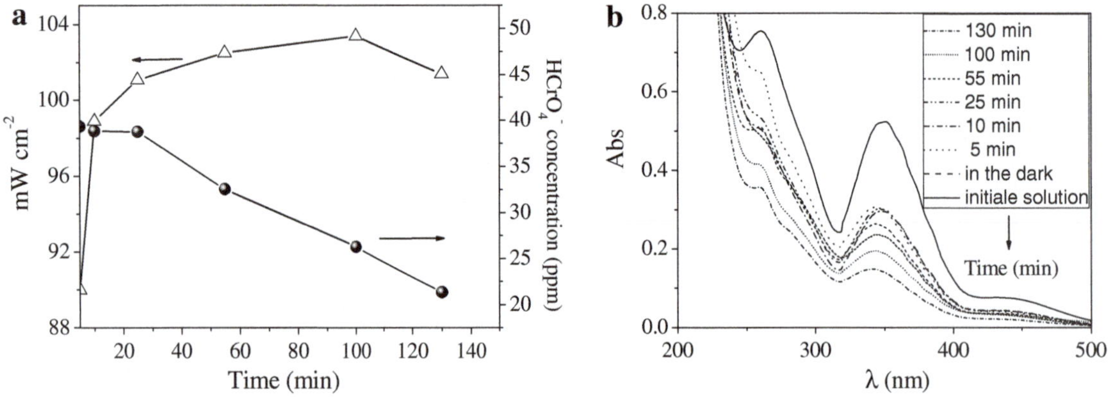

Fig. 10 **a** The effect of the light intensity on the Cr(VI) reduction. **b** The UV–Vis Spectra of the chromate solutions under sunlight on Ba_2SnO_4 over irradiation time

insulation and no correction being made from the light reflection. The light intensity varies slightly over time (Fig. 10a) and fluctuates between 90 and 103 mW cm^{-2} over the experiment duration. The temperature of the solution reached 33 °C at the end. The quantum yield (η) is defined as the number of $HCrO_4^-$ converted per incident photons[2] where $f(h\nu)$ is the photons flux density:

$$\eta = 3\{\text{number reduced of } HCrO_4^- / \text{ photons flux}\} \quad (13)$$

The number 3 is due to the fact that the reaction requires three electrons; η value of 0.13 % is obtained.

Despite its wide gap, the stannate shows an appreciable Cr(VI) photo reduction (Fig. 10b). However, the tendency to saturation over illumination time is ascribed to the competitive water reduction as observed in our previous work (Gherbi et al. 2011).

Conclusion

Cr(VI) is highly toxic and the ability of the perovskite Ba_2SnO_4 for its reduction to less harmful form is reported for the first time. Ba_2SnO_4 is safe, low cost and stable over the whole pH range. The nitrate route is employed for the preparation of homogeneous powder with fine particles. The oxide has been characterized structurally and photo-electrochemically. The conduction band is appropriately positioned with respect to the chromate level, allowing a spontaneous reduction under UV light. The low doping originating from small oxygen deficiency provides a large depletion layer with the occurrence of photo effect. The pH dependence of the flat band potential is judiciously exploited to have an optimal band bending at the interface. Concomitantly, the holes oxidize oxalic acid, thus inhibiting the charges accumulation. The best performance is achieved in acidic medium and the chromate follows a pseudo-first-order kinetic.

Acknowledgments The authors are thankful to Dr M. Izzerouken for X-ray diffraction and M. Fedailine for thermal analysis. This research is financially supported by the Faculty of Chemistry (US-THB, Algiers).

References

Brahimi R, Bessekhouad Y, Nasrallah N, Trari M (2012) Visible light CrO_4^{2-} reduction using the new $CuAlO_2$/CdS hetero-system. J Hazard Mater 219:19–25

Čerňanský M, Přemysl V, Karel K, Radmila K (2002) A study of $PbTiO_3$ crystallization in pure and composite nanopowders prepared by the sol–gel technique. Chem Monthly 133:799–806

Fang Z, Zhen YR, Neumann O, Polman A, García de Abajo FJ, Nordlander P, Halas NJ (2013) Evolution of light-induced vapor generation at a liquid-immersed metallic nano particle. Nano Lett 13:1736–1742

Gherbi R, Nasrallah N, Amrane A, Maachi R, Trari M (2011) Photocatalytic reduction of Cr(VI) on the new hetero-system $CuAl_2O_4/TiO_2$. J Hazard Mater 186:1124–1130

Hernández-Gordillo A, Tzompantzi F, Gómez R (2012) Enhanced photoreduction of Cr(VI) using $ZnS(en)_{0.5}$ hybrid semiconductor. Catal Commun 19:51–55

Hinatsu Y, Tezuka K (1998) Electron paramagnetic resonance study of Pr^{4+}ions doped in $BaSnO_3$, Ba_2SnO_4, and $Ba_3Sn_2O_7$. J Solid State Chem 138:329–333

Iliev V, Tomova D, Rakovsky S, Eliyas A, Li Puma G (2010) Enhancement of photocatalytic oxidation of oxalic acid by gold modified WO_3/TiO_2 photocatalysts under UV and visible light irradiation. J Mol Catal A Chem 327:51–57

Karunakaran C, Panneerselvam Sujatha M, Mathisankar P (2009) Photoreduction of chromium(VI) on ZrO_2 and ZnS surfaces. Chem Monthly 140:1269–1274

[2] $\eta = \frac{\int_{Eg}^{\infty} f(h\nu)\,d(h\nu)}{\int_{0}^{\infty} h\nu f(h\nu)\,d(h\nu)}$.

Kaur R, Pal B (2012) Size and shape dependent attachments of Au nanostructures to TiO_2 for optimum reactivity of Au–TiO_2 photocatalysis. J Mol Catal A Chem 355:39–43

Ketir W, Bouguelia A, Trari M (2008) Photocatalytic removal of M^{2+} (=Ni^{2+}, Cu^{2+}, Zn^{2+}, Cd^{2+}, Hg^{2+} and Ag^{+}) over new catalyst $CuCrO_2$. J Hazard Mater 158:257–263

Khataee AR, Kasiri MB (2010) Photocatalytic degradation of organic dyes in the presence of nanostructured titanium dioxide: influence of the chemical structure of dyes. J Mol Catal A Chem 328:8–26

Köferstein R, Yakuphanoglu F (2010) Semiconducting properties of Ge-doped $BaSnO_3$ ceramic. J Alloys Compd 506:678–682

Lahmar H, Kebir M, Nasrallah N, Trari M (2012) Photocatalytic reduction of Cr(VI) on the new hetero-system. $CuCr_2O_4$/ZnO. J Mol Catal A Chem 353:74–79

Mekatel H, Amokrane S, Bellal B, Trari M, Nibou D (2012) Photocatalytic reduction of Cr(VI) on nanosized Fe_2O_3 supported on natural Algerian clay: characteristics, kinetic and thermodynamic study. Chem Eng J 200:611–618

Nasrallah N, Kebir M, Koudri Z, Trari M (2011) Photocatalytic reduction of Cr(VI) on the novel hetero-system $CuFe_2O_4$/CdS. J Hazard Mater 185:1398–1404

Noori Sepehr M, Zarrabi M, Amrane A (2012) Removal of Cr(III) from model solutions by isolated *Aspergillus niger* and *Aspergillus oryzae* living microorganisms: equilibrium and kinetic studies. J Taiwan Inst Chem Eng 43:420–427

Omeiri S, Gabes Y, Bouguelia A, Trari M (2008) Photoelectrochemical characterization of the delafossite $CuFeO_2$: application to removal of divalent metals ions. J Electroanal Chem 6140:31–40

Omeiri S, Hadjarab B, Bouguelia A, Trari M (2010) Electrical, optical and photoelectrochemical properties of $BaSnO_{3-\delta}$: applications to hydrogen evolution. J Alloys Compd 505:592–597

Paul M, Pal N, Ali M, Bhaumik A (2012) New mesoporous silicotitaniumphosphate and its application in acid catalysis and adsorption of As(III/V), Cd(II) and Hg(II). J Mol Catal A Chem 330:49–55

Poulios I, Kositzi M, Kouras A (1998) Photocatalytic decomposition of triclopyr over aqueous semiconductor suspensions. J Photochem Photobio A Chem 115:175–183

Puangpetch T, Sreethawong T, Yoshikawa S, Chavadej S (2008) Synthesis and photocatalytic activity in methyl orange degradation of mesoporous-assembled $SrTiO_3$ nanocrystals prepared by sol–gel method with the aid of structure-directing surfactant. J Mol Catal A Chem 287:70–79

Trari M, Doumerc JP, Dordor P, Pouchard M, Behr G, Krabbes G (1994) Preparation and characterization of lanthanum doped $BaSnO_3$. J Phys Chem Solids 55:1239–1243

Wu R, Qu J, He H, Yu Y (2004) Removal of azo-dye acid red B (ARB) by adsorption and catalytic combustion using magnetic $CuFe_2O_4$ powder. Appl Catal B Environ 48:49–56

Zidi N, Omeiri S, Hadjarab B, Bouguelia A, Akroun A, Trari M (2010) Transport properties and photo electrochemical characterization of oxygen-deficient $ASnO_{3-\delta}$ (A = Ca, Sr and Ba). Phys B 405:3355–3359

11

Extracellular synthesis of silver nanoparticles using living peanut seedling

D. Raju · Nikita Paneliya · Urmil J. Mehta

Abstract Synthesis of nanoparticles by environment friendly method is an important aspect of nanotechnology. In the present study, extracellular reduction of silver ions to silver nanoparticles was carried out using living peanut plant. The electron microscopic analysis shows that the formed nanoparticles were of different shapes and sizes. The formed nanoparticles were polydispersed. The shapes of the nanoparticles were spherical, square, triangle, hexagonal and rod. Most of the particles were spherical and 56 nm in size. EDS analysis confirmed the formed nanoparticles were of silver. The crystalline nature of nanoparticles was confirmed by diffraction. This method opens up an exciting possibility of plant-based synthesis of other inorganic nanomaterials. This study confirms the synthesis of extracellular silver nanoparticles by living plant.

Keywords Silver · Nanoparticles · Synthesis · TEM · Shapes

Introduction

There is a greater need to develop reliable, green and eco-friendly efficient methods for the synthesis of nanoparticles. Researchers in the field of nanotechnology have been looking at biological systems for toxic-free synthesis of nanoparticles. The particles of size ranging between 1 and 100 nm are referred as nanoparticles (Willems and van den Wildenberg 2005; Simi and Abraham 2007). Nanoparticles have a higher surface area to volume ratio. The higher specific surface area is relevant to its properties such as antimicrobial activity in silver nanoparticles. Biological activity of silver nanoparticles increases with the increase in surface area (Willems and van den Wildenberg 2005).

The synthesis of nanoparticles by various chemical and physical methods have been employed (Ghorbani et al. 2011; Tien et al. 2008), but these have disadvantages because of the use of toxic chemicals and radiation. Biological synthesis of nanoparticles using microorganisms (Klaus et al. 1999; Konishi et al. 2007; Nair and Pradeep 2002), enzyme (Willner et al. 2006) and plants or plant extracts (Shankar et al. 2004; Raju et al. 2011, 2012, 2013a, b) has been suggested as possible alternative eco-friendly method.

Silver in ionic and nano form is mostly used in the medical industry. These are included in ointments to prevent infection of burns and open wounds (Becker 1999). The metal nanoparticles such as gold, and silver are widely used for human contacting areas, so there is a need to develop environmentally friendly methods of nanoparticle synthesis that do not use toxic chemicals.

There are reports on uptake of gold and silver nanoparticles by plant from agar medium and then transferred to shoot (Gardea-Torresdey et al. 2002, 2003). The formation of stable intracellular gold nanoparticles by *Sesbania* seedlings has been reported (Sharma et al. 2007). We have earlier reported the synthesis of extra- and intracellular gold nanoparticles by peanut and its callus cells (Raju et al. 2012, 2013a). Synthesis of nanoparticles by plant can be advantageous over other biological methods by eliminating the lengthy process of maintaining cell cultures (Shankar et al. 2004).

To screen the reduction potential of peanut seedling, we extended our process of synthesis of nanoparticles to other metal (silver) by living peanut seedling. We successfully

D. Raju · N. Paneliya · U. J. Mehta (✉)
CSIR-National Chemical Laboratory, Plant Tissue Culture Division, Pune 411008, Maharashtra, India
e-mail: uj.mehta@ncl.res.in

synthesized the silver nanoparticles by living peanut plant. To the best of our knowledge, there is no report on synthesis of extracellular silver nanoparticle by this plant.

Materials and methods

Arachis hypogaea (peanut cultivar SB-11) seeds were bought from local market. The seeds were washed under running tap water for 10 min, followed by repeated washings with distilled water. Later the seeds were treated with liquid detergent and 1 % bavistin for a period of 30 min, followed by 4 % savlon (v/v) for 12 min. The seeds were then washed thrice with distilled water and treated with 0.1 % $HgCl_2$ (w/v) for 10 min. Adhering $HgCl_2$ was removed by washing the seeds with sterile distilled water. The testa of seeds was removed aseptically, and the seeds were cultured on Whatman filter paper support in deionized water in test tubes. The pH of the deionized water was adjusted to 5.8 before autoclaving. The cultured seeds were incubated in dark till the emergence of radicle, after which they were incubated in 16 h photoperiod at 25 ± 2 °C.

Seed germination and synthesis of silver nanoparticles

After a period of 15 days, well-grown peanut seedlings were used for the synthesis of silver nanoparticles. Five milliliters of 1 mM $AgNO_3$ solution was added to autoclaved test tube under sterile conditions, to which the peanut seedlings were then transferred.

The silver nanoparticles were characterized by UV–vis spectra and transmission electron microscopy (TEM). Distribution of the particles of various sizes was determined using the Gatan software.

UV analysis

The UV analysis of silver nanoparticles formed by living peanut plants at room temperature was carried out. The samples were scanned from 300 to 800 nm wavelengths with a UV spectrophotometer using a dual beam operated at 1 nm resolution. The reaction mixtures were scanned under the same wavelengths after 24, 48 and 72 h.

Transmission electron microscopy

The silver nanoparticle images were recorded with TEM for determination of the shape and size along with their diffraction. Two to three drops of Ag nanoparticle solution were placed on carbon-coated copper grids and allowed to dry. TEM studies were performed on JOEL model 1200EX instrument operated at an accelerating voltage of 200 kV.

Results and discussion

Seed germination

On dark incubation, the emergence of radicals was observed after a period of 4 days in peanut. The germination frequency of peanut was 92.5 %. The well-grown seedlings were used for the synthesis of nanoparticles.

Synthesis of silver nanoparticles

Visible analysis

The well-grown 15-day-old peanut seedlings were taken and exposed to 0.1 mM $AgNO_3$ solution under sterile conditions. A control was also kept in which plant was exposed to sterile deionized water (Fig. 1). There was a change in colour of the solution where seedling was exposed to 1 mM $AgNO_3$ solution. The colourless solution of $AgNO_3$ changed to light yellow colour after a period of 24 h indicating the formation of nanoparticles. The change in the colour is due to the surface plasmon resonance of silver nanoparticles. Also in earlier report (Ramteke et al.

Fig. 1 Peanut seedling exposed to deionized water (control), peanut seedling exposed to 1 mM $AgNO_3$ solution for 24, 48, and 72 h. *Inset image* shows change in colour of the root of peanut seedling exposed to 1 mM $AgNO_3$ solution. *C* control, *E* exposed

2013), it has been demonstrated that the change to brown colour is due to surface plasmon resonance of silver nanoparticles. The intensity of the colour increased from 24 to 48 h (Fig. 1). After 48 h there was reduction in the colour of the solution, which can be observed in 72 h reaction test tube (Fig. 1).

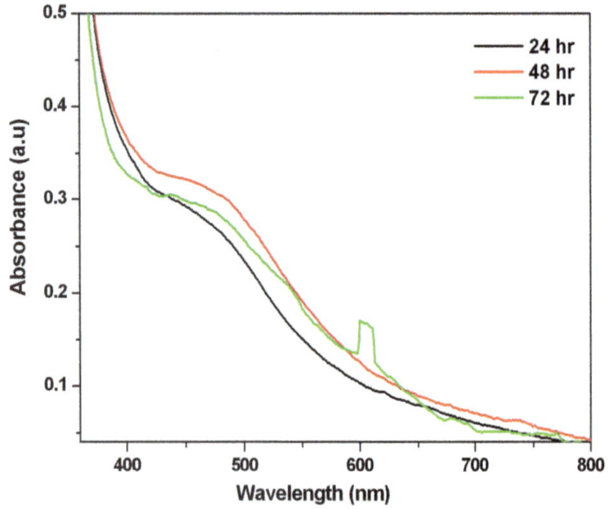

Fig. 2 UV absorbance spectra of 1 mM $AgNO_3$ solution to which the peanut seedling was exposed for different time intervals

After the exposure of the peanut seedlings to $AgNO_3$ solution we observed morphological changes, where no change in the colour of root of control plant could be observed (Fig. 1 inset C), while the root of exposed plant to $AgNO_3$ solution changed to brown colour indicating the synthesis/uptake of silver nanoparticles by the peanut seedling (Fig. 1 inset E). Here, we show the change in the colour of the root to light brown colour. In earlier reports in *A. hypogaea* (peanut) (Raju et al. 2012) and *Sesbania* too, the colour of roots changed to purple when they synthesized intracellular gold nanoparticles and also confirmed the presence of intracellular nanoparticles (Sharma et al. 2007).

UV–visible analysis

After a period of 24 h there was a change in the colour of the reaction solution. An aliquot of 1 ml of light yellow coloured solution was taken aseptically and scanned under UV from 300 to 900 nm. A peak was observed at

Fig. 3 **a, b** TEM images at different magnifications of extracellular silver nanoparticles synthesized by peanut plant, **c** diffraction analysis, and **d** EDS analysis

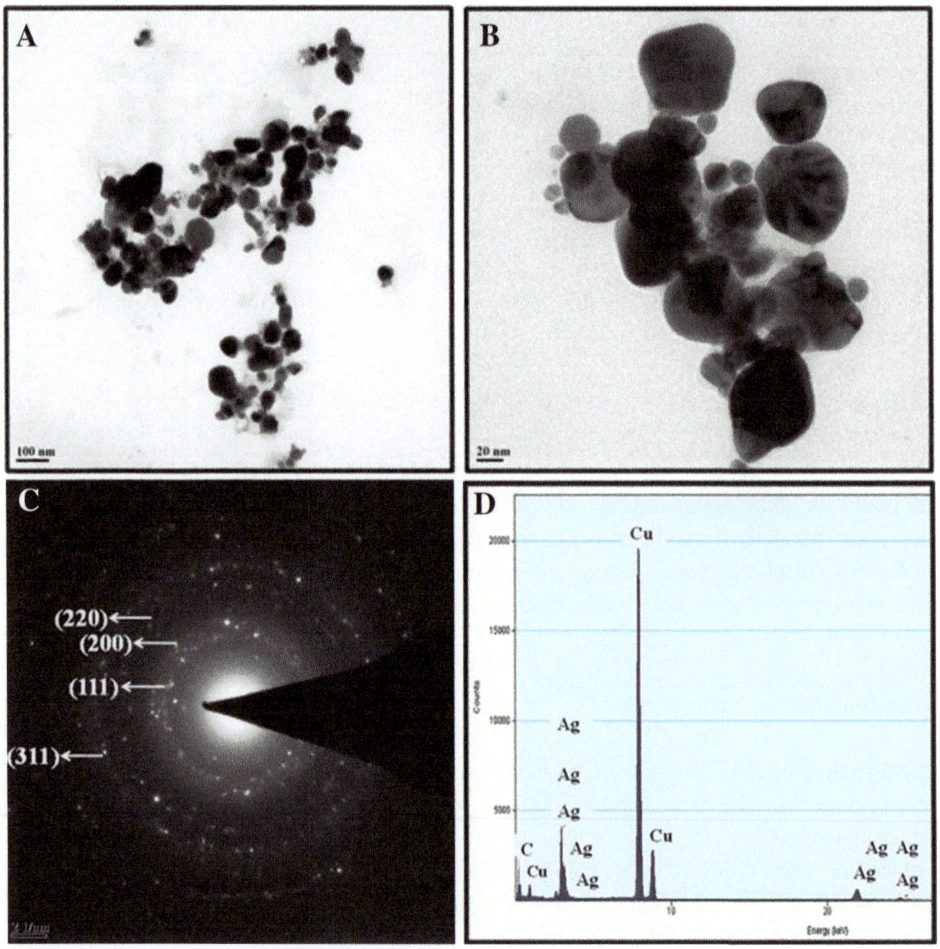

410–450 nm. It indicates the formation of silver nanoparticles (Raja et al. 2012). The UV peak around 600 nm has originated from the hexagonal shaped nanoparticles. This peak could be due to longitudinal oscillation of electrons which can be shifted even up to 1,000 nm in the near-IR region when the ratio of the hexagonal shaped nanoparticles increases (Chen and Carroll 2004). The intensity of peak increased at 48 h and decreased at 72 h (Fig. 2). This could be possibly due to the peanut seedling having uptaken silver nanoparticles from the external solution. However, with these results we cannot conclude exact mechanism for the decrease in intensity of UV absorbance and colour of the reaction solution. A study needs to be conducted for the decreased UV absorbance and colour of the solution.

Transmission electron microscopy (TEM)

The electron microscopy analysis shows that the formed nanoparticles were of different shapes and sizes. They were polydispersed in nature. The silver nanoparticles were well separated and there was no agglomeration. The shapes of nanoparticles were circular, square, hexagonal, triangle and rod (Fig. 3a, b). The selected area electron diffraction (SAED) pattern obtained from silver nanoparticles synthesized by peanut is shown in Fig. 3c. The Scherrer ring pattern characteristic of face centered cubic (fcc) of silver nanoparticles is observed, which confirms the formed nanostructures are crystalline in nature (Fig. 3c). The rings arise due to reflections from (111), (200), (220) and (311) lattice planes of fcc of silver nanoparticles, respectively. Similar SAED pattern was also obtained with silver nanoparticles synthesized using *Aloe vera* extract by Chandran et al. (2006). The EDS attached with TEM shows the peaks of carbon (C), oxygen (O), copper (Cu), silver (Ag). The peak of carbon could be due to biomolecules capped onto the silver nanoparticles (Chandran et al. 2006). Oxygen peak is from the chamber of EDS, while the copper peak is from the grid used for the TEM analysis. The silver peak is from the nanoparticles synthesized by the seedling. The EDS analysis confirms the formed nanoparticles are of silver (Fig. 3d).

Shapes and sizes of nanoparticles

The shapes of the silver nanoparticles were also studied. We could observe different shapes of nanoparticles. Most of the particles were spherical in shape being 68 %, while triangles were 6 %, squares 8 %, hexagonal 8 % and rods 10 % (Fig. 4a). The formation of different shapes of silver nanoparticles at room temperature must be due to the biomolecules secreted by the peanut seedlings. However, with the present results we cannot come to a conclusion for the formation of different shapes of silver nanoparticles.

The distribution of size of nanoparticles was also studied using Gatan software. The size was calculated by measuring 100 numbers of nanoparticles present in the TEM image. The particle size distribution study shows the silver nanoparticles ranged from 30 to 100 nm, most of the particles are 56 nm in size (Fig. 4b).

In previous report, the synthesis of gold nanoparticles with living peanut seedling has been demonstrated, where it was suggested that some of the proteins and phenolics leached from the roots helped in the formation of GNPs (Raju et al. 2012). Possibly these factors might have played role in the reduction of silver nanoparticles also. Further experiments will demonstrate the biomolecules secreted by

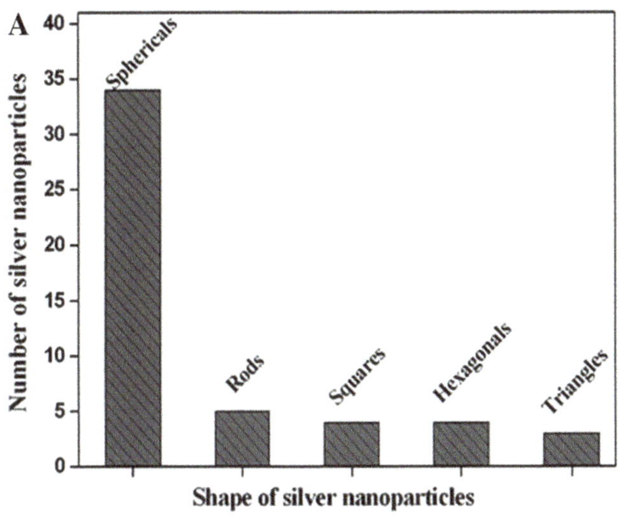

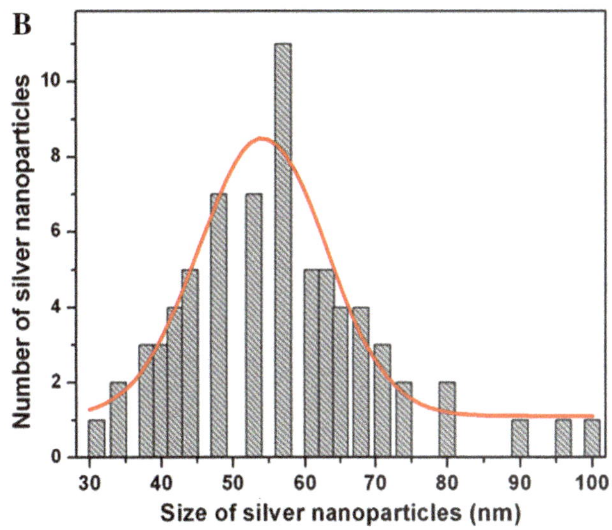

Fig. 4 **a** Various shapes of silver nanoparticles synthesized by peanut seedling, **b** silver nanoparticles size distribution analysis

the peanut seedling responsible for the formation of different shapes of silver nanoparticles.

Conclusion

The development of new methods for synthesis of nanoparticles in respect to economical and environmental challenges is important. There is a need to develop new green methods for synthesis of nanoparticles, to reduce or eliminate the use and generation of hazardous chemicals. We report the green synthesis of extracellular silver nanoparticles with different shapes and sizes using live peanut seedling. The formation of different shapes and sizes could be due to different biomolecules secreted by the seedling. The identification of biomolecules responsible for the formation of silver nanoparticles is under progress. Thus, peanut seedling was found to have the potential ability of synthesizing gold and silver nanoparticles, which can be extended to the study with other metals.

Acknowledgments Authors are grateful to the Council of Scientific and Industrial Research (CSIR), New Delhi for financial support. The authors also thank the Center for Material Characterization, National Chemical Laboratory for TEM analysis.

References

Becker RO (1999) Silver ions in the treatment of local infections. Met Based Drugs 6:311–314

Chandran SP, Chaudhary M, Pasricha R, Ahmad A, Sastry M (2006) Synthesis of gold nanotriangles and silver nanoparticles using *Aloe vera* plant extract. Biotechnol Prog 22:577–583

Chen S, Carroll DL (2004) Silver nanoplates: size control in two dimensions and formation mechanisms. J Phys Chem B 108:5500–5506

Gardea-Torresdey JL, Parsons JG, Gomez E, Peralta-Videa J, Troiani HE, Santiago P, Yacaman MJ (2002) Formation and growth of Au nanoparticles inside live *alfalfa* plants. Nano Lett 2:397–401

Gardea-Torresdey JL, Gomez E, Peralta-Videa JR, Parsons JG, Troiani H, Jose-Yacaman M (2003) *Alfalfa* sprouts: a natural source for the synthesis of silver nanoparticles. Langmuir 19:1357–1361

Ghorbani HR, Safekordi AA, Attar H, Rezayat Sorkhabadib SM (2011) Biological and non-biological methods for silver nanoparticles synthesis. Chem Biochem Eng Q 25(3):317–326

Klaus T, Joerger R, Olsson E, Granqvist CG (1999) Silver-based crystalline nanoparticles, microbially fabricated. Proc Natl Acad Sci USA 96:13611–13614

Konishi Y, Ohno K, Saitoh N, Nomura T, Nagamine S, Hishida H, Takahashi Y, Uruga T (2007) Bioreductive deposition of platinum nanoparticles on the bacterium *Shewanella algae*. J Biotechnol 128:648–653

Nair B, Pradeep T (2002) Coalescence of nanoclusters and formation of submicron crystallites assisted by *Lactobacillus* strains. Cryst Growth Des 2:293–298

Raja K, Saravanakumar A, Vijayakumar R (2012) Efficient synthesis of silver nanoparticles from *Prosopis juliflora* leaf extract and its antimicrobial activity using sewage. Spectrochim Acta A 97:490–494

Raju D, Mehta UJ, Hazra S (2011) Synthesis of gold nanoparticles by various leaf fractions of *Semecarpus anacardium* L. tree. Trees-Struct Funct 25:145–151

Raju D, Mehta UJ, Ahmad A (2012) Phytosynthesis of intracellular and extracellular gold nanoparticles by living peanut plant (*Arachis hypogaea* L.). Biotechnol Appl Biochem 59:471–478

Raju D, Mehta UJ, Ahmad A (2013a) Biosynthesis of highly stable intra and extracellular gold nanoparticles by using live peanut *(Arachis hypogaea)* callus cells. Curr Nanosci 9:107–112

Raju D, Mehta UJ, Hazra S (2013b) Phytosynthesis of silver nanoparticles by *Semecarpus anacardium* L. Mater Lett 102–103:5–7

Ramteke C, Chakrabarti T, Sarangi BK, Pandey R (2013) Synthesis of silver nanoparticles from the aqueous extract of leaves of *Ocimum sanctum* for enhanced antibacterial activity. J Chem Artic ID 278925:1–7

Shankar SS, Rai A, Ahmad A, Sastry M (2004) Rapid synthesis of Au, Ag, and bimetallic Au core Ag shell nanoparticles using Neem (*Azadirachta indica*) leaf broth. J Colloid Interface Sci 275:496–502

Sharma NC, Sahi SV, Nath S, Parsons JG, Gardea-Torresdey JL, Pal T (2007) Synthesis of plant-mediated gold nanoparticles and catalytic role of biomatrix-embedded nanomaterials. Environ Sci Technol 41:5137–5142

Simi CK, Abraham TE (2007) Hydrophobic grafted and crosslinked starch nanoparticles for drug delivery. Bioprocess Biosyst Eng 30:173–180

Tien DC, Liao CY, Huang JC, Tseng KH, Lung JK, Tsung TT, Kao WS, Tsai TH, Cheng TW, Yu BS, Lin HM, Stobinski L (2008) Novel technique for preparing a nano-silver water suspension by the arc-discharge method. Rev Adv Mater Sci 18:750–756

Willems, van den Wildenberg (2005) Roadmap report on nanoparticles. W&W Espana sl, Barcelona

Willner I, Baron R, Willner B (2006) Growing metal nanoparticles by enzymes. Adv Mater 18:1109–1120

12

Studies of graphene–chitosan interactions and analysis of the bioadsorption of glucose and cholesterol

E. Chigo Anota · A. Torres Soto · Gregorio H. Cocoletzi

Abstract Molecular simulations were performed to investigate the structural and electronic properties of graphene (G) nanosheet interacting with the monomer of chitosan (MCh) ($C_6H_{13}O_5N$). The G nanosheet with the $C_{54}H_{18}$ chemical composition is modeled according to the armchair edge and is functionalized with boron atoms. The interaction between the nanosheet and the MCh is investigated to search for better bio-sensing characteristics. Simulations are done within the density functional theory, the generalized gradient approximation is applied to deal with the exchange–correlation energies, and the *all-electron* basis set with double polarization is used. To determine the structure stability, the minimum energy criterion is applied for the G + MCh system in seven different geometries; in addition, it is checked with the non-complex vibration frequency. Results show chemical interactions between the G nanosheets and the MCh in the ground-state geometry. In this geometry, the monomer is oriented perpendicular to the G nanosheet at a distance of 3.9 Å with the nanosheet remaining unchanged. The nanosheet functionalization with boron (to form an epoxy group) and interaction with the monomer yield improved adsorption conditions with a bond length of C_{mesh}–B–N_{Amine} = 3.19 Å and the formation of B–N (boron attached to graphene–amine of the monomer) bond of length 1.57 Å. The polarity of the G + B and G + B + MCh systems displays ionic characteristics contrary to G behavior. The (HOMO–LUMO) energy difference is 1.30 eV for the G + B system and 0.75 eV for the G + B + MCh. Finally, the G + B + MCh system is investigated when D-(+)-glucose and cholesterol are adsorbed. Results show chemisorptions, which suggest the system to be used in biosensor devices.

Keywords DFT theory · Graphene · Functionalization · Chitosan · D-(+)-glucose · Cholesterol

E. Chigo Anota (✉) · A. Torres Soto
Facultad de Ingeniería Química, Benemérita Universidad Autónoma de Puebla, Ciudad Universitaria, San Manuel, Puebla 72570, Mexico
e-mail: echigoa@yahoo.es

G. H. Cocoletzi
Instituto de Física 'Luís Rivera Terrazas', Benemérita Universidad Autónoma de Puebla, Apartado Postal J-48, Puebla 72570, Mexico

Introduction

The functionalization of graphene (G) surface (Novoselov et al. 2004) has given rise to the investigation of new materials such as graphane (Sofo et al. 2007; Elias et al. 2009), graphene oxide (GO) (Dikin et al. 2007; Li et al. 2009), graphone (Zhou et al. 2009), graphanol (Wang and Kaxiras 2010), fluorographene (FG) (Nair et al. 2010), bromination of graphene (Yaya et al. 2011), hydrographene (HG) (Ezawa 2011), chlorination of graphene (Li et al. 2011), GraPOSS (graphene + polyhedral oligomeric silsesquioxanes) (Valentini et al. 2011) and GO-POSS (Xue et al. 2012). In addition, it has motivated the surface modifications of graphene with polymers or other functional groups to vary the properties and find new future applications in optoelectronic devices and/or medicine. Chitosan is an abundant linear biopolymer; it occupies the third place in abundance as compared with cellulose and chitin. Moreover, it displays excellent characteristics to form membranes. Chitosan has good adhesion, excellent mechanical properties, good biocompatibility and good susceptibility for chemical modifications, as it contains a large amount of OH groups and amine functional groups (Kurita 2001; Masuko et al. 2005). One motivation to explore the graphene–chitosan system is the possibility of recovering materials from sea waste as shrimp skeleton and then to obtain chitosan (Hernández Cocoletzi et al. 2009).

Studies have been recently reported by Chigo Anota et al. (2013a) on the hexagonal boron nitride nanosheet functionalized with chitosan (hBN + MCh). Results show chemical interactions (chemisorption) with the ground-state geometry exhibiting a parallel configuration. When defects are generated in the mesh (double vacancy) of the hBN surface, the adsorption is slightly increased. Moreover in the interactions of the monomer with the BN nanotube of different chirality, the monomer is chemically adsorbed (Rodriguez Juárez et al. 2013). The adsorption process on the surface is improved with the presence of vacancies in the tubular systems.

On the other hand, Kang et al. (2009) have studied the electronic properties of the graphene–chitosan compound system. Results suggest that the system may be applied to glucose biosensor devices. These nanosheets have been manufactured using the McAllister et al. (2007) and Schniepp et al. (2006) methods with sheets being suitable for forming the composites. In this work it is invoked the model used by Chigo Anota et al. (2012, 2013b, 2013c) to represent the graphene or boron nitride nanosheet as a CnHm-like cluster to study the adsorption of the monomer of chitosan, represented by the monomeric unit ($C_6H_{13}O_5N$), on the graphene surface (G + MCh). As another case, the adsorption of MCh on the G nanosheet functionalized with boron (G + B) is explored. The structure contains on one side of the central hexagon of the G nanosheet the functional group epoxy and a new B–N bond with the amine group of the monomer. In this way, the binding energy is increased and suggests the possible use of the system as a biosensor device. Moreover, the system has been tested as an adsorbent of glucose and cholesterol biomolecules. In this work, we report changes in the structural properties before and after the adsorption process of the polysaccharides: dipolar moment, chemical reactivity (chemical potential), adsorption energy and the energy difference of HOMO–LUMO frontier orbitals.

Calculation models and method

First principles total energy calculations are performed to study the following systems: graphene–chitosan (G + MCh), graphene doped with boron–chitosan (G + B + MCh), graphene doped with boron–chitosan D-(+)-glucose (G + B + MCh + glucose) and graphene doped with boron–chitosan–cholesterol (G + B + MCh + cholesterol). The exchange-correlation energies are treated with the Hamprecht–Cohen–Tozer–Handy (HCTH) (Boese and Handy 2001) functionals within the generalized gradient approximation (GGA) and the basis set DNP with double polarization (Delley 1990) (that is, *p*-orbitals of H and *d*-orbitals of boron, carbon, nitrogen and oxygen are considered) is applied as implemented in the quantum chemistry software $DMol^3$ code (Delley 1990, 2000). The choice of the HCTH functional for the calculations is done because this allows describing adequately the interaction between the CnHm-like fragment and the graphene sheet as reported by Araujo et al. (2012).

The armchair nanosheets are of height 1.41 nm and of width 1.34 nm; they are mono-hydrogenated at edges (passivated). In this way, the CnHm-like clusters form a mesh with a total of 72 atoms (in 19 hexagons). The chitosan is modeled with the monomeric unity having the chemical composition $C_6H_{13}O_5N$ with 25 atoms. A total of seven configurations for the G + MCh system were explored: three geometries oriented to the chitosan OH groups, one oriented to the oxygen of the piranosic cycle, one perpendicular to the amine group and two parallel geometries to the carbon mesh. In all cases, charge neutrality was considered. The orbital cut radius is 0.37 nm (for the monomer of chitosan and cholesterol) and 0.47 nm (for the G, G + MCh, G + B + MCh, G + B + MCh + glucose and G + B + MCh + cholesterol) on the base function with a tolerance of 1.0×10^{-6} Ha for the energy convergence. The structural stability is achieved following the minimum criterion and the non-complex vibration frequencies (Foresman et al. 1996).

To validate the models, the cohesion energy [$Ecoh = [nE(C) + mE(H) - E(CnHm)/(n + m)]$ of naphthalene ($C_{10}H_8$), phirene ($C_{16}H_{10}$), coronene ($C_{24}H_{12}$) and the cluster $C_{55}H_{18}$ was calculated, obtaining a value of 3.82 a.u./atom for all systems.

To improve the bioadsorption of D-(+)-glucose and cholesterol, the graphene sheet is functionalized with boron at the epoxy-kind structure and then joined to the monomer of chitosan at the nitrogen atom of the amine group. In this work, we report the structural changes (prior and after adsorption), polarity (dipole moment magnitude) and chemical reactivity using the formula $\mu = (HOMO + LUMO)/2$, energy gap (HOMO–LUMO frontier orbitals difference) and adsorption energy [$E(ad)_1 = E(graphene + MCh) - E(graphene) - E(MCh)$, $E(ad)_2 = E(graphene + B + MCh) - E(graphene) - E(B) - E(MCh)$, $E(ad)_3 = E(graphene + B + MCh + X) - E(graphene) - E(B) - E(MCh) - E(X)$; X = D-(+)-glucose and cholesterol]. The molecular electrostatic potential (MEP) is determined as described in the literature (Chigo Anota et al. 2013c).

Results and discussion

Studies are reported of the following systems: graphene–monomer of chitosan, boron-doped graphene–monomers of chitosan, and boron-doped graphene–monomer of chitosan–glucose and boron-doped graphene–monomer of chitosan–cholesterol interactions (Fig. 1). For the presentation of results, we *first* consider the interaction of pristine graphene with the monomer of chitosan (G + MCh) in configurations: 1 the molecule is oriented perpendicular to the nanosheet, the interaction is through the oxygen of the piranosic group (C_5O), 3 where the orientations are defined by the OH functional group, 1 perpendicular through the amine group (NH_2) and 2 where the central hexagon of the graphene is parallel to the monomer. The *following step* considers the graphene functionalization with a boron atom to form an epoxy group with the (G + B) mesh; this is done to improve the G functionalization with chitosan.

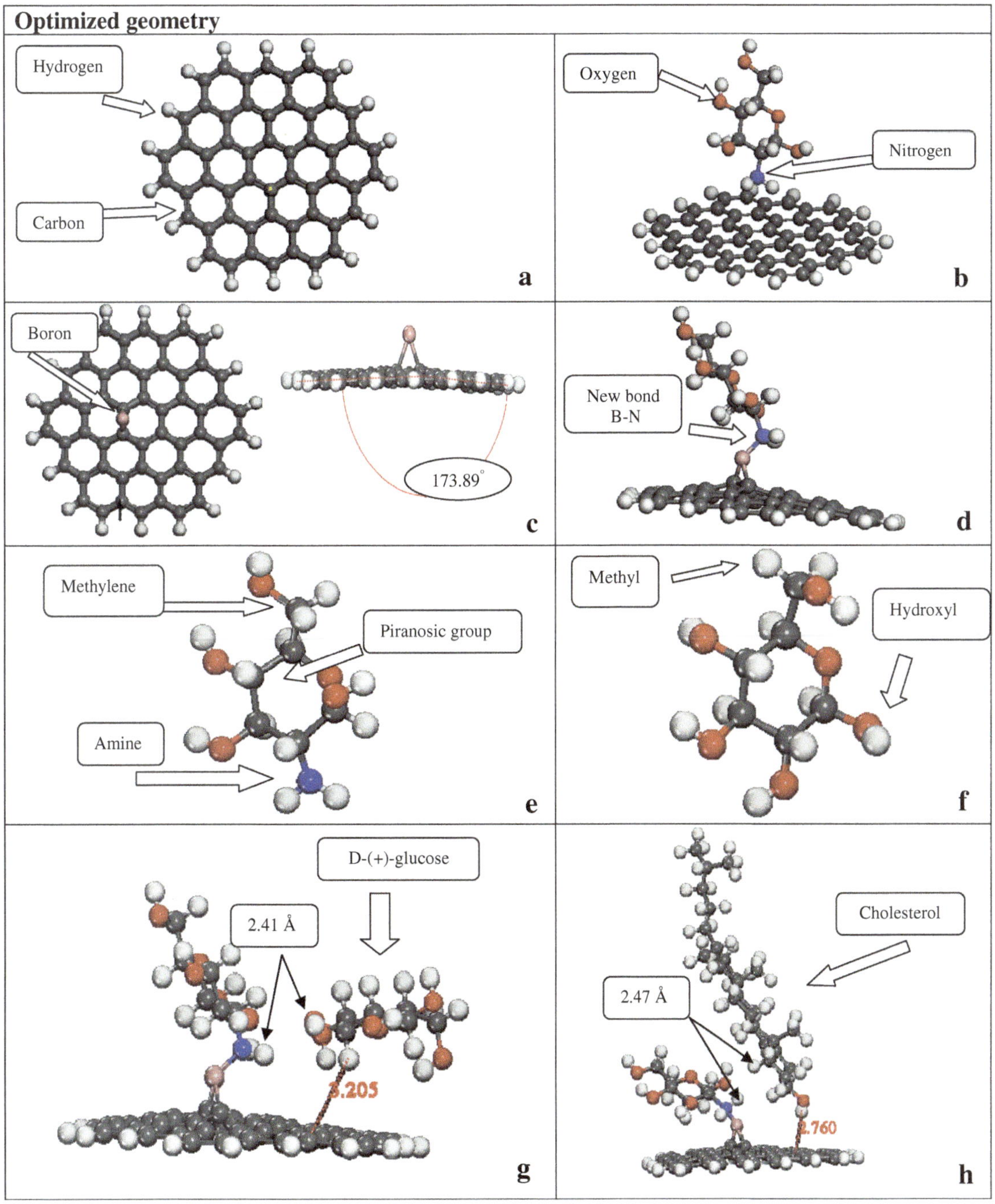

Fig. 1 We depict the relaxed geometries of G (**a**), G + MCh (**b**), G + B (**c**), G + B + MCh (**d**), the monomer of chitosan (**e**), D-(+)-glucose (**f**), G + B + MCh + glucose (**g**) and G + B + MCh + cholesterol (**h**)

Table 1 The optimized parameters of G, G + MCh, G + B + MCh and G + B + MCh–X

System	Bond length (Å)	Dipolar moment (Debye)	Gap (HOMO–LUMO) (eV)	Adsorption energy (eV)
Graphene ($C_{54}H_{18}$) (Chigo Anota et al. 2013b)		2.9×10^{-3}	1.94	
Graphene–MCh	3.9 (C_{mesh}–N of Amine)	3.49	1.30	−8.22
Graphene–boron–MCh	1.87, 1.82 (C_{mesh}–Boron)	1.84	0.75	−8.60
	1.57 (New bond B–N_{amine})			
Graphene–B–MCh–glucose	2.41 (NH_2 of MCh–OH of glucose)	10.78	1.13	−1.62
	2.31 (H of glucose–C_{mesh})	1.85 (glucose)		
Graphene–B–MCh–cholesterol	2.47 (NH_2 of MCh–H of cholesterol)	11.13	0.99	−2.58
	2.76 (OH of Cholesterol–C_{mesh})	1.51 (cholesterol)		

X glucose or cholesterol

Finally, the adsorption of glucose ($C_8H_{12}O_6$) and cholesterol ($C_{27}H_{46}O$) is explored on the G + B + MCh system to test for the sensing capability.

Graphene–chitosan interaction

It is well known that graphene displays extraordinary properties such as high thermal conductivity, mechanical flexibility and good biocompatibility. These make the system suitable for applications in biosensor devices of the electrochemical type (Pumera 2009, 2010; Allen et al. 2010; Long et al. 2012) and moreover allows the formation of devices free of metals (Kuila et al. 2011). Chitosan may be easily obtained from waste shrimp skeleton (Hernández Cocoletzi et al. 2009) and exhibits properties which include good biocompatibility and capability to form membranes suitable for applications (Kurita 2001; Masuko et al. 2005). Consequently in this work, we perform non-periodic molecular simulations to investigate the structural and electronic properties of graphene interacting with chitosan. The minimum energy criterion and the non-negative vibration frequency guarantee an energy minimum of the global potential energy surface. The process of finding the minimum energy is done for all structures. Results show that the lowest minimum energy corresponds to the structure where the monomer of chitosan is oriented perpendicular to the graphene nanosheet, with the interactions being through the amine group of the monomer and the central hexagon of the carbon mesh. The binding energy is −8.22 eV (chemisorption, Table 1) and a bond length of 3.9 Å (Fig. 1b). These results agree well with those obtained in the interactions of chitosan with the BN hexagonal nanosheet (Chigo Anota et al. 2013b).

G nanosheet structural changes induced by the interaction with MCh show that C–C bond lengths take values in the interval of 1.39–1.42 Å, while at the same time the flat surface characteristic remains. In the monomer, the bond angle of the amine group H–N–H is 107.54° which is somewhat contracted as compared with the pristine case (102.15) (Chigo Anota et al. 2013d). In the C–C–O and C–C–C bond angles,contractions of 2.28° and 2.84°, respectively, are obtained again as compared with the pristine cases. Finally, the methyl group (CH_2) exhibits an angle reduction of 0.51°.

The G nanosheet polarity is 2.9×10^{-3} and the MCh polarity is 2.13 D; on the other hand, the G + MCh polarity is 3.49 D, which indicates that the monomer induces ionic characteristics in the G + MCh system, that is, produces the charge localization. The HOMO (with energy −4.73 eV) and LUMO (with energy −2.79 eV) isosurfaces with contributions from the hybridized carbon p_z and s atomic orbitals are concentrated on the nanosheet (Fig. 2a, b). Calculations of the HOMO–LUMO energy difference (1.30 eV, Table 1) indicates that the system is a semiconductor, which is similar to the pristine graphene (Chigo Anota et al. 2013b). This means that the monomer makes no contribution to the HOMO–LUMO energy difference.

The analysis of the vibrations yields stretching oscillations at the wave number 4,092.46 cm^{-1} and bending at 4,181.10 cm^{-1} of the amine group of the monomer of chitosan. Stretching modes at wave numbers 3,655.83, 3,641.61 and 3,662.93 cm^{-1} of the piranosic hydrogen atoms are obtained. Stretching (at 3,699.15 cm^{-1}), anti-stretching (at 2,048.63 cm^{-1}) and scissor (at 2,265.29, 2,346.87, 2,338.16, and 2,472.9 cm^{-1}) modes of the carbon and hydrogen atoms are found in the monomer, in agreement with the isolated G nanosheet oscillations where an anti-stretching mode at 264.63 cm^{-1} is also observed.

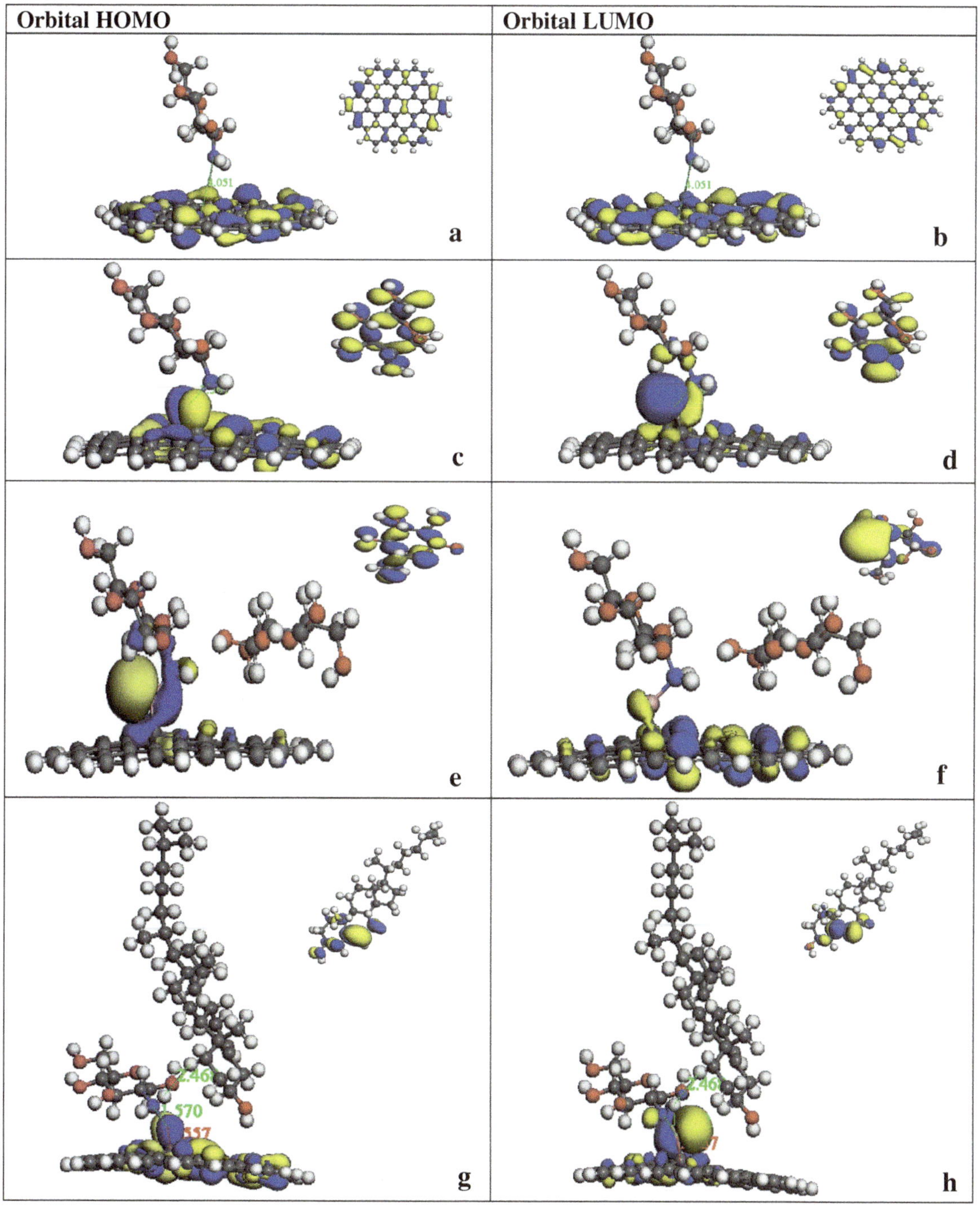

Fig. 2 The HOMO and LUMO molecular orbitals of the systems: G + MCh (**a**, **b**), G + B + MCh (**c**, **d**), G + B + MCh + glucose (**e**, **f**) and G + B + MCh + cholesterol (**g**, **h**). In the *upper right* corner, we display the orbitals of G, the monomer of chitosan, glucose and cholesterol

Boron-doped graphene–chitosan interaction

It has been reported recently the formation of graphene-chitosan films exhibit increased conductivity (attributed to the large surface-to-volume ratio and high conductivity of graphene) and electronic transference processes (Kang et al. 2009); therefore we explore conditions to improve the chitosan adsorption on the G nanosheet. To achieve this, nanosheet functionalization with boron atoms is proposed. This process allows the adsorption of chitosan near the impurity atom at one side of the central hexagon forming an epoxy-like configuration. The structure of minimum energy displays a curvature at the carbon mesh of 6.11° (Fig. 1c), and the bond distance between the G and boron,

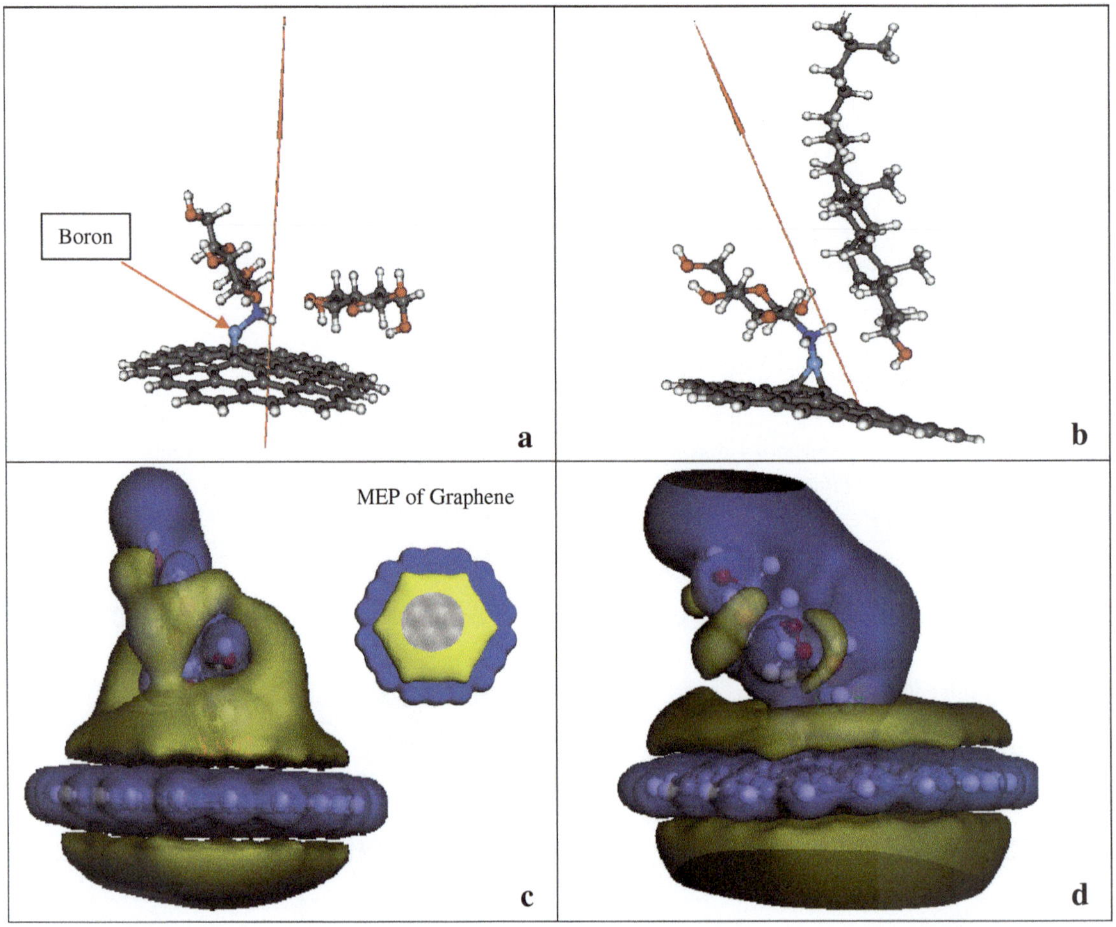

Fig. 3 The dipole moment of G + B + MCh + X system: X glucose (**a**) and X cholesterol (**b**). These values were calculated using the code Gaussian09 (Frisch et al. 2009). **c** The molecular electrostatic potential of G + MCh and G; **d** the corresponding electrostatic potential of G + B + MCh

C_{mesh}–B, takes asymmetry values of 1.87 and 1.82 Å, respectively. A protrusion is formed at the zone and the hybridization changes from sp^2 to sp^3.

When the monomer of chitosan is adsorbed, a new B–N bond (with the nitrogen of the amine group of the monomer) is formed of length 1.57 Å. This value is similar to that reported in the adsorption of the amine group on the hBN nanosheet (Chigo Anota et al. 2013d) and in the monomer adsorption on BN nanotubes (Rodriguez Juárez et al. 2013) with different chirality through the boron atom. The adsorption energy is −8.60 eV (chemisorption, Table 1) which is increased as compared to the previous case (G + MCh); this may suggest the system to be used as a biomolecule sensor, which is due in part to the increment in the charge transfer as shown by the MEP surface.

When the G nanosheet is doped with boron, the molecular orbitals are localized at the vicinity of the C_{mesh}–B bond and at the new B–N_{amine} bond (Fig. 2c, d). These orbital localizations indicate a charge transfer to the polysaccharide (Fig. 2e–h), which induces a better adsorption process. The charge distribution is indicated by the dipole moment direction (Fig. 3a, b) and also shown by the graphs of MEPs (Fig. 3c, d), where the yellow zone is for the negative charge. The large adsorption energy of the monomer on the graphene sheet is validated by the MEP surface which exhibits a charge distribution in the entire interaction zone (Fig. 3c), in a similar fashion as in the pristine graphene (Fig. 3c, inset depicted at the right upper corner).

This charge localization may be measured by Raman experiments, provided the vibrations are increased. The polarity suggests (as a measure of the charge transfer) that graphene functionalized with boron (Ling et al. 2010), deposited on chitosan, may be of practical use, provided the interaction with the biomolecules is favored.

It is important to realize that the molecular HOMO and LUMO frontier orbitals of the polysaccharides (insets at the upper corner of Fig. 2c–h) make no contribution to the G + B + MCh. The structural change induced by the bonding of the monomer with the boron atom of the nanosheet is evident at the H–N–H bond angle of the amine which is reduced by an amount of 1.18° with respect to the boron free structure. The monomer hexagon (piranosic group) is affected by changes in the bond angles: an increase of 2.25° at the C–C–C fragment and 0.63° at the C–C–O fragment. The methyl group experiences a small variation in the bond angle (of the order of 10^{-2}).

Boron-doped graphene–monomer of chitosan–glucose and cholesterol

One possible application of the G + B + MCh system is as a biosensor (receptor); therefore in this section, we explore the glucose adsorption (Fig. 1g) which displays structural and chemical characteristics similar to the monomer of chitosan. Different geometries have been tested. In the *first*, the adsorbate interacts with the OH group, in the *second*, the interaction is through the methyl group; in the *third* and *fourth*, the glucose is parallel to the monomer of chitosan. The cholesterol (Fig. 1h) is in one case perpendicularly bound through the OH group of both systems, two configurations are with parallel orientations and in the last the cholesterol OH fragment is oriented toward the hexagon (piranosic group) of the monomer of chitosan. Minimum energy results yield the monomer of chitosan parallel to the structure with an adsorption energy of −1.62 eV which favors glucose detection (Fig. 1g; Table 1) and an adsorption energy of −2.58 eV for the cholesterol through the polysaccharide group (Fig. 1h; Table 1). The interaction distance is 2.41 Å in the G + B + MCh−Glucose (this is measured from one of the hydrogen atoms of the amine group in the monomer to the oxygen of the OH group in the glucose) and 2.47 Å (this distance measured from one of the hydrogen atoms in the amine group to one of the hydrogen atoms in the cholesterol) for the G + B + MCh + cholesterol and 2.76 Å for the cholesterol and G system. Structural changes are manifested mainly in the monomer of the chitosan, which exhibits a large inclination (Fig. 1g) with the monomer being attached to the G + B when cholesterol is adsorbed, and the amine group is contracted by an angle of 2.11°, the piranosic group at the C–C–O fragment is contracted by 0.49° and the C–C–C is contracted by 1.95°. Moreover, the bond angle at the methyl group is reduced by 0.18°. In the glucose case (Fig. 1h), in the methyl group the bond angle is increased by 0.12°, the C–C–O bond angle is increased by 0.39°, but the C–C–C bond angle is contracted by 1.16° and finally the bond angle at the amine group is reduced by 1.16°.

The vibrational modes in the G + B + MCh + glucose structure stretch at 37.05 cm^{-1} (this is from the polysaccharide toward the amine group), at 345.52 cm^{-1} for the vibration of the H–N fragment in the amine group toward one of the OH groups in the glucose, at 457.37 and 519.45 cm^{-1} in the bending vibration of the outer carbons. There are anti-stretching vibrational modes at 1,676.86 cm^{-1} in the vibrations between hydrogen atoms of the amine group, stretching modes at 1,924.94 and 3,722.79 cm^{-1} of the two hydrogen atoms in the amine group, and stretching modes at 2,372.34 cm^{-1} of the hydrogen atom bonded to the carbon in the C–C–C fragment of the piranosic group. Scissor vibration modes are at 2,561.88 cm^{-1} of the hydrogen atoms of the amine group in the monomer, and anti-stretching modes from 3,779.66 to 3,794.47 cm^{-1} of the outer hydrogen atoms. There are bending vibrations at 3,795.08, 3,801.09 and 3,813.39 cm^{-1} of the hydrogen atoms in the amine group and scissor vibration modes at 3,929.96 cm^{-1} of the methyl group in the monomer chitosan.

In the G + B + MCh + cholesterol structure, at 4,355.90 cm^{-1} the interaction of the cholesterol OH functional group with the carbon mesh occurs (stretching along the *z*-direction), at 50.67 cm^{-1} (stretching) the vibration of the chitosan monomer toward cholesterol is obtained, at 69.37 cm^{-1} the motion is of cholesterol toward MCh. At 4,114.13 cm^{-1}, there is a stretching mode and at 4,114.13 cm^{-1} there is a bending mode of the amine group vibration which interacts with the polysaccharide. At 3,679.13 and 3,662.51 cm^{-1}, there are stretching vibration modes of the amine group hydrogen atoms in the MCh toward the G nanosheet. This interaction is favored by the inclination of the monomer when cholesterol is adsorbed. Finally, the carbon stretching and anti-stretching mesh vibration modes are on the nanosheet plane.

Conclusions

We have presented studies of the interaction between the monomer of chitosan (MCh) and carbon nanosheets (graphene), the G + MCh system. The system has been modified by the functionalization of the carbon mesh with boron atoms (G + B) to form an epoxy-type group which bonds with the nitrogen of the amine group (this forms a new bond) to improve conditions in the detection of polysaccharides such as glucose and cholesterol. The interaction induces structural changes mainly in the monomer. The functionalized graphene nanosheet with chitosan (G + B + MCh) may be a better biosensor than the G + MCh system. This is corroborated by the larger value of the adsorption energy and by the vibrational motions of the amine group which bonds to the boron atom. This is further supported by the dipole moment direction which indicates charge transfer as both polysaccharides are bioadsorbed.

Acknowledgments This work was partially supported by projects: VIEP-BUAP (CHAE-ING13-G), Cuerpo Académico Ingeniería en

Materiales (BUAP-CA-177), Cuerpo Académico Física Computacional de la Materia Condensada (BUAP-CA-191) and VIEP-BUAP-EXC11-G.

References

Allen MJ, Tung VC, Kaner RB (2010) Honeycomb carbon: a review of graphene. Chem Rev 110:132

Araujo V, Peñaranda M, Castellano O, Soscun H (2012) Propiedades estructurales, energéticas y electrónicas del complejo molecular formado por la interacción entre benceno y grafeno extendido: Investigación basada en la teoría del funcional de la densidad DFT. Ciencia 20(special issue):128–136

Boese AD, Handy NC (2001) A new parametrization of exchange-correlation generalized gradient approximation functionals. J Chem Phys 114:5497

Chigo Anota E, Ramírez Gutierrez RE, Escobedo Morales A, Hernández Cocoletzi G (2012) Influence of point defects on the electronic properties of boron nitride nanosheets. J Mol Model 18(5):2175–2184

Chigo Anota E, Escobedo Morales A, Salazar Vilanueva M, Vazquez Cuchillo O, Rubio Rosas E (2013b) On the influence of point defects on the structural and electronic properties of graphene-like sheets: a molecular simulation study. J Mol Model 19(2):839–846

Chigo Anota E, Ramírez Gutiérrez RE, Pérez Sánchez FL, Sánchez Ramírez JF (2013c) Structural characteristics and chemical reactivity of doped graphene nanosheets. Graphene 1(1):31–36

Chigo Anota E, Rodríguez Juárez A, Miguel Castro, Hernández Cocoletzi H (2013d) A density functional theory analysis for the adsorption of the amine group on graphene and boron nitride nanosheets. J Mol Model 19:321–328

Chigo Anota E, Hernández Rodríguez LD, Hernández Cocoletzi G (2013a) Influence of point defects on the adsorption of chitosan on graphene-like BN nanosheets. Graphene.

Delley B (1990) An all-electron numerical method for solving the local density functional for polyatomic molecules. J Chem Phys 92:508–517

Delley B (2000) From molecules to solids with the $DMol^3$ approach. J Chem Phys 113:7756–7764

Dikin DA, Stankovich S, Zimney EJ, Piner RD, Dommett GHB, Evmenenko G, Nguyen ST, Ruoff RS (2007) Preparation and characterization of graphene oxide paper. Nature 448:457–460

Elias DC, Nair RR, Mohiuddin TMG, Morozov SV, Blake P, Halsall MP, Ferrari AC, Boukhvalov DW, Katsnelson MI, Geim AK, Novoselov KS (2009) Control of graphene's properties by reversible hydrogenation: evidence for graphene. Science 323: 610–614

Ezawa M (2013) Quantum percolation transition from graphene to graphane: graph theoretical approach. Nanomat Nanotechnol 3:1–6

Foresman JB, Frisch Æ. (1996) Exploring chemistry with electronic structure methods. 2nd Edn. Gaussian Inc., USA, p 70

Frisch MJ, Trucks GW, Schlegel HB, Scuseria GE et al. (2009) Gaussian09, Revision C.01-SMP. Gaussian Inc., Pittsburgh

Hernández Cocoletzi H, Águila Almanza E, Flores Agustín O, Viveros Nava EL, Ramos Cassellis E, Lee YC (2005) Obtención y caracterización de quitosano a partir de exoesqueletos de camarón. Sup y Vac 21(3):57–60

Kang X, Wang J, Wu H, Aksay IA, Liu J, Lin Y (2009) Glucose oxidase–graphene–chitosan modified electrode for direct electrochemistry and glucose sensing. Biosens Biolectron 25:901–905

Kuila T, Bose S, Khanra P, Mishra AK, Hoon Kim N, Le Hee J (2011) Recent advances in graphene-based biosensors. Biosens Bioelectron 26:4637–4648

Kurita K (2001) Controlled functionalization of the polysaccharide chitin. Prog Plym Sci 26:1921–1971

Li X, Wang H, Robinson JT, Sanchez H, Diankov G, Dai H (2009) Simultaneous nitrogen doping and reduction of graphene oxide. J Am Chem Soc 131:15939–15944

Li B, Zhou L, Wu D, Peng H, Yan K, Zhou Y, Liu Z (2011) Photochemical chlorination of graphene. ACS Nano 5: 5957–5961

Ling X, Xie L, Fang Y, Xu H, Zhang H, Kong J, Dresselhaus MS, Zhang J, Liu Z (2010) Can graphene be used as a substrate for raman enhancement? Nano Lett 10:553–561

Long J, Xie X, Xu J, Gu Q, Chen L, Wang X (2012) Nitrogen-doped graphene nanosheets as metal-free catalysts for aerobic selective oxidation of benzylic alcohols. ACS Catal 2:622–631

Masuko T, Minami A, Iwasaki N, Majima T, Nishimura I, Lee YC (2005) Thiolation of chitosan attachment of proteins via thioether formation. Biomacromolecules 6:880–884

McAllister MJ, Li JL, Adamson DH, Schniepp HC, Abdala AA, Liu J, Herrera Alonso M, Milius DL, Car R, Prud'homme RK, Aksay IA (2007) Single sheet functionalized graphene by oxidation and thermal expansion of graphite. Chem Mater 19:4396–4404

Nair RR, Ren WC, Jalil R, Diaz I, Kravets VG, Britnell L, Blake P, Schedin F, Mayorov AS, Yuan S, Katsnelson MI, Cheng HM, Strupinski W, Bulsheva LG, Okotrub AV, Grigoreva IV, Grigorenko AN, Novoselov KS, Geim AK (2010) Fluorographene: a two-dimensional counterpart of teflon. Small 6(24): 2877–2884

Novoselov KS, Geim AK, Morozov SV, Jiang D, Zhang Y, Dubonos SV, Grigorieva IV, Firsov AA (2004) Electric field effect in atomically thin carbon films. Science 306:666–669

Pumera M (2009) Electrochemistry of graphene: new horizons for sensing and energy storage. Chem Rec 9:211–223

Pumera M (2010) Graphene-based nanomaterials and their electrochemistry. Chem Soc Rev 39:4146–4157

Rodriguez Juárez A, Chigo Anota E, Hernández Cocoletzi H, Flores Riveros A (2013) Adsorption of chitosan on BN nanotubes: a DFT investigation. Appl Surf Sci 268:259–264

Schniepp HC, Li JL, Mcallister MJ, Sai H, Herrera Alonso M, Adamson DH, Prud'homme RK, Car R, Saville DA, Aksay IA (2006) Functionalized single graphene sheets derived from splitting graphite oxide. J Phys Chem B 110:8535–8539

Sofo JO, Chaudhari AS, Barber GD (2007) Graphane: a two-dimensional hydrocarbon. Phys Rev B 75:153401–153404

Valentini L, Cardinali M, Kenny JM, Prato M, Monticelli O (2012) A photoresponsive hybrid nanomaterial based on graphene and polyhedral oligomeric silsesquioxanes. Eur J Inorg Chem 2012:5282–5287

Wang WL, Kaxiras E (2010) Graphene hydrate: theoretical prediction of a new insulating form of graphene. New J Phys 12: 1250121–1250127

Xue Y, Liu Y, Lu F, Qu J, Chen H, Dai L (2012) Functionalization of graphene oxide with polyhedral oligomeric silsesquioxane (POSS) for multifunctional applications. J Phys Chem Lett 3:1607–1612

Yaya A, Ewels CP, Suarez-Martinez I, Wagner PH, Lefrant S, Okotrub A, Bulusheva L, Briddon PR (2011) Bromination of graphene and graphite. Phys Rev B 83:0454111–0454115

Zhou J, Wang Q, Sun Q, Chen XS, Kawazoe Y, Jena P (2009) Ferromagnetism in semihydrogenated graphene sheet. Nano Lett 9(11):3867–3870

13

Growth of carbon nanotube filaments on carbon fiber cloth by catalytic chemical vapor deposition

B. P. Singh · Veena Choudhary · V. N. Singh · R. B. Mathur

Abstract Multiwalled carbon nanotube (MWCNT) filaments were grown by catalytic chemical vapor deposition on 2-D weave carbon fiber (CF) cloth substrates. Two different sets of experiments were carried out to understand the growth mechanism of these filaments. In the first set of experiments where CF cloths were coated with the catalysts particles, CNT filaments having long length (>200 μm) and large diameter (15–25 μm) were obtained. In another set of experiment, where CF cloths without catalyst particles were used, only MWCNTs without any filament formation were obtained. On the basis of the results, a growth mechanism has been proposed. These MWCNT filaments can be used for preparing CNTs reinforced polymer composites having very good structural properties, which are being sought after.

Keywords Carbon filaments · MWCNTs · Chemical vapor deposition

B. P. Singh (✉) · R. B. Mathur
Physics and Engineering of Carbon, Division of Materials Physics and Engineering, CSIR-National Physical Laboratory, New Delhi, India
e-mail: bps@mail.nplindia.org

B. P. Singh · V. Choudhary
Centre for Polymer Science and Engineering, Indian Institute of Technology Delhi, Delhi 16, India

V. N. Singh
Electron and Ion Microscopy Section, CSIR-National Physical Laboratory, New Delhi 110012, India

Introduction

Carbon nanotube (CNT), a fascinating material with outstanding properties has inspired the scientist, engineer and technologist because of its wide range of potential applications (Ajayan and Zhou 2000) in many areas. To utilize their properties for industrial applications, producing high quality CNT on larger scale is necessary. Several methods for the synthesis of CNTs, such as, arc discharge (Choudhary et al. 2013; Mathur et al. 2007), laser ablation (Choudhary et al. 2013; Guo et al. 1995) and chemical vapor deposition (CVD) (Choudhary et al. 2013; Mathur et al. 2008) have been developed. These methods have been further modified to synthesize single walled carbon nanotubes (SWNTs) (Dai et al. 1996; Ebbesen and Ajayan 1992; Thess et al. 1996) and multiwalled carbon nanotubes (MWCNTs) (Chhowalla et al. 2001; Ebbesen et al. 1993). For various specialized applications in electronic devices and composites, CNTs having aligned and controlled structure are required. CVD technique is very useful for producing such structures. Various CVD methods have been developed for synthesizing SWCNTs as well as MWCNT, such as, catalytic CVD or thermal CVD, plasma enhanced CVD (PECVD) (Cheng et al. 1998) and floating catalyst CVD method (Ago et al. 2005; Chen et al. 2006). In the CVD process, organic compounds are decomposed in the presence of catalyst. In the catalytic CVD and PECVD, CNTs growth takes place on the substrate. But in the floating catalyst technique, large quantity of CNTs is generated without use of any substrate. Therefore, the floating catalyst technique is very competitive for producing CNT on large scale for industrial applications. But, controlling the size and composition of the catalyst nanoparticles is a major issue for this technique (Andrews et al. 1999).

Incorporation of carbon nanotubes (CNTs) into a polymer matrix can improve various properties including mechanical (Ruban et al. 2013; Lubineau and Rahaman 2012; Mathur et al. 2010, 2012; Singh et al. 2012a), electrical (Gupta et al. 2013a, b; Pande et al. 2009a, b; Singh et al. 2011, 2013a, b), thermal (Biercuk et al. 2002; Shenogin et al. 2004; Meenakshi and Sudhan 2011; Singh et al. 2008) and biosensing properties (Dhand et al. 2008; Singh et al. 2012b). CNTs have excellent mechanical properties. Tensile strength of 10–500 GPa and Young's modulus of 0.3–1.0 TPa have been reported (Xie et al. 2005). Several works related to CNTs reinforced polymer composites for structural applications have been reported (Jindal et al. 2013; Lubineau and Rahaman 2012; Mathur et al. 2012; Singh et al. 2013a; Xie et al. 2005). Breakthrough in the CNT-polymer composite properties is still awaited, as it had not achieved the values reported for carbon fiber (CF) reinforced polymer composites. There are several reports on the growth of long fibers made from CNT. But most of them are made by spinning of CNT into fibers. In one such study, Zhang et al. (2004) achieved yarn strengths >460 MPa by introducing twist during spinning of MWCNTs. In other study, Ericson et al. (2004) produced well-aligned macroscopic fibers composed of SWNTs by conventional spinning. They used solution spinning to grow macroscopic neat SWNT fibers. Li et al. (2004) spun fibers and ribbons of CNT directly from the CVD synthesis zone of a furnace using a liquid source of carbon and an iron nanocatalyst. The process was realized through the appropriate choice of reactants, control of the reaction conditions and continuous withdrawal of the product using a rotating spindle.

Herein, we report another simple method for producing long MWCNT filaments having length >200 μm, diameter up to 25 μm and consisting of several hundreds of MWCNTs which might be a possible CNTs-based materials to realize the optimum mechanical properties when mixed in polymer matrix.

Experimental

Synthesis

Two different sets of experiments were performed for the growth of MWCNTs on CF cloth (code 6341, woven from T-300 CF tows) obtained from Toray, Japan. In one set of experiments, first, 8 wt% ferrocene was dissolved into toluene by ultrasonication (Choudhary et al. 2013; Mathur et al. 2008). The as received CF cloth having 200 × 50 mm sizes were dipped in the toluene–ferrocene solution to deposit ferrocene on the cloth. Toluene was evaporated by keeping the CF cloth in vacuum oven for 4 h at 80 °C. After drying the cloth, it was placed in the furnace for the deposition of MWCNTs. The reaction zone of the furnace was maintained at 750 °C. Once the temperature was reached, the solution containing a mixture of ferrocene and toluene in particular proportion (0.077 g ferrocene in 1 ml toluene) was injected into the reactor at a point where the temperature was 200 °C. The deposition was carried out for 1 and 2 h, respectively. Argon was used as the carrier gas. After completion of the reaction, the CF cloth was removed from the furnace and weighed for estimating the amount of MWCNT filaments grown on the CF cloth.

In another set of experiments, as received CF cloth (without soaking in the ferrocene and toluene solution) was placed inside the tube furnace followed by injection of toluene–ferrocene mixture for the growth of MWCNTs for 2 h under identical conditions.

Characterization

The as grown CNTs on CF cloth prepared by different technique and duration were characterized using scanning electron microscope (SEM, Zeiss EVO MA-10) to understand the growth mechanism. To further characterize, TEM studies were carried out by scrapping the CNTs from CF cloth, dispersing them in acetone and drop casting one drop of solution on carbon-coated copper grids. TEM studies were carried out using Technai G20-stwin, 300 kV instrument.

Results and discussion

Growth mechanism of MWCNT filament

On the basis of basic concepts of CNTs growth by catalytic CVD technique and SEM measurements, we have proposed a formation mechanism of CNT filaments and is schematically depicted in Fig. 1. Initially ferrocene particles (clusters of ferrocene particles) get deposited in the gap between CF tows. When the desired temperature (750 °C) was reached, nucleation of the CNTs occured in the catalyst clusters and CNT filament formation started. Growth of filament continues because of continuous supply of ferrocene–toluene mixture in the reactor. Growth took place in the filament form because of the presence of ferrocene in the cluster form (not as individual particles). When the reaction time was increased from 1 to 2 h; the filaments started growing in curved or coiled form as shown in the scheme as well as in SEM images.

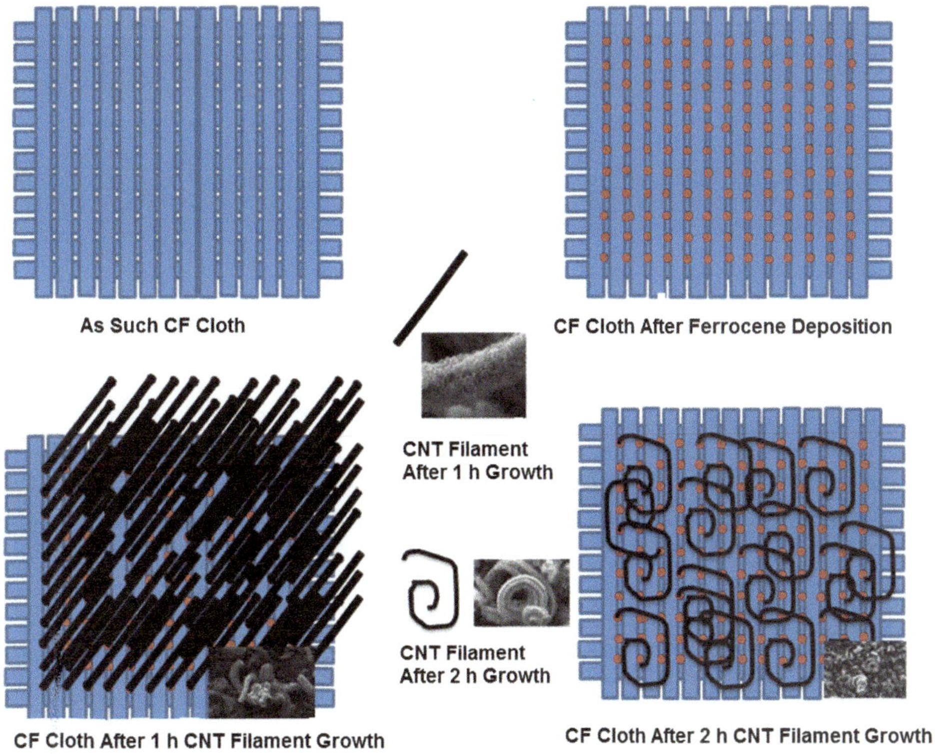

Fig. 1 Schematic representation of growth mechanism of CNT filaments on the surface of CF cloth

SEM/TEM of CNT filaments on CF cloth

The surface morphology of the nanotubes in the form of filaments was investigated using SEM and the results for the samples grown for 1 and 2 h using method 1 are shown in Figs. 2 and 3, respectively. Figure 2a shows the optical microscopic image of plain weave CF cloth and Fig. 2b shows the SEM image of the CNT filaments layer deposited on the CF cloth using method 1 and for 1 h (on CF cloth coated with ferrocene). The SEM image is clearly revealing the copious growth of CNT filaments and is densely packed.

Figure 2c, d shows the magnified SEM images of samples grown for 1 h by method 1. These images clearly reveal the uniform growth of filament. The diameter of the filaments varies from 15 to 25 μm. According to the proposed growth mechanism, the filament growth depends on the presence of ferrocene clusters/particles in the gap between the CF tows. The size of the filament depends on the size of the ferrocene cluster/particles in the gap. The filaments shown in Fig. 2c, d, nucleated from the ferrocene clusters available at various sites of the CF cloth placed inside the furnace at 750 °C. The length of these filaments is very high which cannot be so unless catalyst particles coming through the injection of ferrocene–toluene solution also participate. Further magnified SEM image (Fig. 2e) shows the individual CNT.

The further characterization of the scraped materials from the CF cloth was carried out using TEM. The TEM image is shown in Fig. 2f. The multiwalled structure of CNTs can be clearly seen in the TEM micrograph. These MWCNTs are densely packed. Except the presence of few catalytic particles which are source of growth, MWCNTs were free from impurities.

When the growth was carried out for 2 h, the structure of the filament was quite different from the samples grown for 1 h. Figure 3 shows the structure of CNT filament grown for 2 h on CF cloth. It has been observed that up to 1 h of growth, slightly straight CNT filaments were formed, but after certain time it does not continue in similar fashion and CNT filaments having helical shape were observed. In SEM image, many such filaments are seen. The diameter is almost similar to sample grown for 1 h, i.e., 15–25 μm (Fig. 3a, b). Magnified SEM images (Fig. 3c) show that filaments are made of CNTs having uniform diameter. It is evident that in both the cases (1 h, straight filament and 2 h, coiled filament), filaments are made of very uniform diameter CNTs. The coiled CNTs were further characterized by TEM and the results are shown in Fig. 3d. TEM micrograph is also confirming the presence of MWCNTs. The diameter of CNTs does not depend upon the diameter of ferrocene cluster but it depends upon the size of individual ferrocene particle

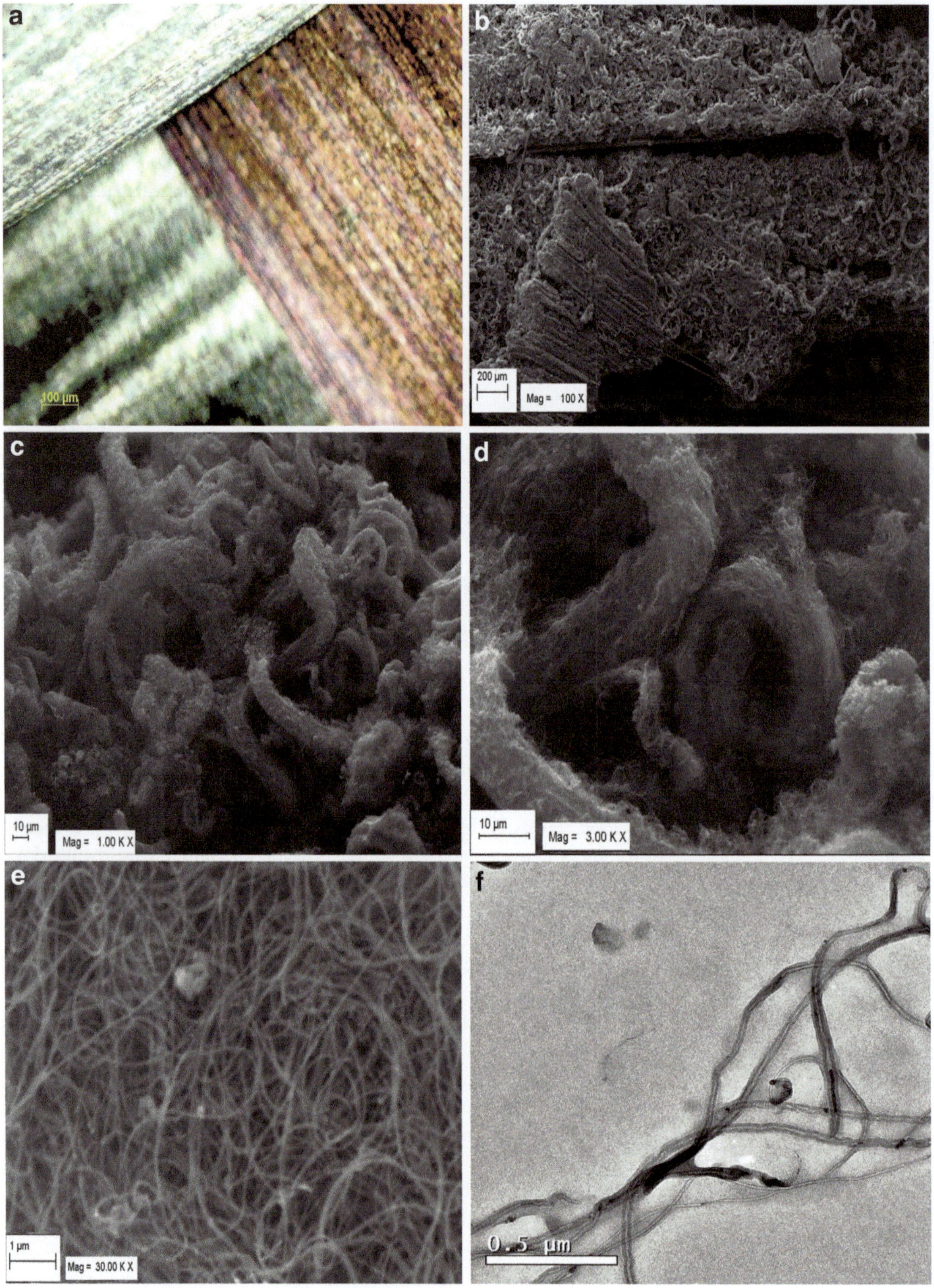

Fig. 2 **a** Optical micrograph image of CF cloth, **b–e** SEM image of CNT filaments grown on CF cloth for 1 h by method 1, and **f** TEM image of MWCNTs scraped from the CF cloth grown for 1 h by method 1

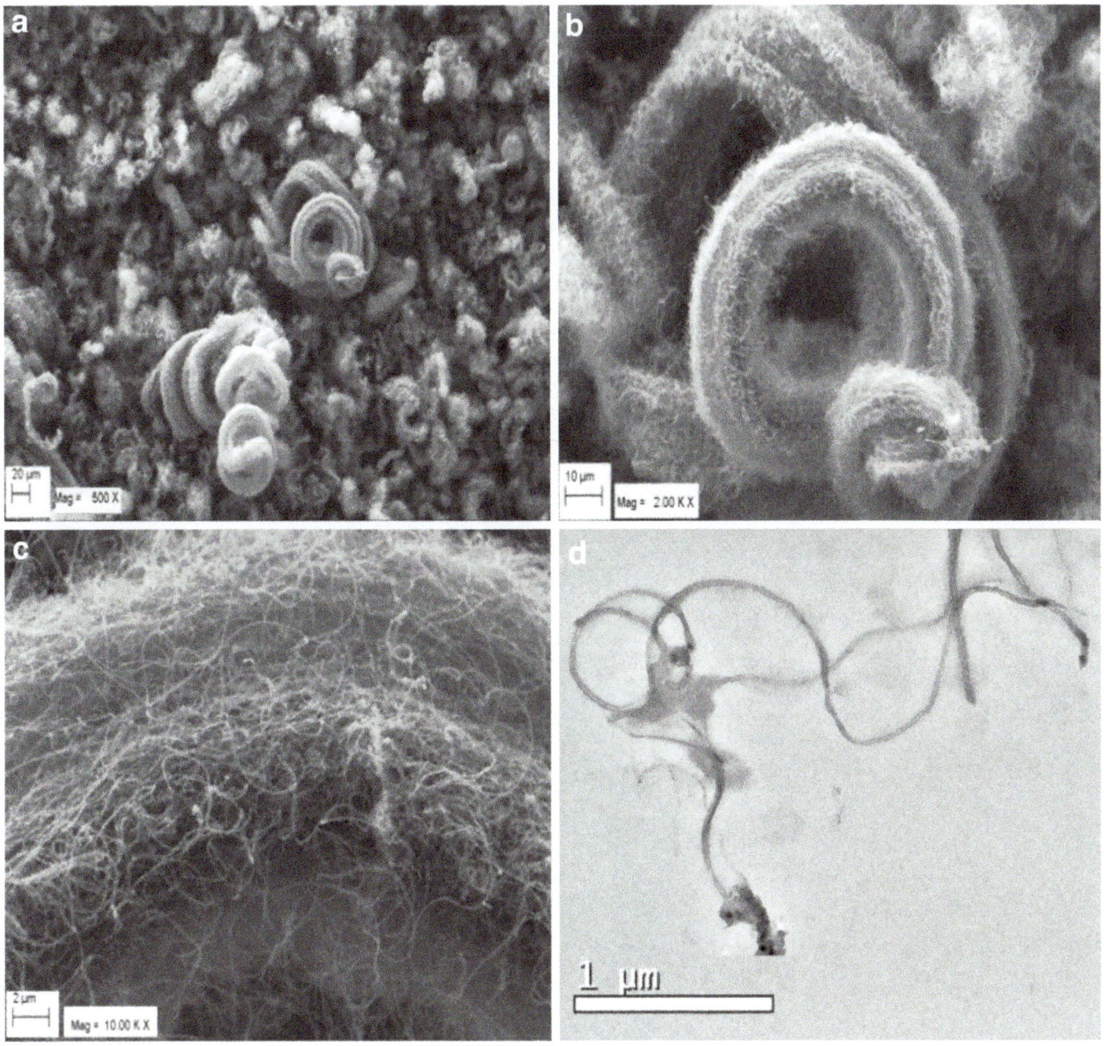

Fig. 3 **a–c** SEM image of CNT filaments grown on CF cloth for 2 h by method 1, **d** HRTEM image of MWCNTs scraped from the CF cloth grown for 2 h by method 1

and, therefore, the diameter of CNTs in the filaments is nearly similar in both the cases (1 and 2 h). The diameters of the individual ferrocene particle in both the cases are similar. Therefore, the growth time is the only parameter to influence the structure of filaments.

When the MWCNTs were grown on CF cloth substrate by method 2 (without coating the CF cloth with ferrocene) for 2 h, the structure is entirely different than sample grown by method 1. In this case, growth of only CNTs took place which was confirmed by SEM (Fig. 4a–c) and TEM images (Fig. 4d). Figure 4a shows that the entire CF cloth substrate was covered with CNTs. The mechanism of the growth of this type of CNTs is different. Here, ferrocene–toluene mixture was passed through the reactor chamber. The ferrocene–toluene mixture produced individual Fe particle and C particle. The individual Fe particles got deposited on the CF substrate and after that carbon particle diffused through the Fe particle and individual tube was formed.

Conclusions

MWCNTs were synthesized on CF cloth by two different methods and for growth times (1 and 2 h). In one method, first ferrocene particles were deposited on the CF cloth and CNTs growth occured with the help of additional available ferrocene + toluene mixture. Using this method, 1 h deposition time was sufficient for the growth of straight CNT filaments having diameter ranging from 15 to 25 µm and length >200 µm. When the samples were grown for 2 h, CNT filaments in the form of coiled structure were obtained. In another method, when CF cloth without ferrocene was used, only CNTs (without filament formation)

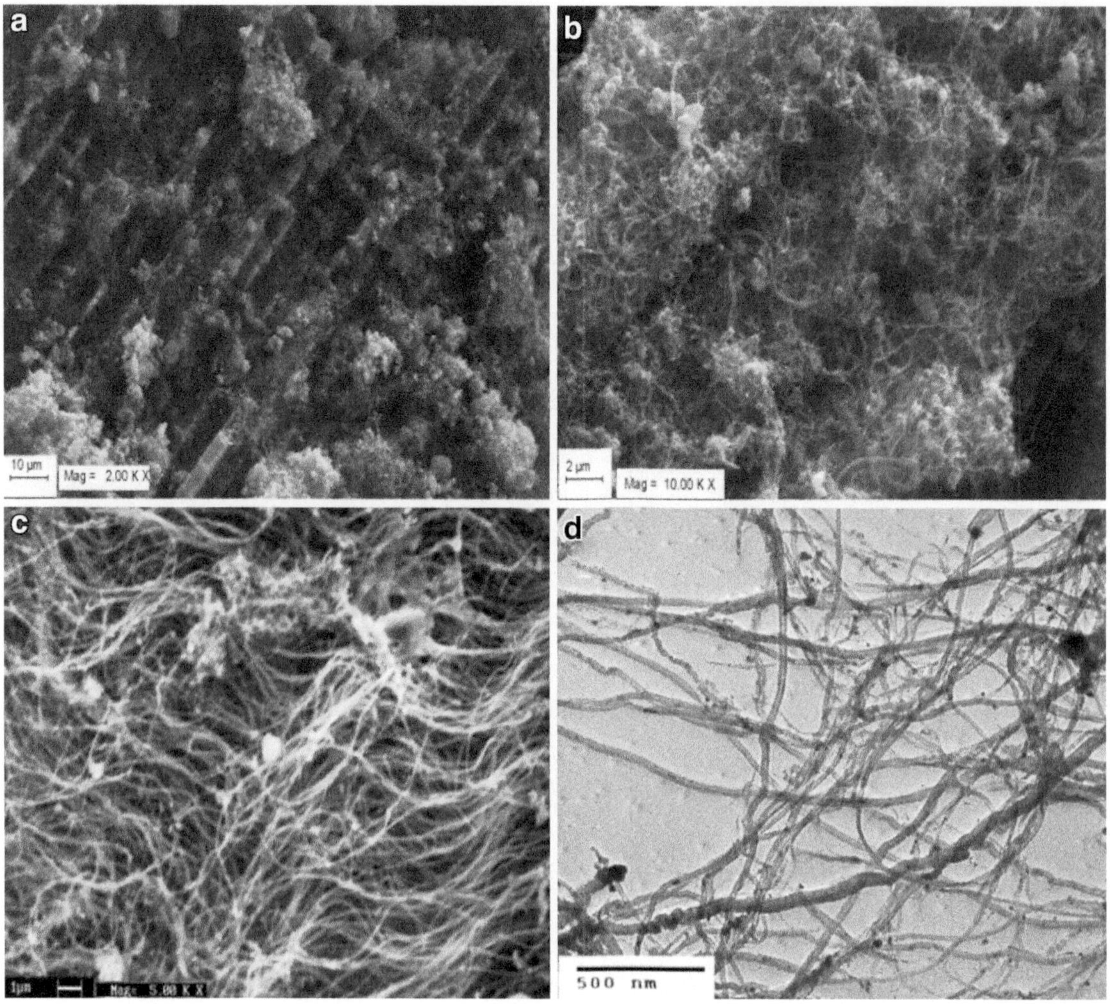

Fig. 4 **a–c** SEM image of CNT filaments grown on CF cloth for 2 h by method 1, **d** HRTEM image of MWCNTs scraped from the CF cloth grown for 2 h by method 2

were obtained. It shows that using ferrocene-coated CF cloth, good quality CNT filaments can be obtained only in 1 h which can be used for enhancing the structural properties when used in CNT-polymer composites.

Acknowledgments The authors wish to express their gratitude to Prof. R.C. Budhani, Director NPL, to accord his permission to publish the results. Authors would like to Mr. K.N. Sood and Mr. Jay Tawale for their support in carrying out SEM.

References

Ago H, Ohshima S, Tsukuagoshi K, Tsuji M, Yumura M (2005) Formation mechanism of carbon nanotubes in the gas-phase synthesis from colloidal solutions of nanoparticles. Curr Appl Phys 5:128–132

Ajayan PM, Zhou OZ (2000) Application of carbon nanotubes. In: D. G, Dresselhaus MS, Avouris P (eds) "Carbon Nanotubes synthesis, structure, properties and applications". Springer, Berlin, pp 391–425

Andrews R, Jacques D, Rao A, Derbyshire F, Qian D, Fan X, Dickey E, Chen J (1999) Continuous production of aligned carbon nanotubes: a step closer to commercial realization. Chem Phys Lett 303:467–474

Biercuk MJ, Llaguno MC, Radosavljevic M, Hyun JK, Johnson AT, Fischer JE (2002) Carbon nanotube composites for thermal management. Appl Phys Lett 80:2767–2769

Chen Y, Sun Z, Li YN, Tay BK (2006) Optimization of carbon nanotube powder growth using low pressure floating catalytic chemical vapor deposition. Mater Chem Phys 98:256–260

Cheng HM, Li F, Su G, Pan HY, He LL, Sun X, Dresselhaus MS (1998) Large-scale and low-cost synthesis of single-walled carbon nanotubes by the catalytic pyrolysis of hydrocarbons. Appl Phys Lett 72:3282–3284

Chhowalla M, Teo KBK, Ducati C, Rupesinghe NL, Amaratunga GAJ, Ferrari AC, Roy D, Robertson J, Milne WI (2001) Growth process conditions of vertically aligned carbon nanotubes using plasma enhanced chemical vapor deposition. J Appl Phys 90:5308–5317

Choudhary V, Singh BP, Mathur RB (2013) Carbon Nanotubes and Their Composites. In: Satoru S (ed) Carbon nanotubes and their composites, syntheses and applications of carbon nanotubes and their composites. InTech, (ISBN: 978-953-51-1125-2)

Dai H, Rinzler AG, Nikolaev P, Thess A, Colbert DT, Smalley RE (1996) Single-wall nanotubes produced by metal-catalyzed disproportionation of carbon monoxide. Chem Phys Lett 260:471–475

Dhand C, Arya SK, Singh SP, Singh BP, Datta M, Malhotra B (2008) Preparation of polyaniline/multiwalled carbon nanotube composite by novel electrophoretic route. Carbon 46:1727–1735

Ebbesen TW, Ajayan PM (1992) Large-scale synthesis of carbon nanotubes. Nature 358:220–222

Ebbesen TW, Hiura H, Fujita J, Ochiai Y, Matsui S, Tanigaki K (1993) Patterns in the bulk growth of carbon nanotubes. Chem Phys Lett 209:83–90

Ericson LM, Fan H, Peng H, Davis VA, Zhou W, Sulpizio J, Wang Y, Booker R, Vavro J, Guthy C, Parra-Vasquez ANG, Kim MJ, Ramesh S, Saini RK, Kittrell C, Lavin G, Schmidt H, Adams WW, Billups WE, Pasquali M, Hwang W F, Hauge RH, Fischer JE, Smalley RE (2004) Macroscopic, neat, single-walled carbon nanotube fibers. Science 305:1447–1450

Guo T, Nikolaev P, Thess A, Colbert DT, Smalley RE (1995) Catalytic growth of single-walled nanotubes by laser vaporization. Chem Phys Lett 243:49–54

Gupta TK, Singh BP, Teotia S, Katyal V, Dhakate SR, Mathur RB (2013a) Designing of multiwalled carbon nanotubes reinforced polyurethane composites as electromagnetic interference shielding materials. J Polym Res 20:1–7

Gupta TK, Singh BP, Dhakate SR, Singh VN, Mathur RB (2013b) Improved nanoindentation and microwave shielding properties of modified MWCNT reinforced polyurethane composites. J Mater Chem A 1:9138–9149

Jindal P, Pande S, Sharma P, Mangla V, Chaudhury A, Patel D, Singh BP, Mathur RB, Goyal M (2013) High strain rate behavior of multi-walled carbon nanotubes-polycarbonate composites. Compos B Eng 45:417–422

Li Y-L, Kinloch IA, Windle AH (2004) Direct spinning of carbon nanotube fibers from chemical vapor deposition synthesis. Science 304:276–278

Lubineau G, Rahaman A (2012) A review of strategies for improving the degradation properties of laminated continuous-fiber/epoxy composites with carbon-based nanoreinforcements. Carbon 50:2377–2395

Mathur RB, Seth S, Lal C, Rao R, Singh BP, Dhami TL, Rao AM (2007) Co-synthesis, purification and characterization of single- and multi-walled carbon nanotubes using the electric arc method. Carbon 45:132–140

Mathur R, Chatterjee S, Singh B (2008) Growth of carbon nanotubes on carbon fibre substrates to produce hybrid/phenolic composites with improved mechanical properties. Compos Sci Technol 68:1608–1615

Mathur RB, Singh BP, Dhami TL, Kalra Y, Lal N, Rao R, Rao AM (2010) Influence of carbon nanotube dispersion on the mechanical properties of phenolic resin composites. Polym Compos 31:321–327

Mathur RB, Singh BP, Tiwari PK, Gupta TK, Choudhary V (2012) Enhancement in the thermomechanical properties of carbon fibre–carbon nanotubes–epoxy hybrid composites. Int J Nanotechnol 9:1040–1049

Meenakshi KS, Sudhan EPJ (2011) Development and study of the thermal and electrical behaviour of TGDDS epoxy nanocomposites for high-performance applications. Appl Nanosci 1:109–115

Pande S, Mathur RB, Singh BP, Dhami TL (2009a) Synthesis and characterization of multiwalled carbon nanotubes-polymethyl methacrylate composites prepared by in situ polymerization method. Polym Compos 30:1312–1317

Pande S, Singh BP, Mathur RB, Dhami TL, Saini P, Dhawan SK (2009b) Improved electromagnetic interference shielding properties of MWCNT–PMMA composites using layered structures. Nanoscale Res Lett 4:327–334

Ruban YJB, Mon SG, Roy DV (2013) Mechanical and thermal studies of unsaturated polyester-toughened epoxy composites filled with amine-functionalized nanosilica. Appl Nanosci 3:7–12

Shenogin S, Bodapati A, Xue L, Ozisik R, Keblinski P (2004) Effect of chemical functionalization on thermal transport of carbon nanotube composites. Appl Phys Lett 85:2229–2231

Singh BP, Singh D, Mathur RB, Dhami TL (2008) Influence of surface modified MWCNTs on the mechanical, electrical and thermal properties of polyimide nanocomposites. Nanoscale Res Lett 3:444–453

Singh BP, Saini P, Gupta TK, Garg P, Kumar G, Pande I, Pande S, Seth RK, Dhawan SK, Mathur RB (2011) Designing of multiwalled carbon nanotubes reinforced low density polyethylene nanocomposites for suppression of electromagnetic radiation. J Nanopart Res 13:7065–7074

Singh BP, Choudhary V, Saini P, Mathur RB (2012a) Designing of epoxy composites reinforced with carbon nanotubes grown carbon fiber fabric for improved electromagnetic interference shielding. AIP Adv 2:022151–022156

Singh K, Singh BP, Chauhan R, Basu T (2012b) Fabrication of amperometric bienzymatic glucose biosensor based on MWCNT tube and polypyrrole multilayered nanocomposite. J Appl Polym Sci 125:E235–E246

Singh BP, Bharadwaj P, Choudhary V, Mathur RB (2013a) Enhanced microwave shielding and mechanical properties of multiwall carbon nanotubes anchored carbon fiber felt reinforced epoxy multiscale composites. Appl Nanosci, pp 1–8.

Singh BP, Choudhary V, Saini P, Pande S, Singh VN, Mathur RB (2013b) Enhanced microwave shielding and mechanical properties of high loading MWCNT–epoxy composites. J Nanopart Res 15:1–12

Thess A, Lee R, Nikolaev P, Dai H, Petit P, Robert J, Xu C, Lee YH, Kim SG, Rinzler AG, Colbert DT, Scuseria GE, Tománek D, Fischer JE, Smalley RE (1996) Crystalline ropes of metallic carbon nanotubes. Science 273:483–487

Xie X-L, Mai Y-W, Zhou X-P (2005) Dispersion and alignment of carbon nanotubes in polymer matrix: a review. Mater Sci Eng R: Rep 49:89–112

Zhang M, Atkinson KR, Baughman RH (2004) Multifunctional carbon nanotube yarns by downsizing an ancient technology. Science 306:1358–1361

14

Stability and electronic properties of isomers of B/N co-doped graphene

Pooja Rani · V. K. Jindal

Abstract We study and compare some of the possible isomers of BN co-doped graphene on the basis of their composition and electronic properties. The effect of doping has been studied theoretically by substituting the C atoms of graphene with an equal amount of B/N with their concentration varying from 4 % (2 atoms of the dopant in 50 host atoms) to 24 % and choosing different doping sites for each concentration. We made use of VASP (Vienna Ab-initio Simulation Package) software based on density functional theory to perform all calculations. While the resulting geometries do not show much of distortion on doping, the electronic properties show a transition from semimetal to semiconductor with increasing number of dopants as in the case of individual B and N doping. The study shows that the BN doping introduces the band gap at the Fermi level unlike individual B and N doping which causes the shifting of Fermi level. High value of the cohesive energy indicates the stability of the resulting heterostructures. Isomers formed by choosing different doping sites differ significantly in relative stability and band gap introduced and these aspects, to a great extent, depend upon position of B and N atoms in the heterostructure.

Keywords Graphene · Isomer · Doping · Stability · Band gap · Density functional theory

P. Rani (✉) · V. K. Jindal
Department of Physics, Panjab University, Chandigarh 160014, India
e-mail: pgoyal0510@gmail.com

Introduction

Graphene is the name given to a single layer of sp^2 bonded carbon atoms arranged in a honeycomb pattern. It is a thermodynamically stable two-dimensional material, in which the valence and conduction bands touch at the so-called Dirac point and so is always referred as a semi-metal or zero-gap semiconductor. Since its successful fabrication in 2004 by micromechanical cleavage of graphite at the University of Manchester, UK, by the group of Geim (Novoselov et al. 2004), it has attracted great interest due to its fascinating properties and potential applications. Graphene has unique physical properties (Novoselov et al. 2004, 2005; Geim and Novoselov 2007; Castro Neto et al. 2009) like massless relativistic fermions that satisfy the Dirac equation, high electrical conductivity (even greater than silver) at room temperature, anomalous quantum Hall effect, optical transparency (Nair et al. 2008), finite conductivity even at zero charge carrier concentration and ballistic transport which originate from its hexagonal honeycomb lattice structure and make it a promising material for application in future nanoelectronics (Novoselov et al. 2006; Yang et al. 2010; Huang et al. 2007; Morozov et al. 2008), sensors (Romero et al. 2009) and optoelectronic applications (Bonaccorso et al. 2010). The linear dispersion curve at the Dirac point gives rise to exciting elementary electronic properties.

The special character of its charge carriers comes from the intersection of the π/π^* electronic bands occurring at the corners of its hexagonal Brillouin zone (Fig. 1). *P6/mmm* symmetry of the space group results in a band degeneracy at the Dirac points (K and K') in the hexagonal Brillouin zone (BZ) leading to zero band gap. This absence of a sizable band gap in graphene sets limitations on its practical applications. So it is of crucial importance to find

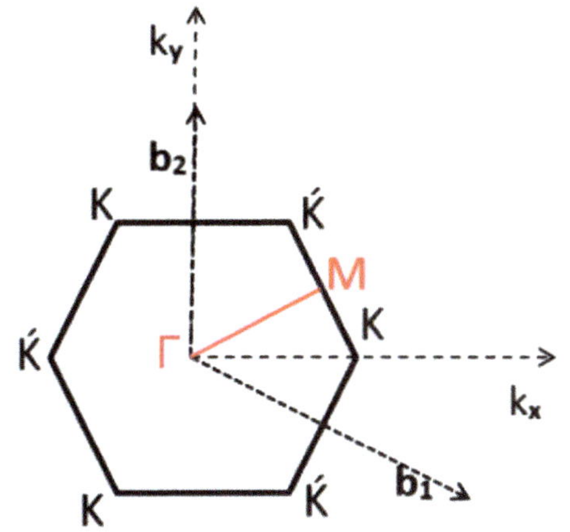

Fig. 1 Hexagonal Brillouin zone of graphene

Fig. 2 Optimized geometries and band structures of different isomers▶ in increasing order of dopant concentration [4 % (**A**), 8 % (**B**), 12 % (**C**), 16 % (**D**), 20 % (**E**) and 24 % (**F**)], respectively. In isomer a, B and N are at ortho position, in isomer b same type of dopants (B or N) are placed at adjacent positions, in isomer c, B and N are at para position (isomer c) and in isomer d, BN pair is at ortho position but distance between two BN pairs is large. Due to higher doping concentration the category (d) cannot be possible in 2 (**E**) and 2 (**F**)

methods to effectively tune the band gap of graphene for applications in nanoelectronics and optoelectronics.

A great amount of effort has been devoted to open a tunable band gap in graphene and different methods have been proposed: chemical functionalization (Wang et al. 2009; Bekyarova et al. 2009; Sofo et al. 2007; Zanella et al. 2008; Choi et al. 2009), doping with heteroatoms (Denis 2010; Dai et al. 2009), using electric fields (Avetisyan et al. 2009; Mak et al. 2009) and depositing graphene on substrates like SiO_2, SiC (Shemella and Nayak 2009; Peng and Ahuja 2008).

Doping with heteroatoms, especially B and N is a widely used technique to modulate the electronic properties (Panchakarla et al. 2009; Wu et al. 2010; Faccio 2010; Tang et al. 2012; Gebhardt et al. 2013) in graphene. B and N are neighboring elements of carbon in the periodic table and have nearly same size and mass that of carbon.

It has been shown by Wu et al. (2010) and in (Rani and Jindal 2013), that doping with B or N individually is an efficient method to open the band gap around the Dirac point, but it introduces the gap below or above the Fermi level. Further, as compared to B–B or N–N bond lengths, B–N bond length is comparable to C–C bond length, which makes the combination of B–N as a better choice to replace a C–C bond without causing much alteration of 2-D lattice. However, introduction of BN in graphene breaks the symmetry of graphene unit cell (Bhandary and Sanyal 2012), which can result in the opening up of a band gap in graphene and can be exploited for band gap-related applications.

It has been pointed out by Deng et al. (2011) that in case of BN co-doping, the gap is introduced at the Fermi level due to combination of pull–push of boron and nitrogen. Fan et al. (2012) have investigated the effect of concentration by varying the number of host atoms, and Manna and Pati (2011) have shown that the patching of graphene and h-BN sheet with semiconducting and/or insulating BxNy(Cz) nanodomains of various geometrical shapes and sizes (h-BN sheet) can significantly change the electronic and magnetic properties. The comparison of stability and band gap of the isomers of B–C–N composites have been studied by Bhandary and Sanyal (2012). However, the detailed analysis of varying concentration (changing the number of BN dopants) and position of B–N pairs and hence the interaction between them and their effect on the stability and band gap of resulting structures is yet to be done.

Therefore, in this paper we find it worthwhile to study the structural stability and the effect on band opening by incorporation of BN with different concentrations by increasing the number of BN dopants and also taking into account different possible doping sites.

Models and computational parameters

Structure minimization and electronic structure calculations have been performed using the VASP (Vienna Ab-initio Simulation Package) (Kresse and Furthmüller 1996; Kresse and Joubert 1999) code which uses plane waves as basis set. The approach is based on an iterative solution of the Kohn–Sham equation (Kohn and Sham 1965) of the density functional theory (DFT). That DFT methods underestimate the band gap of materials, is well known. However, standard DFT methods correctly describe the electronic properties of both graphene and BN (Liu et al. 2003). Therefore, DFT can be fairly used to describe electronic properties of graphene doped by BN, though underestimation of the band gap is possible.

In our calculations, core electrons are replaced by the projector-augmented wave pseudo-potentials and the Perdew–Wang (PW91) (Blöchl 1994) exchange–correlation (XC) functional of the generalized gradient approximation (GGA) is adopted. The plane-wave cutoff energy was set to 450 eV. The calculations of lattice constants and the atomic coordinates at different doping concentrations are made by the minimization of the total energy. A graphene sheet model of 5 × 5 supercell consisting of 50 atoms have been considered for the calculations. The vertical separation between the two sheets has been fixet at 15 Å to avoid the interplanar interactions. We have doped equal amount of B and N concentration thus generating equal number of C–B and C–N bonds. This results in practically identical

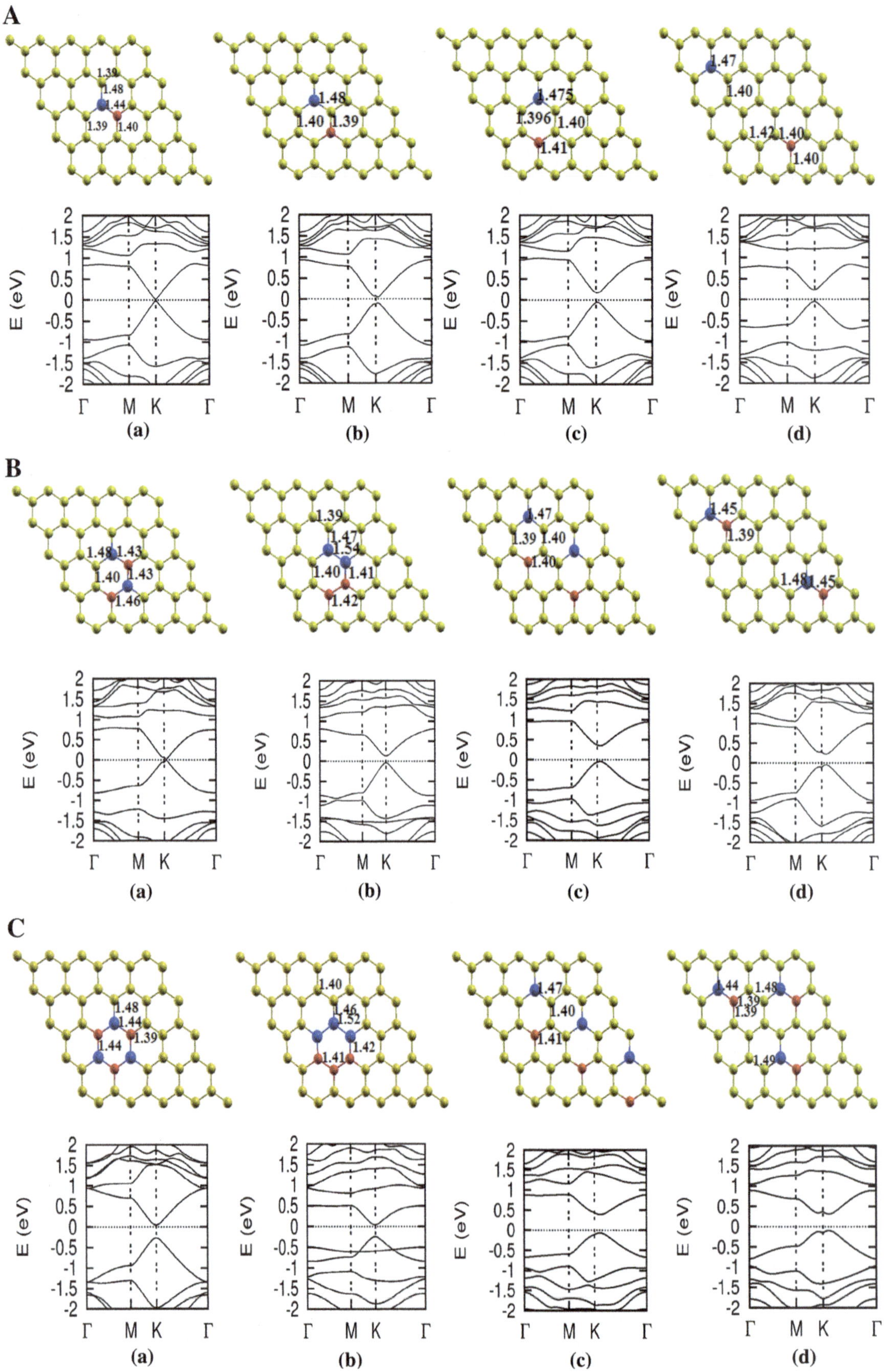
A
B
C
(a)
(b)
(c)
(d)
E (eV)
Γ M K Γ

average atom–atom distance. The Monkhorst–Pack scheme is used for sampling the Brillouin zone. In the calculations, the structures are fully relaxed with a Gamma-centered 7 × 7 × 1 k-mesh. During all calculation processes, the partial occupancies were treated using the tetrahedron methodology with Blöchl corrections (Blöchl et al. 1994). For band structure calculation, partial occupancies for each wavefunction were determined using the Gaussian smearing method with a broadning of 0.01 eV. For geometry optimizations, all the internal coordinates were relaxed until the Hellmann–Feynman forces were <0.008 eV/Å. In order to give a numerical account of the stability of the doped systems, cohesive energy has been calculated using the following expression:

$$^{*}E_{coh}=[E_{tot}-n_iE_i]/n(i = \text{C, B, N}),$$

where E_{coh} is the cohesive energy per atom of BN-doped configuration and E_{tot} and E_i represent total energies of a structure and of individual elements present within the same supercell, respectively. n_i is the number of ith species present in the configuration, while n is the total number of atoms present in a supercell ($n = 50$). A pure graphene sheet was optimized and its bond length and band structure (Geim and Novoselov 2007) are in accordance with literature which confirms the reliability of the simulation method used.

Results and discussion

We carry out our calculations with different concentrations of BN doping (4, 8, 12, 16, 20 and 24 %) in graphene and also take into account different sites of doping for the same concentration. Although we analyze several possible isomers for different concentrations of dopants, for the sake of brevity, we present here only a few isomers with the configurations of mainly four doping sites, i.e. when B and N are at ortho position (isomer a), when same type of dopants (B or N) are at adjacent positions (isomer b), para position (isomer c) and when BN pair is at ortho position but distance between two BN pairs is large (d). The structures of different isomers corresponding to different dopant concentrations are given in Fig 2.

Comparative stability of the isomers

The replacement of two carbon atoms with B–N introduces several interesting features. First, there is very small difference between B–N bond (1.45 Å) and C–C bond length (1.42 Å), but C–B (1.49 Å) bond is 15 % longer than C–N (1.35 Å) bond. So, this will cause distortion in the doped structures, which is going to lower its stability than pure graphene. Second, as mentioned earlier, this will break sublattice symmetry in graphene and bring a significant change in electronic properties. Having these points in mind, in this section we will discuss the stability for various isomers at different concentrations.

We observe that for the low doping concentrations the isomers are more or less energy equivalent, as in the case of 4 % doping concentration (Fig 2A). This is due to the fact that aromaticity and π-band formation of the C-ring do not change much as a whole, which makes those structures iso-energetic. But as the doping concentration goes on increasing, the energy difference becomes significant. This is due to increase in number of B–C, N–C, B–B and N–N bonds and the mismatch leads to strain in the structure. So the structures with more number of these bonds are significantly less stable than other isomers with same doping concentration [see category (c) of Fig 2B–F]. Also the stability for the same doping concentration is greater as compared to the separate B or N doping. This may be due to the equal number of C–B and C–N bonds.

The second observation is that the isomers in which BN is placed as a pair (ortho positions), have considerably larger cohesive energy and hence more stable than their counterparts (para positions) in which B and N are apart. This is due to more stable B–N bond than B–C or N–C bond. When the bond energies are numerically taken into account, the ordering of the bonds is the following (Bhandary and Sanyal 2012; Nozaki and Itoh 1996):

$$\begin{aligned} &\text{B} - \text{N (4.00 eV)} > \text{C} - \text{C (3.71 eV)} > \\ &N - \text{C (2.83 eV)} > \text{B} - \text{C (2.59 eV)} > \\ &B - \text{B (2.32 eV)} > \text{N} - \text{N (2.11 eV)} \end{aligned}$$

So in a way the isomers having more number of C–C or B–N bonds will have more stability over the others.

Thus there is apparent competition between structural stress and chemical bond strength that can be described by calculating the internal energies of the different geometries which are given in Table 1.

Another important aspect is the charge transfer. N is the most electronegative and B is the least one among the three atoms. In case of single BN pair-doped graphene the net charge on N is −1.54 e and +1.82 e on B. This introduces an ionic character in the bond formation. So presence of a B–B or N–N bond makes the structure unstable due to electrostatic repulsion. If the positive and negative charges are situated in an alternative manner, as in case of B–N, binding energy is increased providing stability to the structure [see category (a) of the isomers]. The results show that the B and N binding together is energetically more favorable than being separate in the hexagonal network.

The analysis of structural properties shows the shrinkage of C–C bond lengths which are just adjacent to boron atoms [see isomers (b) in Fig 2] dopants. This possibly results from longer C–B bond lengths because of large

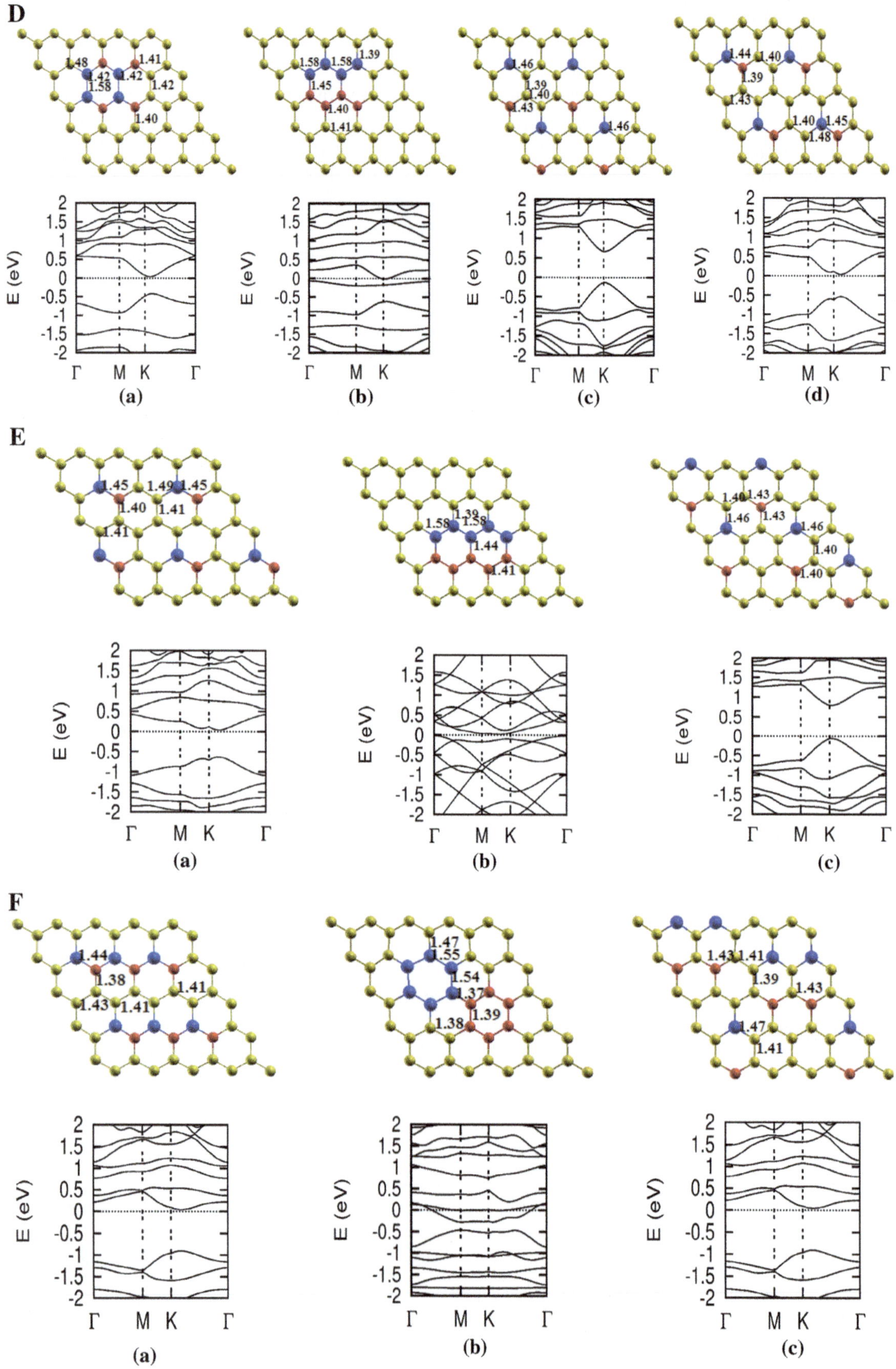

Fig. 2 continued

Table 1 Cohesive energy and band gap for increasing doping concentration and various possible configurations a, b, c, d as described in Fig. 2

BN concentration (%)	Configuration (as shown in Fig. 3)	E^{*}_{coh} (eV/atom)	Band gap (eV)
4	a	−9.13	0.20
	b	−9.11	0.00
	c	−9.11	0.25
	d	−9.10	0.28
8	a	−9.09	0.09
	b	−9.04	0.18
	c	−9.06	0.40
	d	−9.10	0.43
12	a	−9.04	0.31
	b	−8.93	0.28
	c	−8.98	0.56
	d	−9.02	0.58
16	a	−8.98	0.56
	b	−8.80	0.20
	c	−8.95	0.73
	d	−8.87	0.79
20	a	−8.93	0.83
	b	−8.67	0.20
	c	−8.82	0.85
24	a	−8.94	1.06
	b	−8.39	0.48
	c	−8.71	1.08

covalent radius of B than C and preservation of the structure. This also results in large strain in the lattice when the dopants are placed at adjacent positions. The data regarding changes in bond lengths are presented in the structural part of the figures presented in Fig 2.

Electronic properties

The redistribution of surface charge which breaks the local symmetry of graphene is the main reason of band gap opening of graphene on BN doping, which leads to the separation of the valence band and conduction band of graphene at the Dirac point. Since B doping leads to band gap opening below the Fermi level and N doping, above the Fermi level, the counterbalance of both creates a gap at the Fermi level. In addition, the coulomb dipole from BN molecule may also be important to the breaking of symmetry of graphene structure and thus to the band gap opening.

The electronic structure analysis shows that band gap in general increases with increase in concentration of doping. The values of the band gap corresponding to different doping concentrations and isomers are given in Table 1. We give in Fig. 3 a plot of this variation of band gap with % concentration of dopants for the largest band gap for each concentration. We see that the variation of band gap with dopant concentration is almost linear.

As shown by Bhandary and Sanyal (2012), the relative positions of B and N around an all C hexagon is an important factor that determines the electronic properties of the doped isomer. We observe that the band gap is more in the structure in which B and N are at para-positions than the isomer in which they are at ortho positions. This tells us that the interaction between B/N pair also plays an important role in the formation of the band gap.

The width of band gap goes on increasing with the increase in distance between B and N atoms, as shown in Fig. 2 [category (c)], which can be explained by the sum of the effects of Boron and Nitrogen doping into graphene, whereas single Boron or Nitrogen atom doping will break the π bond with inducing a gap in the valence or conduction band only. This also explains the higher band gap as compared to the individual B and N doping. Also when distance between two BN pairs increases, the ortho-para aspect has no effect, i.e. the isomers with ortho and para arrangement of BN pairs have almost same band gap because ortho-para effect has more pronounced effect in case of single hexagon.

From the Fig. 2 it can be observed that for [configuration (b)] corresponding to different concentrations, containing B–B and N–N at adjoint positions, band structure is highly deformed and linear dispersion relation at the Dirac point is also affected. This can be explained by the fact that linear dispersion in graphene is due to its crystal symmetry. So when the dopants are placed at adjacent positions as in isomers. (a), the symmetry is highly distorted which results into deformation of band structure. Also the band gap is minimum in this case.

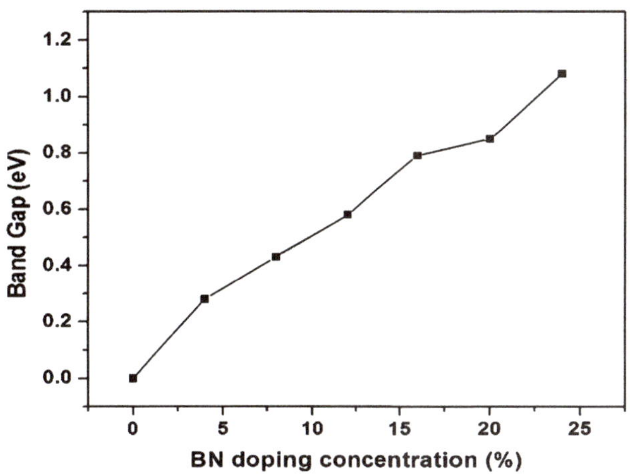

Fig. 3 Variation of band gap with increasing concentration of BN dopants. The *plot* is drawn for the maximum value of band gap for each concentration

Conclusion

Ab-initio calculations were performed to analyze the geometry and electronic structure of isomers of BN-doped graphene sheet. We observe transition of graphene from semimetal to semiconductor with increasing number of dopants. It has been observed that isomers formed by choosing various doping sites differ in the stability, bond length and band gap introduced, depending upon the positioning of the dopant atoms. Thus, through our detailed analysis, we suggest a theoretical approach to design heterographenes with different amount and composition of either B or N doping (Rani and Jindal 2013) or codoping of both (BN). Both mechanisms result in tuning of band gap in a wide range, for device applications. While B or N doping results in band gap at Dirac point with shifting of Fermi Level, BN codoping creates the gap at Fermi level, thus being more efficient. Due to their tunable band structure properties, the resulting materials can be used in applications in various sectors, e.g., nano electronics, gas sensing and hydrogen storage. BN-doped graphene can act as a good electrocatalyst for oxygen reduction reaction (ORR) due to high electron density present on the N atom in BCN heterostructure. Some experimental study in this direction was performed recently by Shuangyin et al. to prepare BN co-doped graphene with tunable composition which has superior electrocatalytic activity over commercial Pt/C electrocatalysts for the oxygen (Wang et al. 2012). The study supports our results and motivates to do further work in this direction.

Acknowledgments We express our gratitude to VASP team for providing the code, the HPC facilities at IUAC (New Delhi) and the departmental computing facilities at Department of Physics, PU, Chandigarh. PR gratefully acknowledges financial support from UGC. VKJ acknowledges support from CSIR as Emeritus Scientist.

References

Avetisyan AA, Partoens B, Peeters FM (2009) Electric field tuning of the band gap in graphene multilayers. Phys Rev B 79:035421

Bekyarova E, Itkis ME, Ramesh P, Berger C, Sprinkle M, De Heer WA, Haddon RC (2009) Chemical modification of epitaxial graphene: spontaneous grafting of aryl groups. J Am Chem Soc 131:1336–1337

Bhandary S, Sanyal B (2012) Graphene-boron nitride composite: a material with advanced functionalities. In: Ning Hu (ed) Composites and their properties. InTech, Rijeka, pp 1–14

Blöchl PE (1994) Projector augmented-wave method. Phys Rev B 50:17953

Blöchl PE, Jepsend O, Andersen OK (1994) Improved tetrahedron method for Brillouin-zone integrations. Phys Rev B 49: 16223–16233

Bonaccorso F, Sun Z, Hasan T, Ferrari AC (2010) Graphene photonics and optoelectronics. Nat Photonics 4:611–622

Castro Neto AH, Guinea F, Peres NMR, Novoselov KS, Geim AK (2009) The electronic properties of graphene. Rev Mod Phys 81:109–162

Choi J, Kim K-J, Kim B, Lee H, Kim S (2009) Covalent functionalization of epitaxial graphene by azidotrimethylsilane. J Phys Chem C Lett 113:9433–9435

Dai J, Yuan J, Giannozzi P (2009) Gas adsorption on graphene doped with B, N, AL, S: a theoretical study. Appl Phys Lett 95:232105

Deng X, Wu Y, Dai J, Kang D, Zhang D (2011) Electronic structure tuning and band gap opening of graphene by hole/electron codoping. Phys Lett A 375:3890–3894

Denis PA (2010) Band gap opening of monolayer and bilayer graphene doped with aluminium, silicon, phosphorus and sulfur. Chem Phys Lett 492:251–257

Faccio R, Fernández-Werner L, Pardo H, Goyenola C, Ventura ON, Mombrú AW (2010) Electronic and structural distortions in graphene induced by carbon vacancies and boron doping. J Phys Chem C 114:18961–18971

Fan X, Shen Z, Liu AQ, Kuo JL (2012) Band gap opening of graphene by doping small boron nitride domains. Nanoscale 4:2157–2165

Gebhardt J, Koch RJ, Zhao W, Hofert O, Gotterbarm K, Mammadov S, Papp C, Gorling A, Steinruck H-P, Seyller Th (2013) Growth and electronic structure of boron-doped graphene. Phys Rev B 87:155437

Geim AK, Novoselov KS (2007) The rise of graphene. Nat Mater 6:183–191

Huang B, Yan QM, Zhou G, Wu J, Gu BL, Duan WH, Liu F (2007) Appl Phys Lett 91:253122

Kohn W, Sham LJ (1965) Self-consistent equations including exchange and correlation effects. Phys Rev A 140:1133–1138

Kresse G, Furthmüller J (1996) Efficient iterative schemes for *ab - initio* total energy calculations using a plane-wave basis set. Phys Rev B 54:11169–11186

Kresse G, Joubert D (1999) From ultrasoft pseudopotentials to the projector augmented-wave method. Phys Rev B 59:1758–1775

Liu Lei, Feng YP, Shen ZX (2003) Structural and electronic properties of h-BN. Phys Rev B 68:104102

Mak KF, Lui CH, Shan J, Heinz TF (2009) Observation of an electric-field-induced band gap in bilayer graphene by infrared spectroscopy. Phys Rev Lett 102:256405

Manna AK, Pati SK (2011) Tunable electronic and magnetic properties in BxNyCz Na nohybrids: effect of domain segregation. J Phys Chem C 115:10842–10850

Morozov SV, Novoselov KS, Katsnelson MI, Schedin F, Elias DC, Jaszczak JA, Geim AK (2008) Giant intrinsic carrier mobilities in graphene and its bilayer. Phys Rev Lett 100:016602

Nair RR, Blake P, Grigorenko AN, Novoselov KS, Booth TJ, Stauber T, Peres NMR, Geim AK (2008) Fine structure constant defines visual transparency of graphene. Science 320:1308

Novoselov KS, Geim AK, Morozov SV, Jiang D, Zhang Y, Dubonos SV, Grigorieva IV, Firsov AA (2004) Electric field effect in atomically thin carbon films. Science 306:666–669

Novoselov KS, Geim AK, Morozov SV, Jiang D, Katesnelson MI, Grigoreva V, Dubonos SV, Firsov AA (2005) Two-dimensional gas of massless Dirac fermions in graphene. Nature (London) 438:197–200

Novoselov KS, McCann E, Morozov SV, Fal'ko VI, Katsnelson MI, Zeitler U, Jiang D, Schedin F, Geim AK (2006) Unconventional quantum Hall effect and Berry's phase of 2π in bilayer graphene. Nature Phys 2:177–180

Nozaki H, Itoh S (1996) Structural stability of BC_2N. J Phys Chem Solids 57:41–49

Panchakarla LS, Subrahmanyam KS, Saha, Achutharao Govindaraj SK, Rishnamurthy HR, Waghmare UV, Rao CNR (2009) Synthesis, structure, and properties of boron- and nitrogen-doped graphene. Adv Mat 21:4726–4730

Peng X, Ahuja R (2008) Symmetry breaking induced bandgap in epitaxial graphene layers on SiC. Nano Lett 8:4464–4468

Rani Pooja, Jindal VK (2013) Designing band gap of graphene by B and N dopant atoms. RSC Adv 3:802–812

Romero HE, Joshi P, Gupta AK, Gutierrez HR, Cole MW, Tadigadapa SA, Eklund PC (2009) Adsorption of ammonia on graphene. Nanotechnology 20:245501

Shemella P, Nayak SK (2009) Appl Phys Lett 94:032101

Sofo JO, Chaudhari AS, Barber GD (2007) Graphane: a two-dimensional hydrocarbon. Phys Rev B 75:153401

Tang YB, Yin LC, Yang Y, Bo XH, Cao YL, Wang HE, Zhang WJ, Bello I, Lee ST, Cheng HM, Lee CS (2012) Tunable band gaps and p-type transport properties of boron-doped graphenes by controllable ion doping using reactive microwave plasma. ACS Nano 6:1970–1978

Wang X, Li X, Yoon Y, Weber PK, Wang H, Guo J, Dai H (2009) N-doping of graphene through electrochemical reactions with ammonia. Science 324:768–771

Wang S, Zhang L, Xia Z, Roy A, Chang DW, Baek JB, Dai L (2012) BCN graphene as efficient metal-free electrocatalyst for the oxygen reduction reaction. Angew Chem 124:4285–4288

Wu M, Cao C, Jiang JZ (2010) Light non-metallic atom (B, N, O and F)-doped graphene: a first-principles study. Nanotechnology. 21:505202

Yang XB, Liu GX, Balandin AA, Mohanram K (2010) Triple-mode single-transistor graphene amplifier and its applications. ACS Nano 4:5532–5538

Zanella I, Guerini S, Fagan SB, Mendes J, Filho AG, Souza Filho (2008) Chemical doping-induced gap opening and spin polarization in graphene. Phys Rev B 77:073404

Flow and heat transfer analysis of Williamson nanofluid

S. Nadeem · S. T. Hussain

Abstract In the present paper, we have examined the two-dimensional flow of Williamson fluid over a stretching sheet under the effects of nano-sized particle also described as nano Williamson fluid. The boundary layer equations of nano Williamson fluid model along with energy and nanoparticle volume fraction are presented and simplified with the help of useful transformations. Governing equations are somewhat different from the ones present in literature (reason is explained in the introduction section). The expressions for coefficients of skin friction and Nusselt number have been computed. The physical features of non-dimensional Williamson parameter, Lewis number, Schmidt number and nano particle parameters (diffusivity ratio and heat capacities ratio) have been discussed by plotting the graphs of velocity, temperature and nanoparticle volume fraction.

Keywords Nano particles · Williamson fluid model · Stretching sheet · Heat transfer

Introduction

Nanofluids are colloidal suspension of base fluid and nanoparticles (1–100 nm). Nowadays these fluids are focus of research because these small size particles can enhance the coefficient of heat transfer several times as compared with base fluid. In some cases researchers have reported about 40 % increase in thermal conductivity. This feature makes them very much suitable for cooling and solidification systems. In polymer industry solidification is an important phenomenon during polymers extrusion. Nanoparticles can play an important role in heat transfer during solidification. Also in extrusion of packaging films clay nanocomposites (Durmus and Kasgoz 2007) and silicate nanoparticles can be used, whose incorporation can create barrier for gases and increases the reliability of packaging films. Heat transfer enhancement due to these small size particles was first time reported by Masuda et al. (1993) This term was first introduced by Choi (1995); he defined nanofluid a liquid containing dispersed submicronic solid particles (nanoparticles). Three models have been widely used to describe the convective transport of nanofluids as discussed by Buongiorno (2006). As the first two models had drawbacks see Buongiorno (2006), he presented a third model and discussed several slip mechanisms and finally incorporated the effects of Brownian motion and thermophoretic diffusion into the equations. In the present article we will stick with Buongiorno model. Nield and Kuznetsov (2009) studied the thermal instability in a porous medium layer saturated by a nanofluid. They introduced the velocity similarity transformations dependent on thermal diffusivity of the porous medium. In another paper (Nield and Kuznetsov 2011) they considered the double diffusion: solute and nanoparticle diffusion in a base fluid. This time again, they introduced the velocity similarity transformations dependent on thermal diffusivity of fluid. Because of these similarity transformations Lewis number (*Le*) appears instead of Schmidt number (*Sc*) in their nanoparticle volume fraction equations. If we look in Buongiorno's paper (Buongiorno 2006), in nanoparticle volume fraction equation, Schmidt number appears instead of Lewis number. Later on several authors (Makinde and Aziz

S. Nadeem · S. T. Hussain (✉)
Department of Mathematics, Quaid-I-Azam University 45320, Islamabad 44000, Pakistan
e-mail: s_thussain@yahoo.com

S. Nadeem
e-mail: snqau@hotmail.com

2011; Khan and Pop 2010; Bachok et al. 2010; Kandasamya et al. 2011) citing the work of Nield and Kuzentsov introduced the Lewis number in nanoparticle volume fraction equation without considering the fact that their velocity transformations do not contain thermal diffusivity. These authors mistakenly reported the Schmidt number as Lewis number in their articles. If we do not choose velocity and stream functions dependent on thermal diffusivity then Lewis number will only appear in heat equation.

Stretching flows study has not lost its appeal yet due to its wide range of applications in polymer, glass, copper wire drawing and plastic industry. Sakiadis (1961) was the first one to study the boundary layer flow over a stretching sheet. Tsou et al. (1967) discussed the heat transfer effects on the boundary layer flow over a stretching sheet. Erickson et al. (1966) extended the work of Tsou et al. for mass transfer. Afterwards large numbers of theoretical studies have been carried out by numerous authors (Liu 2005; Rosali et al. 2012; Kelson and Farrell 2001; Kumaran and Ramanaiah 1996; Ali 1995).

Nadeem et al. (2013) were the first ones who developed the two-dimensional boundary layer equations for the flow of Williamson fluid past a stretching sheet. In the present article we are presenting the two-dimensional flow of nano Williamson fluid (Williamson 1929; Lyubimov and Perminov 2002; Nadeem 2010; Dapra 2007) over a stretching sheet. The governing boundary layer transport equations are first simplified using the suitable similarity transformations. Resulting equations are solved using the Homotopy analysis method (Liao 2003, 2004; Nadeem et al. 2010; Nadeem and Hussain 2009; Ellahi and Riaz 2010; Abbasbandy 2006, 2007; Hayat and Qasim 2010). To observe the convergence of obtained solution h-curve and convergence table are drawn. Graphs have been plotted to observe the impact of various physical parameters on transport phenomenon. Finally, tables are drawn to analyse the effects of important parameters on heat transfer coefficient and nanoparticle volume fraction gradient.

Mathematical formulation

Let us consider the two-dimensional steady flow of an incompressible nano Williamson fluid over a stretching surface. The plate is stretched along x-axis with a velocity Bx, where $B > 0$ is stretching parameter. The fluid velocity, temperature and nanoparticle concentration near surface are assumed to be U_w, T_w and C_w, respectively. The general transport equations for nanofluid are given by (Buongiorno 2006)

$$\operatorname{div} \mathbf{V} = 0, \tag{1}$$

$$\rho \frac{d\mathbf{V}}{dt} = \operatorname{div} \mathbf{S} + \rho \mathbf{b}, \tag{2}$$

$$\rho c\left(\frac{\partial T}{\partial t} + \mathbf{V}\cdot\nabla T\right) = \nabla\cdot k\nabla T + \rho_p c_p \times \left(D_B \nabla\phi\cdot\nabla T + D_T \frac{\nabla T\cdot\nabla T}{T_\infty}\right), \tag{3}$$

$$\frac{\partial C}{\partial t} + \mathbf{V}\cdot\nabla C = \nabla\cdot\left(D_B\nabla C + D_T\frac{\nabla T}{T_\infty}\right), \tag{4}$$

where $\mathbf{V}(u(x, y), v(x, y), 0)$ is the velocity vector, ρ is nanofluid density, $\mathbf{S}$ is Cauchy stress tensor, $\mathbf{b}$ is body force vector, ρc and $\rho_p c_p$ are heat capacities of nanofluid and nanoparticles, respectively, T is temperature, k is nanofluid thermal conductivity, D_B is Brownian diffusion coefficient, C is nanoparticle volumetric fraction, D_T is thermophoretic diffusion coefficient and T_∞ is the ambient fluid temperature. For Williamson fluid model Cauchy stress tensor $\mathbf{S}$ is defined in (Dapra 2007) as

$$\mathbf{S} = -p\mathbf{I} + \tau, \tag{5}$$

$$\boldsymbol{\tau} = \left[\mu_\infty + \frac{(\mu_0 - \mu_\infty)}{1 - \Gamma\dot{\gamma}}\right]\mathbf{A}_1, \tag{6}$$

where $\boldsymbol{\tau}$ is extra stress tensor, μ_0 is limiting viscosity at zero shear rate and μ_∞ is limiting viscosity at infinite shear rate, $\Gamma > 0$ is a time constant, $\mathbf{A}_1$ is the first Rivlin–Erickson tensor and $\dot{\gamma}$ is defined as follows:

$$\dot{\gamma} = \sqrt{\frac{1}{2}\pi}, \quad \pi = \operatorname{trace}(\mathbf{A}_1^2), \tag{7}$$

Here we considered the case for which $\mu_\infty = 0$ and $\Gamma\dot{\gamma} < 1$. Thus Eq. (6) can be written as

$$\boldsymbol{\tau} = \left[\frac{\mu_0}{1 - \Gamma\dot{\gamma}}\right]\mathbf{A}_1, \tag{8}$$

or by using binomial expansion we get

$$\boldsymbol{\tau} = \mu_0[1 + \Gamma\dot{\gamma}]\mathbf{A}_1. \tag{9}$$

Making use of Eqs. (5) and (9) in Eqs. (1) to (4), the two-dimensional boundary layer equations governing the flow are given by

$$\frac{\partial u}{\partial x} + \frac{\partial v}{\partial y} = 0, \tag{10}$$

$$u\frac{\partial u}{\partial x} + v\frac{\partial u}{\partial y} = \nu\frac{\partial^2 u}{\partial y^2} + \sqrt{2}\nu\Gamma\frac{\partial u}{\partial y}\frac{\partial^2 u}{\partial y^2}. \tag{11}$$

$$u\frac{\partial T}{\partial x} + v\frac{\partial T}{\partial y} = \alpha\frac{\partial^2 T}{\partial y^2} + \frac{\rho_p c_p}{\rho c}\left(D_B\frac{\partial C}{\partial y}\frac{\partial T}{\partial y} + \frac{D_T}{T_\infty}\left(\frac{\partial T}{\partial y}\right)^2\right), \tag{12}$$

$$u\frac{\partial C}{\partial x} + v\frac{\partial C}{\partial y} = D_B\frac{\partial^2 C}{\partial y^2} + \frac{D_T}{T_\infty}\frac{\partial^2 T}{\partial y^2}, \tag{13}$$

where $u(x, y)$ and $v(x, y)$ are horizontal and vertical components of velocity, ν is kinematic viscosity and α is nanofluid thermal diffusivity. The corresponding boundary conditions are given by

$$u = U_w;\ v = 0;\ T = T_w,\ C = C_w \quad \text{at } y = 0,$$
$$u \to 0;\ T = T_\infty,\ C = C_\infty \quad \text{as } y \to \infty. \tag{14}$$

Since the surface is stretched with velocity Bx, thus $U_w = Bx$. Introducing the following transformations in above equations

$$u = Bxf'(\eta), \quad v = -\sqrt{B\nu}f(\eta), \quad \eta = \sqrt{\frac{B}{\nu}}y,$$
$$\theta = \frac{T - T_\infty}{T_w - T_\infty}, \quad \phi = \frac{C - C_\infty}{C_w - C_\infty}, \tag{15}$$

with the help of above transformations, Eq. (10) is identically satisfied and Eqs. (11) to (13) along with boundary conditions (14) take the following form:

$$f''' - f'^2 + ff'' + \lambda f'' f''' = 0, \tag{16}$$

$$\theta'' + Prf\theta' + \frac{N_c}{Le}\phi'\theta' + \frac{N_c}{Le \times N_{bt}}\theta'^2 = 0, \tag{17}$$

$$\phi'' + Scf\phi' + \frac{1}{N_{bt}}\theta'' = 0, \tag{18}$$

where f, θ and ϕ are functions of η and prime denotes derivatives w.r.t η. The corresponding boundary conditions are

$$f = 0, f' = 1,\ \theta = 1,\ \phi = 1 \quad \text{at } \eta = 0,$$
$$f' = 0,\ \theta = 0,\ \phi = 0 \quad \text{as } \eta \to \infty. \tag{19}$$

In above transformed equations the following non-dimensional parameters are introduced:

$$\lambda = \Gamma x\sqrt{\frac{2B^3}{\nu}} \text{ (Non Newtonian Williamson parameter)},$$

$$Pr = \frac{\upsilon}{\alpha} \text{ (Prandtl number = momentum diffusivity/ nanofluid thermal diffusivity)},$$

$$Le = \frac{\alpha}{D_B} \text{ (Lewis number = nanofluid thermal diffusivity /Brownian diffusivity)},$$

$$Sc = \frac{\upsilon}{D_B} \text{ (Schmidt number = momentum diffusivity /Brownian diffusivity)}.$$

$$N_c = \frac{\rho_p c_p}{\rho c}(C_w - C_\infty) \text{ (Heat capacities ratio = nano particles heat capacity/nanofluid heat capacity)},$$

$$N_{bt} = \frac{D_B T_\infty (C_w - C_\infty)}{D_T (T_w - T_\infty)} \text{ (Diffusivity ratio = Brownian diffusivity/thermophoretic diffusivity)}.$$

If we put $\lambda = 0$, our problem reduces to the one for Newtonian nano and for $D_B = D_T = 0$ in Eq. (12) our heat equation reduces to the classical boundary layer heat equation in the absence of viscous dissipation. Physical quantities of interest are Local skin friction coefficient c_f, Local Nusselt number Nu and Local Sherwood number Sh.

$$c_f = \frac{\tau_w}{\rho U_w^2}, \quad Nu = \frac{-x}{T_w - T_\infty}\frac{\partial T}{\partial y}\bigg|_{y=0},$$
$$Sh = \frac{-x}{C_w - C_\infty}\frac{\partial C}{\partial y}\bigg|_{y=0}, \tag{20}$$

or by introducing the transformations (15), we have

$$\sqrt{Re}c_f = \left[\left(f'' + \frac{\lambda}{2}f''^2\right)\right]_{\eta=0}, \quad \frac{Nu}{\sqrt{Re}} = -\theta'(0),$$
$$\frac{Sh}{\sqrt{Re}} = -\phi'(0), \tag{21}$$

where $Re = Bx^2/\nu$ is local Reynolds number. Physical parameters of interest will be discussed later in the results section.

Solution technique

Solutions of Eqs. (16–18) are obtained with the help of well-known Homotopy analysis technique (HAM). HAM is a strong analytic technique to solve linear and non-linear, ordinary and partial differential equations. HAM was developed by Liao in 1992. HAM can be equally applied to weak and strong nonlinear problems because it is independent of small physical parameter restriction. It also provides a way to check and adjust the convergence of obtained solution with the help of auxiliary parameters and base functions. The initial guess and operators are taken as

$$f_0(\eta) = 1 - \exp(-\eta),$$
$$\theta_0(\eta) = \exp(-\eta), \tag{22}$$
$$\phi_0(\eta) = \exp(-\eta),$$

$$L_f(f) = \frac{d^3 f}{d\eta^3} - \frac{df}{d\eta},$$
$$L_\theta(\theta) = \frac{d^2\theta}{d\eta^2} + \frac{d\theta}{d\eta}, \tag{23}$$
$$L_\phi(\phi) = \frac{d^2\phi}{d\eta^2} + \frac{d\phi}{d\eta}.$$

The convergence of Homotopy analysis solution greatly depends on the choice of auxiliary parameter h_f, h_θ, h_ϕ.

Results and discussion

In order to check the convergence of obtained solutions, combine h-curve is plotted. It can be observed from Fig. 1

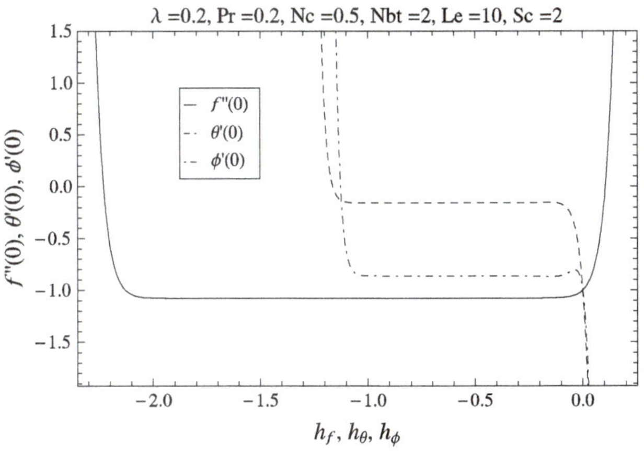

Fig. 1 Combine plot for h curves

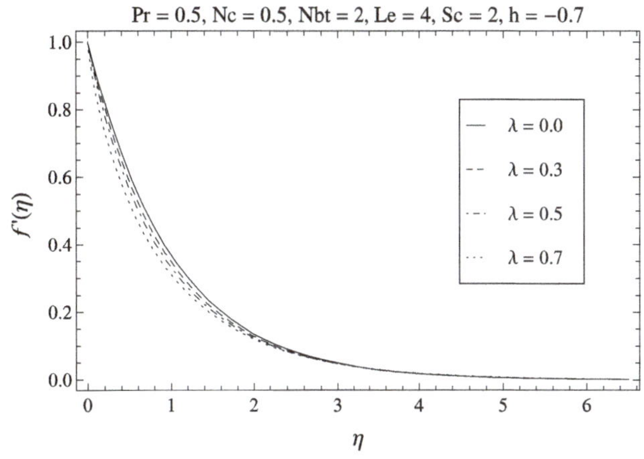

Fig. 2 Velocity variation against different values of λ

Table 1 Convergence of HAM solution for different order of approximation when $Pr = 0.5$, $N_c = 0.5$, $N_{bt} = 2$, $Le = 4$, $Sc = 2$, $\lambda = 0.2$ and $h = h_f = h_\theta = h_\phi = -0.7$

Order of approximation	$-f''(0)$	$-\theta'(0)$	$-\phi'(0)$
1	1.047	0.409	0.650
5	1.076	0.302	0.835
10	1.076	0.308	0.824
18	1.076	0.308	0.824
25	1.076	0.308	0.824
30	1.076	0.308	0.824

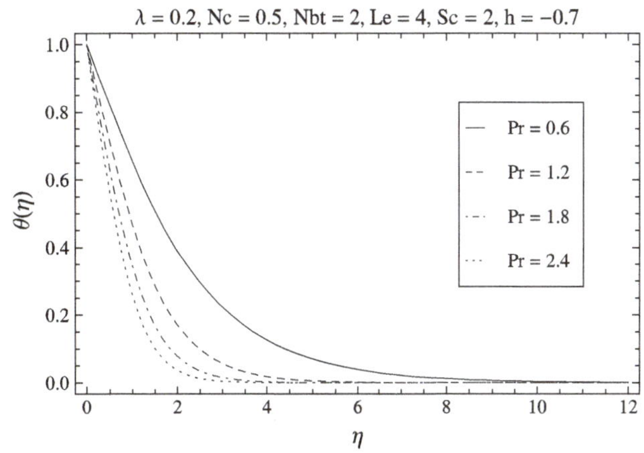

Fig. 3 Temperature variation against different values of Pr

that the admissible range for h_f, h_θ, h_ϕ is $-1.9 \le h_f \le -0.3, -0.9 \le h_\theta \le -0.3, -1.0 \le h_\phi \le -0.3$. Convergence table has been drawn for $f''(0), \phi'(0), \theta'(0)$ when $h = h_f = h_\theta = h_\phi = -0.7$. It is found that the convergence is achieved at 10th order of approximation (Table 1). All the tables and graphs are plotted at 22nd order of approximation.

From Fig. 2 it is observed that for a nanofluid the velocity as well as the boundary layer thickness decreases with the increase in non-Newtonian parameter λ. With increase in Prandtl number Pr, temperature as well as thermal boundary layer thickness decreases, see Fig. 3. It is also noted that the temperature and thermal boundary layer thickness decrease with increase in lewis number Le and N_{bt} (Figs. 4, 5). Since N_{bt} is the ratio of Brownian to thermophoretic diffusivities, increase in N_{bt} means greater activity of nanofluid particles. While they increase with increase in λ and N_c (see Figs. 6, 7). To observe the effects of different parameters on nanoparticle volume fraction, graphs have been plotted against λ, Sc and N_{bt}. Nanoparticle volume fraction decreases with increase in Schmidt number Sc and N_{bt}, while it decrease with increase in λ. (See Figs. 8, 9, 10). It can also be observed from Fig. 11 that nanoparticle volume fraction (close to the boundary) increases with increase in Pr. The behaviour shift of graph away from wall greatly depends on the values of N_{bt}. Since Eq. (18) is a second-order differential equation in both θ and ϕ, here the only thing which can make difference is the value of N_{bt}. Also if the value of N_{bt} is very large then ϕ hardly depends on θ. Therefore, effects of Prandtl number almost diminish for very large values of N_{bt}. It can also be observed from the graphs of velocity, temperature and Nanoparticle volume fraction that nanoparticle volume fraction and temperature boundary layers survive longer as compared with the velocity boundary layer. The importance of nanofluid study is because of heat transfer enhancement. To observe the effects of effective parameters on heat transfer closed to the wall, we plotted the graphs for $-\theta'(\eta)$. It can be observed from Figs. 12 and 13 that the heat transfer in fluid increases with the increase in N_{bt} and Le; this enhancement is very significant in the region very close to the wall but the parameter effect is almost negligible on heat transfer away from the wall. Table 2 is drawn to compare our results for the viscous case in the absence of nanoparticles. These results are

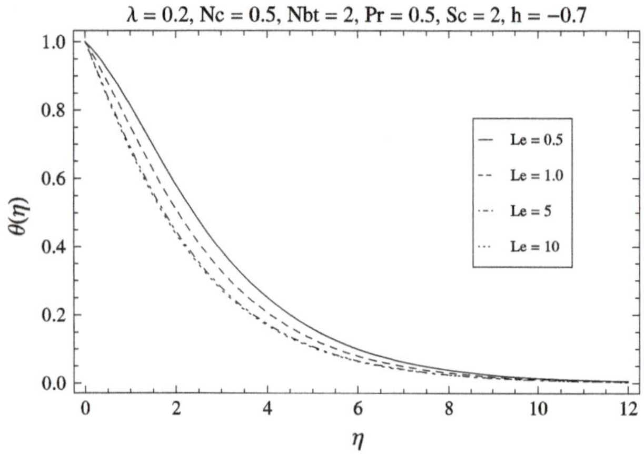

Fig. 4 Temperature variation against different values of *Le*

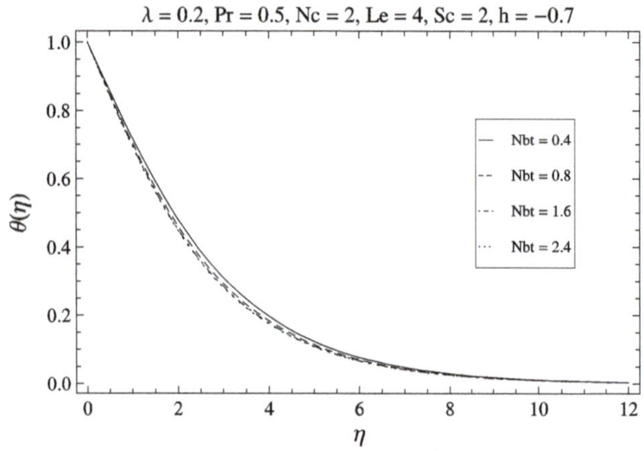

Fig. 5 Temperature graph for different values of N_{bt}

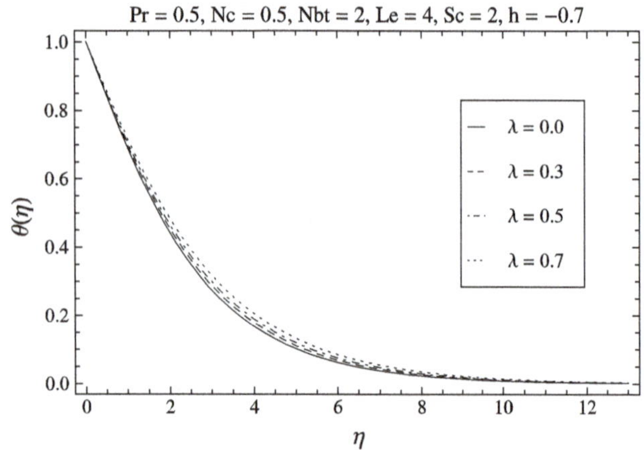

Fig. 6 Temperature variation against different values of λ

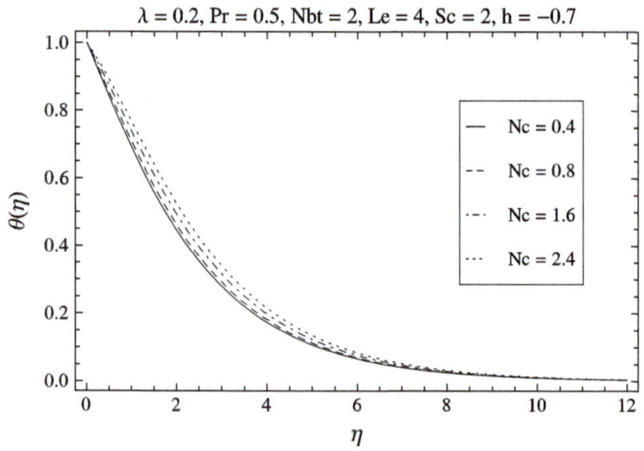

Fig. 7 Temperature variation against different values of N_c

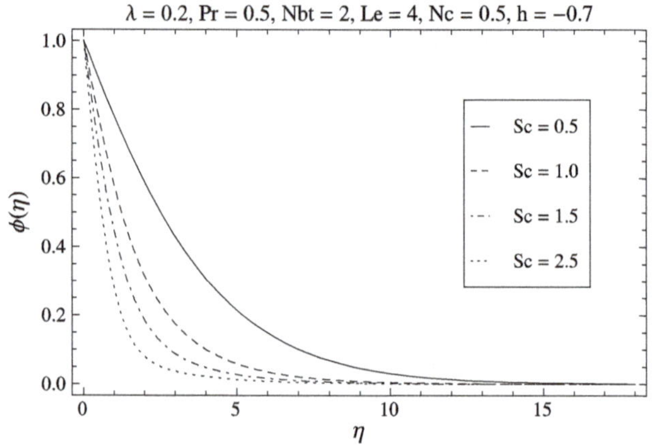

Fig. 8 Nanoparticles volume fraction variation against different values of *Sc*

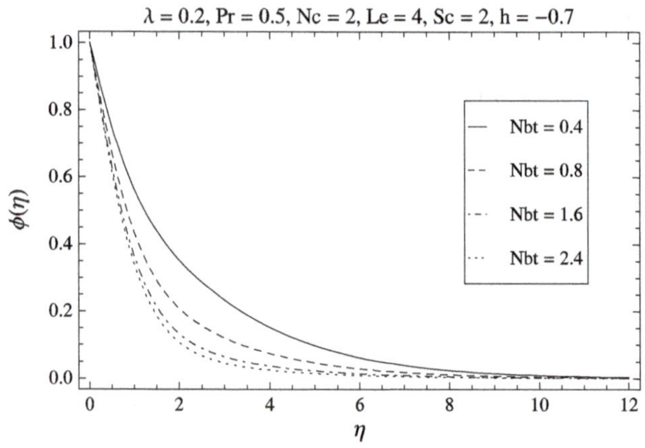

Fig. 9 Nanoparticle volume fraction against different values of N_{bt}

found to be in good agreement. Tables 3 and 4 show wall temperature gradient and wall nanoparticle volume fraction gradient respectively. It is observed that the wall temperature gradient decreases with the increase in λ and N_c. Here *Le* and N_{bt} are important heat transfer parameters for nanofluids; it is observed that the heat transfer increases with the increase in both parameters. Also wall temperature gradient increases with the increase in Prandtl number. From Table 4, it can be observed that wall nano particle volume fraction gradient decreases with the increase in λ and *Le* while it increases with the increase in N_{bt} and *Sc*.

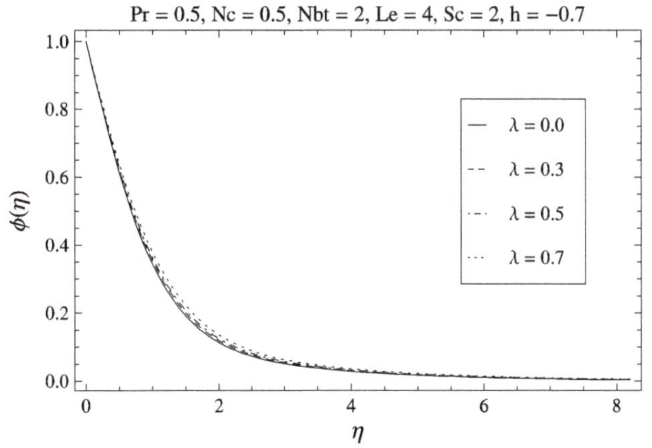

Fig. 10 Nanoparticle volume fraction variation against different values of λ

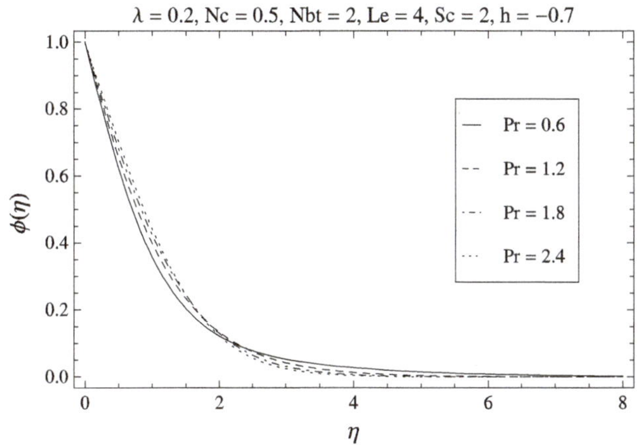

Fig. 11 Nanoparticle volume fraction against different values of *Pr*

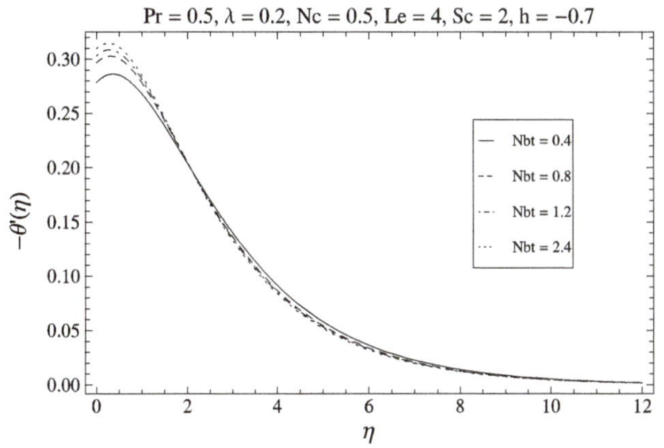

Fig. 12 Heat transfer in fluid against different values of N_{bt}

Concluding remarks

In the present paper we tried to analyse the nano particle effect on boundary layer flow of Williamson fluid over a stretching surface. The governing non-linear equations are solved analytically using HAM. The important findings of the paper are as follows:

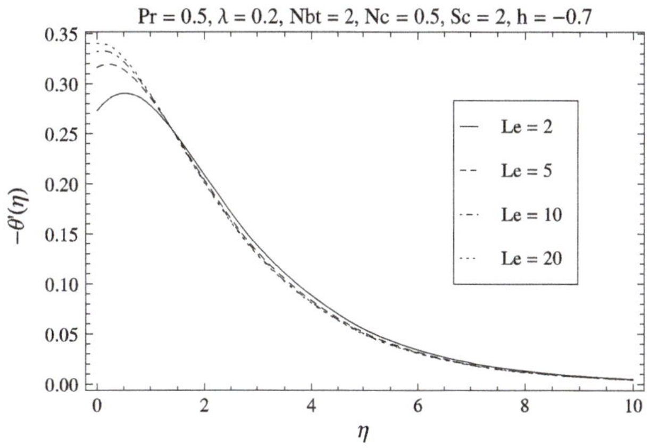

Fig. 13 Heat transfer in fluid against different values of *Le*

Table 2 Comparison table for viscous case ($-\theta'(0)$)

Pr	Present results	Khan and Pop (2010)	Wang (1989)	Golra and Sidawi (1994)
0.07	0.066	0.066	0.066	0.066
0.20	0.169	0.169	0.169	0.169
0.70	0.454	0.454	0.454	0.454
2.0	0.911	0.911	0.911	0.911

Table 3 Values of wall temperature gradient $-\theta'(0)$, when $h = -0.7$

λ	*Le*	N_{bt}	N_c	*Pr*	$-\theta'(0)$
0.0	4.0	2.0	0.5	0.5	0.314
0.2					0.309
0.4					0.302
0.2	4.0	2.0	0.5	0.5	0.309
	10				0.332
	20				0.340
0.2	4.0	0.5	0.5	0.5	0.286
		1.0			0.301
		2.0			0.309
0.2	4.0	2.0	0.5	0.5	0.309
			1.0		0.273
			2.0		0.213
0.2	4.0	2.0	0.5	0.2	0.144
				0.6	0.355
				1.2	0.588

- Lewis number will appear in nano particle volume fraction equation for the case when velocity

Table 4 Values of wall nano particle volume fraction gradient $-\phi'(0)$, when $h = -0.7$

λ	Le	N_{bt}	Sc	$-\phi'(0)$
0.0	4.0	2.0	0.5	0.833
0.2				0.824
0.4				0.811
0.2	4.0	2.0	0.5	0.824
	10			0.815
	20			0.812
0.2	4.0	0.5	0.5	0.620
		1.0		0.751
		2.0		0.823
0.2	4.0	2.0	0.5	0.228
			1.0	0.475
			2.0	0.824

transformation depends on thermal diffusivity; otherwise, Schmidt number will appear.

- The parameters have strong impact on heat transfer very close to the wall and are almost negligible slightly away from wall.
- Wall temperature gradient increases with increase in Le and N_{bt}.
- Wall nano particle fraction gradient increases with N_{bt} and Sc and decreases with Le.

References

Abbasbandy S (2006) The application of homotopy analysis method to nonlinear equations arising in heat transfer. Phys Lett A 360(1):109–113

Abbasbandy S (2007) The application of homotopy analysis method to solve a generalized Hirota–Satsuma coupled KdV equation. Phys Lett A 361(6):478–483

Ali ME (1995) On thermal boundary layer on a power law stretched surface with suction or injection. Int J Heat Mass Transf 16:280–290

Bachok N, Ishak A, Pop I (2010) Boundary-layer flow of nanofluids over a moving surface in a flowing fluid. Int J Therm Sci 49:1663–1668

Buongiorno J (2006) Convective transport in nanofluids. J Heat Transf 128:240–250

Choi SUS (1995), Enhancing thermal conductivity of fluids with nanoparticles. In: The proceedings of the 1995 ASME International Mechanical Engineering Congress and Exposition. ASME, San Francisco, USA, FED 231/MD 66, pp 99–105

Dapra, Scarpi G (2007) Perturbation solution for pulsatile flow of a non-Newtonian Williamson fluid in a rock fracture. Int J Rock Mech Min Sci 44:271–278

Durmus, Kasgoz A, Macosko CW (2007) Linear low density polyethylene (LLDPE)/clay nanocomposites. Part I: structural characterization and quantifying clay dispersion by melt rheology. Polymer 48:4492–4502

Ellahi R, Riaz A (2010) Analytical solutions for MHD flow in a third-grade fluid with variable viscosity. Math Comput Model 52(9–10):1783–1793

Erickson LE, Fan LT, Fox VG (1966) Heat and mass transfer in the laminar boundary layer flow of a moving flat surface with constant surface velocity and temperature focusing on the effects of suction/injection. Ind Eng Chem Fundam 5:19–25

Gorla RSR, Sidawi I (1994) Free convection on a vertical stretching surface with suction and blowing. Appl Sci Res 52:247–257

Hayat T, Qasim M (2010) Influence of thermal radiation and Joule heating on MHD flow of a Maxwell fluid in the presence of thermophoresis. Int J Heat Mass Transf 53(21–22): 4780–4788

Kandasamya R, Loganathanb P, Puvi Arasub P (2011) Scaling group transformation for MHD boundary-layer flow of a nanofluid past a vertical stretching surface in the presence of suction/injection. Nucl Eng Des 241:2053–2059

Kelson NA, Farrell TW (2001) Micropolar flow over a porous stretching sheet with strong suction or injection. Int Commun Heat Mass Transf 28(4):479–488

Khan WA, Pop I (2010) Boundary-layer flow of a nanofluid past a stretching sheet. Int J Heat Mass Transf 53:2477–2483

Kumaran V, Ramanaiah G (1996) A note on the flow over a stretching sheet. Acta Mech 116(1–4):229–233

Liao S (2003) Beyond perturbation: introduction to the homotopy analysis method. Chapman and Hall/CRC, Boca Raton, pp 99–102

Liao S (2004) On the homotopy analysis method for nonlinear problems. Appl Math Comput 147:499–513

Liu C (2005) A note on heat and mass transfer for a hydromagnetic flow over a stretching sheet. Int Commun Heat Mass Transf 32(8):1075–1084

Lyubimov V, Perminov AV (2002) Motion of a thin oblique layer of a pseudoplastic fluid. J Eng Phys Thermophys 75(4):920–924

Makinde OD, Aziz A (2011) Boundary layer flow of a nanofluid past a stretching sheet with a convective boundary condition. Int J Therm Sci 50:1326–1332

Masuda H, Ebata A, Teramae K, Hishinuma N (1993) Alteration of thermal conductivity and viscosity of liquid by dispersing ultra-fine particles. Netsu Bussei 7(4):227–233

Nadeem S, Akbar NS (2010) Numerical solutions of peristaltic flow of Williamson fluid with radially varying MHD in an endoscope. Int J Numer Methods Fluids 66(2):212–220

Nadeem S, Hussain A (2009) MHD flow of a viscous fluid on a nonlinear porous shrinking sheet with homotopy analysis method. Appl Math Mech 30(12):1569–1578

Nadeem S, Hussain A, Vajravelu K (2010) Effects of heat transfer on the stagnation flow of a third order fluid over a shrinking sheet. Zeitschrift für Naturforschung 65(a):969–994

Nadeem S, Hussain ST, Lee C (2013) Flow of a Williamson fluid over a stretching sheet. Braz J Chem Eng 30(3):619–625

Nield DA, Kuznetsov AV (2009) Thermal instability in a porous medium layer saturated by a nanofluid. Int J Heat Mass Transf 52:5796–5801

Nield DA, Kuznetsov AV (2011) The onset of double-diffusive convection in a nanofluid layer. Int J Heat Fluid Flow 32:771–776

Rosali H, Ishak A, Pop I (2012) Micropolar fluid flow towards a stretching/shrinking sheet in a porous medium with suction. Int Commun Heat Mass Transf 39(6):826–829

Sakiadis C (1961) Boundary layer behavior on continuous solid flat surfaces. Am Inst Chem Eng J 7:26–28

Tsou FK, Sparrow EM, Goldstein RJ (1967) Flow and heat transfer in the boundary layer on a continuous moving surface. Int J Heat Mass Transf 10:219–235

Wang Y (1989) Free convection on a vertical stretching surface. J Appl Math Mech 69:418–420

Williamson RV (1929) The flow of pseudoplastic materials. Ind Eng Chem 21(11):1108–1111

16

Synthesis, characterization and evaluation of effect of phytogenic zinc nanoparticles on soil exo-enzymes

K. Sri Sindhura · T. N. V. K. V. Prasad · P. Panner Selvam · O. M. Hussain

Abstract Nanobiotechnology, the bio-branch of nanotechnology is considered to be one of the fastest emerging research fields. Biosynthesis of metallic nanoparticles is currently under exploitation. Use of plant and plant materials for the synthesis of Zinc nanoparticles is relatively new and exciting research field. The biogenic zinc nanoparticles were synthesized using the leaves of *Parthenium hysterophorous* by green synthesis route. UV–VIS absorption spectroscopy was used to monitor the quantitative formation of zinc nanoparticles. The characteristics of the synthesized zinc nanoparticles were studied using scanning electron microscopy and nanoparticle analyzer. Zinc nanoparticles were observed to be spherical in shape with size range of 16 to 108.5 nm. The measured zeta potentials varied from 100.4 to 117.20 mV indicate high dispersion of the zinc nanoparticles. The synthesized zinc nanoparticles showed good enzymatic activity and microbial activity. The physiological parameters increased from 30 to 60 days of sowing when compared to control.

K. Sri Sindhura · O. M. Hussain (✉)
Department of Physics, Sri Venkateswara University, Tirupati 517 502, Andhra Pradesh, India
e-mail: hussainsvu@gmail.com

T. N. V. K. V. Prasad
Department of Nanotechnology, Institute of Frontier Technology, Regional Agricultural Research Station, Acharya Nagarjuna Ranga Agricultural University, Tirupati 517 502, Andhra Pradesh, India

P. Panner Selvam
Department of Soil Science and Agricultural Chemistry, Indian Institute of Horticultural Research, Bangalore 560 089, India

Keywords Zinc nanoparticles · Green synthesis · *Parthenium hysterophorous* · *Arachis hypogea* L · Enzymatic activity · Microbial activity

Introduction

Nanoscience is about creating new chemical and biological nanostructures, uncovering and understanding their novel properties, and ultimately about learning how to organize these new structures into larger and more complex functional structures and devices. The applications of nanomaterials in agriculture and food sector make all the more relevance and importance since the Indian economy is predominantly depending on agriculture to feed the large population. The nano-sized particles increase the potency of active ingredients and potentially reduce the quantity to be applied. Nano-structured particles/matrix make soil binder similar to natural crop and plant residues providing dust control, soil moisture retention and re-vegetation stimulant for seeding efforts, which simultaneously increase the production while lowering product cost. Indeed, over the past several years, plants, algae, fungi, bacteria, and viruses have been used for production of low-cost, energy-efficient, and non-toxic metallic nanoparticles (Parikh 2010).

Green Synthesis of metallic nanoparticles is eco-friendly. Biosynthetic methods employing either biological microorganisms or plant extracts have emerged as a simple alternative to physical and chemical methods. Plant and plant materials have become a potential source for the synthesis of metallic nanoparticles recently. A number of researchers have reported on synthesis of silver and gold nanoparticles using different plant materials (Sharma et al. 2007, Badri Narayanan and Sakthivel 2008; Badri Narayanan and Sakthive 2010; Philip 2010; 2011). Synthesis of Zinc oxide nanoparticles is also in progress (Kumar et al. 2011; Singh et al. 2011; Gunalan et al. 2012). But till now there is no report on the synthesis of zinc nanoparticles using plant materials. To the best of our knowledge this is the first report on the green synthesis of zinc nanoparticles. The reason for selecting zinc is—zinc, an essential micronutrient plays an important role in the physiological functions of plants. Plants can easily uptake zinc (rhizosphere). Utilization of zinc in tissues is called internal efficiency and uptake of zinc is called external efficiency. Particle size may affect agronomic effectiveness of zinc fertilizers (Prasad et al. 2012).

Parthenium hysterophorous is a species of flowering plant in the aster family, Asteraceae. In India, it is locally known as Congress weed. Coming to the toxicity of the leaves, contact with this plant causes dermatitis and respiratory malfunction in humans, dermatitis in cattle and domestic animals, due to the presence of toxin parthenin. Parthenin plays a pivotal role as root exudate in allelopathic interference with surrounding plants (Belz et al. 2007). Although Parthenium is considered as toxic plant, it is useful for many medicinal, allelopathic and industrial applications (Oudhia 2001).

The main objective of the proposed work is to synthesize the zinc nanoparticles in biological route and the application of these biogenic zinc nanoparticles on agricultural crops by conducting the pot culture experiments to observe the variations in enzyme activity, microbial activity and physiological traits of the plants by taking the soil from the pot cultures and to make a comparative study on the application of zinc nanoparticles applied plants and the plants grown under normal crop conditions (by maintaining controls) to test the potential of green synthesized zinc nanoparticles.

Materials and method

Collection of plant materials

Fresh leaves of *Parthenium hysterophorous* were collected from the fields of Regional Agricultural Research Station, Acharya Nagarjuna Ranga Agricultural University, Tirupati, India. The leaves were thoroughly washed for 2–3 times with distilled water and then kept shade dried for 5 days to free from moisture. After shade drying, the leaves were kept in hot air oven at 50° C for 5–10 min. Then the leaves were subjected to grinding.

Preparation of leaf extract and samples

The ground powder was again sieved (0.5 mm) for three times. After getting the fine powder plant decoction was made by adding distilled water. The decoction was heated under hot plate for 15 min at 60° C. Then the decoction was filtered using Whatmann No.1 filter paper to get the clear filtered solution of the plant extract. To the plant extract, 1 molar zinc nitrate solution was added in the ratio of 1:19. Two samples were prepared by diluting the first sample (sample-1, 100 %) to its half (sample-2, 50 %). These solutions were kept for heating up to 10–15 min and a change in the color was observed from pale yellow color to dark honey color at 65 °C. After the color change, the samples were purified by centrifugation at 18G for 15 min.

Recording localized surface plasmon resonance (LSPR)

UV–Visible Spectrophotometer (Schimadzou 4650) was used to record LSPR of zinc nanoparticles. Wavelength verses absorption spectra were recorded for the two samples. Surface Plasmon Resonance (SPR) in nanometer-sized structures is called localized surface plasmon resonance. SPR is the basis of many standard tools for measuring adsorption of material onto planar metal surfaces or onto the surface of metal (Zeng et al. 2011). It is the fundamental principle behind many color-based biosensor applications. LSPRs are collective electron charge oscillations in metallic nanoparticles that are excited by light. They exhibit enhanced near-field amplitude at the resonance wavelength.

Particle size and zeta potential measurement

The zeta potentials were recorded using Particle size analyzer (Horiba Nanoparticula 100). This apparatus uses Dynamic Light Scattering (DLS) phenomenon. The sample holder temperature was maintained at 25° C, a technique used to determine the size distribution profile of small particles in solution. This measurement depends on the size of the particle core, the size of surface structures, particle concentration, and the type of ions in the medium.

Fourier transform infrared spectroscopic (FTIR) studies

FTIR spectrum (Bruker Tensor 27) was recorded in mid IR region in the range of 400–4,000 wavenumber (cm^{-1}). A drop of the sample solution (sample-1) was carefully added to the potassium bromide crystals using 1 ml micropipette and the spectrum was recorded in transmittance (%) mode. From the FTIR analysis, the presence of functional groups can be identified. The biological components interact with the metal salts via these functional groups and undergo bio-reduction into nanoparticles (Ganesh Babu and Gunasekaran 2009).

Scanning electron microscope (SEM) studies

The particle size and shape can be identified using the scanning electron microscope (Carlziess EVO 50). For the preparation of sample for SEM, sample-1 was coated on a thick aluminum sheet cut into 1 × 1 cm width. This sheet was kept on hot plate and the temperature was maintained at 65° C. 0.5 ml of sample was taken into the micropipette and carefully poured on to the aluminum sheet. This step was repeated for three times. Like this, a smear was obtained on the sheet. This was taken for the SEM analysis.

Pot culture experiment

Arachis hypogea L, one of the important oil seeds crops of India grown in rainfed soils was selected for pot culture. The soil pH was measured as 6.8. 20 pots (15 × 12 × 12 cm dimension for each pot) were raised as five replications, for which, 15 pots were added with sample-1 in three different variations in three treatments and five pots were maintained as controls without adding any sample. The bacteria, fungi and actinomycetes population were analyzed from soil samples using serial dilution plating technique after 60 days of sowing. The population was expressed as cfu g^{-1} of oven dry soil. Phosphatase and dehydrogenase enzymatic activities were estimated by adopting standard methods, after 30 and 60 days of sowing.

Physiological parameters like plant height, number of leaves and leaf surface ratio (leaf length and breadth) were calculated in two time intervals of 30 and 60 days of sowing. Similarly, root length, shoot length, fresh and dry weight of roots and shoots and total biomass (dry weight of root and shoot) were estimated after 60 days of sowing. The total biomass was calculated after removing the plant from all the pots and by separating each plant into its root and shoot. This was followed by measuring the root length and shoot length and weighing of fresh and dry weights of all the roots and shoots. The fresh weights were measured immediately after removing the plant from the pots and the dry weights were measured after incubating the roots and shoots in an oven for 36 h at 60 °C.

Results and discussion

Ultraviolet–visible spectroscopy

Zinc nanoparticles were characterized by studying the absorption spectra recorded from the Ultraviolet–visible spectral analysis for the two samples (Table 1). Formation of zinc nanoparticles in aqueous colloidal solution were confirmed using Ultraviolet–visible spectral analysis. Zinc nanoparticles normally show a broad peak in the UV–vis spectrum in the range of 230–330 nm (Revina et al. 2007). The optical transitions have been observed at 327.50 and 330.00 nm (Fig. 1a, b) corresponds to the formation of zinc nanoparticles. The absorbance values at different wavelengths are given in Table 1. Between the two samples, highest absorbance of 0.373 A % at 330.00 nm was observed for sample-1 because of higher concentration compared to sample-2. The bio matrix present in the plant extracts may leads to the change in the absorbance of UV–vis spectra.

Table 1 Absorption verses wavelength, particle size and zeta potential values for sample-1 and sample-2

S·No.	Wavelength	Absorbance	Particle size	Zeta potential
	(nm)	(A %)	(nm)	(mV)
Sample-1	330.00	0.373	16.10	117.20
Sample-2	327.50	0.358	58.60	100.40

Sample-1: 95 ml of 1M zinc nitrate solution +5 ml of *Parthenium hysterophorous* leaf extract, Sample-2: Half of the sample-1 diluted to its half (50 ml of sample-1 +50 ml of double distilled water)

The results show promise for the development of a "green" process for nanoparticle synthesis. Biomolecules found in plants induce the reduction of Zn^{2+} ions into Zn nanoparticles. The process of reduction is extracellular and fast leading to the development of easy biosynthesis of zinc nanoparticles. At the time of breakdown of zinc nitrate into zinc and nitrate ions, a nitrate reductase enzyme makes the reaction to take place.

$$ZnNo_3 \xrightarrow{\text{Nitratereductase}} Zn^{2+} + No_3^{2-}$$

The zinc content was estimated from Inductively Coupled Plasma-Optical Emission Spectrophotometer (ICP-OES—Leeman Labs Prodigy XP) as 65 ppm (Wolf et al. 2009).

Nanoparticle analyzer and SEM studies

From the nanoparticle analyzer, the zeta potential was measured as 117.20 mV for sample-1 and 100.40 mV for sample-2 (Fig. 2a, b) and particle size was measured as 16.10 nm for sample-1 and 58.60 nm for sample-2 (Fig. 3a, b). The significance of zeta potential is that its value can be related to the stability of colloidal dispersions. The zeta potential indicates the degree of repulsion between adjacent and similarly charged particles in dispersion. Zeta potential indicates the stability of the nanoparticles. Here, the zeta potential value indicated good stability with high potential.

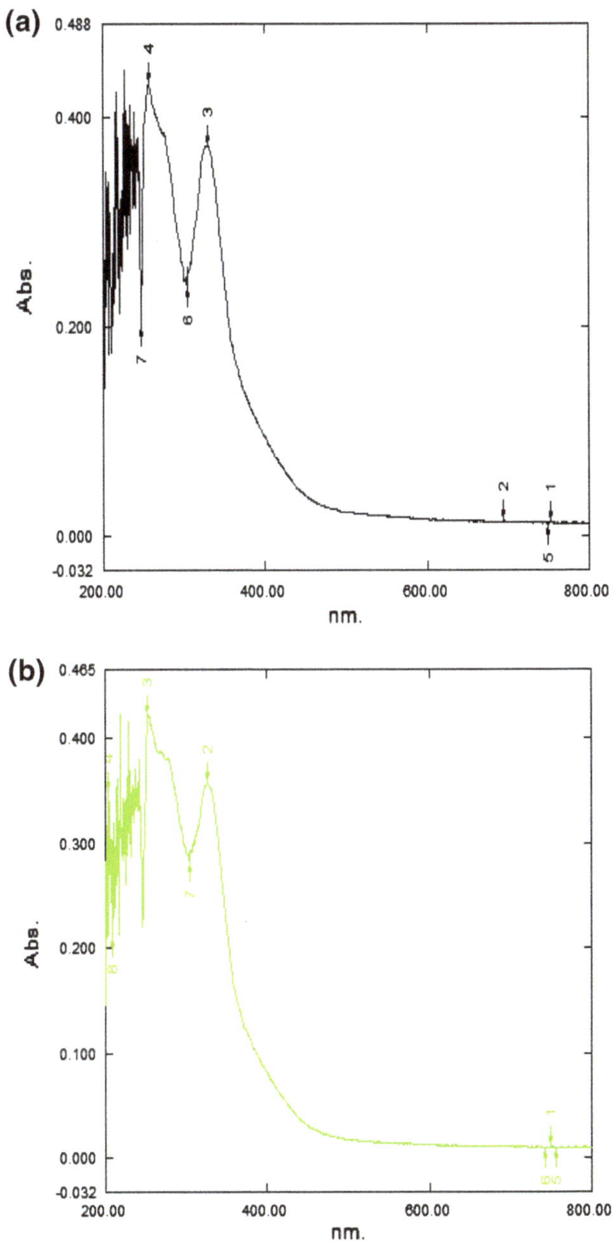

Fig. 1 Wavelength verses absorbance spectra. **a** Sample-1 **b** Sample-2

From Table 1, highest zeta potential of 117.20 mV and least particle size of 16.10 nm were observed for sample-1 (higher concentration than sample-2 as sample-2 is the diluted form of sample-1) (Table 1). From the scanning electron microscopic image of sample-1, zinc nanoparticles were observed to be spherical in shape with a size of 108.5 nm (Fig. 4).

FTIR Studies

The FTIR spectra were recorded for the *Parthenium hysterophorous* leaf extract (Fig. 5a) and sample-1 (Fig. 5b). The FTIR spectrum of *Parthenium hysterophorous* leaf extract showed the presence of functional groups of alcohols, phenols, alkenes, alkanes, carbonyls, aromatics, alkyl halides and alkynes. The spectrum showed the bands for the functional groups located at 3,779.57, 3,691.36, 2,910.73, 2,851.62, 1,672.48, 1,417.53, 1,275.29, 1,155.78 and 583.50 cm^{-1}. The strong band of –C = O– stretch (carbonyls) was recorded at 1,672.48 cm^{-1}. The medium bands of –C–C– (in ring) stretch (aromatics) and $-CH_2-X$ stretch (alkyl halides) were recorded at 1,417.53 and 1,155.78 cm^{-1}. The FTIR spectrum of sample-1 showed the presence of functional groups of alcohols, phenols, alkenes, nitriles, carbonyls, aromatics, aliphatic amines, alkyl halides, alkenes and alkynes (Fig. 5). The IR spectrum of zinc nanoparticles showed the bands for the functional groups located at 3,673.15, 3,117.36, 2,388.24, 1,675.46, 1,410.28, 1,267.29, 1,162.84, 1,011.83, 800.57, 586.77 cm^{-1}. The strong band of –C = O– stretch (carbonyls) was recorded at 1,675.46 cm^{-1}. The medium bands of –C–C– (aromatics) and $-CH_2-X$ (alkyl halides) were recorded at 1,410.28, and 1,162.84 cm^{-1}. From the

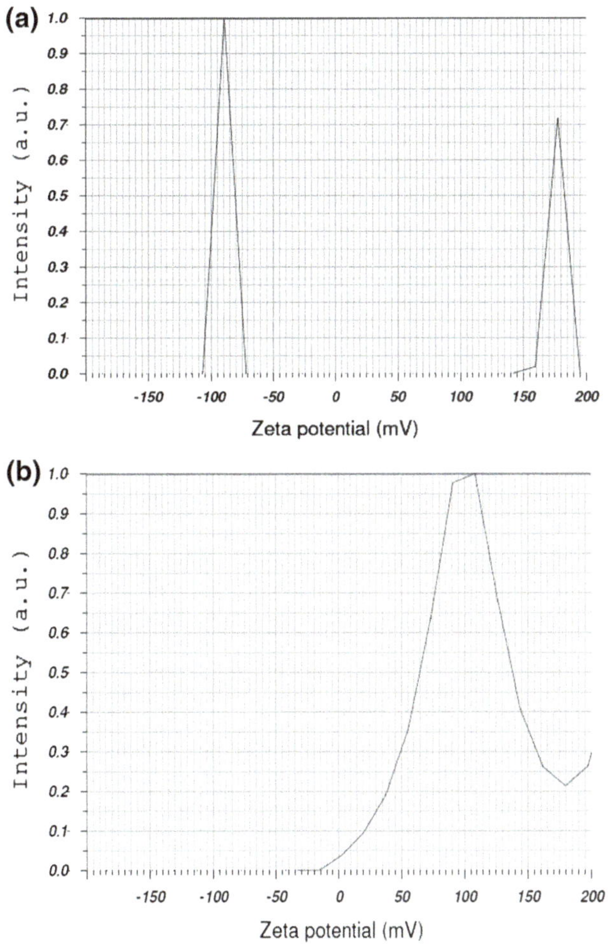

Fig. 2 Zeta potential spectra. **a** Sample-1 **b** Sample-2

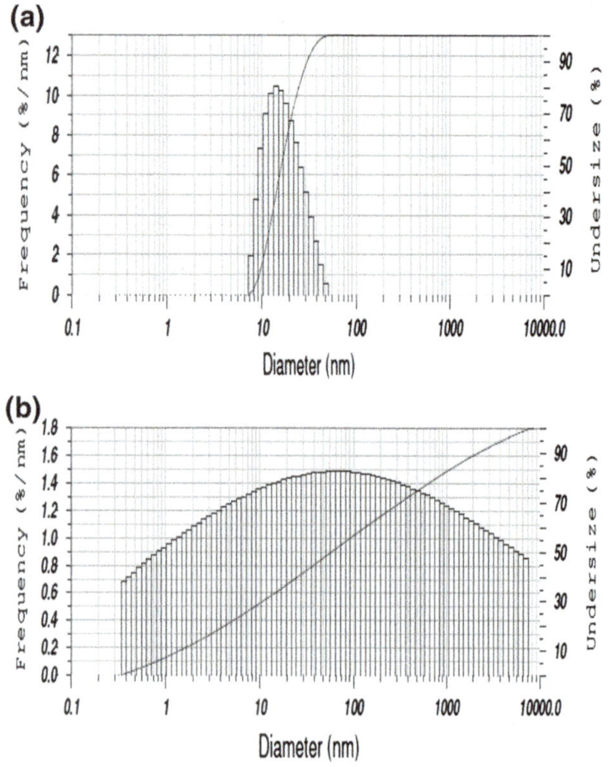

Fig. 3 Particle size distribution spectra. **a** Sample-1 **b** Sample-2

FTIR spectra of *Parthenium hysterophorous* leaf extract and sample-1, the change in wave number of the functional groups was due to the reduction and stabilization of the metal group—Zn. When the spectrum of leaf extract was compared with the spectrum of sample-1, the functional groups of nitriles and aliphatic amines were observed from the spectrum of sample-1 due to the stretching of nitrile (C ≡ N) and amine ($-NH_2$) bands.

Enzymatic activity

For the enzymatic activity analysis, acidic phosphatase activity, alkaline phosphatase activity and dehydrogenase activity were estimated for the pot culture, in which, sample-1 was added in three treatments, against the control in two regular time intervals of 30 and 60 days of sowing (Table 2). Sample-1 showed good results compared to sample-2. So sample-1 was taken for the enzyme and microbial activities and also to estimate the physiological parameters.

Treatment-1: 2.5 ml of sample-1 + 2.5 ml of distilled water.
Treatment-2: 1.0 ml of sample-1 + 1.0 ml of distilled water.
Treatment-3: 0.5 ml of sample-1 + 0.5 ml of distilled water.
Control: Without adding any sample.

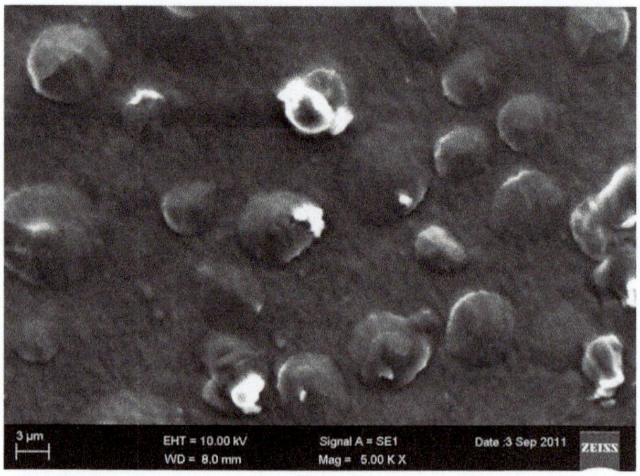

Fig. 4 Scanning electron microscopic image of spherical shaped zinc nanoparticles

The soil enzyme activities viz., dehydrogenase and phosphatase activities were performed after 30 and 60 days of sowing. The phosphatase and dehydrogenase enzyme activities were performed to test the enzymatic efficiency of the applied zinc nanoparticles. An overview of soil exo-enzymes is as follows—Soil phosphatase activity (Acid and Alkaline) was inhibited by the citrate phosphate buffer. The phosphatase activity was based on the determination of p-nitrophenol released after the incubation of oil with p-nitrophenol phosphate for 1 h at 37 °C. Acid and alkaline phosphatase activities differ only in terms of pH. Dehydrogenase activity was based on the estimation of the triphenyl tetrazolium chloride reduction rate to triphenyl formazan in soils after incubation at 30 °C for 24 h, where the tri phenyl tetrazolium is an artificial electron acceptor.

Enzyme activities were increased in response to all treatments when compared to the control confirmed that the zinc nanoparticles applied treatments showed significant variations in acid and alkaline phosphatase activity and dehydrogenase activity among the nanoparticles treatment after 30 and 60 days of sowing and also increased from 30 to 60 days time period against control when compared to Sunghyun et.al (2011). The measured phosphatases and dehydrogenases were cell bound and were not the extracellular activities. The slight raise in values for treatment-1 compared to treatment 2 and 3 is because sample-1 was applied in greater volume for treatment-1 (5 ml) compared to treatment-2 (2 ml) and treatment-3 (1 ml). The reason for performing the enzymatic activity analysis was to report that the applied zinc nanoparticles induce the phosphatase (acid and alkaline) and dehydrogenase enzyme activities of the pot culture experiment against the control. Our experiment revealed that the zinc nanoparticles induced the enzyme activity. Hence the green synthesized zinc nanoparticles were potential and non-toxic

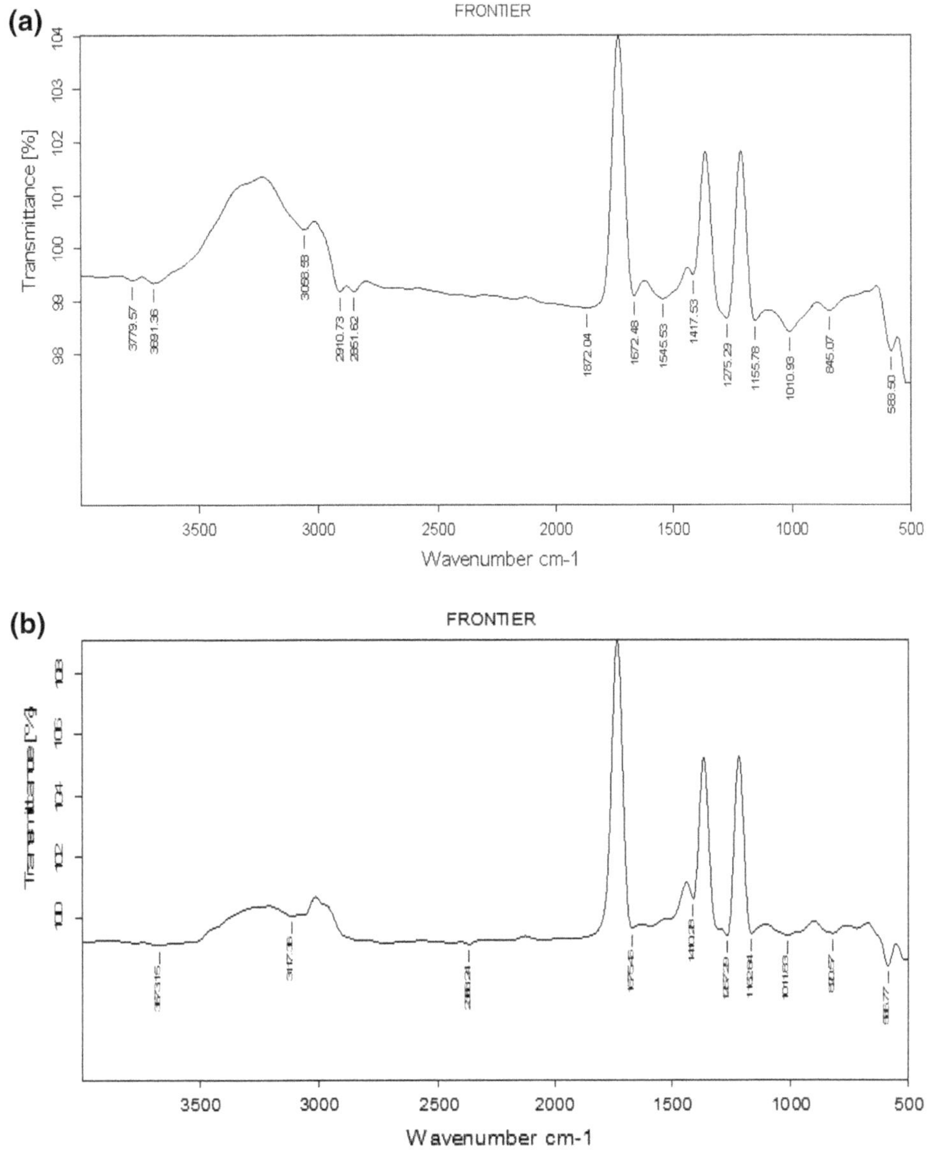

Fig. 5 FTIR Spectra. **a** *Parthenium hysterophorous* leaf extract **b** Sample-1

Table 2 Effect of zinc nanoparticles on change in phosphatase and dehydrogenase activities in *Arachis hypogea* L. rhizosphere

S.No.	Acidic phosphatase activity (μg of p-nitrophenol released g-1 of soil h-1)		Alkaline phosphatase activity (μg of p-nitrophenol released g-1 of soil h-1)		Dehydrogenase activity (μg of TPF released g-1 of soil h-1)	
	30 days	60 days	30 days	60 days	30 days	60 days
T-1	15.83 ± 0.08	17.87 ± 0.09	21.08 ± 0.12	18.52 ± 0.07	1.49 ± 0.01	9.04 ± 0.05
T-2	15.72 ± 0.08	17.58 ± 0.09	19.47 ± 0.12	14.31 ± 0.07	1.21 ± 0.01	8.01 ± 0.05
T-3	15.51 ± 0.08	15.91 ± 0.09	18.96 ± 0.12	13.45 ± 0.07	1.12 ± 0.01	7.46 ± 0.05
Control	12.64 ± 0.08	14.25 ± 0.09	14.08 ± 0.12	12.86 ± 0.07	1.02 ± 0.01	6.15 ± 0.05
CD (<5 %)	0.18	0.19	0.25	0.14	0.02	0.10

Each value is the ±SE of five replications. T-1: Treatment-1—2.5 ml of sample-1 +2.5 ml distilled water, T-2: Treatment-2—1.0 ml of sample-1 +1.0 ml distilled water, T-3: Treatment-3—0.5 ml of sample-1 +0.5 ml distilled water

Control Without adding any sample, *CD (<5 %)* Critical difference at 5 % level

Table 3 Effect of zinc nanoparticles on change in microbial population in *Arachis hypogea* L. rhizosphere

S.No.	Bacteria (cfu × 10^5 g^{-1} dry soil)	Fungi (cfu × 10^3 g^{-1} dry soil)	Actinomycetes (cfu × 10^3 g^{-1} dry oil)
T-1	7.40 ± 0.04	2.00 ± 0.02	0.85 ± 0.01
T-2	7.20 ± 0.04	2.40 ± 0.02	0.53 ± 0.02
T-3	7.40 ± 0.04	2.00 ± 0.02	1.35 ± 0.02
Control	5.50 ± 0.04	1.10 ± 0.02	0.47 ± 0.02
CD (<5 %)	0.09	0.03	0.03

and also enhanced the phosphatase and dehydrogenase enzyme activities.

Microbial activity

From the microbial population estimation, numbers of colonies were estimated for bacteria, fungi and actinomycetes for the three treatments (Treatment 1–3) along with the control. The results presented in Table 3 indicated that the bacterial population was significantly higher (7.2–7.4 × 10^5 g^{-1} dry soil) in nanoparticle-treated samples as compared to the control (5.5 × 10^5 g^{-1} dry soil). Similarly, the nanoparticle-treated samples had significantly higher fungi (2.0–2.4 × 10^3 g^{-1} dry soil) and actinomycetes (0.53–1.35 × 10^3 g^{-1} dry soil) compared to the controls. The microbial population estimation was performed to test the microbial activity in the soils to which zinc nanoparticles were applied. The results presented in Table 3 shows that the plant extracted zinc nanoparticles induce the rhizosphere microbial population. Compared to the control, treatments 1–3 showed higher number of colonies concluding that the applied zinc nanoparticles induced the growth of bacteria, fungi and actinomycetes. Among the three treatments, treatment-1 showed higher number of colonies as sample-1 is applied in greater volume for treatment-1. The reason for performing the microbial activity analysis was to report that the zinc nanoparticles induced the colony formation of microorganisms by applying the synthesized zinc nanoparticles to the bacterial, fungal and actinomycetes culture media by maintaining controls and incubating the petri-plates under controlled laboratory conditions. The microbial activity revealed that the synthesized zinc nanoparticles induce the growth of microbial population as the number of colonies were increased in response to all treatments, compared to the control.

Physiological traits

From the pot culture experiment, the physiological growth parameters—numbers of leaves, plant height, leaf surface ratio, root length, shoot length, fresh and dry weights of roots and shoots and total biomass were measured for the three treatments against the control. Leaf surface ratio, number of leaves and plant height were recorded after 30 and 60 days of sowing for all the three treatments against control.

Among the three treatments, treatment-1 showed high variations for all the physiological parameters against the control because for treatment-1, zinc nanoparticles (sample-1) were applied in greater volume compared to treatments 2 and 3. From Table 4, from control to treatment-1, leaf length was increased from 1.95 to 2.52 cm (increased by 22.6 %) after 30 days of sowing and from 2.25 to 2.73 cm (17.6 %) after 60 days of sowing. Leaf breadth increased from 1.02 to 1.24 cm (30 days, 17.7 %) and from 1.15 to 1.45 cm (60 days, 20.69 %). Number of leaves increased from 38 to 79 (30 days, 51.9 %) and from 46 to 98 (60 days, 53 %) and the plant height increased from 18.25 to 30.34 cm (30 days, 39.8 %) and from 30.14 to 50.41 cm (60 days, 40.21 %).

Simultaneously, root lengths and shoot lengths, fresh weight and dry weight of roots and shoots and total biomass were recorded for the three treatments against the control after 60 days of sowing (Table 5). Among the three

Table 4 Physiological growth parameters of *Arachis hypogea* L. pot culture by applying the zinc nanoparticles synthesized from *Parthenium hysterophorous* leaf extract

S.No.	30 days				60 days			
	Leaf length (cm/plant)	Leaf breadth (cm/plant)	No. of leaves (/plant)	Plant height (cm/plant)	Leaf length (cm/plant)	Leaf breadth (cm/plant)	No. of leaves (/plant)	Plant height (cm/plant)
T-1	2.52 ± 0.01	1.24 ± 0.01	79 ± 0.33	30.34 ± 0.21	2.73 ± 0.01	1.45 ± 0.01	98 ± 0.57	50.41 ± 0.31
T-2	2.25 ± 0.01	1.21 ± 0.01	63 ± 0.33	29.32 ± 0.21	2.61 ± 0.01	1.42 ± 0.01	79 ± 0.57	40.67 ± 0.31
T-3	2.02 ± 0.01	1.08 ± 0.01	42 ± 0.33	30.34 ± 0.21	2.43 ± 0.01	1.39 ± 0.01	61 ± 0.57	39.14 ± 0.31
Control	1.95 ± 0.01	1.02 ± 0.01	38 ± 0.33	18.25 ± 0.21	2.25 ± 0.01	1.15 ± 0.01	46 ± 0.57	30.14 ± 0.31
CD (<5 %)	0.03	0.01	0.73	0.46	0.03	0.02	1.24	0.68

Table 5 Effect of zinc nanoparticles on the growth of *Arachis hypogea* L under pot culture conditions

S.No	Root length	Shoot length	Root weight		Shoot weight		Total biomass
			Fresh weight	Dry weight	Fresh weight	Dry weight	
T-1	7.29 ± 0.055	44.16 ± 0.23	0.58 ± 0.03	0.23 ± 0.001	13.33 ± 0.09	7.06 ± 0.056	7.29 ± 0.01
T-2	3.39 ± 0.055	42.26 ± 0.23	0.46 ± 0.03	0.15 ± 0.001	10.06 ± 0.09	2.25 ± 0.056	3.39 ± 0.01
T-3	2.37 ± 0.055	40.25 ± 0.23	0.40 ± 0.03	0.14 ± 0.001	6.02 ± 0.09	2.12 ± 0.056	2.26 ± 0.01
Control	1.92 ± 0.055	34.18 ± 0.23	0.26 ± 0.03	0.13 ± 0.001	4.81 ± 0.09	1.78 ± 0.056	1.91 ± 0.01
CD (<5 %)	0.119	0.51	0.007	0.003	0.197	0.121	0.119

treatments, treatment-1 showed high significant variations against the control. From control to treatment-1, root length increased from 1.92 to 7.29 cm (73.7 %), shoot length increased from 34.18 to 44.16 cm (22.6 %), fresh weight of root increased from 0.26 to 0.58 gm (55.2 %), dry weight of root increased from 0.13 to 0.23 gm (43.5 %), fresh weight of shoot increased from 4.81 to 13.33 gm (63.9 %), dry weight of shoot increased from 1.78 to 7.06 gm (74.8 %) and the total biomass increased from 1.91 to 7.29 gm (73.8 %) after 60 days of sowing.

The recorded physiological parameters concluded that the zinc nanoparticles applied plants showed better results when compared to the control. The results presented in Table 5 indicated that the effect of zinc nanoparticles on the growth of the *Arachis hypogea* L. pot culture was significantly high for treatment-1 than treatments 2–3 compared to the control after 30 and 60 days of sowing, and was also increased from 30 to 60 days time period of sowing. The results indicated that the effect of zinc nanoparticles on the root and shoot lengths, fresh and dry weight of roots and shoots and total biomass of the *Arachis hypogea* L. pot culture was also significantly high compared to the control after 60 days of sowing.

Conclusion

Green synthesis promises eco-friendly and non-toxic route to synthesize the zinc nanoparticles. This process can be easily scaled up to produce large quantities of zinc nanoparticles for nanobiotechnology industry and offers solutions to various problems in agriculture such as reducing the use of fertilizer, pesticide and water to improve plant breeding techniques. Here, the zinc nanoparticles were synthesized using *Parthenium hysterophorous* leaf extract by green synthesis route. The optical absorption peaks recorded at 327.50 and 330.00 nm confirms the formation of zinc nanoparticles. The estimated particle size was 16.10–58.60 nm, grain size was 108.50 nm and zeta potential was 100.40–117.20 mV, respectively. Sample-1 (100 %) exhibited highest zeta potential and least particle size. The enzyme and microbial activities and physiological traits were tested using sample-1, which showed significant variations among the nanoparticles treated samples compared to the control. The potential microbial activity of as prepared zinc nanoparticles is exciting. Zinc nanoparticles coupled with microbial activity promises potential applications in agriculture where zinc is one of the essential micronutrients which need to be supplied to the crop plants.

Acknowledgments Mrs. K. Sri Sindhura would like to thank Department of Science and Technology, New Delhi, India, for providing INSPIRE Fellowship and also Department of nanotechnology, Regional Agricultural Research Station (R.A.R.S), Acharya N. G Ranga Agricultural University, Tirupati and Indian Institute of Horticultural Research (IIHR, ICAR) for giving me permission to carry out the necessary laboratory work.

References

Badri Narayanan K, Sakthivel N (2008) Coriander Leaf mediated biosynthesis of gold nanoparticles. Mater Lett 62:4588–4590

Belz RG, Reinhardt CF, Foxcroft LC, Hurle K (2007) Residue allelopathy in *Parthenium hysterophorus* L.—does parthenin play a leading role? Crop Prot 26:237–245.

Ganesh Babu MM, Gunasekaran P (2009) Production and structural characterization of crystalline silver nanoparticles from Bacillus cereus isolate. Colloid Surface A 74:191–195

Gunalan S, Sivaraj Rajeshwari, Rajendran Venckatesh (2012) Green synthesized ZnO nanoparticles against bacterial and fungal pathogens. Progr Natur Sci 22:693–700

Kaushik N, Thakkar MS, Snehit S, Mhatre MS, Rasesh Y, Parikh MS (2010) Biological synthesis of metallic Nanoparticles. Nanomed Nanotechnol 6:257–262

Narayanan KB, Sakthive N (2010) Photosynthesis of gold nanoparticles using leaf extract of Coleus amboinicus Lour. Mater Charact 61:1232–1238

Oudhia P (2001) MEDICINAL Weeds in banana orchids: a boon for small farmers of chhattisgarh (INDIA). Agric. Sci. Digest 21:267–268

Philip Daizy (2010) Green synthesis of gold and silver nanoparticles using Hibiscus rosa sinensis. Phys E 42:1417–1424

Philip Daizy (2011) Mangifera indica leaf-assisted biosynthesis of well-dispersed silver nanoparticles. Spectrochim Acta A Mol Biomol Spectrosc 78:327–331

Prasad TNVKV, Sudhakar P, Sreenivasulu Y, Latha P, Munaswamy V, Raja Reddy K, Sreeprasad TS, Sajanlal PR, Pradeep T (2012) Effect of nanoscale zinc oxide particles on the germination, growth and yield of peanut. J Plant Nutr 35:905–927

Revina AA, Oksentyuk EV, Fenin AA (2007) Synthesis and properties of zinc nanoparticles: the role and prospects of radiation chemistry in the development of modern nanotechnology. Prot Met 43:613–618.

Sharma Nilesh C, Sahi Shivendra V, Nath Sudip, Parsons Jason G, Gardea-Torresdey Jorge L, Pal Tarasankar (2007) Synthesis of plant—mediated gold nanoparticles and catalytic role of bio matrix- embedded nanomaterials. Environ Sci Technol 41:5137–5142

Singh Ravindra P, Shukla Vineet K, Yadav Raghavendra S, Sharma Prashant K, Singh Prashant K, Pandey Avinash C (2011) Biological approach of zinc oxide nanoparticles formation and its characterization. Adv Mat Lett 2:313–317

Sunghyun Kim, Jungeun Kim, Insook Lee (2011) Effects of and ZnO nanoparticles and Zn^{2+} on soil enzyme activity and bioaccumulation of Zn in Cucumis sativus. Chem Ecol 27:49–55

Vinay Kumar B, Bhojya Naik HS, Girija D, Vijaya Kumar B (2011) ZnO nanoparticle as catalyst for efficient green one-pot synthesis of coumarins through Knoevenagel condensation. J Chem Sci 123:615–621

Wolf RE, Todd AS, Brinkman S, Lamothe PJ, Smith KS, Ranville JF (2009) Measurement of total Zn and Zn isotope ratios by quadrupole ICP-MS for evaluation of Zn uptake in gills of brown trout (Salmo trutta) and rainbow trout (Oncorhynchus mykiss). Talanta 80:676–684.

Zeng S, Yong K-T, Roy I, Dinh X-Q, Yu X, Luan F (2011) A review on functionalized gold nanoparticles for biosensing applications. Plasmonics 6:491.

17

Synthesis and structural, optical and thermal properties of CdS:Zn^{2+} nanoparticles

S. Muruganandam · G. Anbalagan · G. Murugadoss

Abstract Undoped and Zn (1–5, 10 %) -doped CdS nanoparticles were successfully synthesized by chemical method and polyvinylpyrrolidone was used as capping agent. The morphology and crystalline structure of the samples were studied by transmission electron microscopy and X-ray diffraction. The average particle size of the spherical nanoparticles determined by these techniques was of the order of 2.5–6 nm. The functional groups of the capping agent on CdS:Zn^{2+} surface were identified by FT-IR study. The band gap of the nanoparticles was calculated using UV–visible absorption spectra and the result showed that the band gap values were dramatically blue shifted from the bulk CdS. The optimum concentration of the doping ions was selected through absorption study. Photoluminescence of the CdS:Zn^{2+} nanoparticle showed strong blue and green emission. The thermal properties of the nanoparticles were analyzed by thermogravimetric-differential thermal analysis.

Keywords CdS · Nanoparticles · Chemical method · XRD · Photoluminescence

Introduction

In recent years, semiconductor nanoparticles have received broad attentions because of their interesting and novel electronic, thermoelectronic and optical properties intrinsically associated with their low dimensionality and the quantum confinement effect (Pan et al. 2001; Duan et al. 2003; Murugadoss 2014). The properties of nanostructured materials are highly affected by the physical parameters that include size, shape, composition, crystallinity and microstructure (Seoudi et al. 2010; Chawla Puja et al. 2010). CdS is an important semiconductor compound of the II–VI group with excellent physical properties and band gap energy of 2.42 eV. It has been extensively studied due to its potential application in field effect transistor, light-emitting diodes, photocatalysis, biological sensors, solar cells and photodegradation of water pollutants (Alivisatos 1996; Colvin et al. 1994; Klein et al. 1997; Ruxandra and Antohe 1998; Romeo et al. 2010; Beecroft and Ober 1997). The electronic properties of CdS nanoparticles doped with transition metals are strongly influenced by the doped transition metals (Liu Kewei et al. 2007). Wageh and Badr (2008) have reported that the $Cd_{1-x}Zn_xS$ nanoparticle is stabilized by the bifunctional organic molecule. Unni (2009) has reported on the size-tunable aqueous synthesis and PL properties of the Zn^{2+}- and Cu^{2+}-doped CdS nanophosphors. There have been only a few reports on Zn^{2+}-doped CdS nanoparticles (Wang et al. 2010; Zhang et al. 2007; Xing et al. 2006; Wang et al. 2012). Transition metals doped with CdS nanoparticles with good structural, magnetic and optical qualities have been reported from a variety of techniques such as surfactant-assisted synthesis, chemical precipitation method, spray pyrolysis, ion implantation and high-energy electron irradiation. In this work, we used the chemical precipitation method. It is the most popular technique that is used in industrial applications because of the cheap raw materials, easy handling and large-scale production (Souici et al. 2006). To improve the uniformity and capability of CdS:Zn nanoparticle, it is

S. Muruganandam · G. Anbalagan (✉)
Department of Physics, Presidency College, Chennai 600 005, Tamilnadu, India
e-mail: anbu24663@yahoo.co.in

G. Murugadoss
Centre for Nanoscience and Technology, Anna University, Chennai 600 025, Tamilnadu, India

capped with water-soluble polyvinyl pyrrolidone (PVP). Through the surface modification with PVP, highly monodispersed CdS nanoparticles were prepared, which exhibited significantly enhanced luminescence and high chemical stability.

Experimental

Materials

To synthesize Zn-doped CdS, the following materials were used. The chemical reagents used were of analytical reagent grade without further purification. Cadmium acetate ($Cd(CH_3COO)_2.2H_2O$), zinc acetate ($Zn(CH_3COO)_2.4H_2O$) and sodium sulfide ($Na_2S.xH_2O$) obtained from Nice Chemical Company were used as precursors. Polyvinyl pyrrolidone (PVP 40,000) was obtained from Aldrich. All the chemicals were above 99 % purity. All the glassware used in this experimental work were acid washed. Ultrapure water was used for all dilution and sample preparations.

In a typical experiment, 2.66 g of cadmium acetate in 100 ml deionised water was mixed with 25 ml of zinc acetate (1–5 and 10 %) solution. The mixture was stirred magnetically at 80 °C until a homogeneous solution was obtained. Then, 1.1 g (0.1 M) of 100 ml Na_2S was added drop by drop to the above mixture. After the Na_2S injection, a yellow voluminous precipitate appeared. It slowly dissolves under formation of CdS:Zn^{2+} nanoparticles under stirring for 30 min at 120 °C. The obtained yellow color colloidal solution was purified by dialysis against de-ionized water and acetone several times to remove impurities. The products were dried in a hot air oven at 80 °C for 2 h. In addition, for synthesis of surfactant (PVP) capped particles, different amounts (0.5–2.5 g) of PVP were added to cadmium acetate solution before the addition of zinc acetate.

Characteristic techniques

The synthesized nanoparticles were characterized using the following techniques: X-ray diffraction (XRD) pattern was collected on an X'Pert PRO diffractometer with Cu Kα radiation ($\lambda = 1.54066$ Å). Transmission electron microscopy (TEM) was performed using HITACHI (Japan) H–7650, 100 kV. The formations of nanoparticles were studied using Fourier transfer infrared spectroscopy (FT-IR) NICOLET AVATAR 3330. The UV–visible absorption spectrum was recorded using a Varian incorporates (U.S) Cary 500 UV–visible spectrometer. Room temperature photoluminescence phenomenon of the samples was studied on Cary-Eclipse spectrometer. Thermal analysis including TGA and DTA was carried out using a simultaneous thermal analyzer SDT Q600 V8.3 Build 101, heating rate 20 °C/min in air atmosphere.

Results and discussion

Structural and morphological study

Figure 1 shows the XRD pattern of undoped, Zn^{2+}- (1–5 and 10 %) doped and PVP-capped CdS nanoparticles. The XRD peaks are found to be very broad indicating very fine

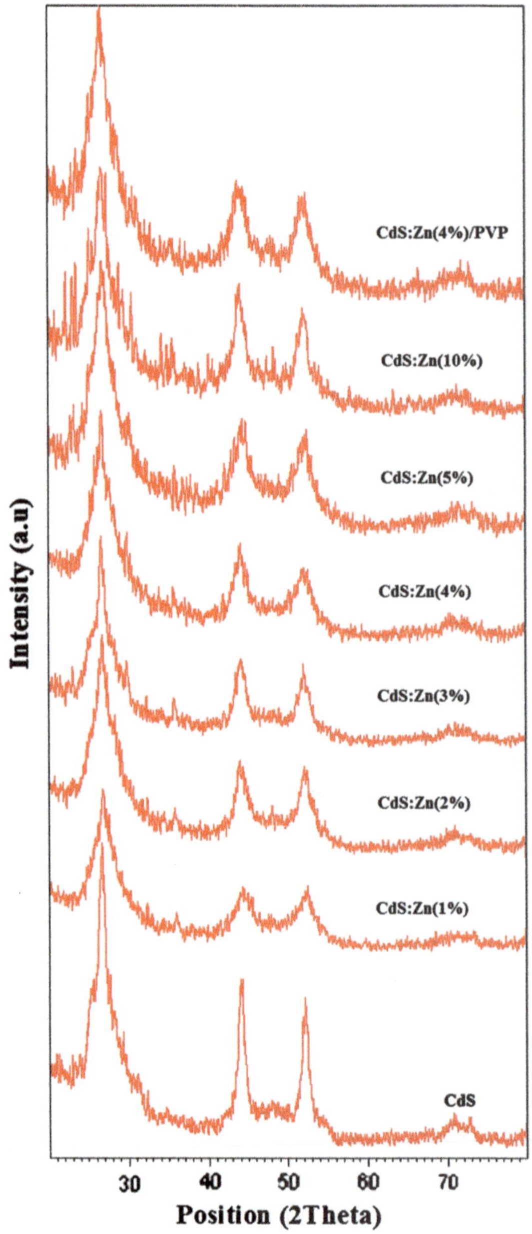

Fig. 1 XRD patterns of the undoped CdS, doped CdS:Zn^{2+} (1-5 and 10 %) and PVP-capped CdS:Zn^{2+} (4 %) nanoparticles

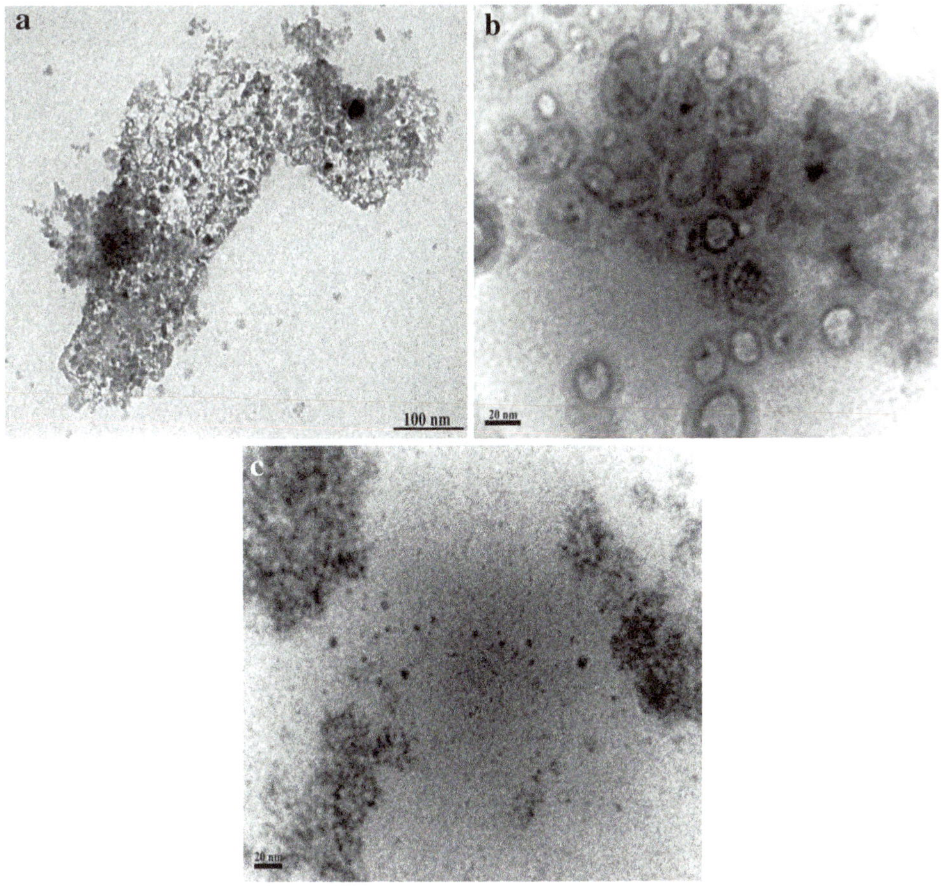

Fig. 2 TEM images show **a** undoped CdS, **b** doped $CdS:Zn^{2+}$ (4 %) and **c** $CdS:Zn^{2+}$ (4 %)/PVP-capped nanoparticles

size of the grains. The XRD pattern exhibits prominent broad peaks at $2\theta = 26.5°$, 43.9° and 52.1° which could be indexed to diffraction from the (1 1 1), (2 2 0) and (3 1 1) planes of cubic CdS, according to JCPDS file (JCPDS no. 10–454). It can be seen that the peak position of the doped samples is not changed with respect to doping (Zn^{2+}) concentrations.

However, the relative intensity is varied. The peak broadness of the surfactant-capped nanoparticles shows decrease in size than that of the free-surfactant particles. It depicts the role of the surfactant in this research work. The average grain size was calculated using Scherrer's formula (Murugadoss et al. 2010).

$$D = k\lambda/\beta \cos\theta \tag{1}$$

where k is particle shape factor (taken as 0.94), λ is the X-ray wavelength (0.1542 nm), β is full width at half maximum (FWHM) and θ is the diffraction angle in degrees (half of the peak position angle). According to the calculation, the average grain size of undoped CdS nanoparticles is 4.5 nm. The size of the Zn^{2+}-doped (4 %) and PVP-capped $CdS:Zn^{2+}$(4 %) nanoparticles is 3.4 and 2.8 nm, respectively. Figure 2a, b shows TEM images of the pure CdS and doped $CdS:Zn^{2+}$(4 %) nanoparticles aggregated with a size of 4.5 and 3.5 nm, respectively. As shown in Fig. 2a, the particles are spherical in shape.

However, the Zn (4 %) -doped CdS (Fig. 2b) nanoparticles showed as core–shell structure. The PVP-capped $CdS:Zn^{2+}$ (Fig. 2c) shows that the particles are well dispersed with an estimated particle size of about 3.0 nm. These values are consistent with the results obtained from X-ray diffraction. From the TEM photographs, one can find that PVP played an important role in controlling the size of $CdS:Zn^{2+}$.

FT-IR study

FT-IR spectra of pure CdS, $CdS:Zn^{2+}$ (1–5 and 10 %) -doped and PVP-capped $CdS:Zn^{2+}$ (4 %) nanoparticle are shown in Fig. 3a–h. The presence of strong absorption band in the region between 600 and 700 cm^{-1} is due to Cd–S stretching vibration (Martin and Schaber 1982). The absorption peak occurs at 927 cm^{-1} due to S–O stretching vibration. The weak absorption band appeared at 1,110 cm^{-1} due to Zn–S stretching mode, and may be due to some amount of Cd^{2+} replaced by Zn^{2+} ions (Murugadoss 2010). The weak peaks at 2,360, 2,343 and 1,645 cm^{-1}, attributed to microstructure formation, may be present in the samples. The strong peaks appearing at 3,433

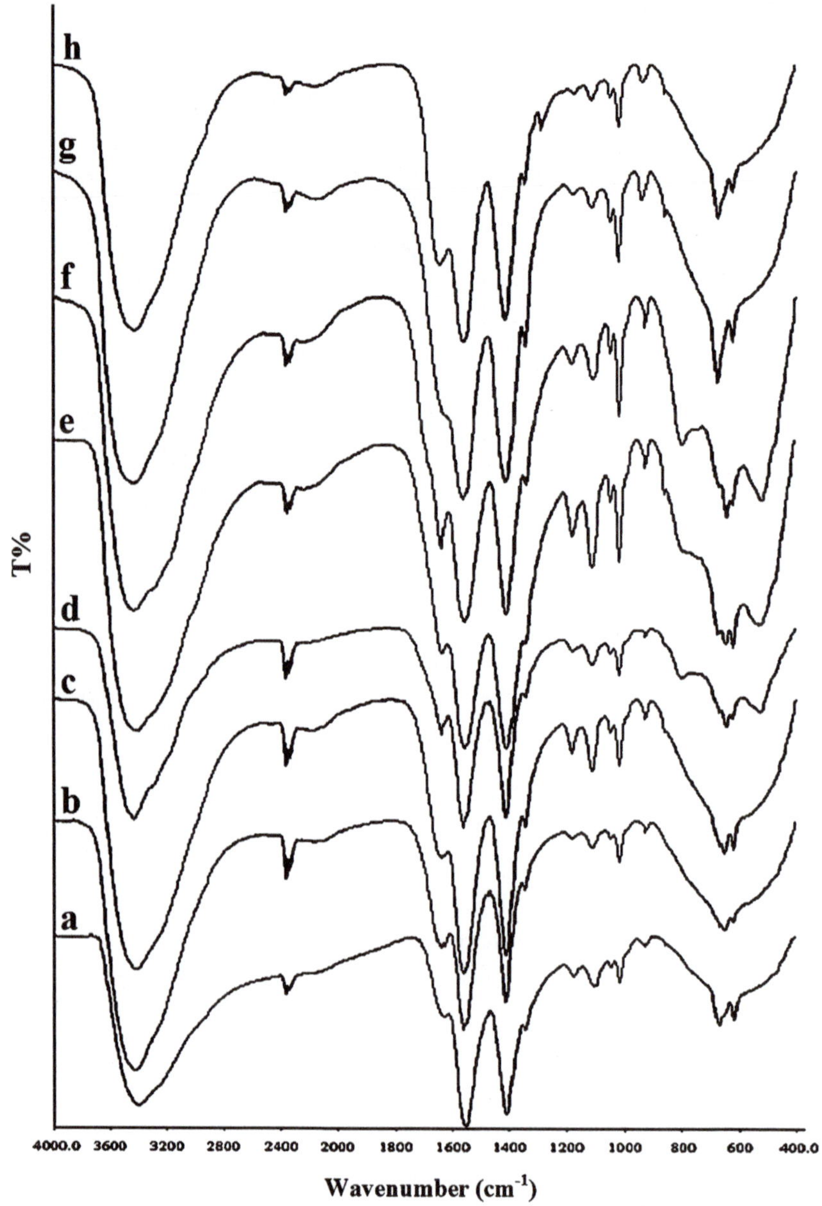

Fig. 3 FT-IR spectra show **a** undoped CdS, **b–g** doped CdS:Zn^{2+}(1–5 and 10 %) and **h** CdS:Zn^{2+}(4 %)/PVP-capped nanoparticles

and 1,560 cm^{-1} are due to O–H stretching vibration and CO_2 adsorbed on the surface of the particles, respectively. In fact, adsorbed water and CO_2 are common to all samples exposed to the atmosphere. As shown in Fig. 3h, the peak appeared at 1,290 cm^{-1} due to C–O stretching vibration. It shows the presence of PVP on the CdS:Zn^{2+} surface.

Optical study

Figure 4 shows UV–visible absorption spectra of undoped CdS:Zn^{2+}(1–5 and 10 %) and PVP-capped CdS:Zn^{2+} nanoparticle in the range of 400–700 cm^{-1}. The absorption values of all samples were blue shifted from bulk CdS. Similar result was observed by Unni et al., for Zn^{2+} (5 %) -doped CdS nanoparticle. This result is a direct consequence of the quantum conferment effect, due to the decrease in particle size (Brus 1984). The band gap of the samples was determined by the following equation (Chowdhury et al. 2005) and presented in Table 1:

$$E_{gn} = \mathrm{hc}/\lambda_e \quad (2)$$

Using the calculated band gap value, the average particle size was measured by Brus formula (Brus 1984).

$$E_{np} = E_g + \hbar\pi^2/2R^2\left[1/m_e^* + 1/m_h^*\right] - 1.8e^2/4\pi\varepsilon_0\varepsilon_r R \quad (3)$$

where E_{np} is the band gap of the nanoparticles, E_g is the band gap of bulk CdS (2.42 eV), m_e* is the effective mass of the electron (0.19 m_o), m_h* is the effective mass of the hole (0.8 m_o), m_o is the rest mass of electron, R is the radius of the particle, ε_r is dielectric constant (5.7), ε_o

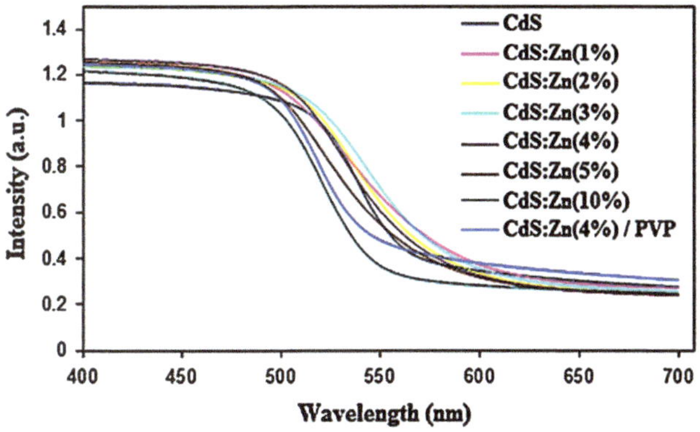

Fig. 4 UV–Vis absorption spectra of undoped CdS, doped CdS:Zn^{2+} (1–5 and 10 %) and CdS:Zn^{2+} (4 %)/PVP-capped nanoparticles

Table 1 Energy gap, absorption and particle size of undoped, doped and PVP-capped CdS: Zn^{2+}(4 %) nanoparticles

Doping concentration	Band gap (E_{gn}) eV	Absorption wavelength (nm)	UV–Vis particle size (nm)	XRD particle size (nm)	TEM particle size (nm)
CdS	2.60	478	4.4	4.5	4.5
CdS: Zn^{2+}(4 %)	2.73	454	3.3	3.4	3.5
CdS:Zn^{2+}(4 %)/ PVP	2.78	447	3.0	2.8	3.0

is the permittivity of free face and 'e' is the charge of electron.

As shows in Table 1, the particle sizes are decreased by increasing the band gap values. This absorption shift is due to the quantum size effect, representing a change in band gap along with exciton features, which can be used as a measure of particle size and size distribution (Beecroft and Ober 1997). The calculated particle sizes are in good agreement with the XRD and TEM results (Table 1).

The PL spectra of synthesized nanoparticles recorded at room temperature for different absorption values are presented in Fig. 5. Figure 5a, b shows the PL spectrum of undoped CdS and Zn^{2+} (1–5 and 10 %) -doped CdS and PVP-capped CdS:Zn^{2+} nanoparticle, respectively. The PL spectrum is observed to be broad in the visible region.

In addition, the peak positions of the doped samples are slightly shifted with respect to the doping concentrations as shown in Fig. 5b. The broad PL spectrum is due to trap state emissions arising from surface defect states (Saunders et al. 2006). When CdS nanoparticles absorb photons, the electrons are excited from valence band to the conduction band and are trapped by defects. The recombination of the defects and excitation states induced by Zn^{2+} provides modified surface trap states, which enhance and shift the luminescence band (Yang et al. 2002). Both samples (undoped and doped CdS) are found to exhibit strong emission bands at 527 and 542 nm in the visible region. The peak position of PL spectra is nearly same for all the samples. However, the relative intensities were changed. The emission peak of the PVP-capped CdS:Zn^{2+} nanoparticle is more enhanced than the uncapped particles. It shows that PVP is not only used to control the particle size, but also to reduce the surface defects. Enhanced PL emission is observed for the PVP-capped CdS:Zn^{2+} nanoparticles as shown in Fig. 5b.

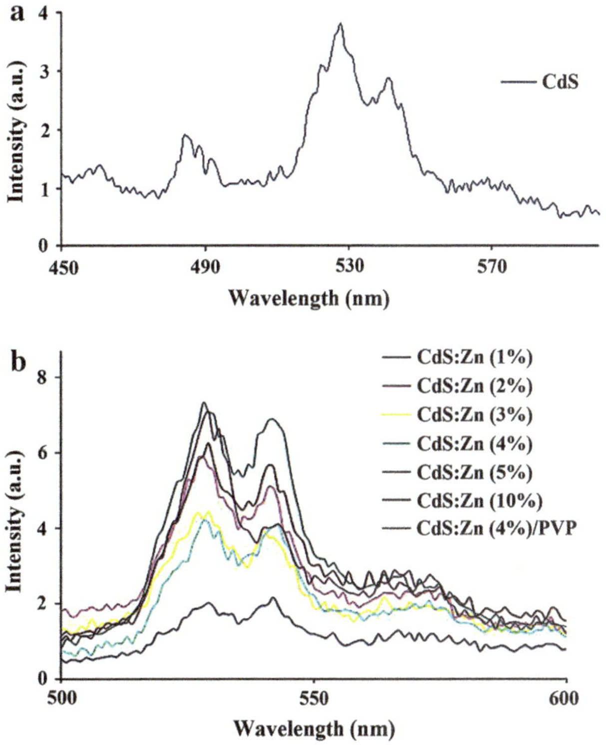

Fig. 5 **a** Photoluminescence spectra of the CdS nanoparticles, **b** CdS:Zn^{2+}(1–5 and 10 %) and CdS:Zn^{2+}(4 %)/PVP-capped nanoparticles

Thermal study

Thermogravimetric (TG) and differential thermal analysis (DTA) were used to study the thermal decomposition of undoped and Zn^{2+}-doped CdS nanoparticles. The synthesized specimens were heated from room temperature to 1,000 °C with an increment of 20 °C/min in air. The combined plots of TG and DTA curves show the undoped (Fig. 6a) and Zn^{2+} (4 %) -doped CdS (Fig. 6b) nanoparticles. From TG data plots, it is noticed that the weight loss of the nanoparticles is found to take place up to 800 °C. The weight loss of TG curves of both the samples up to 100 °C mainly corresponds to evaporation of water, residual solvent and organic molecules.

The DTA curve of Fig. 6a shows a strong exothermic peak at 328 °C, probably corresponding to the lattice

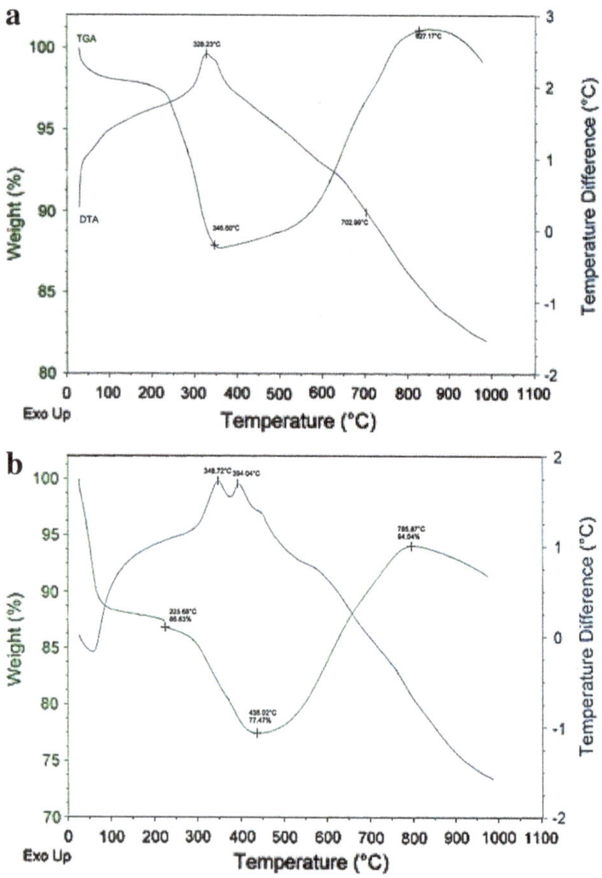

Fig. 6 **a** TG and DTA curves of undoped CdS nanoparticles and **b** CdS:Zn^{2+} (4 %) nanoparticles

deformation of CdS or release of sulfur ion from CdS. In addition, two exothermic peaks were observed at 348 and 394 °C for Zn-doped CdS nanoparticles (Fig. 6b). It may be due to sequential events of the lattice deformation and release of the dopant ion, respectively. The corresponding weight loss is observed as shown in Fig. 6a, b in TG curves. On further increasing the temperature (>400 °C), a significant weight gain is observed at 850 °C for CdS and 800 °C for CdS:Zn as shown in the TGA curves (Fig. 6a, b). These may be due to oxidation of the cadmium ions in the air atmosphere.

Conclusions

CdS and CdS:Zn nanoparticles are potential candidate for various applications like optoelectronic devices and solar cells; hence, they were studied in the present work. TEM results showed that the synthesized particles were spherical in shape with 2.5–6 nm range size. The presence of the capping agent (PVP) on CdS:Zn was identified through FT-IR. The position of the absorption peaks was shifted toward higher energy with a decrease in the nanoparticles size. The optimum concentration of the doping ion was selected as 4 % by optical study. Thermal stability and decomposition of the CdS and Zn-doped CdS nanoparticles were discussed by TG-DTA analysis.

References

Alivisatos AP (1996) Semiconductor clusters, nanocrystals, and quantum dots. Science 271:933–937

Beecroft LL, Ober CK (1997) Nanocomposite materials for optical applications. Chem Mater 9:1302–1317

Brus LE (1984) Electron–electron and electron-hole interactions in small semiconductor crystallites: the size dependence of the lowest excited electronic state. J Chem Phys 80:4403–4409

Chowdhury S, Ahmed GA, Mohanta D, Dolui SK, Avasthi DK, Choudhury A (2005) Luminescence study of bare and coated CdS quantum dots: effect of SHI irradiation and ageing. Nucl Instr Meth Phys Res B 240:690–696

Colvin VL, Schlamp MC, Alivisatos AP (1994) Light-emitting diodes made from cadmium selenide nanocrystals and a semiconducting polymer. Nature 370:354–357

Duan XF, Huang Y, Agarwal R, Lieber CM (2003) Single nanowire electrically driven lasers. Nature 421:241–245

Kewei L, Zhang JY, Xiaojie W, Li B, Li B, Youming L, Fan X, Shen D (2007) Fe-doped and (Zn, Fe) co-doped CdS films: could the Zn doping affect the concentration of Fe and the optical properties? Physica B 38:248–251

Klein DL, Roth R, Lim AKL, Alivisatos AP, McEuen PL (1997) A single-electron transistor made from a cadmium selenide nanocrystal. Nature 389:699–701

Martin TP, Schaber H (1982) Matrix isolated II-VI molecules: sulfides of Mg, Ca, Sr, Zn and Cd. Spectrochem Acta A 38:655–660

Murugadoss G (2010) Synthesis and optical characterization of PVP and SHMP-encapsulated Mn^{2+}-doped ZnS nanocrystals. J Lumin 130:2207–2214

Murugadoss G (2014) Synthesis and study of optical and thermal properties of multi layers coated CdS core–shell nanocomposites. J Lumin 146:430–434

Murugadoss G, Rajamannan B, Ramasamy V (2010) Synthesis, characterization and optical properties of water-soluble ZnS:Mn^{2+} nanoparticles. J Lumin 130:2032–2039

Pan ZW, Dai ZR, Wang ZL (2001) Nanobelts of semiconducting oxides. Science 291:1947–1949

Puja C, Lochab SP, Singh N (2010) Photoluminescence, thermoluminescence and Raman studies of CdS nanocrystalline phosphor. J Alloy Comp 492:662–666

Romeo N, Bosio A, Romeo A (2010) An innovative process suitable to produce high-efficiency CdTe/CdS thin-film modules. Solar Energy Mater Soller Cells 94:2–7

Ruxandra V, Antohe S (1998) The effect of the electron irradiation on the electrical properties of thin polycrystalline CdS layers. J Appl Phys 84:727–733

Saunders AE, Popov I, Banin U (2006) Synthesis of hybrid CdS–Au colloidal nanostructures. J Phys Chem B 110:25421–25429

Seoudi R, Shabaka A, Eisa WH, Anies B, Farage NM (2010) Effect of the prepared temperature on the size of CdS and ZnS nanoparticle. Physica B 405:919–924

Souici AH, Keghouche N, Delaire JA, Remita H, Mostafavi M (2006) Radiolytic synthesis and optical properties of ultra-small stabilized ZnS nanoparticles. Chem Phys Lett 422:25–29

Unni C, Philip D, Smitha SL, Nissamudeen KM, Gopchandran KG (2009) Aqueous synthesis and characterization of CdS, CdS:Zn^{2+} and CdS:Cu^{2+} quantum dots. Spectrrochimica Acta Part A 72:827–832

Wageh S, Badr MH (2008) $Cd_{1-x}Zn_xS$ nanoparticles stabilized by a bifunctional organic molecule. Physica E 40:2810–2813

Wang L, Wang W, Shang M, Yin W, Sun S, Zhang L (2010) Hydrothermal synthesis of monodisperse ZnxCd1-xS spheres and their photocatalytic properties. Int J Hydrogen Energy 35:19–25

Wang J, Li B, Chen J, Li N, Zheng J, Zhu Z (2012) Enhanced photocatalytic H2- production activity of CdxZn1-xS nanocrystals by surface loading MS (M = Ni Co., Cu) species. Appl Surf Sci 259:118–123

Xing C, Zhang Y, Yan W, Guo L (2006) Band structure-controlled solid solution of Cd1- xZnxS photocatalyst for hydrogen production by water splitting. Int J Hydrogen Energy 31:2018–2024

Yang P, Lu MK, Song CF, Xu D, Yuan DR, Cheng XF, Zhou GJ (2002) Luminescence of Cu^{2+} and In^{3+} co-activated ZnS nanoparticles. Opt Mater 20:141–145

Zhang K, Jing D, Xing C, Guo L (2007) Significantly improved photocatalytic hydrogen production activity over Cd1-xZnxS photocatalysts prepared by a novel thermal sulfuration method. Int J Hydrogen Energy 32:4685–4691

18

Metal oxide-based nanoparticles: revealing their potential to enhance oil recovery in different wettability systems

Luky Hendraningrat · Ole Torsæter

Abstract This paper presents systematic studies of hydrophilic metal oxide nanoparticles (NPs) dispersed in brine intended to reveal their potential to enhance oil recovery (EOR) in various rock wettability systems. The stability in suspension (nanofluid) of the NPs has been identified as a key factor related to their use as an EOR agent. Experimental techniques have been developed for nanofluid stability using three coupled methods: direct visual observation, surface conductivity and particle size measurements. The use of a dispersant has been investigated and has been shown to successfully improve metal oxide nanofluid stability as a function of its concentration. The dispersant alters the nanofluid properties, i.e. surface conductivity, pH and particle size distribution. A two-phase coreflood experiment was conducted by injecting the stable nanofluids as a tertiary process (nano-EOR) through core plugs with various wettabilities ranging from water-wet to oil-wet. The combination of metal oxide nanofluid and dispersant improved the oil recovery to a greater extent than either silica-based nanofluid or dispersant alone in all wettability systems. The contact angle, interfacial tension (IFT) and effluent were also measured. It was observed that metal oxide-based nanofluids altered the quartz plates to become more water-wet, and the results are consistent with those of the coreflood experiment. The particle adsorption during the transport process was identified from effluent analysis. The presence of NPs and dispersant reduced the IFT, but its reduction is sufficient to yield significant additional oil recovery. Hence, wettability alteration plays a dominant role in the oil displacement mechanism using nano-EOR.

L. Hendraningrat (✉) · O. Torsæter
Department of Petroleum Engineering and Applied Geophysics, Norwegian University of Science and Technology, NTNU, 7491 Trondheim, Norway
e-mail: luky.hendraningrat@ntnu.no

Keywords Enhanced oil recovery · Metal oxide nanoparticles · Characterisation · Nanofluids stability · Wettability alteration

Introduction

In recent years, most investigations have shown that nanoparticles (NPs) offer promise for future enhanced oil recovery (EOR) processes in which silica-based NPs are most commonly used (Ju et al. 2006; Miranda et al. 2012; Roustaei et al. 2012; Hendraningrat et al. 2013a). Hydrophilic silica NPs were chosen because they consist of greater than 99 % silicon oxide, which is the main constituent of sandstone reservoirs (Hendraningrat et al. 2012). Moreover, silicon oxide is not a chemical that harms the environment. However, inhalation of dried silica NPs into the human body is harmful over a long period. Miranda et al. (2012) mentioned other benefits of using silica NPs for EOR purposes. Silica NPs are an inorganic material that is easy to manipulate in terms of physical chemistry properties. In addition, silica also can be easily surface functionalised from a hydrophobic to hydrophilic state via silanization with hydroxyl groups or sulphonic acid. If the NPs do not aggregate to a larger size, they have the ability to flow through a pore size (typically measured in microns). Hence, the stability of the nanofluid becomes a critical parameter.

The rapid evolution of nanotechnology has created various types of NPs, i.e. metal oxides for multi-purpose applications in many sectors. Currently, metal oxides play an essential role in many areas of chemistry, physics and materials science (Fernandez-Garcia and Rodriguez 2007). Metal oxide-based nanofluids are frequently used for thermal conductivity and heat transfer (Fedele et al. 2011; Keblinski et al. 2008; Yu et al. 2008). The most widely used nanometals are Al_2O_3, MgO, ZrO_2, CeO_2, TiO_2, ZnO and Fe_2O_3, and these metal oxide NPs exhibit unique physical and chemical properties (Fernandez-Garcia and Rodriguez 2007). In the oil and gas industries, the use of metal oxide NPs was recently investigated for trapping and controlling fine migration (Ahmadi et al. 2013; Ogolo et al. 2013) and reducing heavy oil viscosity by acting as a catalyst (Li et al. 2007; Hascakir et al. 2008; Shokrlu and Babadagli 2010).

Because the nanofluid is injected through a typical porous medium with a micron-sized pore throat, it is important to maintain the stability of the nanofluids in terms of nanometer size. Most NPs tend to aggregate once they are hydrated (Keller et al. 2010). The NPs aggregate easily because they have a large ratio of surface area to volume (due to the small particle size); therefore, possess high surface energies and consequently create an aggregate form to minimise these surface energies (Wu et al. 2008). The surface-to-volume ratio is an essential aspect for reactivity, which is defined as the rate at which a chemical reaction will proceed. Materials with a high surface-to-volume ratio react at much faster rates because additional surfaces are available to react. If the NPs create an aggregate form, it may lead to pore throat blocking/plugging and alter the rock properties (Ogolo et al. 2012; Hendraningrat et al. 2013b, c; Ehtesabi et al. 2014).

The classical framework theory used to describe the stability of colloids in suspension (i.e. NPs dispersed in brine as the base fluid) is the DLVO theory (named after Derjaguin, Landau, Verwey and Overbeek). The state of the NPs dispersed in brine is determined by the interactions between the NPs and between the NPs and the base fluid (Rao 2010). This theory primarily addresses the balance between two opposing forces, i.e. van der Waals (VDW) attraction and electrostatic repulsion, to explain why certain colloidal systems agglomerate while others remain discrete and dispersed in suspension. The VDW attraction is the result of forces between individual molecules in each colloid. The electrostatic repulsion becomes significant when two colloids approach each other due to charge interactions, and their double layers begin to interfere. Energy is required to minimise this repulsion by creating an aggregate form. Increasing the repulsive forces by increasing the surface charge between particles provides better stability in the suspension, because it maintains discrete particles and prevents them from gathering into larger sizes in the process known as agglomeration. Each suspension carries the same electrical charge, which produces a force of mutual electrostatic repulsion between adjacent particles. The surface conductivity is closely related to the surface charge. If the charge is sufficiently high, the colloids will remain discrete and dispersed in suspension. Decreasing the charge has the reverse effect. The particle surface charge can be controlled by modifying the suspension liquid, i.e. changing pH, shifting the ionic environment or adding a stabiliser (Rao 2010).

The actual oil reservoirs can be water-, intermediate- or oil-wet. The initial rock wettability will affect oil recovery after waterflooding and plays a vital role in oil production (Morrow 1990). In a water-wet system, the most oil will be produced before water breakthrough after which the rate will decline. Indeed, the water breakthrough will occur earlier in intermediate- or oil-wet systems than in a water-wet system. The wettability affects the relative permeability, which controls the flow and spatial distribution of fluids in a porous medium (Anderson 1987). However, the wettability is not the only parameter that controls the relative permeability curves. Anderson (1987) stated that the relative permeability curves will also be affected by pore geometry, fluid distribution, saturation, and saturation history.

The oil displacement mechanism with use of NPs is not yet clearly understood. Wasan and Nikolov (2003), Chengara et al. (2004), and Wasan et al. (2011) concluded that the structural disjoining pressure is important. The structural disjoining pressure is correlated to the fluid's ability to spread along the surface of a substrate due to an imbalance of the interfacial forces among the solid, oil and aqueous phases (Chengara et al. 2004). The NPs create ordered structures near the three-phase contact line (wetting wedge) of a drop on a solid surface, which promotes spreading of the nanofluid along the surface as monolayer particles. The zero contact angle of the aqueous phase means that the nanofluid has completely spread and has a notably large surface area. Therefore, the fluid possesses a high surface energy with adsorption and alters the surface energy and wettability of the system (Giraldo et al. 2013).

The use of metal oxide nanofluids for EOR purposes that involve stability analysis as well as coreflood experiments with light crude oil and Berea sandstones in various wettability systems has not yet been reported. This paper describes investigation of metal oxide NPs in various initial rock wettability systems to evaluate these materials as alternative solutions for future EOR. Experimental studies on nanofluid stabilities were performed using three techniques: direct visual observation, surface conductivity measurement and particle size analysis. The use of a dispersant in metal oxide NPs was investigated at various concentrations and temperatures. A two-phase coreflood experiment was conducted by injecting those stable nanofluids as a tertiary process (nano-EOR) through Berea sandstone core plugs with various wettability systems ranging from water- to oil-wet. The reason for using different wettabilities in this research was to identify the most suitable conditions for each type of NPs. Therefore, the relationship between oil recovery, metal oxide NPs, and initial rock wettability could be identified. The fluid–fluid and rock–fluid interactions were analysed via contact angle measurement, interfacial tension (IFT) measurement and effluent analysis to observe possible oil displacement mechanisms.

Experiments

Fluids

Degassed crude oil from a field in the North Sea was used as the oleic phase in this study. This oil has a viscosity of approximately 5.1 cP at 22 °C. For the base aqueous phase, a synthetic reservoir brine was prepared from sodium chloride (NaCl) at 3.0 wt. % (approx. 30,000 ppm) and deionised water. The density and viscosity were measured using a pycnometer and a Brookfield rotating viscometer (model LVDV-II + P), respectively. The pH and surface conductivity were measured using a digital pH meter Metroohm (serial 827). This brine was used as the dispersion fluid for the NPs, and the fluid properties are summarised in Table 1. Adding NPs into the brine slightly decreased its density and pH value, whereas the viscosity and surface conductivity were increased.

Table 1 Fluid properties at room condition

Fluid	Density (g/cm^3)	Viscosity (cP)	pH	Surface Conductivity (mV)
Degassed crude oil	0.826	5.10	–	–
Brine, NaCl 3 wt %	1.022	1.001	6.76	14.2
Al_2O_3 nanofluid 0.05 wt %	1.020	1.03	6.551	29.0
TiO_2 nanofluid 0.05 wt %	1.016	1.08	6.413	36.6
SiO_2 nanofluid 0.05 wt %	1.020	1.02	6.220	49.9

Table 2 Nanoparticles properties

Properties	Al_2O_3	TiO_2	SiO_2
Specific surface area, BET (m^2/g)	65 ± 10	50 ± 15	45–60
Declared primary particle size (nm)	17	21	40
Bulk density (g/L)	50	130	90–110
pH (4–5 % in dispersion)	4.5–6	3.5–4.5	3.8–4.8

Nanoparticle preparation and characterisation

Three fumed hydrophilic oxide NPs, i.e. aluminium (Al_2O_3), titanium (TiO_2), and silica (SiO_2), were employed in this study. These NPs were selected due to similar specific surface areas (Brunauett–Emmett–Teller/BET) of approximately 45–65 m^2/g. The average primary particle sizes of those NPs were declared as 17 nm for Al_2O_3, 21 nm for TiO_2 and 40 nm for SiO_2. Table 2 lists the properties of the NPs. Characterisation and elemental analysis of NPs were carried out using a scanning electron microscope (SEM, Hitachi Model S-5500) and integrated energy-dispersive X-ray spectroscopy (EDX), as shown in Figs. 1, 2, 3. The carbon appears because the NPs were placed on carbon conductive tape. Based on this characterisation, the NPs took on a secondary form after aggregation under SEM that is much larger than their primary particle size. As explained in the previous section, the NPs easily aggregate because they have a large surface-to-volume ratio (due to the small particle size) and therefore possess high surface energies. The NPs consequently create

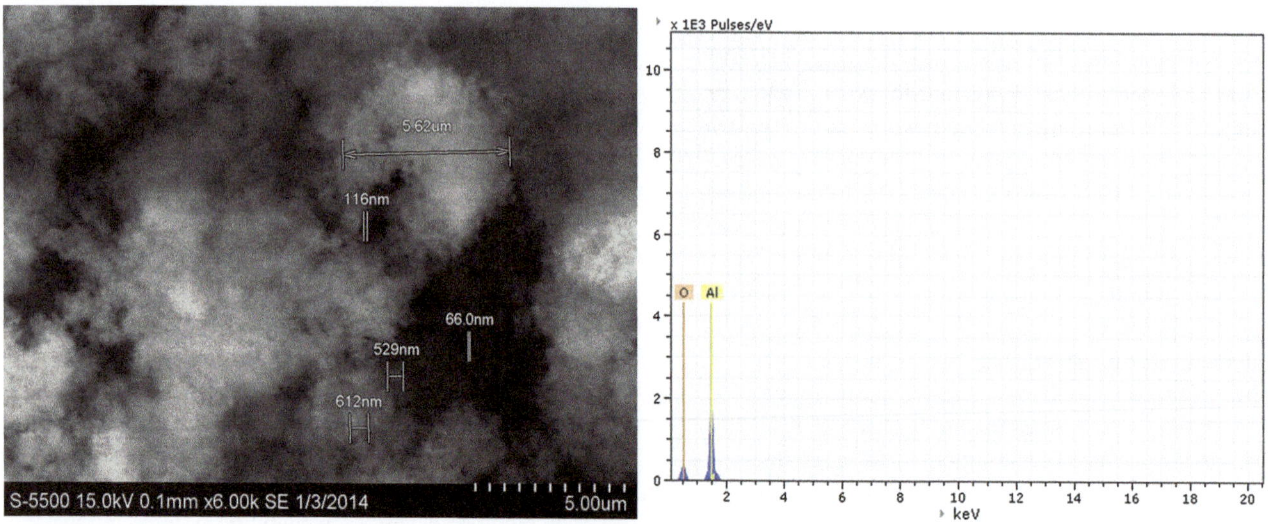

Fig. 1 Morphology and elemental analysis of Al_2O_3 NPS from SEM-integrated EDX instrument

an aggregate form to minimise this surface energy. In addition, materials with high surface-to-volume ratios react at much faster rates because additional surfaces are available to react. To obtain a more precise particle size characterisation, the NanoSight instrument was used, which operates based on a conventional optical microscope concept but uses a laser light source to illuminate nano-sized particles within a sample introduced to the viewing unit via a disposable syringe, and the particles subsequently scatter the light. The detector measures the NPs in a particular area (Nanosight 2009).

Nanofluid preparation and characterisation

Each type of NP was synthesised using brine with a weight concentration of 0.05 wt. %. This concentration was selected based on our previous concentration optimisation study of silica NPs (Hendraningrat et al. 2013c). The nanofluids were synthesised using high-speed magnetic stirring and ultra-sonication at 60–100 % amplitude. The fluid properties of the nanofluids are provided in Table 1. Figure 4 shows brine (NaCl 3 wt. %) containing NPs at 0.05 wt. % at the initial condition. Of the group, the TiO_2 nanofluid is the milkiest in appearance due to original colour from the particles. The characterisation of NP sizes and distributions dispersed in brine were measured using nanoSight. Figure 5 shows the NP distribution measurements dispersed in brine, ranging from 0 nm to greater than 1,000 nm in particle size. The means of the NP diameters measured from NanoSight are 98 nm for Al_2O_3, 106 nm for TiO_2 and 53 nm for SiO_2.

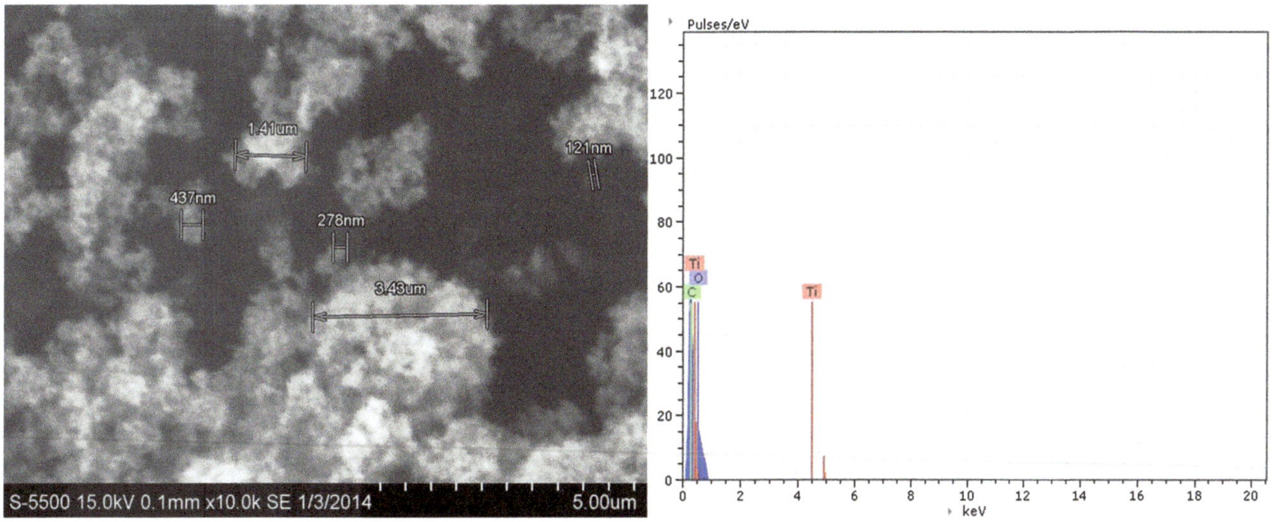

Fig. 2 Morphology and elemental analysis of TiO_2 NPS from SEM-integrated EDX instrument

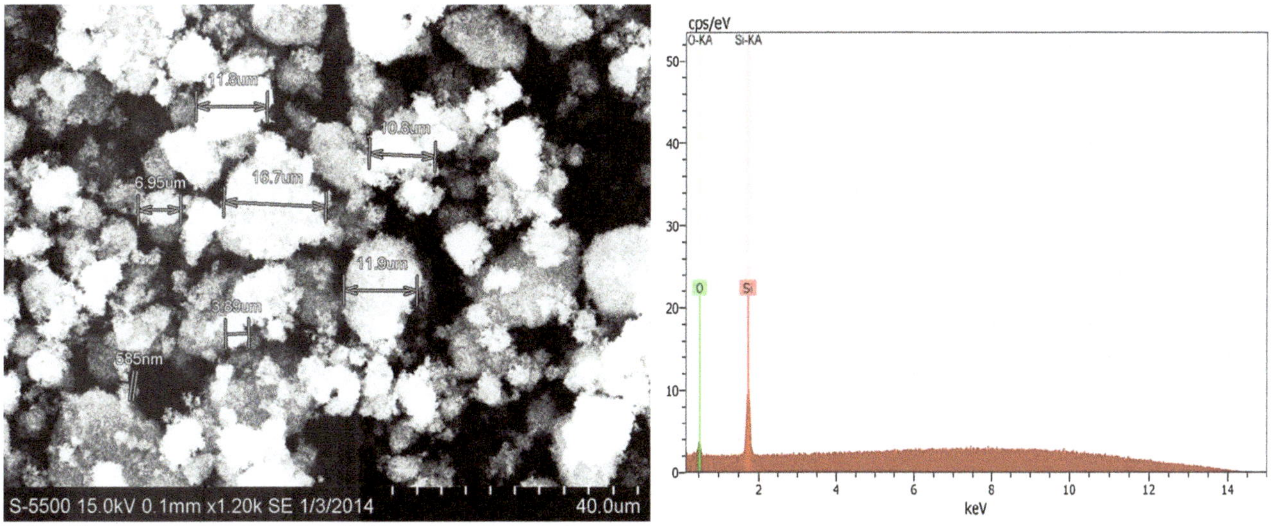

Fig. 3 Morphology and elemental analysis of SiO_2 NPS from SEM-integrated EDX instrument

Nanofluid stabilising agent

A polyvinylpyrrolidone (PVP) stabiliser known as Povidone is used in this study. This material has an ultrafine amorphous form, a size of approximately 50–250 μm and a bulk density of 400–600 g/L (Folttmann and Quadir 2008). The molecular form of PVP is $(C_6H_9NO)_n$, and the chemical structure of PVP is shown in Fig. 6. An advantage of PVP is its universal solubility both in hydrophilic and hydrophobic solvents (Folttmann and Quadir 2008), and it is a non-toxic stabiliser that is safe for the environment. Currently, PVP is widely used as a binder for capsules and granules (especially in pharmaceuticals), a film former in medical plastics, a taste masker in chewable tablets and a toxicity reducer (Folttmann and Quadir 2008). In this study, PVP type K30 was added to avoid earlier NP aggregation by improving the metal oxide nanofluid stability. The K30 refers to the *K* value of this PVP in the range 27.0–32.4. The *K* value characterises the mean molecular weight of PVP, and the higher the *K* values, the higher the molecular weight.

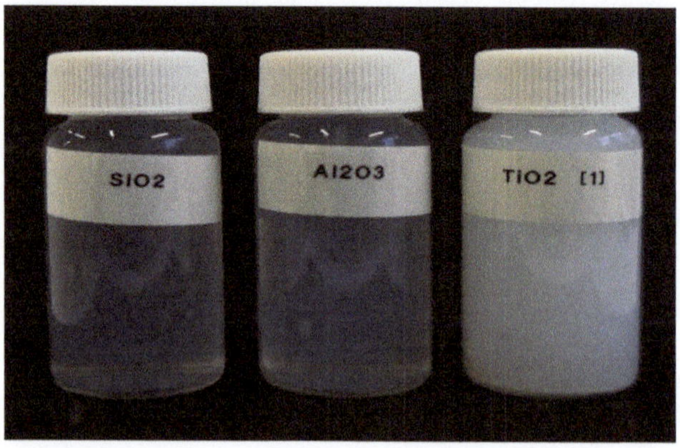

Fig. 4 Brine (NaCl 3 wt %) containing nanoparticles 0.05 wt % at initial condition

Core preparation and characterisation

All cores used in this study consisted of Berea sandstone, which is commonly used as a sandstone core porous medium in most laboratory research. The core has an average diameter and length of 3.8 and 4.0 cm, respectively. All cores were first cleaned with methanol and dried in a heating cabinet at 80 °C for 2–3 days until no dry weight alteration was observed before and after cleaning based on standard laboratory procedure (Torsæter and

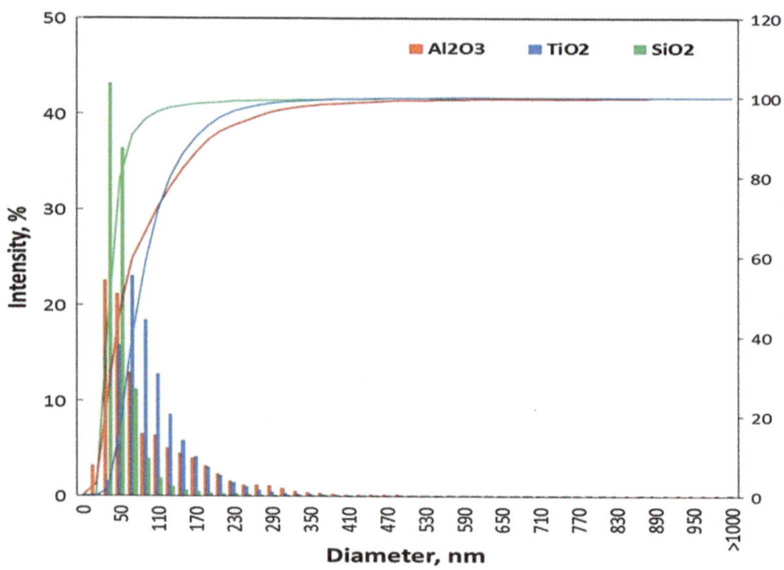

Fig. 5 NanoSight measurement of nanoparticles size distribution for brine (NaCl 3 wt %) containing 0. 05 wt % nanoparticles at initial condition

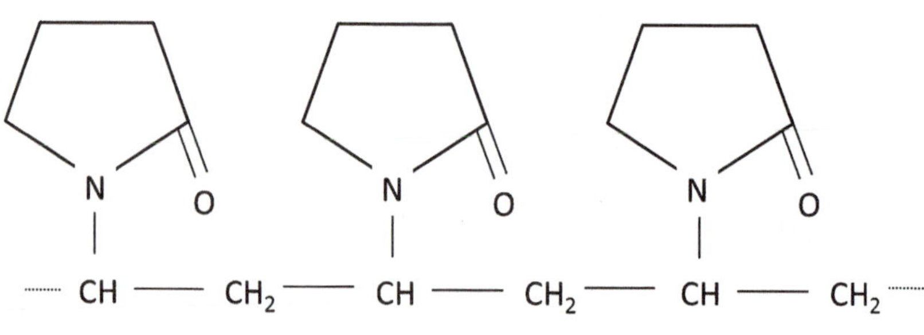

Fig. 6 Chemical structure of PVP (Adapted from Folttmann and Quadir 2008)

Abtahi 2003). The porosities were measured using a helium porosimeter, resulting in values from 15 to 17 %. Selected cores were cut into smaller specimens for characterisation purposes. Others were used in further ageing processes to alter the original wettability from water-wet to intermediate- and oil-wet systems. The pore size of Berea sandstone was observed using SEM, which reported micron-sized pores, as shown in Fig. 7.

The core mineralogy was characterised using two methods: energy-dispersive X-ray (EDX) and X-ray diffraction (XRD). Figure 8 shows the EDX elemental analysis of areas A and B. The only elements present in both locations are Si, O, Al and K. Reactive clays such as 'Na^{+}', 'Ca^{+2}', 'Cl^{-}', or 'Mg^{+2}' were not identified based on this analysis, but a minor amount of those elements may be present. Table 3 provides the XRD analysis data from four samples and shows consistent results. The core was dominated by quartz, which is composed of elements Si and O (91–93 %). The potassium (K^{+}) and aluminium (Al^{+3}) may originate from either alkali feldspar ($KAlSi_3O_8$) or residues. The minor residue is identified as pyroxene, which has the general formula XYZ2O6, where $X = Mg^{+2}$, Fe^{+2}, Mn^{+2}, Li^{+}, Ca^{+2}, Na^{+}; $Y = Al^{+3}$, Fe^{+3}, Cr^{+3}, Cr^{+3}, Ti^{+4}, Mg^{+2}, Fe^{+2}, Mn^{+2}; and $Z =$ Si, Al^{+3}, Fe^{+3}. The XRD results are consistent with the EDX results. Hence, it was assumed that the obtained mineralogy is representative for all core plugs studied in this work.

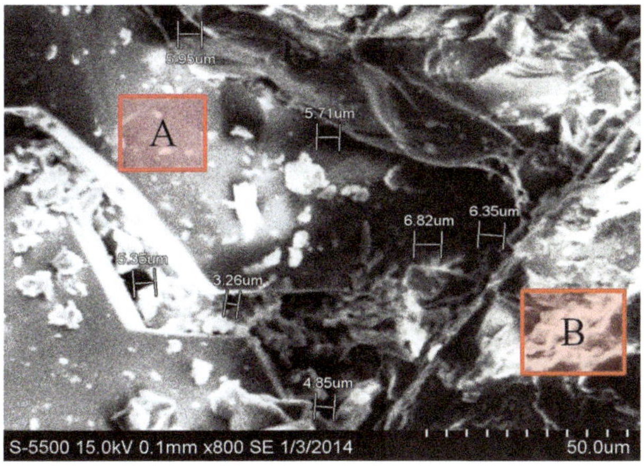

Fig. 7 Morphology of Berea sandstone SEM image with pore size identification

Quartz plate preparation

A series of rectangular transparent plates of quartz was used as the solid substrates in this study. Each quartz plate has average dimensions of 2.8 cm in length, 1.5 cm in width and 0.19 cm in height. The quartz plates were first cleaned with a detergent solution and rinsed with methanol followed by distilled water. The plates were dried in a heating cabinet at 50–60 °C for 3–5 days. The original wettability of the quartz plate was observed by measuring the contact angle between brine and crude oil and is considered as water-wet (Hendraningrat et al. 2013a).

Nanofluid stability analysis

Nanofluid stability analysis was carried out using three coupled techniques: direct visual observation, surface conductivity measurement and NanoSight. The visual

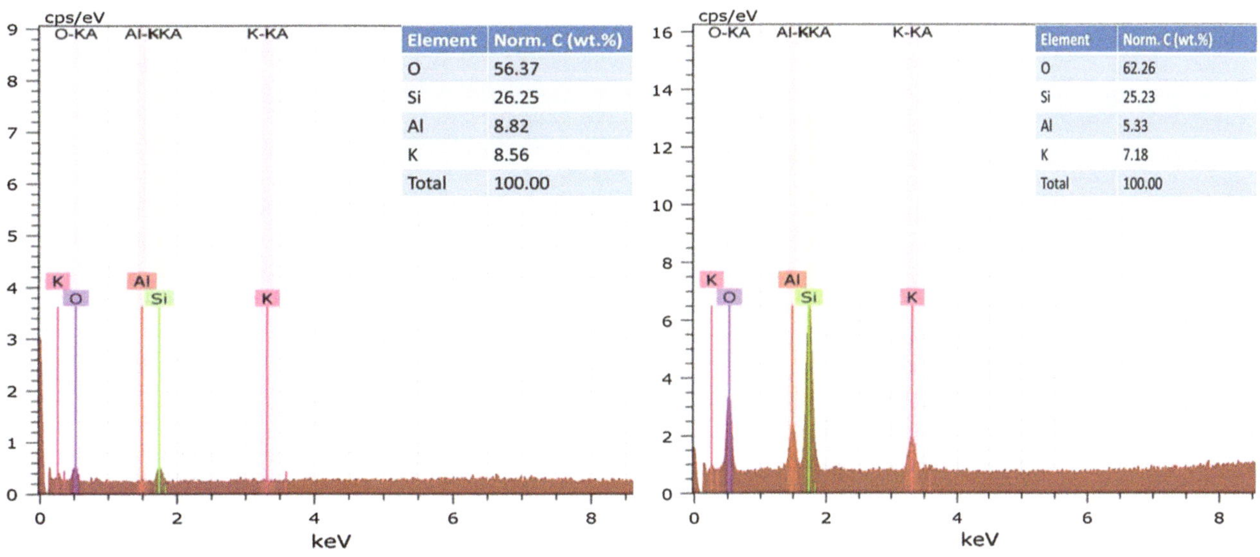

Fig. 8 Quantitative mineral elements analysis using EDX integrated with SEM at area *A* (*left*) and *B* (*right*)

Table 3 XRD mineral analysis of Berea core plugs (% mass)

Mineral	Sample-1	Sample-2	Sample-3	Sample-4
Quartz	93	91	92	93
Alkali feldspar	6	8	8	7
Pyroxene	<1	<1	<1	<1

observation was conducted at various temperatures: room temperature, 50 and 80 °C. The dispersant PVP was used and studied to avoid early metal oxide NP aggregation. Three different concentrations of PVP were studied: 0.1, 0.5 and 1.0 wt. %. The surface conductivity of each nanofluid solution was measured using a digital pH meter. Each colloid carries a "like" surface charge that produces a force of mutual electrostatic repulsion between adjacent particles. The surface conductivity is closely related to the surface charge. If the charge is sufficiently high, the colloids will remain in discrete form and exhibit good stability in suspension. Decreasing the surface charge has the reverse effect. The particle stability also can be observed using NanoSight. The nanofluids will be unstable if particles aggregate and create a secondary form that is larger than the primary particle size. NanoSight is a useful technique for identifying the particle size and its distribution in solution.

Wettability alteration via an ageing process

The original wettability of the cores and quartz plates was altered from water-wet to intermediate- and oil-wet via an ageing process. The wettability of porous rock was determined using the Amott test, which combines spontaneous and forced displacement at room conditions with a wettability index (I_W) formula as follows:

$$I_w = \frac{V_{o1}}{V_{o1} + V_{o2}} - \frac{V_{w1}}{V_{w1} + V_{w2}} \quad (1)$$

where V_o describes the oil volume produced from the imbibition process, V_w describes the water volume from the drainage process, subscript '1' denotes a spontaneous displacement process and '2' indicates a forced displacement process.

The wettability index (I_W) of a porous medium can be correlated with the contact angle obtained from a smooth surface by the inversed cosine (arccos) of the contact angle. A water-wet system is represented by an I_W from +0.25 to +1.0 (contact angle from 0° to 75°), intermediate-wet by an I_W from −0.25 to +0.25 (contact angle from 75° to 105°) and oil-wet by an I_W from −0.25 to −1.0 (contact angle from 105° to 180°). The original wettabilities of the Berea sandstone cores show an Amott–Harvey wettability index (I_W) ranging from +0.7 to +0.9 and are considered as strongly water-wet (Hendraningrat et al. 2013d). Ageing in crude oil altered the Berea core plugs to intermediate-wet. The ageing process took approximately 4 weeks at temperatures of 80–90 °C and altered the core wettability I_W to an average of +0.05 (Li et al. 2013). At the same time, a mixture of 70 % (V/V) Pentane (C_5–H_{12}) and 30 % (V/V) Surfasil™ (C_8–H_{24}–Cl_2–O_3–Si_4) was used to alter the core and quartz plate wettability from water-wet to oil-wet. Surfasil™ can be used for surface treatment purposes. The latter process took approximately 6 days at room temperature. Based on the Amott test, the substrate has an I_W in the range from −0.5 to −0.6 and is considered oil-wet.

Contact angle measurement

The contact angle was measured directly on transparent quartz plates using a Goniometry KSV CAM instrument at ambient conditions. The system consists of a quartz plate as the solid, a degassed crude oil from a field in the North Sea as the oleic phase and nanofluids as the aqueous phase, as shown in Fig. 9. The measurement is based on the Young–Laplace formula as follows:

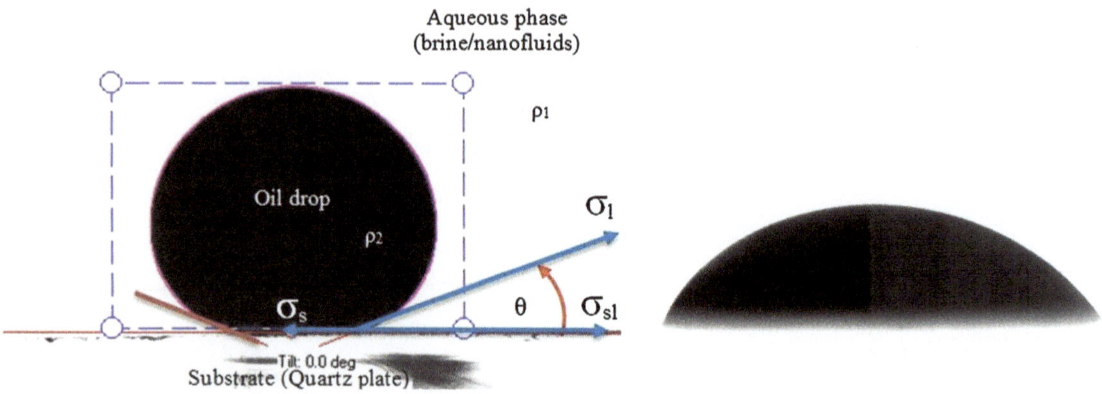

Fig. 9 Contact angle formation on quartz glass of crude oil against brine/nanofluids: water-wet system (*left*) and oil-wet (*right*)

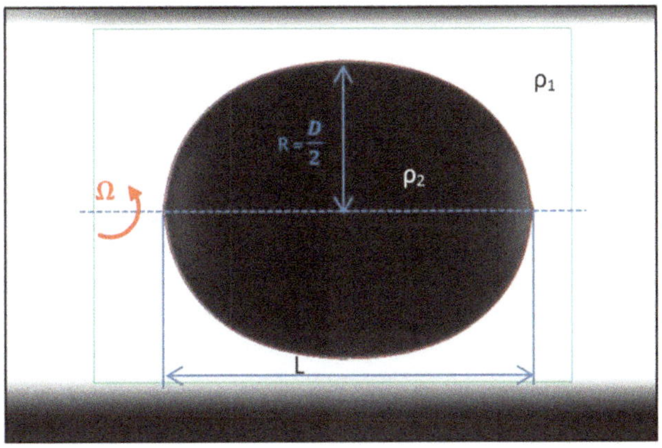

Fig. 10 Interfacial tension measurement between crude oil and brine/nanofluid

$$\sigma_s = \sigma_{sl} + \sigma_l \cos\theta \qquad (2)$$

where σ describes the interfacial tension components of phase, indices s and l stand for the solid and liquid phases, respectively, σ_{sl} represents the interfacial tension between the two phases and θ is the contact angle corresponding to the angle between vectors σ_l and σ_{sl}.

Treiber et al. (1971) defined the contact angle in a three-phase system (water, oil and rock surface) as follows: water-wet in the range of 0°–75°, intermediate/neutral-wet in the range of 75°–105° and oil-wet in the range of 105°–180°. A zero contact angle indicates that the denser fluid completely wets the solid.

Interfacial tension measurement

The interfacial tension (IFT) of degassed crude oil was measured against brine/nanofluids as the aqueous phase using a SVT20 spinning drop video tensiometer at room temperature, as shown in Fig. 10. This measurement involves crude oil with brine and silica-based and Al_2O_3-based nanofluids with PVP. Due to poor clarity, the IFT of the TiO_2-based NPs with PVP nanofluid could not be measured. The rotational speed was maintained near 5,000–6,000 rpm. The formula used to measure IFT is as follows (Than et al. 1988):

$$\sigma = \frac{\Delta\rho \times \Omega^2 \times (D_{app})^3}{8n^3 J_D(\frac{L}{D})} \qquad (3)$$

where σ is the interfacial tension (dyn/cm), $\Delta\rho$ is the density difference (g/cm^3), Ω is the rotational rate of the cylinder (s^{-1}), D_{app} is the measured drop diameter (cm), n is the refractive index of the heavy fluid, D is the true diameter of the drop ($D = \frac{D_{app}}{n}$), J_D is a correction factor and a function of L/D, and L/D is the aspect ratio (e.g. the ratio of the drop length to its diameter).

Coreflood procedure

All cores were fully saturated with brine in a vacuum container for approximately 1–2 days. The liquid (brine) permeability of cores was measured with a range of 100–400 mD. The two-phase oil–water system coreflooding instrument is shown in Fig. 11. The sleeve pressure was set to a constant 20 bar in the Hassler core cell. The drainage process was started by injection of degassed crude oil at a flowrate from 1 to 40 cm^3/min with approximately 8–10 of pore volume (PV) until no more brine was produced and thus the initial water saturation was established. As a first forced imbibition process, brine (denoted as WF1) was injected at a constant rate of 0.2 cm^3/min with a range of 3–5 PV into the core plugs until no more oil was produced and thus residual oil saturation was established due to waterflooding. The injection rate was chosen to approximate the typical reservoir velocities. Next, the injection was continued at a similar injection rate and pore volume of nanofluids (silica and metal oxides) as a tertiary recovery process (denoted as NF). In the final stage, brine was re-injected (denoted as WF2) in a range of 3–5 PV under the same conditions as those of previous flooding. The recovered oil was accumulated into a gravity-settling separator and calculated based on material balance. A camera was installed to record the oil recovery during the flooding process, and the effluents were collected in a water accumulator. For each sequential flooding process, the effluent was collected at the last PV to analyse its pH and surface conductivity alterations to avoid contamination from the previous sequence. A differential pressure transmitter in the range of 0–30 bar was connected to the inlet and outlet ends of the core holder to measure and record the pressure drop across the core during the flooding process. The data were recorded into a computer in real time. As a comparison, brine with dispersant alone was used in the tertiary flooding and is discussed in the following section.

Results and discussion

Nanofluid stability analysis

Providing a stable nanofluid is challenging, because NPs have a tendency to aggregate due to forces of different natures that interact amongst particles and may lead to agglomeration (Fedele et al. 2011). Decreasing the stability of nanofluids will reduce their advantages. We observed that nanofluid stability has been observed as a critical issue prior to use as EOR agents (Hendraningrat et al. 2013b). The stability of nanofluids is a function of its concentration as dispersed in brine. During the injection process, the tendency of NPs to agglomerate will affect the porosity and

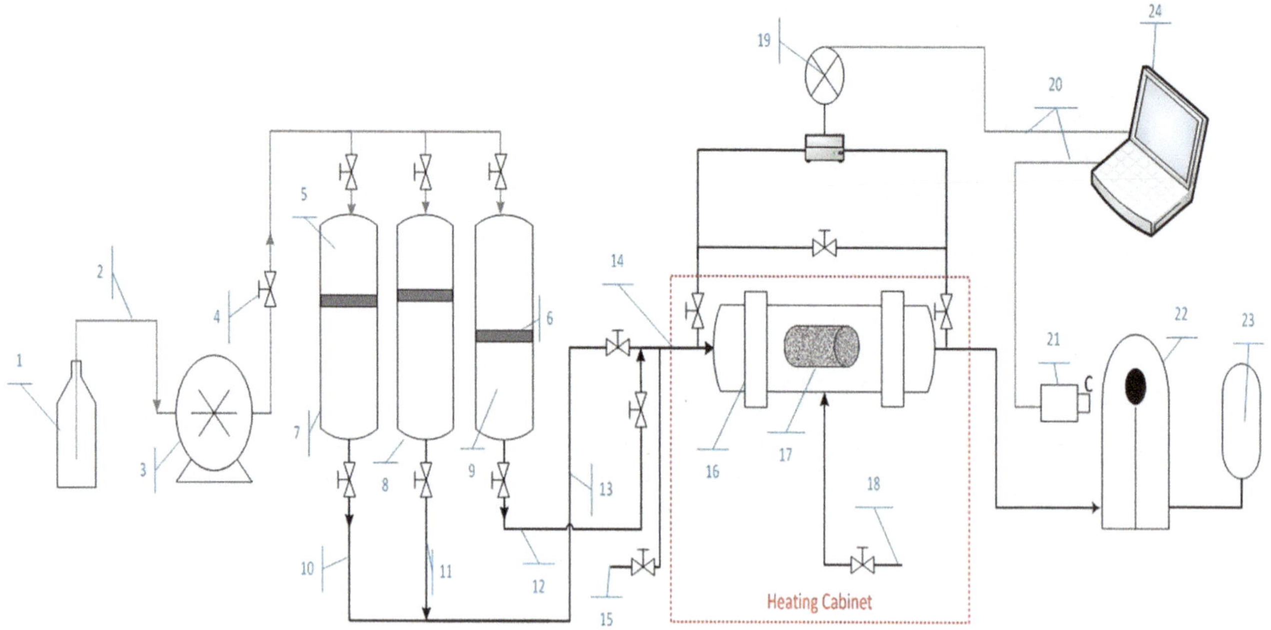

Fig. 11 Experimental setup schematic *1* pump fluid (Exxol D60), *2* injection pipe (ID 1.5 mm), *3* pump, *4* valve, *5* pump fluid in vessel-A, *6* piston plate, *7* brine in vessel-A, *8* nanofluid in vessel-B, *9* crude oil in vessel-C, *10* brine pipeline, *11* nanofluid pipeline, *12* oil pipe, *13* water-phase pipeline, *14* inlet pipeline, *15* bypass valve, *16* Hassler core cell inside oven, *17* core plug inside cell, *18* sleeve pressure, *19* pressure transmitter, *20* connection cable to PC, *21* camera recorder, *22* oil–water separator, *23* water accumulator, *24* Computer/PC as controller and data acquisition

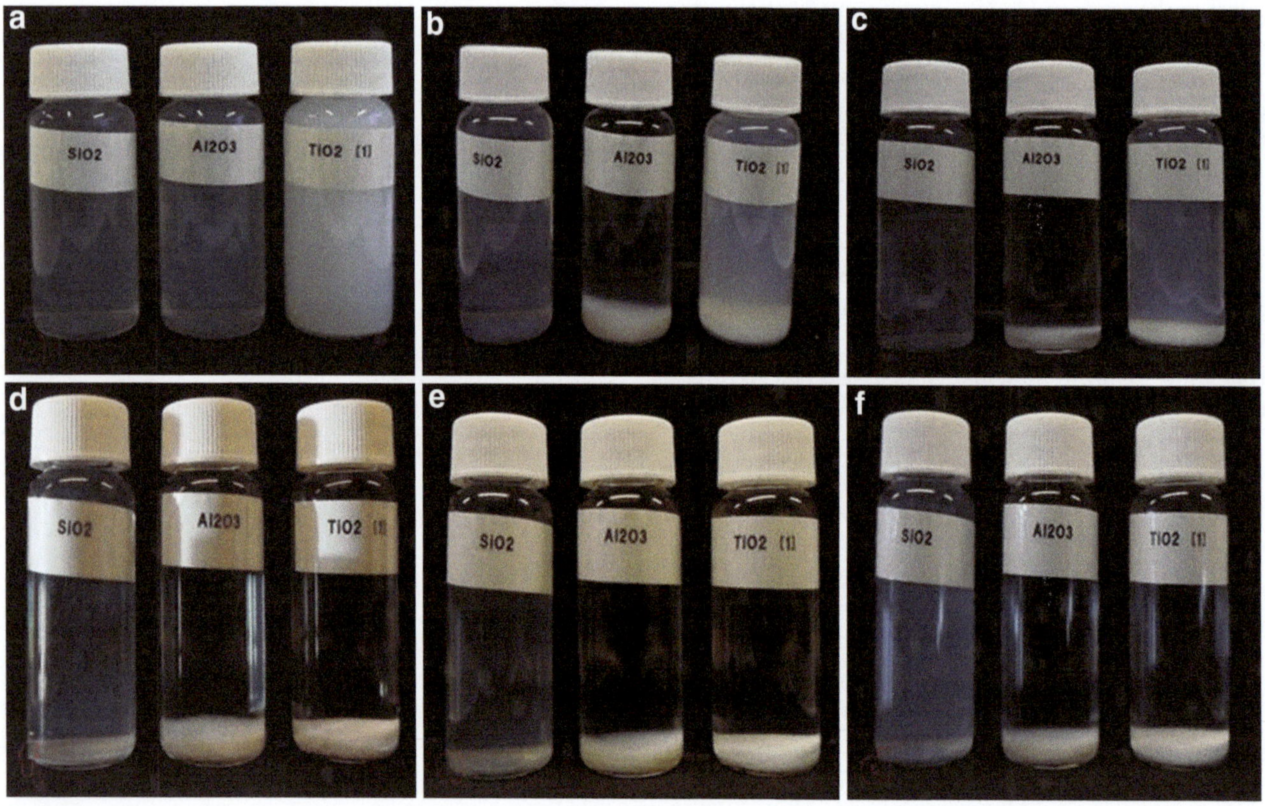

Fig. 12 Untreated nanofluids stability observations at room condition: **a** initial condition, **b** $t = 1$ h, **c** $t = 3$ h, **d** $t = 24$ h, **e** $t = 48$ h, **f** $t = 168$ h

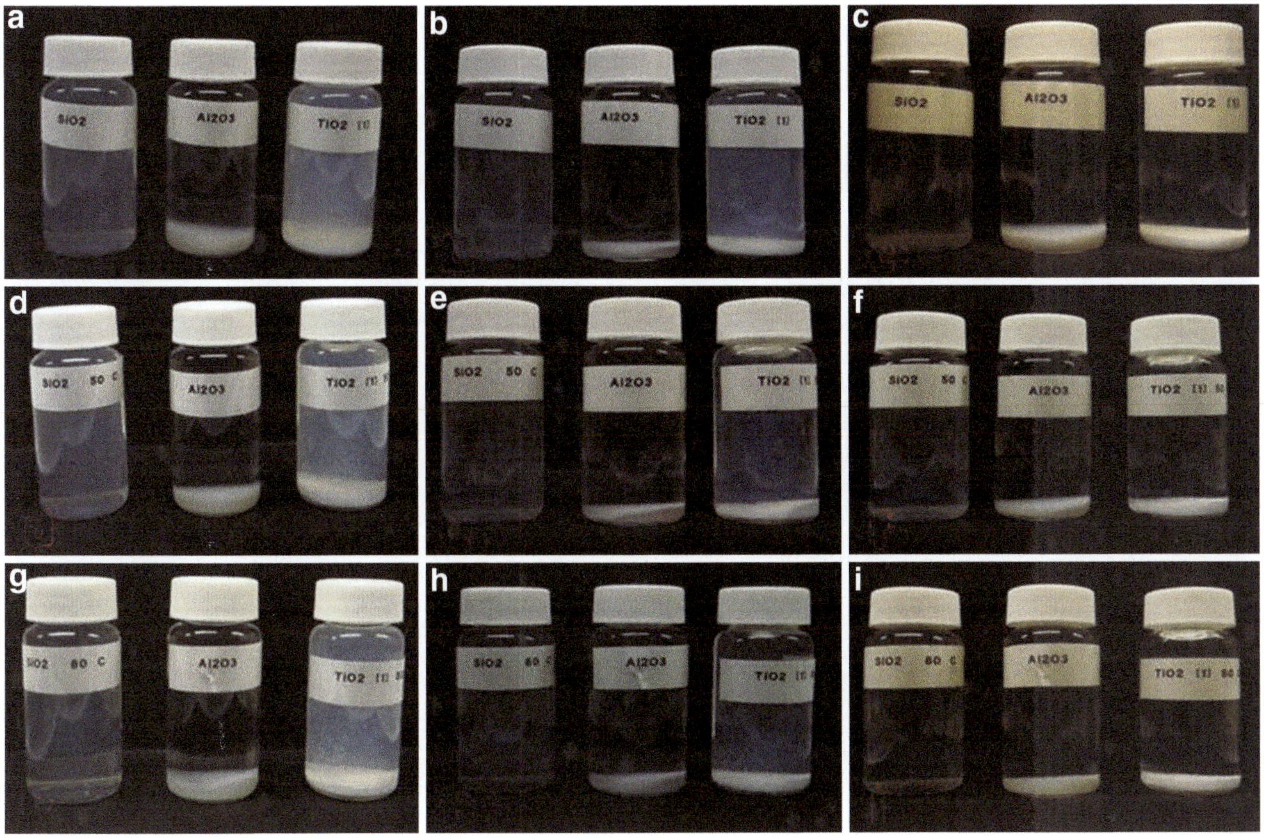

Fig. 13 Untreated nanofluids stability observations at various temperatures: 25 °C (*top row*), 50 °C (middle row), 80 °C (*bottom row*). The observation were taken at; 1 h (*first column*); 3 h (*second column*) and 8 h (*third column*)

impair permeability. It was also observed that oil recovery is not proportional to the nanofluid concentration, which means that oil recovery is enhanced if the nanofluid concentration increases but decreases if the concentration is higher than a critical value. For oil recovery, the critical value seems to be an optimum hydrophilic nanofluid concentration of 0.05 wt. %, and we selected this concentration in this study. The nanofluid stability analysis was developed using three techniques: direct visual observation, surface conductivity measurement and particle size analysis.

Based on our visual observation of stability over 168 h, as shown in Figs. 12, 13, untreated hydrophilic metal oxide NPs (0.05 wt. %) were much easier to precipitate than silica dioxide when dispersed in brine solution. As shown in Fig. 12, these particles began to precipitate in less than 1 h at room conditions. The A_2O_3-based nanofluid was completely precipitated after approximately 3 h, and the TiO_2-based nanofluid agglomerated slightly later (Fig. 13). The SiO_2-based nanofluids show much better stability over approximately 24–48 h. Thermal stability was also observed at room temperature, 50 and 80 °C, as shown in Fig. 14, to determine the stability in reservoir temperature conditions. The behaviour appears analogous for metal oxides at room conditions, which means that the higher temperature does not significantly affect the NP aggregation. These visual stability observations are consistent with the surface conductivity measurements, as shown in Table 1. The SiO_2-based nanofluid has the highest surface conductivity (49.9 mV) among the nanofluids, whereas the Al_2O_3-based nanofluid has the lowest (29 mV). The surface conductivity is inversely related to the pH value, and the higher the particle charge potential, the more stable is the nanofluid. The surface conductivity is closely related to the surface charge. If the charge is sufficiently high, the colloids will remain discrete and dispersed in suspension.

Determination of the mean particle size dispersed in brine was carried out using NanoSight, as shown in Fig. 5, and the results are summarised as follows: 98 nm for Al_2O_3, 106 nm for TiO_2 and 53 nm for SiO_2. Although the SiO_2 nanoparticle has the largest primary particle size (40 nm) among the NPs used in this study (see Table 2), its mean diameter is the smallest when dispersed in brine (Fig. 15). The Al_2O_3 NPs have the highest degree of aggregation, almost six times larger than the primary particle size. These results are consistent with those from two previous techniques in which Al_2O_3-based nanofluid precipitated fastest and had the lowest surface conductivity.

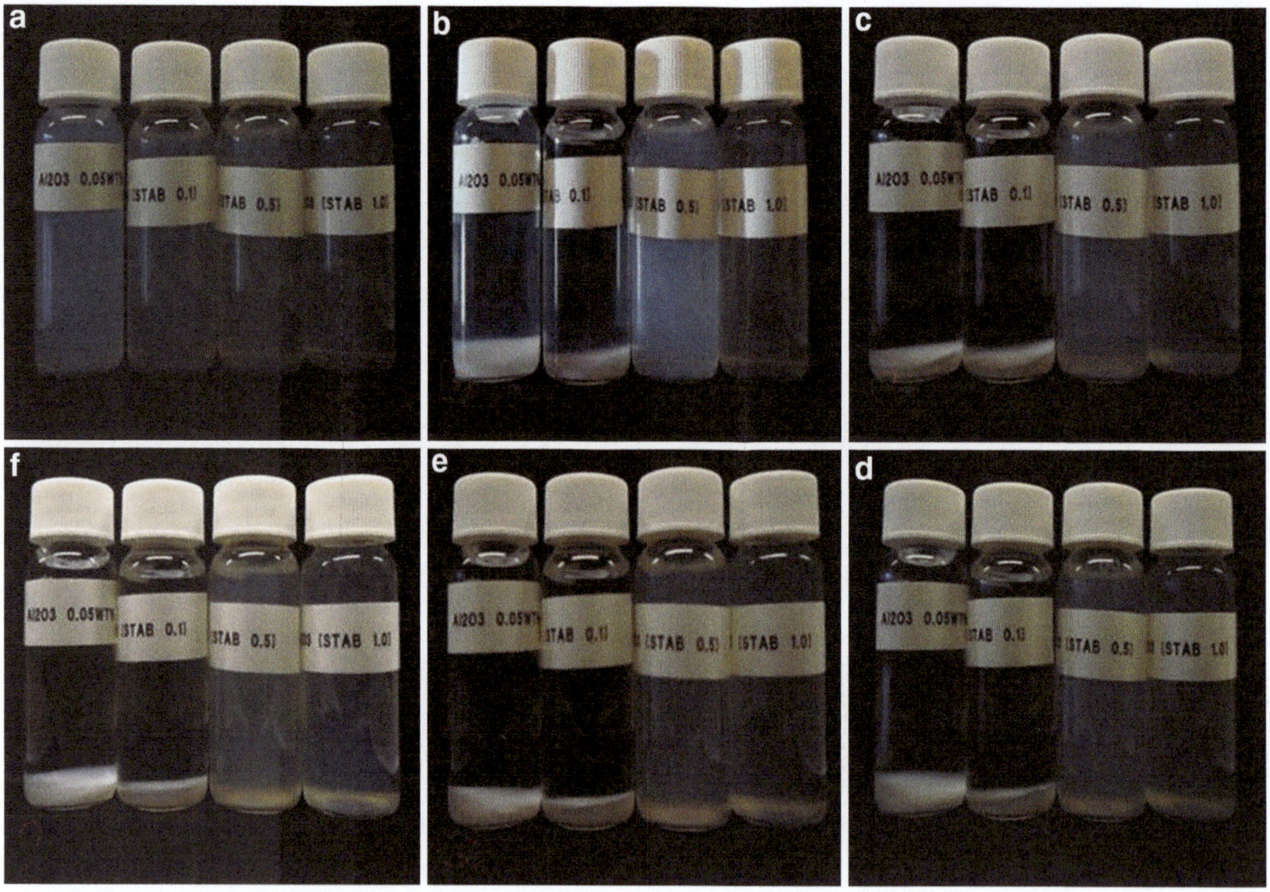

Fig. 14 PVP treated Al2O3 nanofluids stability observations at 50 °C at different concentrations: 0.0; 0.1; 0.5 and 1.0 wt %. **a** initial condition, **b** $t = 1$ h, **c** $t = 24$ h, **d** $t = 72$ h, **e** $t = 96$ h, **f** $t = 168$ h

The development of three coupled techniques was successful in identifying nanofluid stability.

Effect of adding a stabiliser

To improve metal oxide nanofluid stability, PVP (Povidone K30) is added at three different weight concentrations of 0.1, 0.5 and 1.0 wt. % but was not added to the silica-based nanofluid because this suspension was considered stable with a mean particle size distribution close to the average single particle size. Visual stability was observed at 50 °C during a 168-hour period, as shown in Figs. 14, 16. It was observed that PVP provided better stability at particular concentrations. At a low concentration of 0.1 wt. %, PVP does not provide any effect on the metal oxide nanofluid stability. Increasing the PVP concentration to 1.0 wt. % provides better stability, and at 1.0 wt. % of PVP, the nanofluid stability was investigated for more than 96 h for the Al_2O_3-based nanofluid but over shorter time period for the TiO_2-based nanofluid, approximately 48 h. However, both conditions showed good stability at higher temperatures, i.e. 50 °C.

Addition of PVP also altered such nanofluid properties as pH and surface conductivity. Figure 17 shows how PVP significantly changed these nanofluid properties, but density and viscosity were not significantly altered. Higher PVP concentrations caused lower pH values and higher surface conductivities on metal oxide nanofluids, higher than those of SiO_2 nanofluids. At higher PVP concentrations (0.5 and 1 wt. %), the surface conductivity of Al_2O_3 nanofluids exceeded that of TiO_2 nanofluids. The highest conductivity of Al_2O_3 is similar only to that of brine with PVP (1 wt. %). These results are consistent with the visual stability observation in which the Al_2O_3 nanofluid with PVP showed better stability than TiO_2 nanofluids at the given high PVP concentration. For nanofluid stability, it may be of interest to find a suitable dispersant for each metal oxide NP together with its optimum concentration. Fedele et al. (2011) observed that PEG 600 (Alfa Aesar) is a suitable dispersant for TiO_2-based nanofluids.

A dispersant (PVP) with 1 wt. % concentration in metal oxide NPs was selected for the particle size distribution study, as shown in Figs. 18 and 19. The presence of PVP successfully increased the intensity of the mean particle

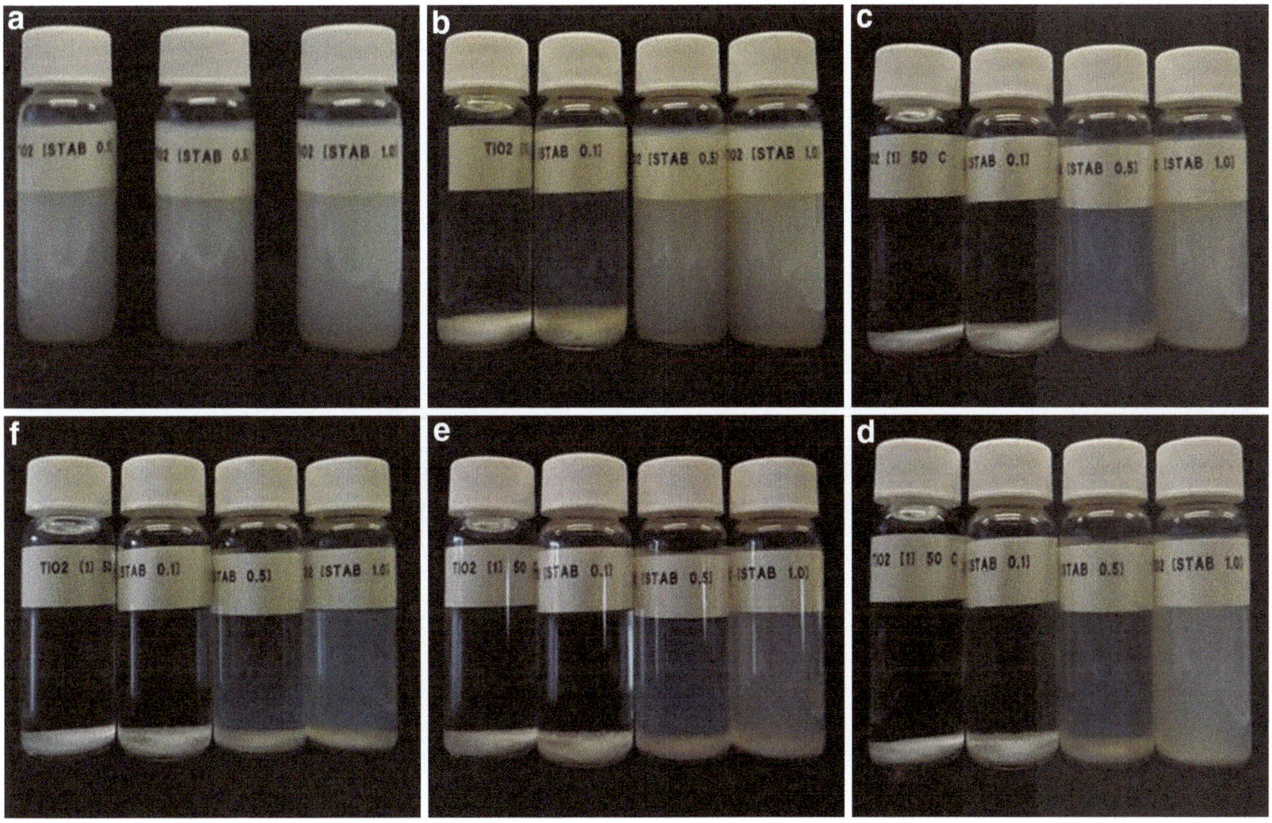

Fig. 16 PVP treated TiO2 nanofluids stability observations at 50 °C different concentrations: 0.0; 0.1; 0.5 and 1.0 wt %. **a** initial condition, **b** $t = 1$ h, **c** $t = 24$ h, **d** $t = 72$ h, **e** $t = 96$ h, **f** $t = 168$ h

Fig. 17 The pH and surface conductivity alteration of nanofluids at various PVP concentrations at room condition

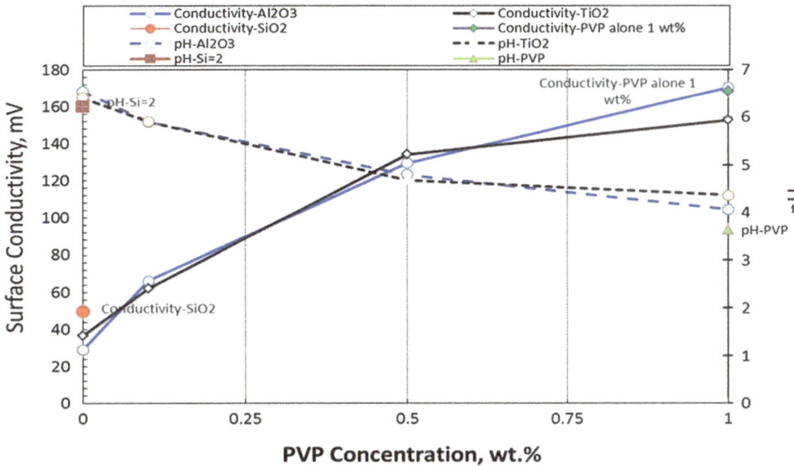

diameter, closer to the primary particle size of 26 nm for Al_2O_3 NPs and 29 nm for TiO_2 NPs. Stable nanofluids were identified by increasing the population of small-sized NPs dispersed in brine. The mean particle size grows larger with time (7 days) due to aggregation but their diameter is identified as equal to or a slightly greater than the size of the nanofluid without PVP in the range of 80–150 nm, which is much smaller than the size of the pore throat (typically in micron size). These observations demonstrate the potential for EOR processes in Berea sandstone, because the nanofluid may easily flow through a core pore throat of micron size.

Coreflood experiment

The coreflood experiment results from several Berea core plugs are summarised in Table 4. A dispersant (PVP) at 1 wt. % was added to the metal oxide-based nanofluid based

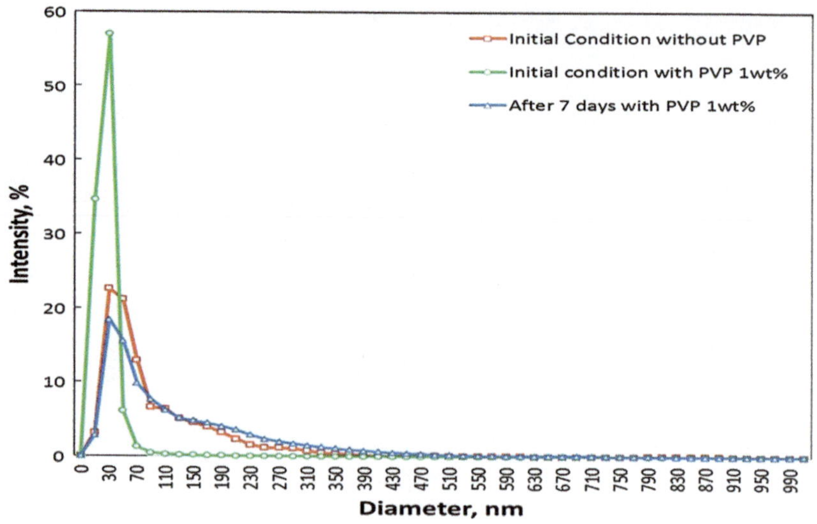

Fig. 18 NanoSight measurement of nanoparticles size distribution for brine (NaCl 3 wt %) containing 0.05 wt % Al_2O_3-based NPs

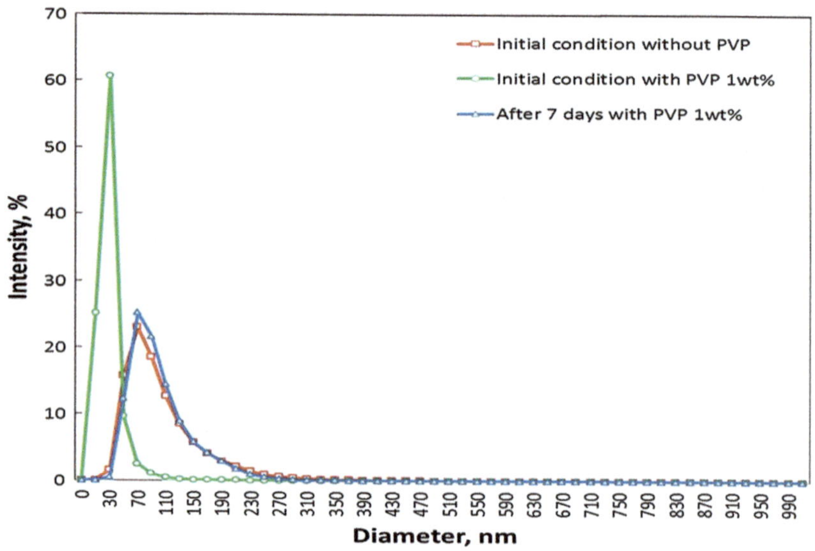

Fig. 19 NanoSight measurement of nanoparticles size distribution for brine (NaCl 3 wt %) containing 0.05 wt % TiO_2-based NPs

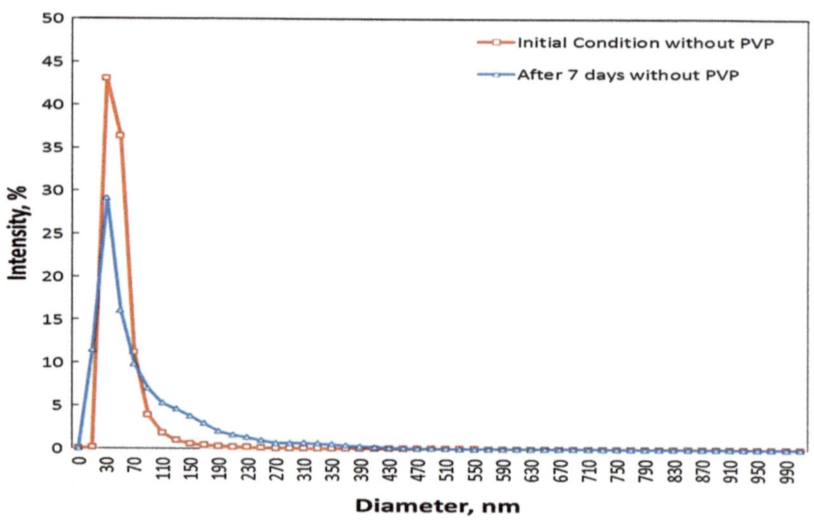

Fig. 15 NanoSight measurement of nanoparticles size distribution for brine (NaCl 3 wt %) containing 0.05 wt % SiO_2-based NPs

Table 4 Summary of coreflood experiments

Core #	Initial water saturation, % PV	Oil recovery, % OOIP			Residual oil saturation, % PV			Total recovery, % OOIP	Case NF
		WF1	NF	WF2	WF1	NF	WF2		
1-WW	19	52	8	6	32	26	23	66	Al_2O_3 with PVP
2-WW	20	57	11	9	29	22	16	77	TiO_2 with PVP
3-WW	32	53	2	19	35	33	19	74	SiO_2 with PVP
4-WW	16	55	1	2	30	29	28	58	PVP alone
5-IW	32	54	5	6	32	28	24	65	Al_2O_3 with PVP
6-IW	30	60	4	6	28	25	21	70	TiO_2 with PVP
7-IW	31	53	2	16	32	31	20	71	SiO_2 with PVP
8-OW	22	50.9	0.4	2.3	38.2	37.9	36.1	53.5	Al_2O_3 with PVP
9-OW	25	42	6	21	43	39	23	69	TiO_2 with PVP
10-OW	27	46	2	25	39	38	19	73	SiO_2 with PVP

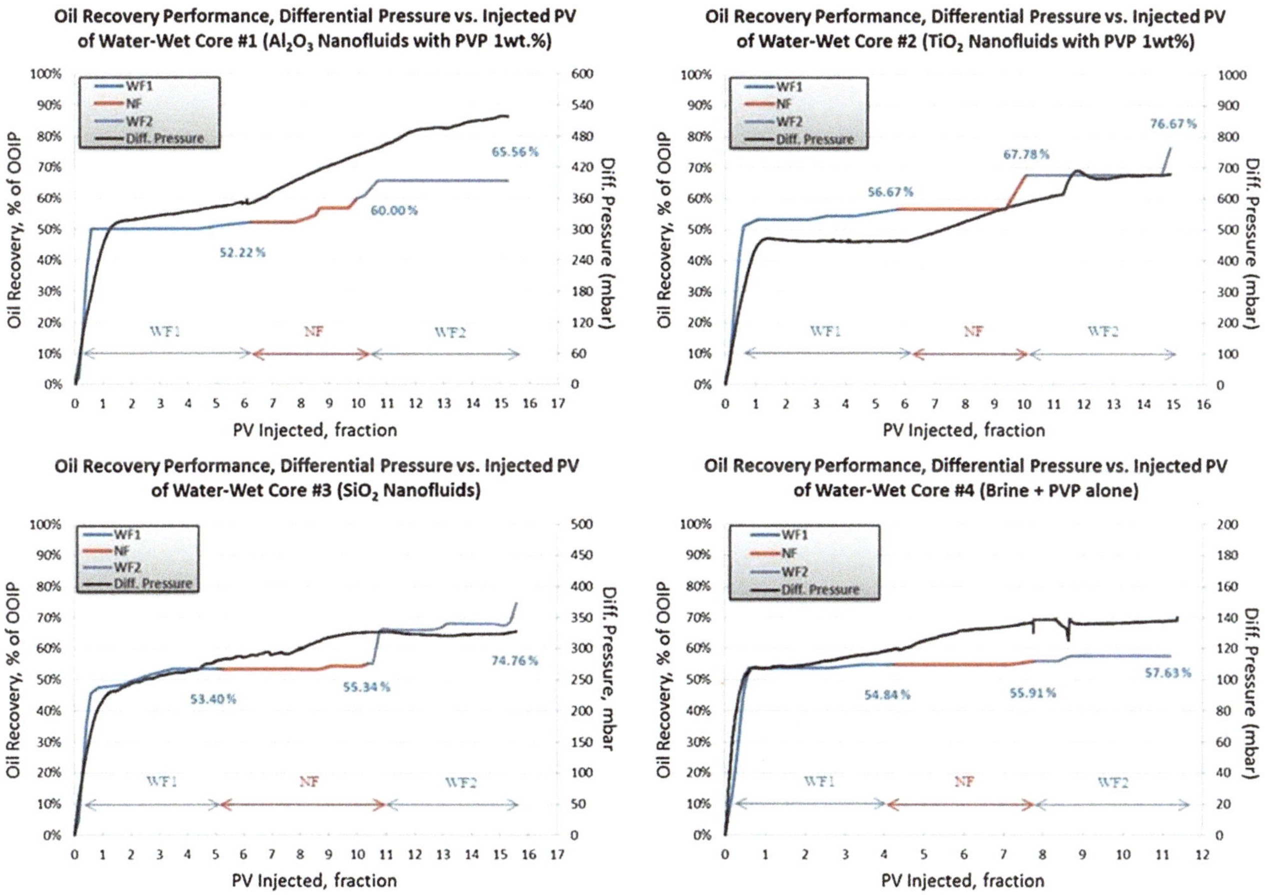

Fig. 20 Oil recovery profile, differential pressure vs. injected PV of water-wet cores

on previous stability analysis. The average oil recoveries after waterflooding with brine (WF1) are in the range of 52–57 % of original oil in place (OOIP) for a water-wet system, 53–60 % of OOIP for an intermediate-wet system and 42–51 % of OOIP for an oil-wet system. Morrow (1985) observed that oil recovery reached a maximum for cores in an intermediate-wet system, which is likely related to disconnection and trapping of the oil phase. After a water cut of 100 %, nanofluids were injected as a tertiary recovery process (nano-EOR, denoted as NF). The nanofluids successfully reduced residual oil saturation. However, as shown Figs. 20, 21, 22, the nanofluids

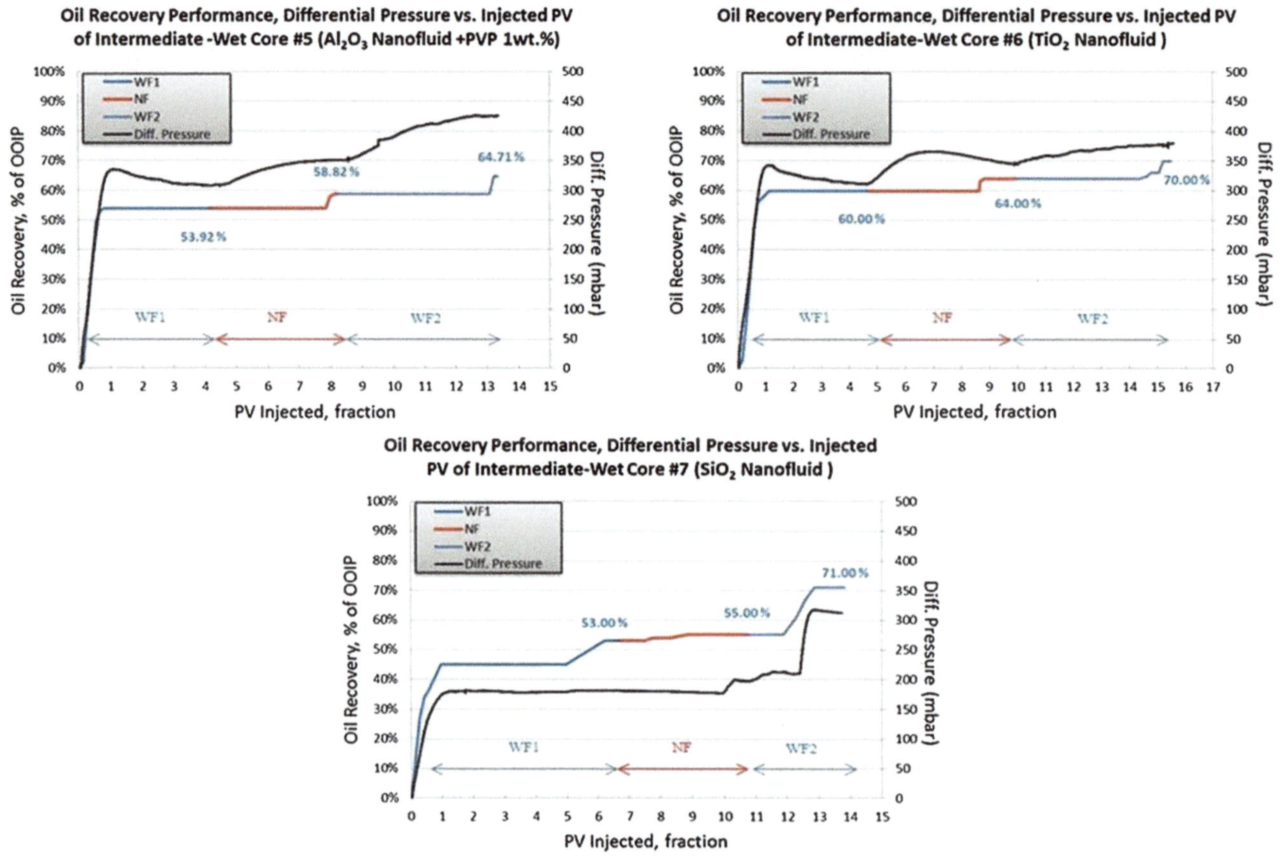

Fig. 21 Oil recovery profile, differential pressure vs. injected PV of intermediate-wet cores

required a certain amount of time to displace additional oil, at least 0.5 PV during NF, likely due to physicochemical interactions between the NPs and the grain surfaces. The additional oil recovery from the NF sequence of both metal oxide-based nanofluids is much greater than that of silica-based nanofluid, especially in a water-wet system. The additional oil recovery (NF) of metal oxide-based nanofluids decreased as the wettability changed from water- to intermediate- and oil-wet systems. However, the TiO_2-based nanofluid (with PVP) showed the best performance in this sequence (NF) in all wettability systems. The absence of metal oxide NPs in brine with dispersant solution had no significant impact on the additional oil recovery. Hence, the combination of metal oxides and dispersant shows great potential for EOR. In the last sequence (WF2), brine was re-injected to flush out any remaining particles and might have reduced residual oil in the core. The most interesting result came from the silica-based nanofluid. As shown, the residual oil saturation decreased dramatically from 11 to 19 % points of PV, whereas the oil recovery increased from 19 to 25 % points of OOIP in all wettability systems, and the trend appears identical in all silica-based nanofluid cases. Metal oxide NPS have a greater ability to adsorb into the pore throat than silica NPs, and pH plays significant role in the trapping mechanism (Ogolo et al. 2013). The effectiveness of particle trapping is greater at lower pH. Both metal oxide-based nanofluids (with PVP) provide much lower pH values than the silica-based nanofluid and tend to adsorb inside the pore throat rather than washing out during the NF and WF2 sequences. Silica oxide NPs have a low adsorption capacity and are easily washed-out (due to negative charges) and displace more oil. The NP adsorption during transport in a porous medium involves physicochemical interactions between NPs and grain surfaces (Zhang et al. 2013).

Figure 23 shows the filtration mechanism of NPs in Berea sandstone from observation of the inlet point of cores after the WF2 sequence. It is clear that there were no surface nano-cakes formed during NF flooding in all cases (metal oxides with PVP and silica-based nanofluids). Increasing the injection rate may affect the accumulation of NPs at the core inlet (Hendraningrat et al. 2013d). The particle velocity through the porous medium might be lower than the nanofluid injection rate and will create nano-caking (increasing pressure drop) at an injection rate of 0.8 cm^3/min, as shown in Fig. 23d. If the nanofluid remains stable at nanometer size, core plugging could be avoided and a physicochemical adsorption mechanism might occur between the particles and the porous medium.

Displacement mechanism

In this study, several possible measurements, i.e. contact angle, IFT and effluent analyses, were conducted to identify the oil displacement mechanism using metal oxide NPs. Karimi et al. (2012) observed that ZrO_2 plays an important role in the wettability alteration of a carbonate reservoir rock, whereas it could also change the wettability of the rock from a strongly oil-wet to strongly water-wet condition. Giraldo et al. (2013) observed that alumina-based nanofluids altered the wettability of cores from an oil-wet to a water-wet condition. Ehtesabi et al. (2014) reported a similar phenomenon in which the rock wettability changed from an oil- to water-wet condition after treatment with TiO_2 NPs. Therefore, metal oxide-based nanofluids may change the rock wettability and affect the oil displacement mechanism.

Figure 24 shows the contact angle measurements for different aqueous phases in a water-wet system (a) and an oil-wet system (b). Once the hydrophilic metal oxide (with PVP) is introduced into the brine, it reduces the contact angle from 54° to 21°, which means that metal oxide-based nanofluids have rendered the quartz plate more strongly water-wet. In contrast, PVP alone rendered the quartz weakly water-wet. The highest reduction in contact angle was achieved with the TiO_2-based nanofluid, consistent with the oil recovery results obtained from the TiO_2-based nanofluid applied as tertiary recovery (NF) in a water-wet system. Due to the poor clarity of the TiO_2-based nanofluid in the oil-wet system, the measurements could not be performed in addition to IFT measurements. Similar to the water-wet system, the presence of hydrophilic NPs in brine tends to change the quartz plate to a weakly oil-wet system. However, the silica-based nanofluid is more effective in an oil-wet system than an Al_2O_3-based nanofluid, a result consistent with the observation that additional oil recovery from SiO_2-based nanofluid is greater than with Al_2O_3-based nanofluid in an oil-wet system.

The IFT was measured at room temperature, as shown in Table 5, and a brine-crude oil system is used with the base value of 19.2 mN/m and a pH value of 6.76. The presence of NPs reduced the IFT between the oleic phase and the aqueous phase as well as the presence of PVP as dispersant. The lowest IFT was achieved with metal oxide NPs, although they do not have the lowest pH. As shown in Fig. 17, the lowest pH was obtained from brine with PVP solution (approx. pH 3.6), and metal oxide-based nanofluids have a pH in the range 4–5. Additional oil recovery due to a silica-based nanofluid is greater than that of a brine with PVP solution, even though silica-based nanofluids have a higher IFT than a brine with PVP solution. It can be concluded that the degree of IFT reduction is not a main parameter in oil displacement in nano-EOR.

In addition, the pH and surface conductivity of the effluent were also analysed, and the results are summarised in Table 6. Based on the measurements, the pH value and surface conductivity of injected brine were altered in all cores before and after injection of NF. The effluent became more acidic (lower pH) and the surface charge increased. This alteration may have occurred due to adsorption of NPs during NPs transport through the porous medium, resulting in wettability alteration. This particle retention during NP transport in porous media has been observed in a previous study (Hendraningrat et al. 2013b). The trend of the pH value of Al_2O_3-based nanofluids after the WF2 sequence is lower than in the NF sequence and has the highest capacity for adsorption into the pore throat amongst the other nanofluids due to the lowest pH. It is observed that pH plays a significant role in the trapping mechanism, and Al_2O_3 is identified as one of the most effective NPs for trapping and migrating in fine sand (Ogolo et al. 2013). However, the pH reduction is not the main cause of the displacement mechanism, as demonstrated in the case of brine with PVP alone. This solution has the lowest pH but does not have a significant effect on additional oil recovery for either the NF or WF2 sequences. Although pH induced an IFT reduction, the relationship is not proportional to the additional oil recovery. Hence, wettability alteration plays a dominant role in the oil displacement mechanism using nano-EOR.

Concluding remarks and future work

A systematic study of hydrophilic metal oxide nanoparticles (NPs) for EOR purposes was performed, and the stability of metal oxide-based nanofluids was successfully improved by adding PVP at 1 wt. %. The PVP altered the fluid properties, i.e. pH and surface conductivity, and provided better nanofluid stability. The combinations of metal oxide nanofluids and PVP improved oil recovery to a greater extent than silica-based nanofluid and dispersant alone, and the TiO_2-based nanofluid was especially effective for all wettability systems. The additional oil recovery (NF) of metal oxide-based nanofluids decreased as the wettability changed from water- to intermediate- and oil-wet systems. The contact angle, interfacial tension (IFT) and effluent were measured and analysed in the displacement mechanism observations. It was observed that all nanofluids rendered the quartz plate more water-wet, and the TiO_2-based nanofluid was the most effective wettability-altering fluid. This result is consistent with and proportional to the additional oil recovery from the coreflood

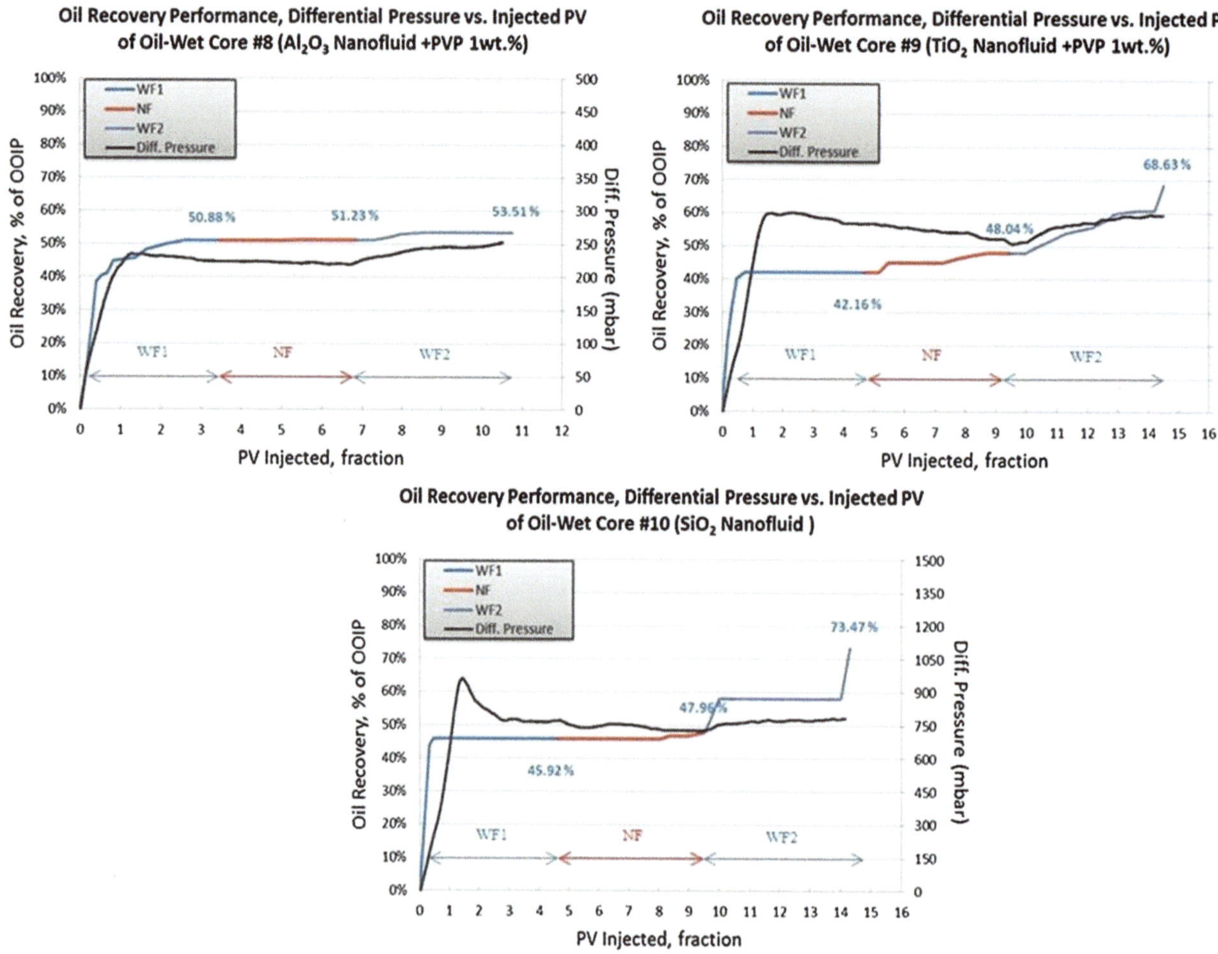

Fig. 22 Oil recovery profile, differential pressure vs. injected PV of oil-wet cores

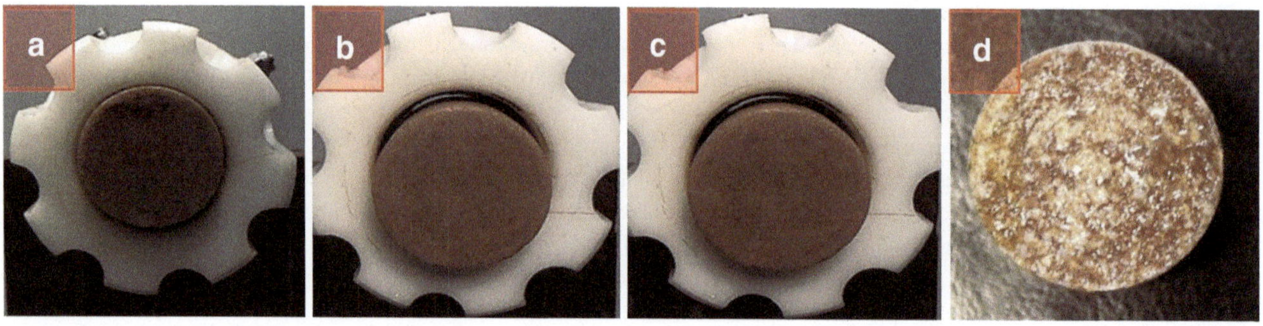

Fig. 23 Core inlet point observation after nanofluid flooding: **a** Al_2O_3-based, **b** TiO_2-based **c** SiO_2-based, **d** SiO_2-based at higher injection rate ($0.8\ cm^3/min$)

experiment. The particle adsorption during the transport process was identified from the effluent analysis. The presence of NPs reduced the IFT between the oleic phase and aqueous phase as did the presence of PVP as dispersant, but the degree of IFT reduction is not proportional to the additional oil recovery. The lowest IFT was achieved by metal oxide NPs even though this material does not have the lowest pH. The additional oil recovery due to silica-based nanofluid is greater than that of brine with PVP solution, even though the silica-based nanofluid has a higher IFT than the brine with PVP solution. Therefore, wettability alteration plays a dominant role in the oil

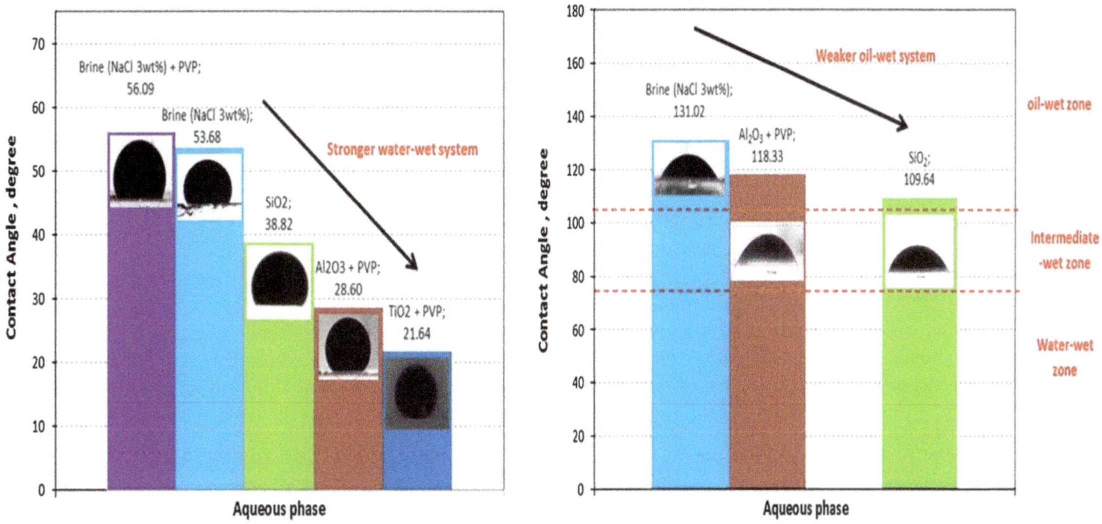

Fig. 24 Contact angle measurement of aqueous phase on quartz plate at room temperature: **a** water-wet and **b** oil-wet system

Table 5 Interfacial tension measurement for aqueous phase against crude oil (at room temperature)

Fluid	(mN/m)
Brine (NaCl 3 wt. %)	19.2
Al_2O_3-based (0.05 % wt.) + PVP 1 wt. %	12.8
TiO_2-based (0.05 % wt.) + PVP 1 wt. %	n.a
Silica-based (0.05 % wt.)	17.5
Brine (NaCl 3 wt. %) + PVP 1 wt. %	15.7

Table 6 Effluent analysis

Core #	pH			Surface conductivity (mV)		
	WF1	NF	WF2	WF1	NF	WF2
1	6.245	5.846	5.656	46.5	68.8	79.5
2	6.097	5.049	5.876	47.2	109.8	62.7
3	6.525	6.144	6.161	27.8	49.1	48.2
4	5.839	5.271	5.330	51.3	81.4	78.2
5	5.560	5.434	5.291	80.1	87.2	95.2
6	5.690	4.987	5.552	72.6	112.2	82.2
7	6.241	6.068	6.129	46.7	52.1	47.1
8	5.541	4.616	n.a	63.6	112.6	n.a
9	4.748	4.52	4.53	108	119.9	119.5
10	4.871	4.723	4.869	96.1	101.7	96.3

displacement mechanism using nano-EOR. Finding a suitable stabiliser for each nanofluid and its optimum concentration might be important in future work as well as repeating the experiment with representative formation water at reservoir conditions.

SI metric conversion factors

$$\text{mD} \times 9.869233 \quad \text{E}-04 = \mu\text{m}^2$$
$$\text{cP} \times 1.000000 \quad \text{E}+00 = \text{mPa.s}$$
$$\text{dyn/cm} \times 1.000000 \quad \text{E}+00 = \text{mN/m}$$
$$\text{bar} \times 1.000000 \quad \text{E}-02 = \text{kPa}$$

Acknowledgments The authors gratefully acknowledge Evonik Industries AG and Elkem Silicon Materials for providing nanoparticles and the Department of Petroleum Engineering and Applied Geophysics, Norwegian University of Science and Technology (NTNU) for financial support necessary to complete this study and permission to publish this article.

References

Ahmadi M, Habibi A, Pourafshary P (2013) Zeta-potential investigation and experimental study of nanoparticles deposited on rock surface to reduce fines migrations. SPE J 18(3):534–544 (SPE-142633-PA)

Anderson WG (1987) Wettability literature survey-part 5: the effects of wettability on relative permeability. J Pet Tech 39(11):1453–1468 (SPE-16323-PA)

Chengara A, Nikolov A, Wasan DT, Trokhymchuck A, Henderson D (2004) Spreading of nanofluids driven by the structural disjoining pressure gradient. J Colloid Interface Sci 280:192–201

Ehtesabi H, Ahadian MM, Taghikhani V, Ghazanfari MH (2014) Enhanced heavy oil recovery in sandstone cores using TiO nanofluids. Energy Fuels 28(1):423–430

Fedele L, Colla L, Bobbo S, Barison S, Agresti F (2011) Experimental stability analysis of different water-based nanofluids. Nanoscale Res Lett 6:300

Fernandez-Garcia M, Rodriguez JA (2007) Metal oxide nanoparticles. Brookhaven National Laboratory, New York

Folttmann H, Quadir A (2008) Polyvinylpyrrolidone (PVP)–One of the most widely used excipients in pharmaceuticals: an overview. Drug Deliv Technol 8(6):22–27

Giraldo J, Benjumea P, Lopera S, Cortés FB, Ruiz MA (2013) Wettability alteration of sandstone cores by alumina-based nanofluids. Energy fuel 27:3659–3665

Hascakir B, Babadagli T, Akin S (2008) Experimental and numerical modeling of heavy-oil recovery by electrical heating. Paper SPE/PS/CHOA 117669-MS presented at the international thermal operations and heavy oil symposium, Calgary, Canada, 20–23 October 2008

Hendraningrat L, Shidong L, Torsæter O (2012) A glass micromodel experimental study of hydrophilic nanoparticles retention for EOR project. Paper SPE 159161-MS presented at the SPE Russian oil & gas exploration & production technical conference and exhibition, Moscow, Russia, 16–18 October 2012

Hendraningrat L, Li S, Torsæter O (2013a) A coreflood investigation of nanofluid enhanced oil recovery. J Petrol Sci Eng 111:128–138

Hendraningrat L, Engeset B, Suwarno S, Li S, Torsæter O (2013b) Laboratory investigation of porosity and permeability impairment in Berea sandstones due to hydrophilic nanoparticle retention. Paper SCA2013-062 presented at the international symposium of the society of core analysts held in Napa Valley, California, USA, 16–19 September 2013

Hendraningrat L, Li S, Torsæter O (2013c) A coreflood investigation of nanofluid enhanced oil recovery in low-medium permeability Berea sandstone. Paper SPE 164106-MS presented at SPE international symposium on oilfield chemistry, The Woodlands, Texas, 8–10 April 2013

Hendraningrat L, Li S, Torsæter O (2013d) Effect of some parameters influencing enhanced oil recovery process using silica nanoparticles: an experimental investigation. Paper SPE-165955 presented at SPE Reservoir Characterisation and Simulation Conference and Exhibition, Society of Petroleum Engineers, Abu Dhabi, UAE, 16–18 September 2013

Ju B, Tailiang F, Mingxue M (2006) Enhanced oil recovery by flooding with hydrophilic nanoparticles. China Particuology 4:41–46

Karimi A, Fakhroueian Z, Bahramian A, Khiabani NP, Darabad JB, Azin R, Arya S (2012) Wettability alteration in carbonates using zirconium oxide nanofluids: EOR Implications. Energy Fuels 26(2):1028–1036

Keblinski P, Prasher R, Eapen J (2008) Thermal conductance of nanofluids: is the controversy over? Nanopart Res 10:1089

Keller AA, Wang H, Zhou D, Lenihan HS, Cherr G, Cardinale BJ, Miller R, Ji Z (2010) Stability and aggregation of metal oxide nanoparticles in natural aqueous matrices. Environ Sci Technol 44(6):1962–1967

Li W, Zhu J, Qi J (2007) Application of nano-nickel catalyst in the viscosity reduction of Liaohe extra-heavy oil by aqua-thermolysis. J Fuel Chem 35(2):176–180

Li S, Kaasa AT, Hendraningrat L, Torsæter O (2013). Effect of silica nanoparticles adsorption on the wettability index of Berea sandstone. Paper SCA2013-059 presented at the international symposium of the society of core analysts held in Napa Valley, California, USA, 16–19 September 2013

Miranda CR, De Lara LS, Tonetto BX (2012) Stability and mobility of functionalized silica nanoparticles for enhanced oil recovery application. Paper SPE 157033-MS presented at SPE international oilfield technology conference, Noordwijk, The Netherlands, 12–14 June 2012

Morrow NR (1985) A review of the effects of initial saturation, pore structure, and wettability on oil recovery by waterflooding. In: Proceeding of the north sea oil and gas reservoirs seminar. Graham and Trotman, Ltd: 179-191, Trondheim, Norway, 2–4 December 1985

Morrow NR (1990) Wettability and its effect on oil recovery. J Pet Tech 42(12):1476–1484 (SPE-21621-PA)

Nanosight (2009) Operating manual. NanoSight Ltd., UK

Ogolo NA, Onyekonwu MO Akaranta O (2013) Trapping mechanism of nanofluids on migrating fines in sand. Paper SPE 167502-MS presented at the Nigeria annual international conference and exhibition, Lagos, Nigeria, 30th July–1st August 2013

Ogolo NA, Olafuyi OA, Onyekonwu MO (2012) Enhanced oil recovery using nanoparticles. Paper SPE 160847-MS presented at the SPE Saudi Arabia section technical symposium and exhibition, Al-Khobar, Saudi Arabia, 8–11 April 2012

Rao Y (2010) Nanofluids: stability, phase diagram, rheology and applications. Particuology 8:549–555

Roustaei A, Saffarzadeh S, Mohammadi M (2012) An evaluation of modified silica nanoparticles' efficiency in enhancing oil recovery of light and intermediate oil reservoirs. Egypt J Petrol 22(3):427–433

Shokrlu YH, Babadagli T (2010) Effects of nano-sized metals on viscosity reduction of heavy oil/bitumen during thermal applications. Paper CSUG/SPE 137540-MS presented at the Canadian unconventional resources and international petroleum conference, Calgary, Canada, 19–21 October 2010

Than P, Preziosi D, Joseph D, Arney M (1988) Measurement of interfacial tension between immiscible liquids with the spinning rod tensiometer. J Colloid Interface Sci 124(2):552–559

Torsæter O, Abtahi M (2003) Experimental reservoir engineering laboratory workbook. Department of Petroleum Engineering and Applied Geophysics, Norwegian University of Science and Technology (NTNU), Trondheim

Treiber LE, Duane LE, Archer L, Owens WW (1971) A laboratory evaluation of the wettability of fifty oil-producing reservoirs. SPE J 12:531–540

Wasan DT, Nikolov A (2003) Spreading of nanofluids on solids. J Nat 423:156–159

Wasan DT, Nikolov A, Kondiparty K (2011) The wetting and spreading of nanofluids on solids: role of the structural disjoining pressure. Curr Opin Colloid Interface Sci 16:344–349

Wu W, He Q, Jiang C et al (2008) Magnetic iron oxide nanoparticles: synthesis and surface functionalization strategies. Nanoscale Res Lett 3:397–415

Yu W, France DM, Routbort JL, Choi SUS (2008) Review and comparison of nanofluid thermal conductivity and heat transfer enhancements. Heat Transfer Eng 29:432

Zhang T, Murphy M, Yu H, Bagaria H, Nielson B, Bielawski C, Johnston K, Huh C, Bryant S (2013) Investigation of nanoparticle adsorption during transport in porous media. Paper SPE 166346-MS presented at the SPE annual technical conference and exhibition, New Orleans, Lousiana, 30th September–2nd October 2013

19

Hydraulic classifier system for fractionation of nano $CaCO_3$ particles

H. F. Aly · M. A. Akl · Hesham M. A. Soliman · Aref M. E. AbdEl-Rahman · A. I. Abd-Elhamid

Abstract A laboratory scale hydraulic classifier system was developed for calcium carbonate nanoparticles fractionation. The system is based on the differences in the settling velocity of particles in aqueous fluid at different dynamic viscosities along different settling stages. Different factors affecting the fractionation process were studied, such as the effect of water volume, L, terminal (settling) velocity in different stages, v_s, $CaCO_3$ feed concentration, g/L and flow rate (L/h) of the dispersed fluid solution. The particles obtained were characterized using SEM and showed that the developed system can fractionate particles within the size range 25–33 nm. A simple model for the results obtained is developed and discussed in terms of the different parameters affecting particles size is given. Further, the calcium carbonate used was characterized before and after fractionations using Vibratory sieve shaker, SEM, EDS, XRD and FTIR.

Keywords Calcium carbonate · Nanoparticles · Fractionation · Hydraulic classifier

H. F. Aly
Hot Laboratories Center, Atomic Energy Authority, Nasr 13759, Egypt

M. A. Akl
Faculty of Science, Mansoura University, Mansoura, Egypt

H. M. A. Soliman · A. M. E. AbdEl-Rahman ·
A. I. Abd-Elhamid (✉)
Advanced Technology and New Materials Research Institute, City for Scientific Research and Technology Applications, P. O. Box 21934, SRTA, Egypt
e-mail: ahm_ch_ibr@yahoo.com

Introduction

Nanoparticles are now an important research area because they are widely used in numerous technological and medical applications (Guo et al. 2007). Owing to their very small size and large surface area-to volume ratio, fractionation of these powders is very attractive. Ultrafine powders can be obtained by mechanical attrition or by self-assembly using chemical reactions. In most cases, fractionation and classification of particles obtained are required.

Fractionation is one of the major techniques used for separation of nanoparticles. Many types of powder classifiers are available. These exemplified by cyclone type separators (Bryczkowski and Chmielniak 2001), cross flow air type classifiers (Wang et al. 2001), Rotating vibrating conical disk separator (Yamamoto et al. 1998) and impeller wheel type classifier (Galk et al. 1999). These classifiers address fractionation of powder particles within the micro range.

When concerning nanoparticles, a review on strategies for size and/or shape selective purification of nanoparticles was recently published by Kowalczyk et al. (2011). This review present and discuss different methods for separation of nanoparticles. Among these, methods based on the density gradient centrifugation are reported and discussed. In this system, particles are accelerated in a centrifugal field and particles of different sizes and/or shapes move with different velocities in the medium provided which is the basis for particles separation into distinct bands. To obtain selective separation between different particles sizes, the technique is modified by density gradient centrifugation in which a liquid column providing the density gradient required. This technique was successfully applied by Sun et al. (2009), to fractionate gold nanoparticles. The main advantages of these techniques are the speed and

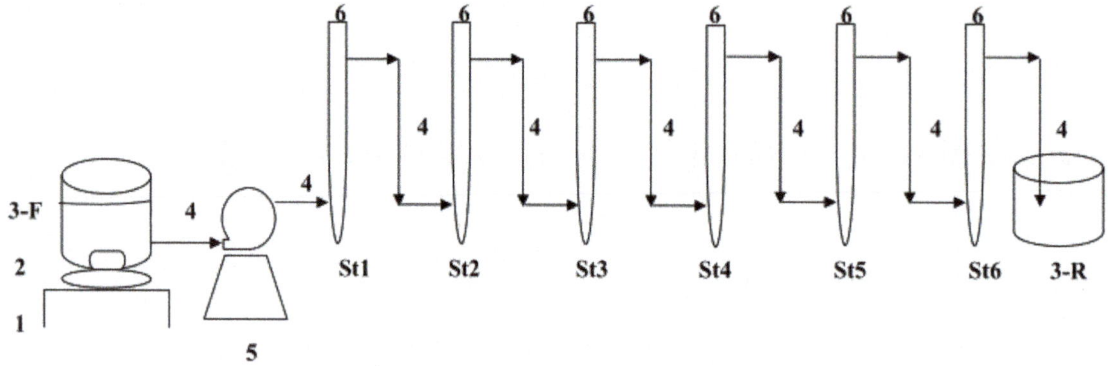

Fig. 1 Schematic diagram of hydraulic classifier system. 1) *magnetic stirrer*, 2) *Magnet*, 3-F & 3-R) *Glass beaker (2000 ml)*, 4) *plastic pipe*, 5) *Pump*, 6) *50 ml Plastic tubes*

selectivity of separation, provided that a suitable liquid is used to give the density gradient medium required without disturbing the nanoparticles to be fractionated.

In this contribution, a simple laboratory scale hydraulic classifier system was developed based on the kinetic gravity separation and settling velocity. The system is simple, economic (where we use one pump and water as fractionation medium), continuous and the particles obtained are homogenous and in the nanoscale. In this kind of separation, the particles are separated on the basis of differences in their terminal velocity (Van Kooy et al. 2004). The developed system was used for fractionation of calcium carbonate particles (commercial and artificial samples were used). However, the sample can be fractionated into different sizes at different settling stages. Further, a simple model for the results obtained is given.

Developed particles classifier system

Mechanical attrition is top-down approach for production of nanoparticles, in which milling was used, followed by particles fractionation and classification. In this concern, for particles classification to nanoparticles, a fractionation system was developed. This system is based on sorting particles by specific gravity in stream of hydraulic water that rises at a controlled rate. In this environment, heavier particles will settle down and discriminated while lighter particles will be carried out. This was followed for different stages to fractionate the different particles.

A schematic diagram for the system is given in Fig. 1. The system consists of 2L feed beaker, 3-F, contains the particles to be classified in certain volume of water. These fed particles are dispersed by a magnetic stirrer, 1, with a magnet, 2. After complete dispersion of the particles, the dispersed particles are feeding by a pump, 5, at constant flow rate, through a six stages test tubes (travelling zone),

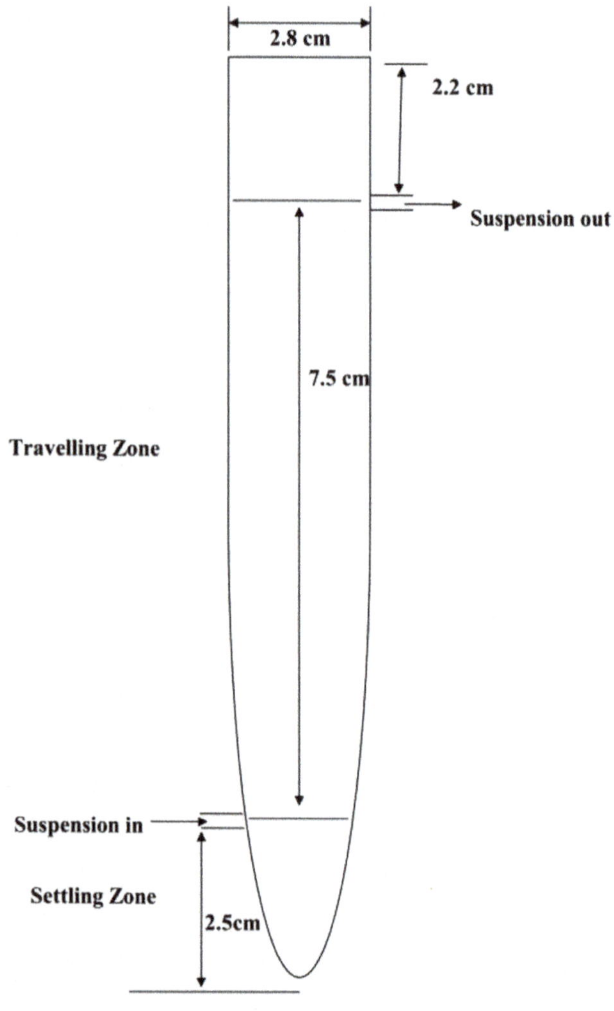

Fig. 2 Schematic diagram of the stage

connected in a series, from the top to the bottom of each tube. Based on the different factors affecting the flow rate of the suspended particles, which will be studied, the particles with different sizes will settle down in the cone part (settling zone) of the test tube of each stage. At the end of

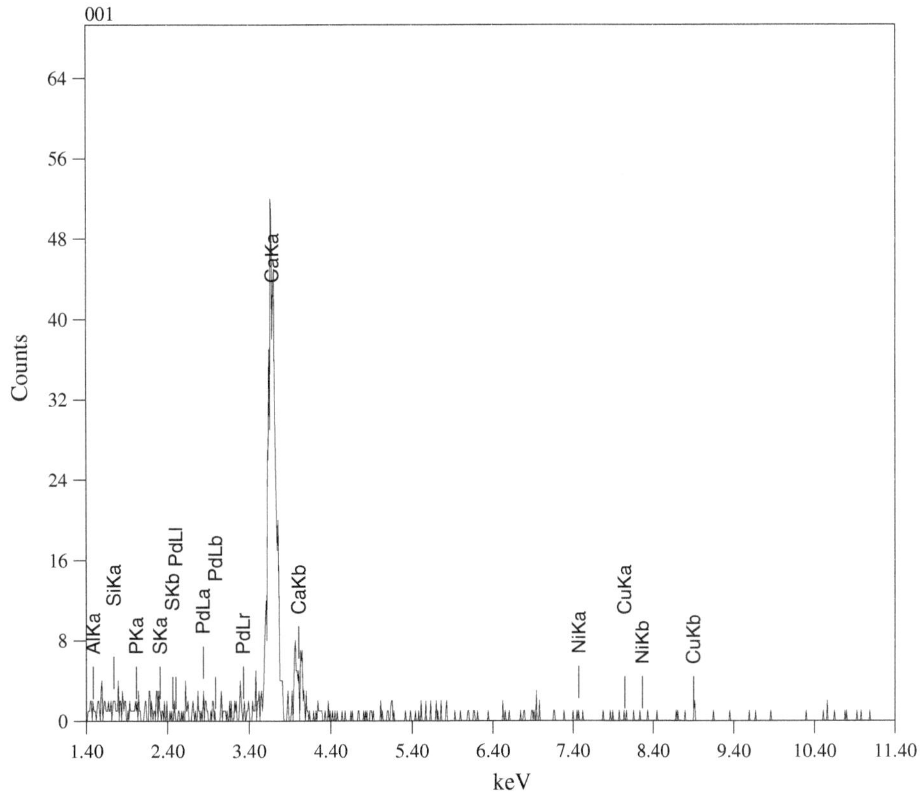

Fig. 3 The EDS analysis for calcium carbonate obtained from target for chemical industry company (S1)

Element	(keV)	mass%	Error%	At%	Compound	mass%	Cation K
C K	0.277	12.01	0.16	18.97			7.9586
O K	0.525	55.83	0.71	66.20			35.3681
Al K	1.486	0.01	0.24	0.01			0.0137
Si K	1.739	0.02	0.22	0.02			0.0286
P K	2.013	0.50	0.20	0.30			0.8193
S K	2.307	0.11	0.16	0.07			0.1956
Ca K	3.690	29.12	0.28	13.78			52.2119
Ni K	7.471	1.09	1.03	0.35			1.5658
Cu K	8.040	0.54	1.37	0.16			0.7427
Pd L	2.838	0.76	0.51	0.14			1.0958
Total		100.00		100.00			

the 6th test tube (St. 6), a receiving beaker, 3-R, is present to collect the end fraction of these particles. A detailed dimension of the tubes used at the different stages is given in Fig. 2. This system was used to fractionate two samples of commercial $CaCO_3$.

Experimental

Chemicals

Two samples of calcium carbonate were subjected to fractionation experiments. The first sample (S1) was obtained from Target Company for Chemical Industry, Alexandria, Egypt, its purity was determined by EDS analysis as seen in Fig. 3. In addition, the second one (S2) was obtained from Riedel–DeHaen Company, Germany (98.5 %).

Instruments

GP 3202 analytical balance (Sartorius, USA), SB 162 hot plate stirrer (Stuart, UK.) and Pump drive S101 (Heidolph, Germany).

AS200 Basic Vibratory Sieve Shaker (Retsch, Germany), JSM-636 OLA Scanning Electron Microscopy (SEM) (Jeol, Japan) with EDS analysis was used to measure the elemental analysis, XRD- 7000 X-Ray Diffraction (Shimadzu, USA), and 8,400 s Fourier transmission infra red spectroscopy (FTIR) Shimadzu, Japan.

Procedures

Sieving of $CaCO_3$

Two procedures were used in case of $CaCO_3$; dry sieving and wet sieving. In case of dry sieving, 100 g obtained from Target for chemical industry company, Alexandria, Egypt, or 5 g of $CaCO_3$ from Riedel-deHaën, Germany, was placed on vibratory sieve shaker for 10 min at amplitude of 100 mm. While in case of wet sieving, 5 g of $CaCO_3$ from (Target for chemical industry company, Egypt or Riedel-deHaën, Germany) was dispersed in 800 ml distill water for 2 h then poured on vibratory sieve shaker for 10 min at amplitude of 100 mm. The sieving for each size fraction was dried at 105 °C for 12 h.

Fractionation procedure

The fractionation procedure was carried out by taken a certain weight of $CaCO_3$ and well mixed in 800 ml of double distilled water using magnetic stirrer for a predetermined period. After stirring, the mixture was left for 10 min for settling of large particle size (< 20 μm). These particles were then separated from suspended particles by decantation and discarded. The separated dispersed particles were subjected to stirring and particle size fractionation by feeding at a predetermined flow rate into the hydraulic classifier apparatus developed in this work (Fig. 1). After a certain period enough to fractionate the particles in different stages, a drop of the settled particles separated in the bottom of the different fractionating stages was placed on a glass slide and subjected to air drying for SEM characterization. The different conditions affecting particle separation was applied on the sample S1 to obtain the optimum condition for particle classifications. The optimum condition for S1 sample was then applied to S2.

Particle size determination

To study the morphology of the particles present in samples, it was first coated with gold using Carbon Sputter Coater SPI, Module control, SPI supplies USA. When this surface is subjected to the electrons beam, the reflected beam of electrons carries a bright and clear image of it. After coating, the sample was placed in the cavity of JEOL JSM-6390 LA, Scanning Electron Microscope (SEM) provided with Smile View software for particle size measurements.

To obtain good statistical measurements for particle size, the work of Vigneau et al. (2000), was consulted. Accordingly, three drops from 6th stage sample was taken on three different slides. Each slide was scanned for at least 20 particles at different eight areas of the slide to give a total of 500 particles for the three slides (Fig. 4d). The particle size distribution for the three slides is shown in Fig. 3d. The mean particle size of the sample was found to equal 31 nm with standard deviation ±2 using the software Smile View.

Results and discussion

Sieving of $CaCO_3$

Two samples of different suppliers of calcium carbonate were studied as given in the experimental. A known weight of each sample was subjected to dry and wet sieving and the weight of each sieved fraction of certain particle size was determined. Tables 1 and 2 give the results of dry sieving, and Tables 3 and 4 are for wet sieving. These results indicated that wet sieving brought finer particles as well as better classifications. This is exemplified for both samples S1 and S2. In the case of sample S1, the largest particle size is in the range 250–125 μm for dry sieving whereby for the same sample, the largest particle size is 125–63 μm for wet sieving. Similar results were obtained for sample S2. For both samples, the smallest particle size was less than 45 μm. This fraction size was taken as feed source for further fractionation using the hydraulic system developed and given in this work.

Basic concept of the particles classifier

The main concept of this particle separator is based on the differences in the settling velocity of different particles in a fluid of different dynamic viscosities along the different stages. In this concern, the particles are transported from one stage to another leaving particle of terminal velocity at each stage.

The terminal velocity of sediment particles (negative velocity) is a defined as the rate at which the sediment settles in still fluid. This velocity is diagnostic of particle size and is also sensitive to the shape and density of particles as well as the viscosity and density of the fluid. All of these are integrated in the transport parameter from stage to another. To predict the particle settling velocities, the main forces acting on a particle in fluid are the particle weight, the particle buoyancy and the drag force of the fluid. The difference between the weight force, F_w, and the buoyancy force, F_b, is the net gravitational force, F_g, given by:

$$F_g = F_w - F_b$$

Since, $F_w = \rho_s.V\, g$
and $F_b = \rho_f.\ V\, g$
Then, $F_g = (\rho_s - \rho_f)\ V\, g$ where ρ_s and ρ_f are the particle and fluid densities, and V is the volume of sediment particles and g is the acceleration of gravity.

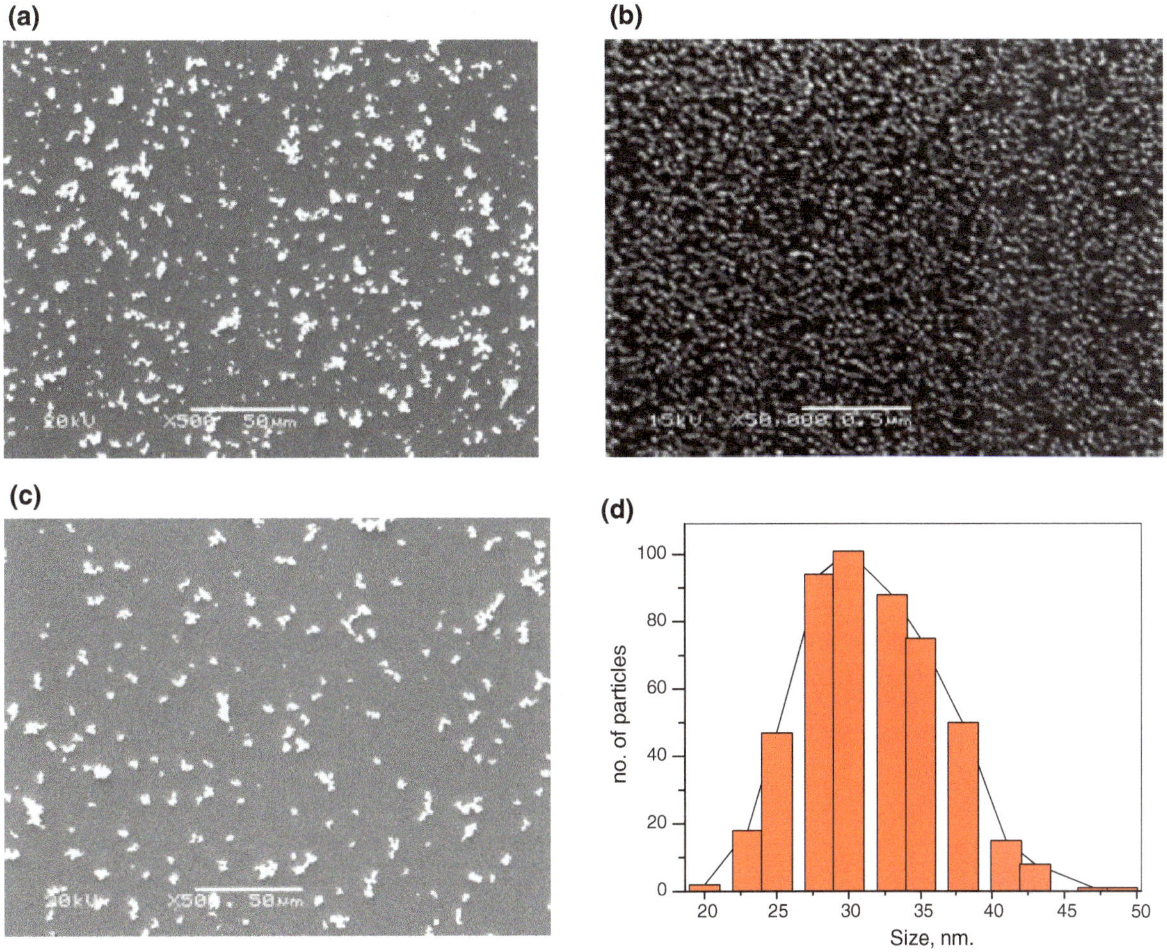

Fig. 4 Effect of water volume (L). 6th Stage, Stirring time: 2 h, Stirrer speed: 650 rpm, [$CaCO_3$]: 5g/800L, V_{water}. **a** 0.5 L, **b** 0.8 L and **c** 1.5 L

Table 1 Weight of different sizes of $CaCO_3$ (S1) after dry sieving

Size (μm)	Weight (g)
250–125	81.30
125–63	18.52
63–45	0.10
<45	0.05

$CaCO_3$ weight = 100 g, amplitude = 100 mm, time = 10 min

Table 2 Weight of different sizes of $CaCO_3$ (S2) after dry sieving

Size (μm)	Weight (g)
500–250	0.14
250–125	3.84
125–63	0.97
63–45	0.08
<45	0

$CaCO_3$ weight = 5 g, amplitude = 100 mm, time = 10 min

Table 3 Weight of different sizes of $CaCO_3$ (S1) after wet sieving

Size (μm)	Weight (g)
125–63	0.03
63–45	0.02
<45	4.2

$CaCO_3$ weight = 5 g, water volume = 800 ml, dispersing time = 2 h, sieving amplitude = 100 mm, sieving time = 10 min

Table 4 Weight of different sizes of $CaCO_3$ (S2) after wet sieving

Size (μm)	Weight (g)
250–125	0.06
125–63	0.05
63–45	0.03
<45	4.7

$CaCO_3$ weight = 5 g, water volume = 800 ml, dispersing time = 2 h, sieving amplitude = 100 mm, sieving time = 10 min

Assuming that the particles are spherical, then;

$$F_g = (\pi/6)\, d^3(\rho_s - \rho_f)\, g \quad (1)$$

where d is the diameter of the particles.

Equation 1 represents the net gravitational force on spherical particle. This force is downward (positive) if the sediment particles are denser than the fluid and upward (less than zero) if the sediment particles are less dense than the fluid.

Imagine a sediment spherical particle initially at rest, once the fluid transports the particles, velocities increase and the particles now moving through the fluid and get a fractional drag force, F_D. This drag force is dependent on the particles size and relative velocity, and the fluid viscosity (Stokes 1891).

The net force of the particle will be the differences between the net gravitational force and the drag force, and the particle will settle down once its velocity equals the settling velocity. This was given by the stocks law, which expresses the drag force in terms of the dynamic viscosity, μ, and the diameter of the particle, d, as well as the settling velocity, v_s, for small particles moving with laminar flow. This law is given by the following relation (Barry and Napier-Munn 2006; Rhods 2008);

$$F_D = 3\pi\, \mu\, v_s d \quad (2)$$

Under these conditions, $F_D = F_g$

Therefore, from Eqs. 1 and 2,

$$(\pi/6)\, d^3(\rho_s - \rho_f)g = 3\pi\, \mu\, v_s d$$

or,

$$v_s = \left(gd^2/18\mu\right)(\rho_s - \rho_f) \quad (3)$$

Equation 3, indicates that the settling velocity is indirectly proportional to the dynamic viscosity μ, as long as $(\rho_s - \rho_f)$ is constant, (http://www.gordonengland.co.uk/conversion/dynamic_viscosity.htm).

Experimentally, μ were calculated from flow rate/vertical travelled distance Kg s^{-1} $m^{-1}$and, at constant flow rate and feed concentration, v_s is directly proportional to $(\rho_s - \rho_f)$, and ρ_s was calculated from feed concentration in each stage/volume, kg m^{-3}, and ρ_f were calculated from feed concentration in each stage/volume, kg m^{-3}.

Further, the settling velocity is directly proportional to the square root of particle diameter at constant dynamic viscosity.

Factors affecting particles classification

Based on the basic concept of the developed fractionation system, the following factors affecting particle diameter, the following factors were investigated.

Effect of water volume (L)

Three different water volumes for the feed solution were used to show the effect of these volumes on the particle size of the $CaCO_3$, while keeping the other parameters constant. Figure 4a–c shows the size of the particles using 0.5, 0.8 and 1.5 L of water, respectively. It is clear that the particles size of average 25–33 nm is obtained when using 0.8 L of water, Figure 4b, while those obtained when using more or less of this volume (0.5 and 1.5 L)were of average size of 1.1–6.5 μm and 1.4–9 μm, Fig. 4a and c, respectively. These variations can be related to the fact that at low volume (500 ml water), the proportion of solids in water increases with subsequent increase in the pulp density and formation of particles aggregate. On the other hand at high dilution, 1,500 ml, the stirring efficiency decreases with subsequent decrease in separation efficiency. This result indicates that for feed concentration of 5 g $CaCO_3$ dispersed in 800 ml distilled water the different particles are well dispersed by the mixing speed used.

Effect of terminal (settling) velocity in different stages

The SEM images, Fig. 5, shows that the particles size range obtained at stages 3, 4, 5 and 6 are between 1.2 and 10.3 μm, Fig. 5a, 0.7–5.7 μm, Fig. 5b, 0.066–0.178 μm, Fig. 5c, and 30–47 nm, Fig. 5d, respectively. This indicates that as the number of stage increases, the size of particle decreases. Accordingly, particles of nano size, 30–47 nm, are separated in the 6th stage of classifier system.

To discuss the variations in the particles size obtained in each stage, equation, 3, was consulted. In this concern, the terminal (settling) velocity at a certain stage n will be given by the following relation:

$$v_{s,n} = (g/18) * \left[(\rho_s - \rho_f)_n / \mu_n\right] * d_n^2 \quad (4)$$

where n is the stage number at a certain stage, n $(\rho_s - \rho_f)_n$ and μ_n are constant. Then,

$$v_{s,n} = k_1 * d_{n,}^2 \quad \text{where } k_1 \text{ is constant} \quad (5)$$

Therefore, the relation between $v_{s,n}$ and d_n^2 will give linear relation, with negative slope, since $v_{s,n}$ is a downward velocity. The experimental results obtained are given in Fig. 6, and showed a linear negative relation with a slope equals –97 s. This support Eq. 6 deduced based on the previous assumptions.

Effect of $CaCO_3$ feed concentration (g/L)

The concentration of calcium carbonate in the feed dispersed solution was investigated in the range 1.5–15 g/800 ml. At all concentration used, the flow rate of the feed

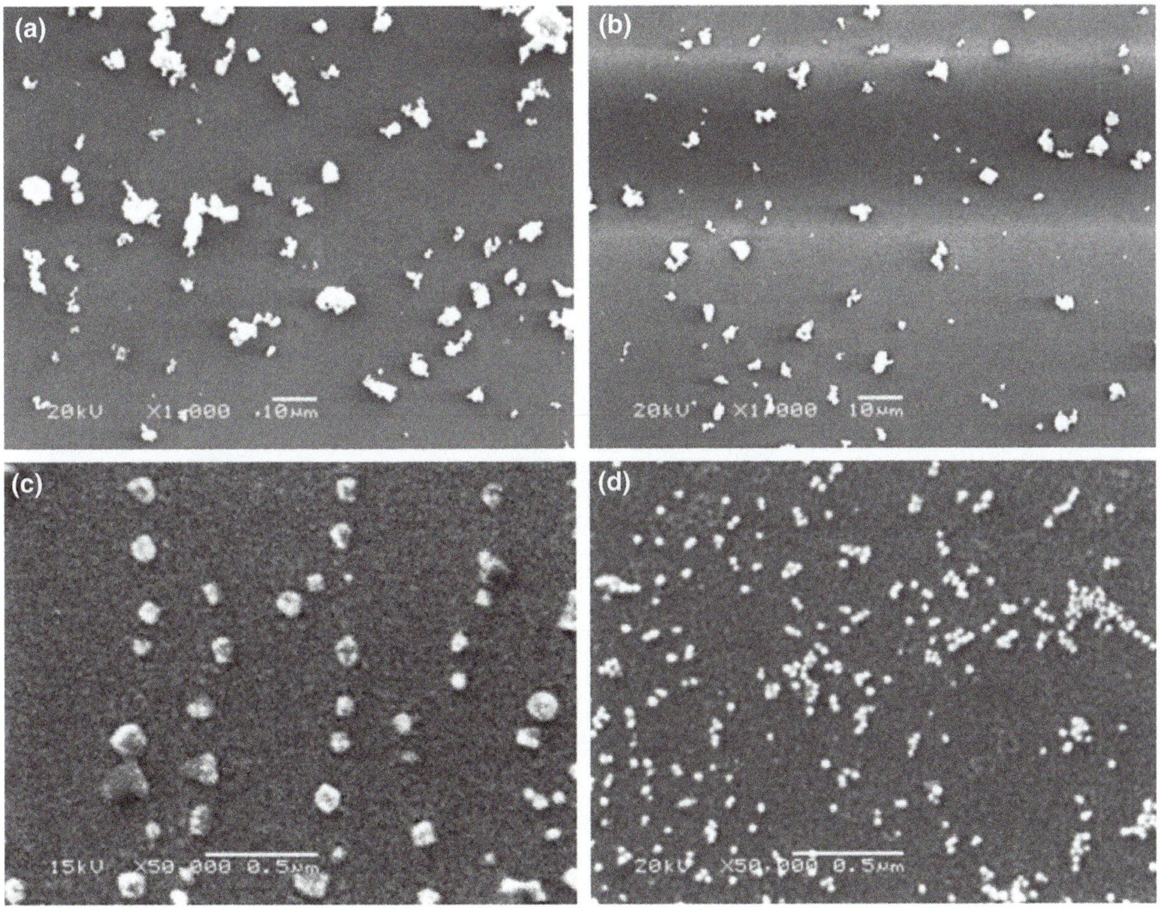

Fig. 5 Effect of terminal (settling) velocity in different stages. [$CaCO_3$]: 1.5g/800 L, Stirring time: 2 h, Stirrer speed: 650 rpm, Flow rate: 0.2 L/h. **a** 3rd stage, **b** 4th stage, **c** 5th stage and **d** 6th stage

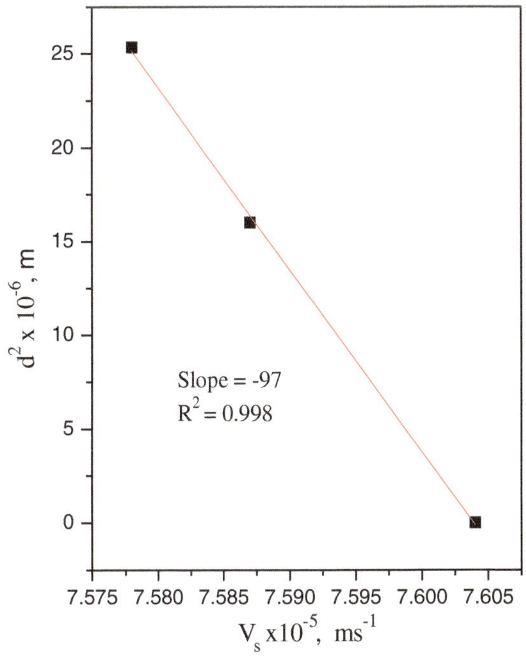

Fig. 6 Effect of the velocity of fluid on the particle size of $CaCO_3$ obtained at 6th stage

dispersed solution was 0.2 L/h, a mixing speed was 650 rpm and the stirring time was 2 h. The separated particles at 6th stage were determined for each feed concentration, Fig. 7a–d. When the feed of $CaCO_3$ concentration was 1.5 g/800 ml, the nanoparticles formed in the 6th stage obtained was in the range 30–47 nm (Fig. 7a). Increasing the concentration of the dispersed particles to 5 g/800 ml the range of the nanoparticles separated decreased to 25–33 nm (Fig. 7b). At higher dispersed $CaCO_3$ concentration, 10 and 15 g/800 ml, the $CaCO_3$ range of particle size obtained at 6th stage increased to 0.137–0.231 µm (Fig. 7c) and 0.156–0.429 µm (Fig. 7d), respectively. The impact of the dispersed phase concentration on particle classification is mainly related to the mechanical viscosity, μ, as long as no aggregate of particles are present in the dispersed phase.

Equation 3 can be modified to relate the particle diameter in terms of dynamic viscosity as follows:

$$d^2 = (18/g) * [v_s/(\rho_s - \rho_f)] * \mu \qquad (6)$$

Then, the particle diameter at certain stage n will give the following relation:

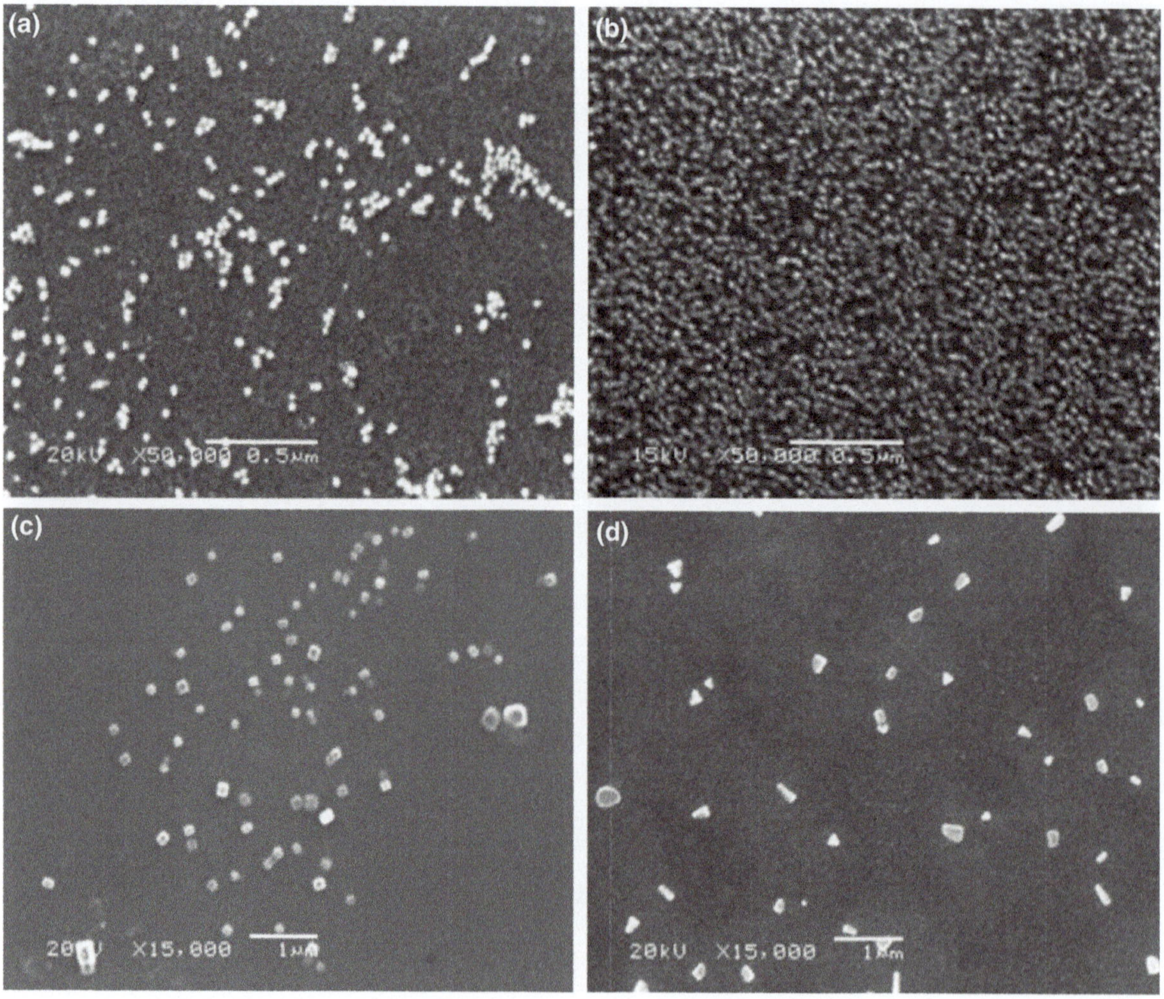

Fig. 7 Effect of $CaCO_3$ feed concentration on $CaCO_3$ particles size at 6th stage. Stirring time: 2 h. Stirrer speed: 650 rpm. Flow rate: 0.2 L/h. Water volume: 800 ml, $wt_{CaCO_3}Wt_{CaCO3}$ **a** 1.5 g, **b** 5 g, **c** 10 g and **d** 15 g

$$d_n^2 = (18/g) * \left[v_{s,n}/(\rho_s - \rho_f)_n\right] * \mu_n \quad (7)$$

For certain concentration and stage $v_s/(\rho_s - \rho_f)$ is constant.

$$\text{Then, } d_n^2 = k_2 * \mu_n \quad (8)$$

Figure 8 shows a linear relation between d_n^2 and μ_n with positive slope 0.22 m^2 s kg^{-1} was obtained. This indicates that the particle diameter is directly proportional to the viscosity of the fluid at certain stage and this support Eq. 8.

Effect of flow rate (L/h) of the fluid

The SEM images of the particles obtained at the 6th stage at different flow rates are shown in Fig. 9. From these images, microparticles were detected by increasing the flow rate. The particle size observed at flow rate 0.2 L/h is 25–33 nm (Fig. 9a), at flow rate 0.3 L/h is 0.11–0.59 μm (Fig. 9b) and at flow rate 0.4 L/h is 0.8–8.6 μm (Fig. 9c).

To illustrate the particle size diversity at different flow rates Eq. 6 is consulted, and

$$\mu = \text{F.R} / L \quad (9)$$

where F.R: flow rate and L: vertical travelled distance. Therefore,

$$d^2 = (18v_s/g) * [1/L(\rho_s - \rho_f)] * \text{F.R} \quad (10)$$

For certain concentration and at 6th stage, the term $[1/L(\rho_s - \rho_f)]$ is constant and for certain particle v_s is constant. Therefore,

$$d^2 = k_3 * \text{F.R} \quad (11)$$

The d^2 was plotted vs. FR, the relation traced in Fig. 10, shows clearly that as the flow rate increases, the particle size increases linearly with the slope around 1.4×10^{-3} m s kg^{-1} which support Eq. 11.

The best conditions of the above factors are [$CaCO_3$ weight (5 g), 0.8 L water, stirring speed (650 rpm), stirring

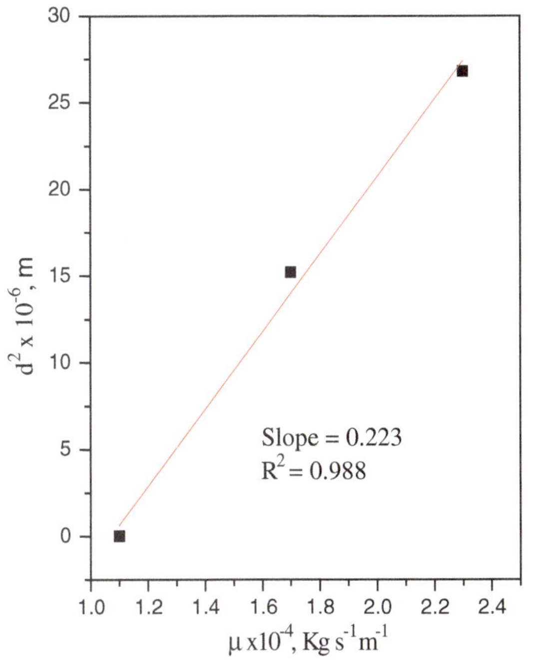

Fig. 8 Effect of dynamic viscosity of fluid particle size of $CaCO_3$ obtained at 6th stage

time (2 h), flow rate (0.2 L/h), and 6th stage] were applied on $CaCO_3$ of sample S2. Figure 11 shows the particles obtained with size ranged between 25 and 48 nm.

Characterization of calcium carbonate

XRD analysis

X-ray diffraction patterns for the two samples S1 and S2 before and after fractionation are given in Figs. 12 and 13, respectively. From the figures, an intense sharp peak at $2\theta = 29°$ is obtained for both samples before and after fractionation. This indicates that $CaCO_3$ have crystal structure, this crystalline polymorph is calcite. The crystal size obtained for S1 (Fig. 12) are $D = 55$ nm and $D = 42$ nm before and after fractionation, respectively. In addition, the crystal sizes obtained for S2 (Fig. 13) are $D = 48$ nm and $D = 42$ nm before and after fractionation, respectively. The crystal size was calculated using Scherrer's equation. These values are comparable with that obtained by SEM; however, SEM

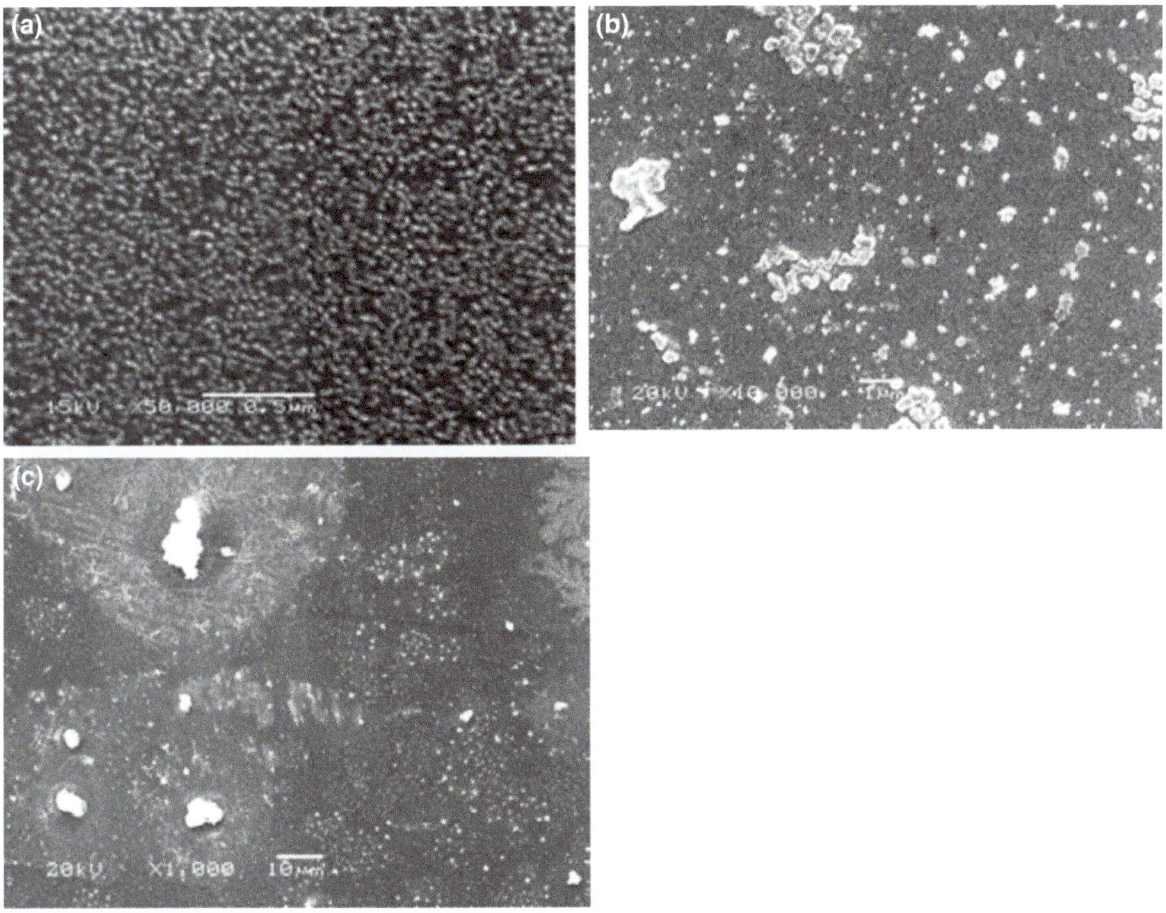

Fig. 9 Effect of flow rate (L/h) of the fluid. 6th stage, Stirring time: 2 h. Stirrer speed: 650 rpm. [$CaCO_3$]: 5 g/800 L. F.R. **a** 0.2 L/h, **b** 0.3 L/h and **c** 0.4 L/h

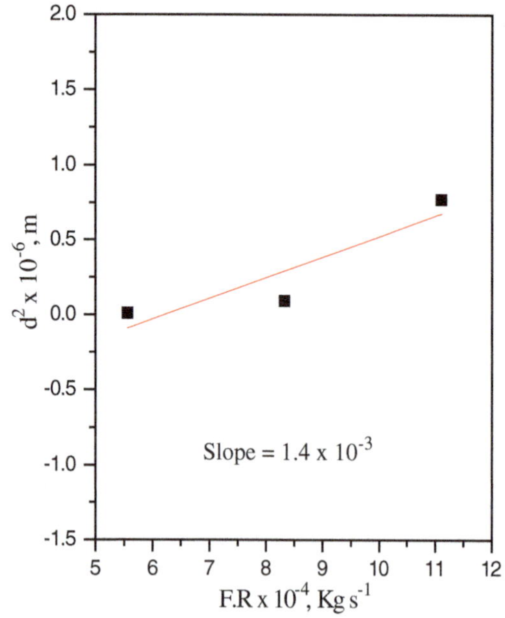

Fig. 10 Effect of flow rate on the $CaCO_3$ particle size at 6th stage

results are more detailed concerning the particle size distribution if statistically examined carefully.

FTIR analysis

The IR spectra of $CaCO_3$ nanoparticles fractionated from S1 and S2, respectively, (a: before fractionation and b: after fractionation for each samples) are given in Figs. 14 and 15. $CaCO_3$ has three main peaks $\nu_3 = 1{,}425\ cm^{-1}$,

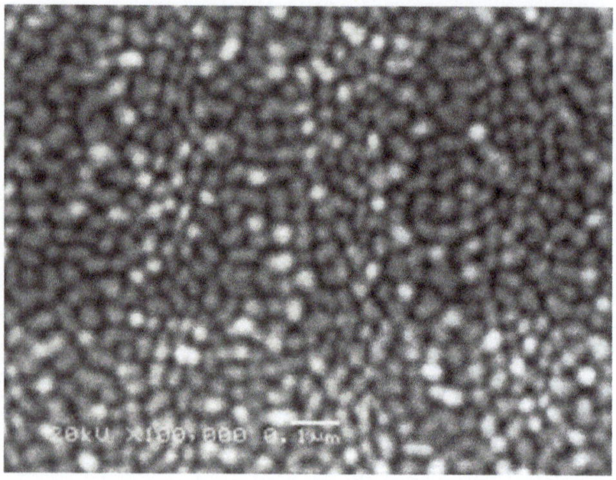

Fig. 11 Image of particles obtained from $CaCO_3$ from Riedel-deHaën (S2). [$CaCO_3$]: 5 g/800 L. Stirring speed: 650 rpm. Stirring time: 2 h, Flow rate: 0.2 L/h. 6th stage, water volume: 0.80 L

$\nu_2 = 874\ cm^{-1}$ and $\nu_4 = 712\ cm^{-1}$ (Yue Lin-hai et al. 2000). The ν_3 band is strong and broad while ν_2 and ν_4 bands are weak and narrow. The main peaks in our samples (S1 and S2) are shown at $\nu_3 = 1{,}465\ cm^{-1}$ (strong and broad band), $\nu_2 = 868\ cm^{-1}$ (weak and broad band) and $\nu_4 = 663\ cm^{-1}$ (weak and broad band). Furthermore, for S1, the band at $1{,}465\ cm^{-1}$ obtained before fractionation, Fig. 14a, is more broader than that obtained after fractionation, Fig. 14b. And the band at $3{,}444\ cm^{-1}$ related to the O–H vibrational mode for the adsorbed water obtained before fractionation, Fig. 14a, is less intense than that obtained after fractionation, Fig. 14a. This may be

Fig. 12 XRD characterization peaks for $CaCO_3$ from Target for chemical industry company (S1); B1: before ($D = 55$ nm) and B2: after fractionation ($D = 42$ nm)

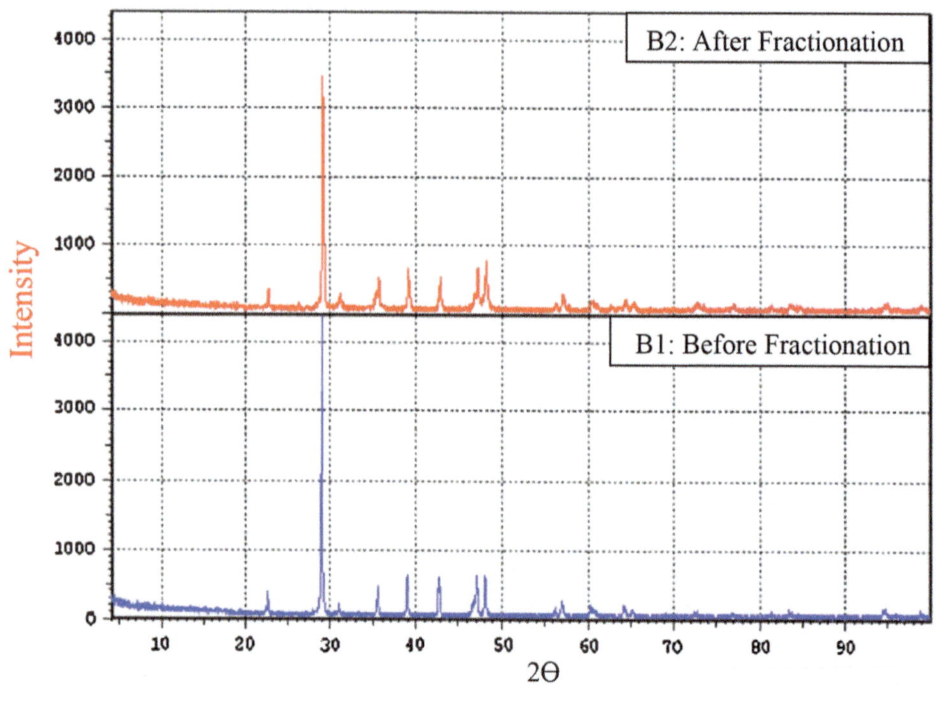

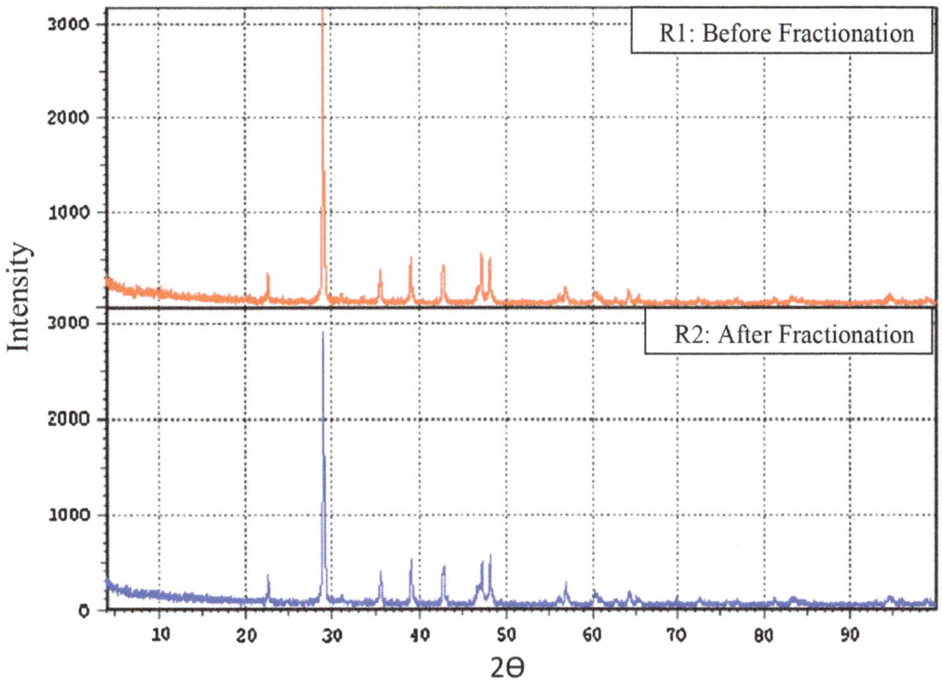

Fig. 13 XRD characterization peaks for $CaCO_3$ from Riedel-deHaën (S2); R1: before(D = 48 nm) and R: after fractionation (D = 42 nm)

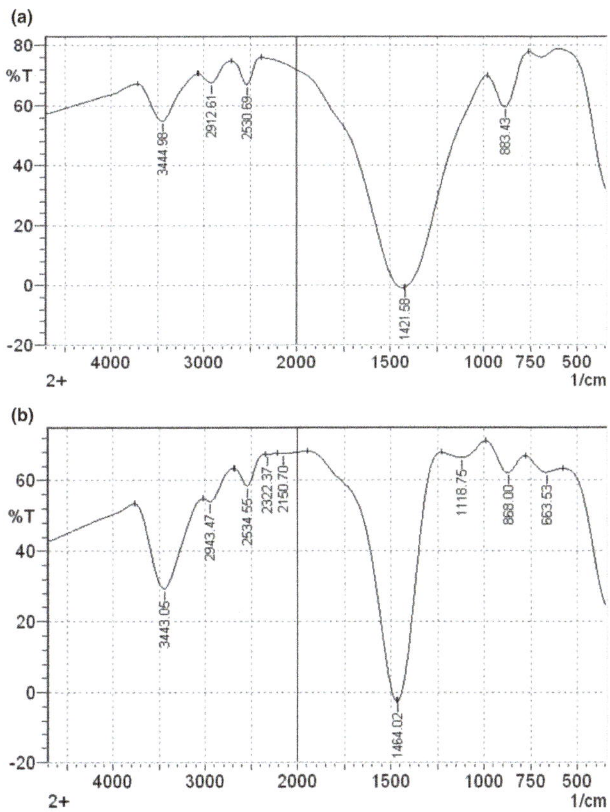

Fig. 14 FTIR spectrum of fractionated $CaCO_3$ nanoparticles from S1: **a** before and **b** after fractionation

referring to the high reactivity of the fractionated $CaCO_3$ nanoparticles. Similarly, for S2 the band at 3,444 cm^{-1} related to the O–H vibrational mode for the adsorbed water obtained before fractionation, Fig. 15a, is less intense than that obtained after fractionation, Fig. 15b.

EDS of fractionated $CaCO_3$ nanoparticles

Figures 16 and 17 revealed to the EDS analysis for the $CaCO_3$ nanoparticles obtained after fractionation form the two samples S1 and S2, respectively. The two figures showed that the $CaCO_3$ nanoparticles obtained have some impurities. By comparison between the EDS analysis for the sample S1 (before; Fig. 3 and after; Fig. 16), we observed that some impurities disappear and another impurities decrease to very low percent. The purity of the sample S2 increases from 98.5 to 99.77 %.

Conclusion

In this work, we developed a laboratory scale hydraulic classifier system for fractionation of calcium carbonate nanoparticles. The used procedure is very simple and relatively of low cost where water was used as fractionation medium, a feeding pump and simple glassware. By this system, we fractionate calcium carbonate of commercial resources to particle diameters in the range 25–33 nm and 25–48 nm by varying the different parameters. The experimental results were modeled in terms of the different parameters affecting particles fractionations.

Although the developed system is simple and easy to manufacture, yet the time required for fractionation is

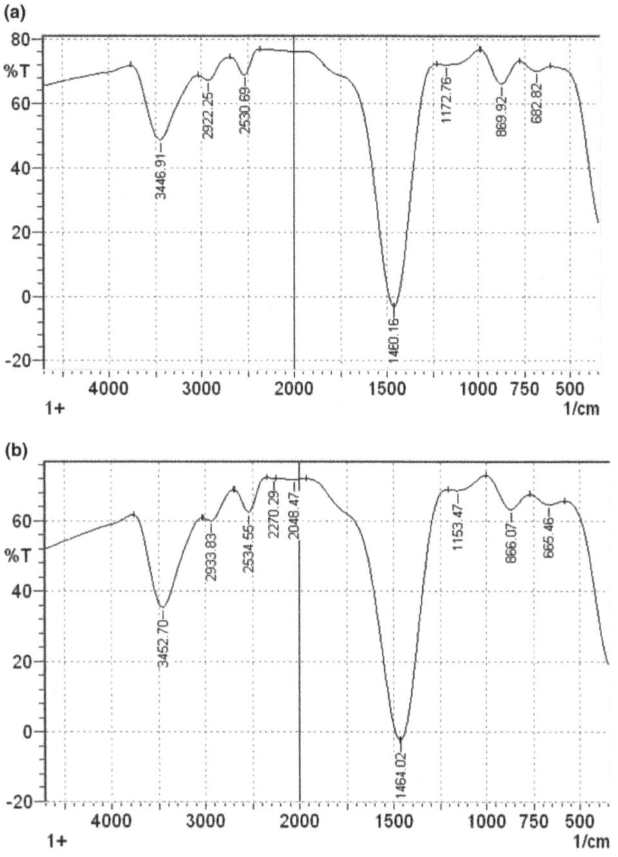

Fig. 15 FTIR spectrum of fractionated $CaCO_3$ nanoparticles from S2: **a** before and **b** after fractionation

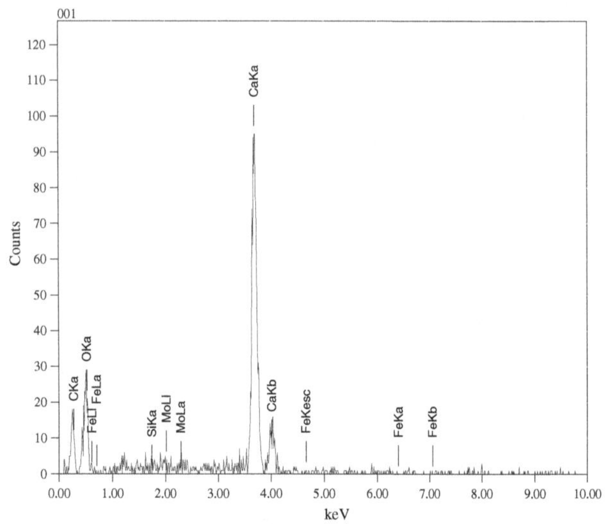

Element	(keV)	mass%	Error%	At%	Compound	mass%	Cation	K
C K	0.277	11.18	0.15	19.26				7.1719
O K	0.525	45.42	0.90	58.73				19.9788
Si K	1.739	0.18	0.20	0.13				0.2079
Ca K	3.690	41.80	0.26	21.57				70.7815
Mo L	2.293	1.43	0.47	0.31				1.8599
Total		100.00		100.00				

Fig. 16 The EDS analysis for calcium carbonate obtained from Riedel–DeHaen Company (S1) after fractionation

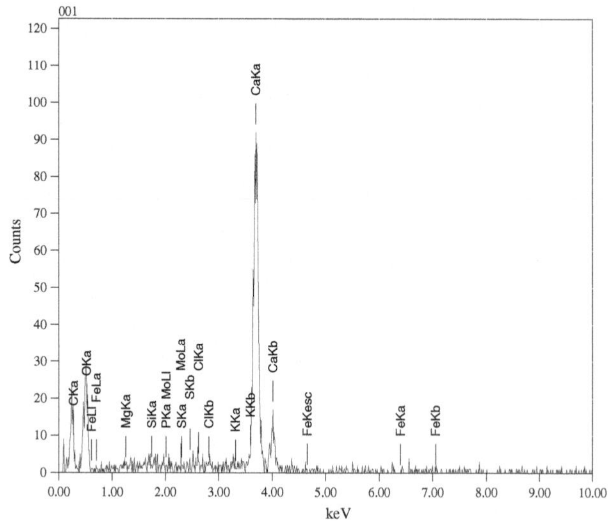

Element	(keV)	mass%	Error%	At%	Compound	mass%	Cation	K
C K	0.277	7.76	0.16	13.38				4.9740
O K	0.525	49.98	0.93	64.73				23.3164
S K	2.307	0.04	0.16	0.03				0.0680
Cl K	2.621	0.34	0.18	0.20				0.5614
Ca K	3.690	41.88	0.28	21.66				71.0802
Total		100.00		100.00				

Fig. 17 The EDS analysis for calcium carbonate obtained from Riedel–DeHaen Company (S2) after fractionation

relatively long compared to density gradient centrifugation. However, the density gradient centrifugation requires a suitable centrifuge and density gradient medium which do not disturb the nanoparticles to be separated further is needs more expensive components.

From the characterization of the fractionated nanoparticles from the two samples S1 and S2, the nanoparticles more reactive, have small crystal size and high purity compared by the original samples used.

References

Barry AW, Napier-Munn T (2006) An introduction to the practical aspects of ore treatment and mineral recovery. Elsevier Science & Technology Books, 17th edition

Bryczkowski A, Chmielniak T (2001) Method of calculation of new cyclone-type separator with swirling baffle and bottom take off of clean gas—part II: experimental verification. Chem Eng Process 40:245–254

Galk J, Peukert W, Krahnen J (1999) Industrial classification in a new impeller wheel classifier. Powder Technol 105:186–189

Guo Feng, Li Ying, Hong-Xia Xu, Zhao Guo-Qing, He Xiu-Juan (2007) Size-controllable synthesis of calcium carbonate nanoparticles using aqueous foam films as templates. Mater Lett 61:4937–4939

Kowalczyk Bartlomiej, IstvanLagzi, Grzybowski BA (2011) Nanoseparations: strategies for size and/or shape-selective purification of nanoparticles. Curr Opin Colloid Interface Sci 16:135–148

Lin-hai Yue, Shui Miao Xu, Zhu-de Xu X (2000) Distortion of crystal lattice and abnormal infra-red behavior in nanocrystalline $CaCO_3$. J Zhejiang Univ (SCIENCE) 1:178–183
Rhods M (2008) Introduction to particle technology Handbok, 2nd edn. Wiley, New York
Stokes GG (1891) Mathematical and physical paper III. Cambridge University Press, Cambridge
Sun XM, Tabakman SM, Seo WS et al (2009) Separation of nanoparticles in a density gradient: FeCo@C and gold nanocrystals. Ange Chem Int Ed 48:939–942
Van Kooy L, Mooij M, Rem P (2004) Kinetic gravity separation. Phys Sep Sci Eng 13:25–32
Vigneau E, Loisel C, Devaux MF, Cantoni P (2000) Number of particles for the determination of size distribution from microscopic image. Powder Technol 107:243–250
Wang Q, Chr Melaaen M, De Silva SR (2001) Investigation and simulation of a cross-flow air classifier. Powder Technol 120:273–280

Yamamoto Ken-ichi, Tohyama Masakazu, Sugimoto Masunori (1998) Continuous separation of differently shaped fine particles on a rotating vibrating conical disk effects of different operating conditions on separation characteristics under various throughputs. Powder Technol 99:1–10

20

Gold nanoparticles assisted characterization of amine functionalized polystyrene multiwell plate and glass slide surfaces

Gunasekaran Dharanivasan · Thangavelu Rajamuthuramalingam · Denison Michael Immanuel Jesse · Nagappan Rajendiran · Krishnan Kathiravan

Abstract We demonstrated citrate-capped gold nanoparticles assisted characterization of amine functionalized polystyrene plate and glass slide surfaces through AuNPs staining method. The effect of AuNPs concentration on the characterization of amine modified surfaces was also studied with different concentration of AuNPs (ratios 1.0–0.0). 3-Aminopropylyl triethoxy silane has been used as amine group source for the surface modification. The interactions of AuNPs on modified and unmodified surfaces were investigated using atomic force microscopy and the dispersibility, and the aggregation of AuNPs was analyzed using UV–visible spectrophotometer. Water contact angle measurement and X-ray photoelectron spectroscopy (XPS) were used to further confirmation of amine modified surfaces. The aggregation of AuNPs in modified multiwell plate leads to the color change from red to purple and they are found to be adsorped on the modified surfaces. Aggregation and adsorption of AuNPs on the modified surfaces through the electrostatic interactions and the hydrogen bonds were revealed by XPS analysis. Remarkable results were found even in the very low concentration of AuNPs (ratio 0.2). This AuNPs staining method is simple, cost-effective, less time consuming, and required very low concentration of AuNPs. These results can be read out through the naked eye without the help of sophisticated equipments.

G. Dharanivasan · T. Rajamuthuramalingam ·
D. Michael Immanuel Jesse · K. Kathiravan (✉)
Department of Biotechnology, University of Madras,
Guindy Campus, Chennai 600 025, Tamil Nadu, India
e-mail: kathir68@rediffmail.com; drkkathiravan@gmail.com

N. Rajendiran
Department of Polymer Science, University of Madras,
Guindy Campus, Chennai 600 025, Tamil Nadu, India

Keywords Polystyrene multiwell plate · Glass slides · Amine functionalization · Gold nanoparticles interaction

Introduction

Immobilization of biomolecules on the solid surface was the key step in the diagnosis of infectious and genetic diseases, analysis of gene expression, forensic science, toxicology, drug discovery, biosensing, and environmental monitoring (Kannoujia et al. 2010; North et al. 2010). In general, polystyrene (PS), polypropylene, polycarbonate, and glass slide (GS) have been used as a solid materials for the immobilization of biomolecules (DNA and protein) because of their versatile chemical, physical, and surface adherence properties. Among these, PS and GS are most commonly used solid materials because they readily adsorb to DNA and proteins (antibodies or antigens) via non-covalent interactions (Chu et al. 2005), excellent optical as well as mechanical properties, and cost-effective (North et al. 2010). However, the non-covalent interactions of DNA and proteins on their surfaces also have limitations like desorption during the time of washing, poor recognizing properties, protein denaturation, loss of bioactivity, and low sensitivity (Rebeski et al. 1999; Butler 2000). To overcome these problems, covalent immobilization protocols have been developed and it ensured with the strong attachment and displaying a reorganization site of interest with respect to its target (Goddard and Hotchkiss 2007). Covalent immobilization of biomolecules on PS and GS surfaces can be achieved using a variety of techniques including wet chemical treatments, silane monolayers, plasmas, flame, and UV irradiation (Goddard and Hotchkiss 2007). Among these, silane monolayer using

3-aminopropylyl triethoxy silane (APTES), 3-aminopropylyl trimethoxy silane (APTMS), N-[3-(trimethoxysilyl) propyl]ethylenediamine (TMSPED), and 3-glycidoxy propyl trimethoxy silane (GOPS) for the functionalization of the amine and the epoxy groups on the PS plate and the GS surfaces (Huang et al. 2012; Moirangthem et al. 2012; Wang and Vaughn 2008; Yang et al. 2005; Lin et al. 2006; Kobayashi et al. 2007; Dixit et al. 2011). APTES functionalized PS plates and GS are used in the immobilization of antibodies for the detection of antigens. Similarly, oligonucleotide probes/DNA were used to detect target DNA sequences and single nucleotide polymorphisms (SNP) typing and assaying in the protein kinases activity (Dixit et al. 2010; Qin et al. 2007; Pack et al. 2007; Hirayama et al. 1996; Nikiforov et al. 1994; Kim et al. 2008).

The presence of functional groups and chemical status of elements, rough and smooth surface, hydrophobic and hydrophilic properties of materials was characterized by using X-ray photoelectron spectroscopy (XPS), atomic force microscopy (AFM), and water contact angle (WCA) measurements (North et al. 2010; Qin et al. 2007; Idage and Badrinarayanan 1998). However, these characterization methods are cost-effective, time consuming and they are not affordable for routine modification process in the view of commercialization. Therefore, an essential need to develop a simple and effective characterization method was considered as an important one. The use of AuNPs to characterize the amine modified GS surfaces through the staining method was recently reported (Carré and Lacarrière 2007). Herein, we have demonstrated citrate-capped AuNPs assisted characterization of amine modified PS plate and GS surfaces based on the AuNPs aggregation, color changes and adsorption. This characterization was completed within 1 h and the results could be read out by the naked eye without the help of any sophisticated equipments.

Materials and methods

Materials

Sodium citrate was obtained from Sisco Research Laboratories Pvt. Ltd (Mumbai, India). PS 96-multiwell plate, nitric acid (HNO_3), sulphuric acid (H_2SO_4), and hydrogen peroxide (H_2O_2) were purchased from Thermo Fisher Scientific India Pvt. Ltd (Mumbai, India). Hydrogen tetrachloroaurate trihydrate ($HAuCl_4{\bullet}4H_2O$) was purchased from Loba Chemie Pvt. Ltd. (Mumbai, India). The glass slides were purchased from HIMedia Laboratories Pvt. Ltd (Mumbai, India), GOPS and APTES were obtained from Sigma Aldrich Co. (St. Louis, USA).

Synthesis of citrate-capped AuNPs

Citrate-capped AuNPs were synthesized based on the early described boiling method (Grabar et al. 1995). In brief, 100 mL of 1 mM $HAuCl_4$ was heated and added 10 mL of 38.8 mM sodium citrate at the boiling point and this solution was stirred continuously for 15 min. The solution was cooled at room temperature and characterized by UV–visible spectrophotometer (Shimadzu UV-1600, Japan) and HR-TEM [Model Fei Technai G^2 F30 S-TWIN with 250 kV high-resolution (UHR) pole piece].

Amine modification on PS plate and GS surfaces

Functionalization of amine group on the surface of PS plate and GS was carried out with earlier described methods (Kaur and Raman Suri 2007; Ji et al. 2011). The plates were treated with freshly prepared HNO_3 and H_2SO_4 mixtures (47:53 ratio, 250 μL/well) at room temperature for 30 min in a fume hood with constant shaking. The plates were washed three times with ddH_2O followed by the addition of 5 % of APTES (250 μL/well, pH 6.9) and incubated at room temperature for 2 h. The excess APTES was washed three times with ddH_2O and incubated at 60 °C for 2 h to enhance the APTES binding over the PS surface. Then the modified plates were kept at 4 °C for overnight before to use.

GS was first treated with piranha solution (H_2O_2/H_2SO_4, 30:70 ratio) for 12 h, washed thoroughly with ddH_2O and dried under a stream of N_2 gas. For the silanization, the treated GSs were immersed in a solution containing 1 % of AEPTS for 30 min at room temperature. After the incubation, the slides were rinsed with ddH_2O, dried under a stream of N_2 gas and the slides were kept for 30 min at 120 °C.

AuNPs staining for the characterization of amine modified PS plate and GS surfaces

Citrate-capped AuNPs solution (250 μL) was added to each well in amine modified PS plate and kept at room temperature for 1 h. The color changes were observed and the optical absorbance was measured using with UV–visible spectrophotometer. The modified GS was immersed in AuNPs solution for 1 h and rinsed with ddH_2O then dried at room temperature. The stained slides were scanned using a cannon flatbed scanner LiDE 110. Modified GS treated with AuNPs solution was also analyzed by UV–visible spectrophotometer along with AuNPs with unmodified GS as a control. Different ratios of AuNPs (ratios: 1.0, 0.8, 0.6, 0.4, 0.2 and 0.0) were prepared with ddH_2O to investigate the effect of AuNPs concentrations on this characterization method. In detail, 250 μL of different ratios of AuNPs solutions was added to each well in amine modified and unmodified PS plate then incubated for 1 h at room temperature. The color changes were observed and corresponding optical absorbance of AuNPs solutions was analyzed by UV–visible spectrophotometer. Similarly, 1 cm × 2.5 cm pieces of amine modified GS were incubated with 3 mL of different ratios of AuNPs solutions (ratio 1.0–0.0) for 1 h at room temperature. After the incubation, GS pieces were removed from the AuNPs solutions then rinsed with ddH_2O and dried at room temperature. Air dried GS pieces were scanned using cannon flatbed scanner. Simultaneously GS pieces treated with AuNPs solutions were analyzed by UV–visible spectrophotometer along with the unmodified GS piece treated with different ratios of AuNPs solutions as control.

WCA measurement

WCA measurement was carried out using the static sessile drop technique. Contact angle measurements were performed at room temperature using a goniometer (GBX Digidrop, France) equipped with a microsyringe helps to control the volume of liquid drop (2 μL). Four water drops were placed at different locations on modified and the unmodified surfaces of PS plates and GS. Average contact angles were measured and images were recorded.

XPS and AFM analyses

Chemical composition of modified and AuNPs stained surfaces of PS plates and GS was revealed by XPS analysis. The samples were prepared in 1 × 0.4 cm size of the GS and the bottom of the well in PS plates. Prepared samples were placed on the sample holder and transferred to analysis chamber under ultra high vacuum (UHV) of 3×10^{-10} mbar. Elements of modified surfaces were analyzed by the OMICRON™ XPS system (Omicron Nano Technology GmbH, Germany) using monochromatic Al Kα X-ray source operated at 300 W (anode at 15 kV and emission at 20 mA). Survey the XPS data were recorded with using pass energy of 50 eV and sleep size 0.5 eV with 0.2 s dwell time. High-resolution scans were recorded with pass energy of 20 eV, sleep size 0.03 eV, and dwell time of 0.2 s with three sweeps. The recorded spectrum data analyses and multiple curve fitting were performed using Casa XPS software. Topology of modified and AuNPs stained PS plate and GS surfaces was analyzed using AFM (Park AFM System XE 70, South Korea) operated in a set frequency of 273.73 Hz with NC-AFM head mode.

Results and discussion

Synthesis and characterization of citrate-capped AuNPs

The amine functionalized PS plate and GS surfaces were characterized by using citrate-capped AuNPs based on their aggregation, color changes and adsorption were illustrated as schematic representation in Fig. 1. AuNPs were synthesized by earlier described boiling method (Grabar et al. 1995). The yellow color reaction solution was gradually changed to red color. It indicates that the formation of AuNPs and the corresponding SPR band of NPs was found at 520 nm in UV–visible spectra as shown in Fig. 2. The inset of the Fig. 2 (TEM image) shows the spherical shape of the NPs with an average diameter of 13 nm. The reduction of Au^{3+} ion to Au^0 metal and control

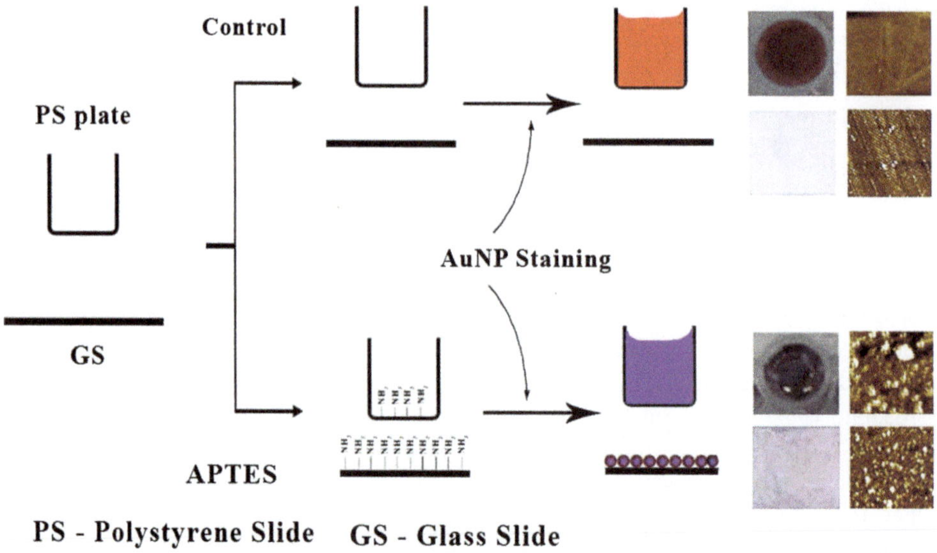

Fig. 1 Schematic representation of the characterization of amine modified PS plate and GS surfaces using citrate-capped AuNPs

the further nucleation growth was achieved using sodium citrate because it acts as reducing and capping agent. The synthesized citrate-capped AuNPs were used for further studies.

WCA measurement of amine modified PS plate and GS surfaces

Amine group was functionalized on PS plate and GS surfaces through the silane monolayer method using APTES as a source of amine group. Nitro group was introduced on the benzene of PS after treated with a nitrating mixture (HNO_3 and H_2SO_4) and GS surfaces were treated with piranha solution. Then the treated surfaces were functionalized with amine functional group through the condensation reaction with organosilane (APTES) (Kaur and Raman Suri 2007; Ji et al. 2011). The hydrophobic and hydrophilic properties of modified and unmodified PS plate and GS surfaces were characterized using with WCA measurement method.

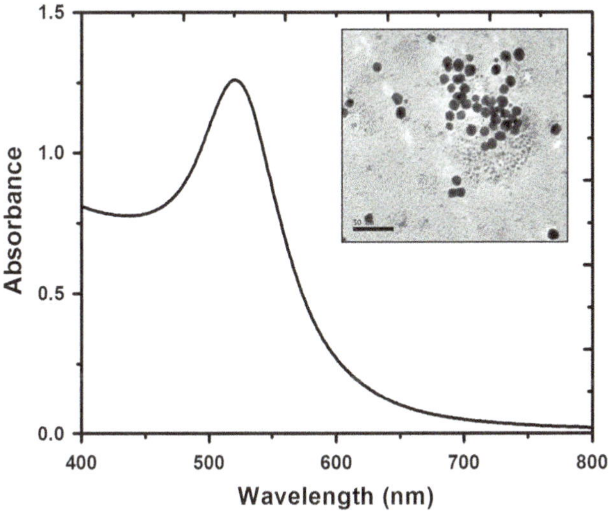

Fig. 2 UV–visible absorption spectra and the corresponding TEM image of synthesized citrate-capped AuNPs in aqueous medium

Figure 3a, b shows that the average WCA of the modified and unmodified PS plate surfaces was found to be 55.5° and 78°, respectively. The contact angle of water on the modified surface was lower than the unmodified surface. This could be due to the decrease of the hydrophobicity of PS plate surfaces by the amine modification, which facilitates the interaction of water molecules with amine group of the modified surfaces through hydrogen bonds. It proves that the PS plate surfaces were modified with the amine group (North et al. 2010). Similarly, Fig. 3c, d shows that the average contact angle of water on the amine modified and unmodified GS surfaces was found to be 49° and 19°, respectively. WCA of modified surface was higher than the unmodified one. This could be due to the increase of the hydrophobicity as well as decrease of the wettability of the surface by the presence of aliphatic carbon chains (ethyl and propyl stretches) as a result of silanation (Qin et al. 2007).

AuNPs staining of amine modified PS plate and GS surfaces

Amine modified PS plate and GS surfaces were investigated through AuNPs staining method. Figure 4a shows SPR band of AuNPs present in the unmodified multiwell PS plate which was found at 525 nm (black line) and SPR band was shifted towards the longer wavelength with the broadening of peak which was found in amine modified plate (red line). This could be due to the absence of AuNPs aggregation in the unmodified multiwell plate as evidenced from the un-changes in the color of the solution. The shift of the SPR band confirms that the AuNPs were getting aggregated in the amine functionalized plates and the color of the solution changes from red to blue as shown in the inset image of Fig. 4a. The mode of AuNPs aggregation in

Fig. 3 Photographic images of water contact angle measurements of modified and unmodified PS plate and GS surfaces. **a** Amine modified PS plate surface, **b** unmodified PS plate surface, **c** amine modified GS surface, and **d** unmodified GS surface

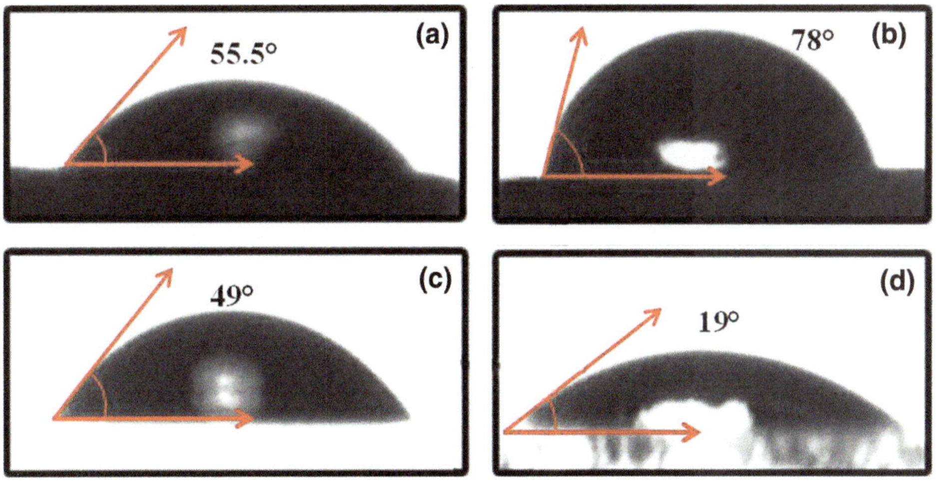

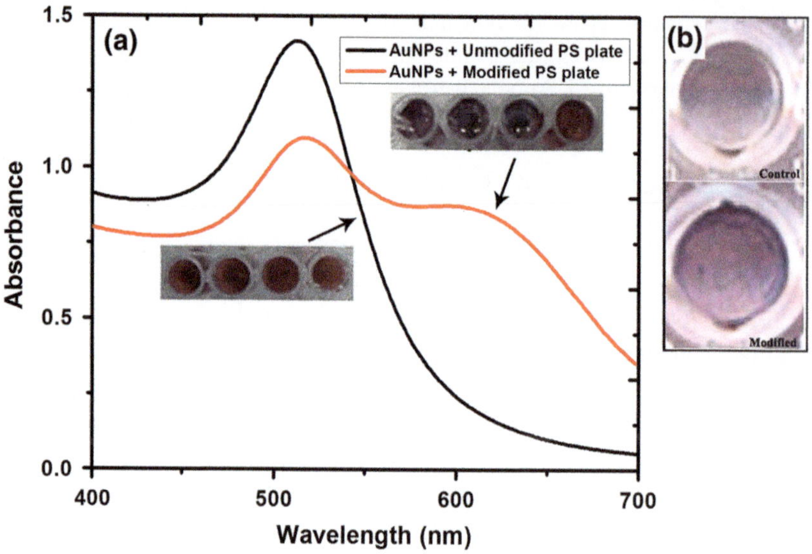

Fig. 4 **a** UV–visible spectra of AuNPs in the unmodified PS plate (*black line*) and amine modified PS plate (*red line*) and **b** scanned image of control and AuNPs stained amine modified PS plate surface, respectively

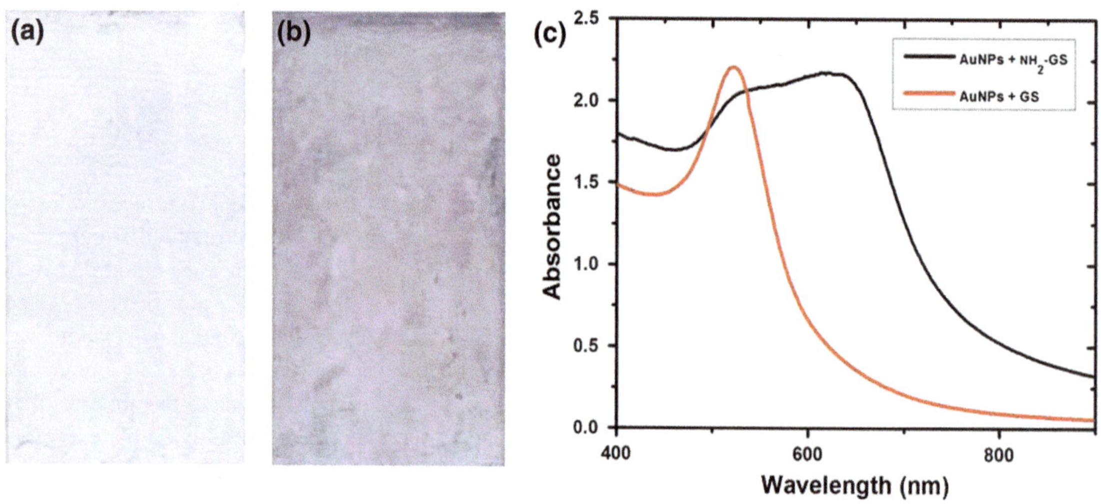

Fig. 5 Scanned images of AuNPs stained unmodified GS surface (**a**) and amine modified GS surface (**b**). (**c**) UV–visible absorbance spectra of AuNPs with modified GS (*black curve*) and unmodified GS (*red curve*)

modified plate might be due to the screening of electronegative repulsion among the NPs by protonated amine groups present on the modified plate surface. Also, AuNPs were weakly adsorbed on the modified surface through the electrostatic interactions and the hydrogen bonds between the positively charged amine groups and the negatively charged AuNPs as shown in Fig. 4b. In the case of unmodified plate, AuNPs were not adsorbed on the surface (Moirangthem et al. 2012; Carré and Lacarrière 2007; Fujiwara et al. 2006). Figure 5a, b shows that the AuNPs stained modified and unmodified GS surface, respectively. Purple staining was observed only on the amine modified GS surface but not on the unmodified one, and corresponding UV–visible absorbance spectra of AuNPs treated with modified and unmodified GS were also taken (Fig. 5c). AuNPs with unmodified GS show a sharp SPR band at 522 nm without any shift towards the longer wavelength region (Fig. 5c curve–red). But, the absorbance spectra of AuNPs with amine modified GS show a broadening peak towards the longer wavelength (above 600 nm) was shown in Fig. 5c (curve – black). These absorbance spectra prove that AuNPs dispersed with unmodified GS and aggregation with modified GS. Aggregation of AuNPs mainly caused by the protonated amine group present on the modified GS surface (Carré and Lacarrière 2007; Fujiwara et al. 2006; Huang et al. 2012).

The effect of AuNPs concentration on the characterization of amine modified surfaces was also revealed. The different concentrations of AuNPs were prepared by the dilution with water (1.0, 0.8, 0.6, 0.4, 0.2, and 0.0) and were treated with amine modified/unmodified PS plate and GS. The color of AuNPs solution was changed from red to purple and blue in amine modified PS multiwell plate while decreasing the concentration of AuNPs from the ratio of 1.0 to 0.2 was illustrated in Fig. 6a. From red to purple color change was observed in the ratio from 1.0 to 0.6 and red to light blue color shift was found in the 0.2 ratio of AuNPs. The intermediate color between purple and blue was found in the ratio of 0.4 and no more color was found in the ratio 0.0 of AuNPs. But the case of AuNPs solutions in the unmodified PS multiwell plate shows that the gradual decrease of intensity of red color without any color changes, while decreasing the ratio of AuNPs from 1.0 to 0.0 (Fig. 6a). These color changes of AuNPs in the amine modified PS plate might be due to the aggregation of AuNPs caused by protonated amine group present on the surface. But in the unmodified PS plate, the AuNPs solution was remained in red color without any color changes and this could be due to the absence of the amine group on the unmodified plate surface as a result AuNPs not aggregated and present as such individual particles. In addition, UV–visible absorbance spectra of AuNPs with unmodified and modified PS plate were also taken for further confirmation. Figure 6b shows the maximum absorbance and a sharp SPR band of different ratios of AuNPs (from 1.0 to 0.0) with the unmodified PS plate were found to be at 522 nm. The intensity of the peak was decreased (OD: 0.76, 0.69, 0.49, 0.25, and 0.12) with decreasing the ratio of AuNPs from 1.0 to 0.2, respectively, and no peak was found in 0.0 ratio. But an optical absorbance spectra of AuNPs from the amine modified PS plate show broadening peak towards the longer wavelengths (more than 650 nm), which corresponds to the purple and blue colors of aggregated AuNPs. In Fig. 6c, UV– visible absorbance spectra show the low intensity peak that was found at 522 nm and high intensity peak at more than 650 nm. These results confirm that AuNPs aggregation occurred only in amine modified PS plate but not in unmodified plate and also proved that even the very low concentration of AuNPs (ratio 0.2) enough to characterize the modified PS plate surface. The color changes can be observed through naked eye and to be revealed by UV–visible spectrophotometer.

Similarly, the amine modified and unmodified GS surfaces were also characterized by gold staining with different ratios of AuNPs (1.0 –0.0). An intensity of AuNPs staining on the amine modified surface was gradually decreased while decreasing the ratio of AuNPs from 1.0 to 0.2 shown in Fig. 7a. Strong pinkish purple color stain was observed on the modified surface treated with 1.0 ratio of AuNPs which was shifted to purple color when decreasing the AuNPs ratio to 0.8 and finally the shift was reached to light blue color in 0.2 ratio of AuNPs. The color of staining was not observed in 0.0 ratio of AuNPs. These could be due to the aggregation and adsorption of AuNPs on the amine modified surfaces by electrostatic interaction

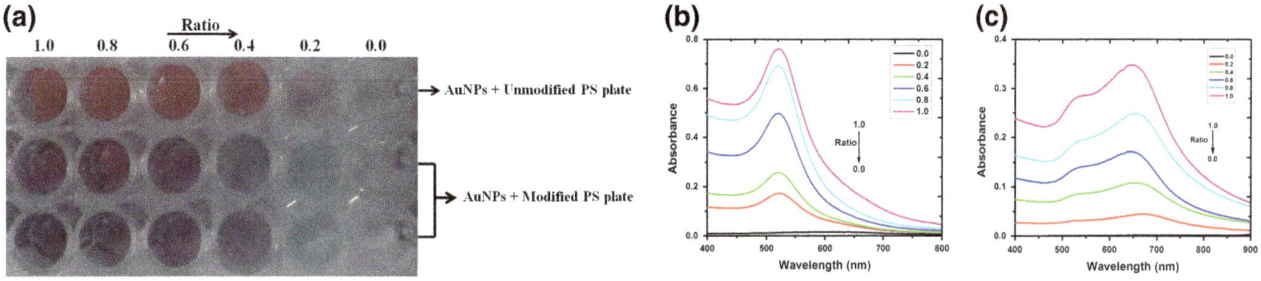

Fig. 6 (**a**) Characterization of amine modified and unmodified PS plate with different ratios of AuNPs (1.0, 0.8, 0.6, 0.4, 0.2, and 0.0). UV–visible absorbance spectra of different ratios of AuNPs with unmodified PS plate (**b**), and amine modified PS plate (**c**)

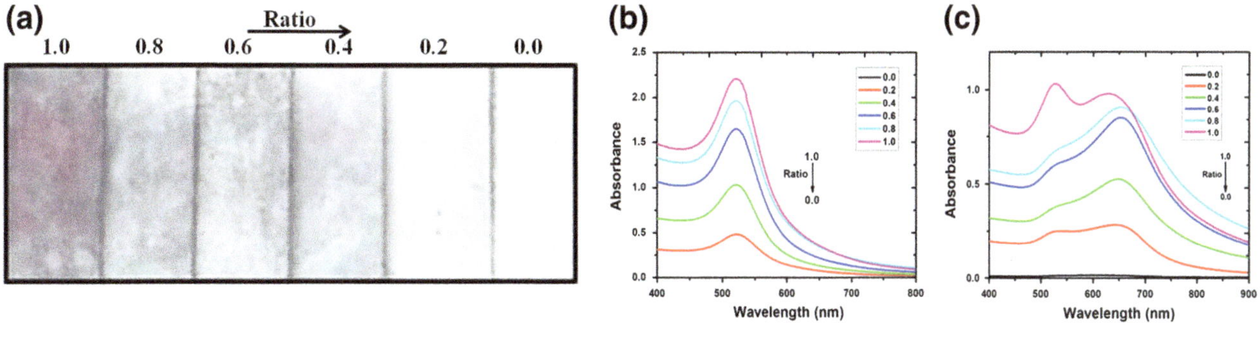

Fig. 7 **a** Characterization of amine modified GS surface using different ratios of AuNPs (1.0, 0.8, 0.6, 0.4, 0.2, and 0.0) through gold staining. UV–visible spectra of different ratios of AuNPs treated with unmodified GS **b**, **c** amine modified GS

between the protonated amine and the negative charged AuNPs. Where the AuNPs ratio is too low, the NPs easily get aggregated and adsorbed on the modified surface due to the presence of protonated amine group. The number of individual particles adsorbed on the modified surface was decreased by decreasing the ratio of AuNPs from 1.0 to 0.2 as a result less number of particles were adsorbed on the modified surface in the lower ratio of AuNPs (ratio 0.2), finally all particles get aggregated. Due to the huge population of NPs at higher concentration (ratio 1.0), complete aggregation of AuNPs could not be possible by defined modified surface of GS as a result more number of individual NPs adsorbed on the surface which make the AuNPs stained GS appeared in pinkish purple color. However, noticeable blue color stain was clearly found up to 0.4 ratios of AuNPs but not in the ratio 0.2 and below.

The optical absorbance of AuNPs solutions with unmodified and modified GS were measured by UV–visible spectrophotometer. In Fig. 7b, UV–visible spectra of different ratios of AuNPs with unmodified GS show the sharp SRP band at 521 nm without any shift. At the same time the peak intensity was decreased with decreasing the ratio of AuNPs from 1.0 to 0.2, and 0.0, respectively. In the case of AuNPs with amine modified GS indicates the shift of the SPR band towards the longer wavelengths (more than 600 nm) with broadening peaks were shown in Fig. 7c. These could be due to the dispersion and aggregation of AuNPs by the absence and presence of the amine group on the unmodified and the modified GS surfaces, respectively. The remarkable spectral shift was found up to 0.2 ratios of AuNPs with modified GS. These results were similar to the findings of different ratios of AuNPs with modified and unmodified PS plate.

XPS characterization of amine modified and AuNPs stained PS plate and GS surfaces

The elements present on amine modified and AuNPs stained PS plates and GS surfaces were analyzed by suing XPS and are summarized in Table 1. Figure 8a depicts the XPS survey spectra of amine modified PS plate surface and it indicates the presence of N, Si, C, and O. High-resolution spectra of C 1s show peaks at 285.2, 286.5, and 288.9 eV which were characteristic features of C–C/C–H, C–N, and C–O, respectively (Fig. 8b). Peak at 400.3 eV in N 1s spectra indicates the presence of protonated amine (NH_3^+) and O 1s spectra show that binding energy peak at 532.4 eV was specific for O–Si chemical bond (Fig. 8c, d). Appearance of binding energy peak at 103.6 eV was characteristic of Si–O/Si–C chemical composition found in Si 2p spectra (Fig. 8e). These characteristic features of XPS spectra prove the presence of the amine group on the PS plate surface. This could be the reason for AuNPs aggregation and adsorption as well as decreases the hydrophobicity of PS plates (Flores-Perez and Ivanisevic 2007; Durdureanu-Angheluta et al. 2012).

XPS survey spectra of AuNPs stained aminated PS plates (Fig. 9a) illustrate the presence of N, C, Si, Au, and O elements and binding energy 1 eV was subtracted for the carbon charge correction from the obtained spectra for curve fitting. The binding energy peaks at 285.1, 287.1, and 288.7 eV were found in C 1s spectra and they were

Table 1 XPS binding energy peaks of amine modified and AuNPs stained PS plate and GS surfaces with their possible assignments

Amine functionalized PS plate surface							
Elements	C 1s			N 1s	O 1s	Si 2p	
Binding energy (eV)	285.2	286.5	288.9	400.3	532.4	103.6	
Types of assignments	C–C/C–H	C–N	C–O	NH_3^+	O–Si	Si–O/Si–C	
Relative atomic concentration (%)	73.05			6.60	16.53	3.83	
AuNPs stained amine functionalized PS plate surface							
Elements	C 1s			N 1s	Si 2p	Au 4f	
Binding energy (eV)	285.1	287.1	288.7	400.5	102.8	83.8	87.4
Types of assignments	C–C/C–H	C–N	C–O	NH_2/NH_3^+–C	Si–O	Au 4f7/2	Au 4f5/2
Amine functionalized GS plate surface							
Elements	C 1s			N 1s	O 1s	Si 2p	
Binding energy (eV)	284.4	285.4	287.0	400.5	532.4	103.3	
Types of assignments	C–C	C–N	HN–CO	NH_3^+	O–Si	Si–O	
Relative atomic concentration (%)	53.67			1.25	24.04	21.04	
AuNPs stained amine functionalized PS plate surface							
Elements	C 1s			N 1s	Si 2p	Au 4f	
Binding energy (eV)	283.9	285.6	288.3	400.0	102.6	84.3	88.0
Types of assignments	C–C	C–N	HN–CO	NH_2/NH_3^+	Si–O	Au 4f7/2	Au 4f5/2

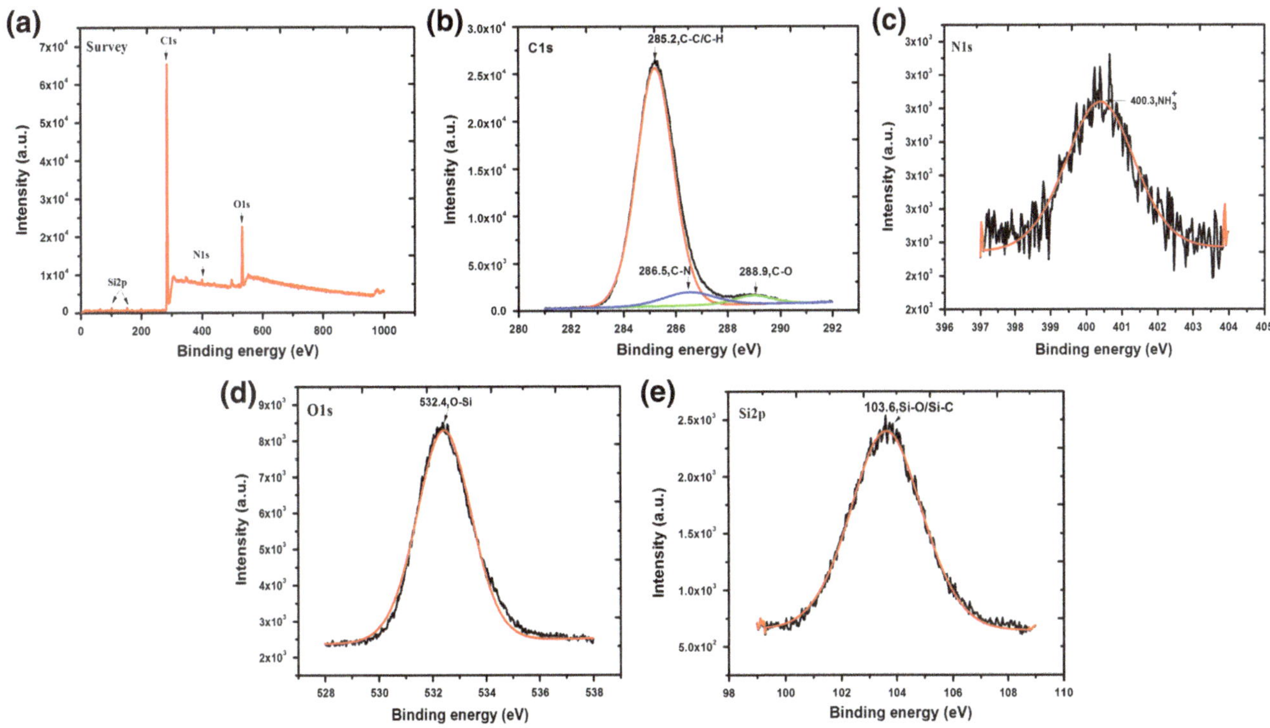

Fig. 8 XPS spectra of amine modified PS plate. **a** Survey scan, high-resolution scan in the **b** C 1s region, **c** N 1s region, **d** O 1s region, and **e** Si 2p region. N 1s, Si 2p and O 1s spectra indicate that the presence of organosilane with the amine functional group on the APTES treated PS plate surface

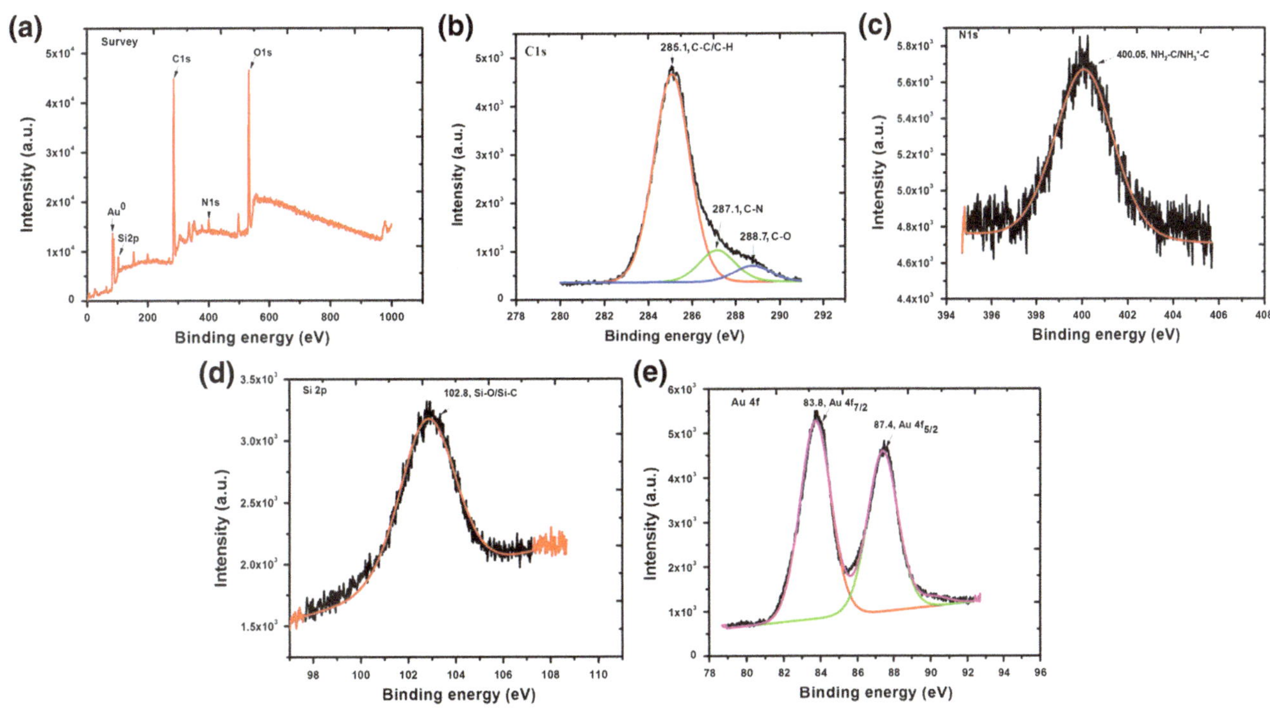

Fig. 9 XPS spectra of AuNPs stained amine modified PS plate. **a** Survey scan, high-resolution scan in the **b** C 1s region, **c** N 1s region, **d** Si 2p region, and **e** Au 4f region. N 1s spectra indicate that the presence of the amine functional group and Au 4f region proves that the presence of metallic gold on the amine modified PS plate surface

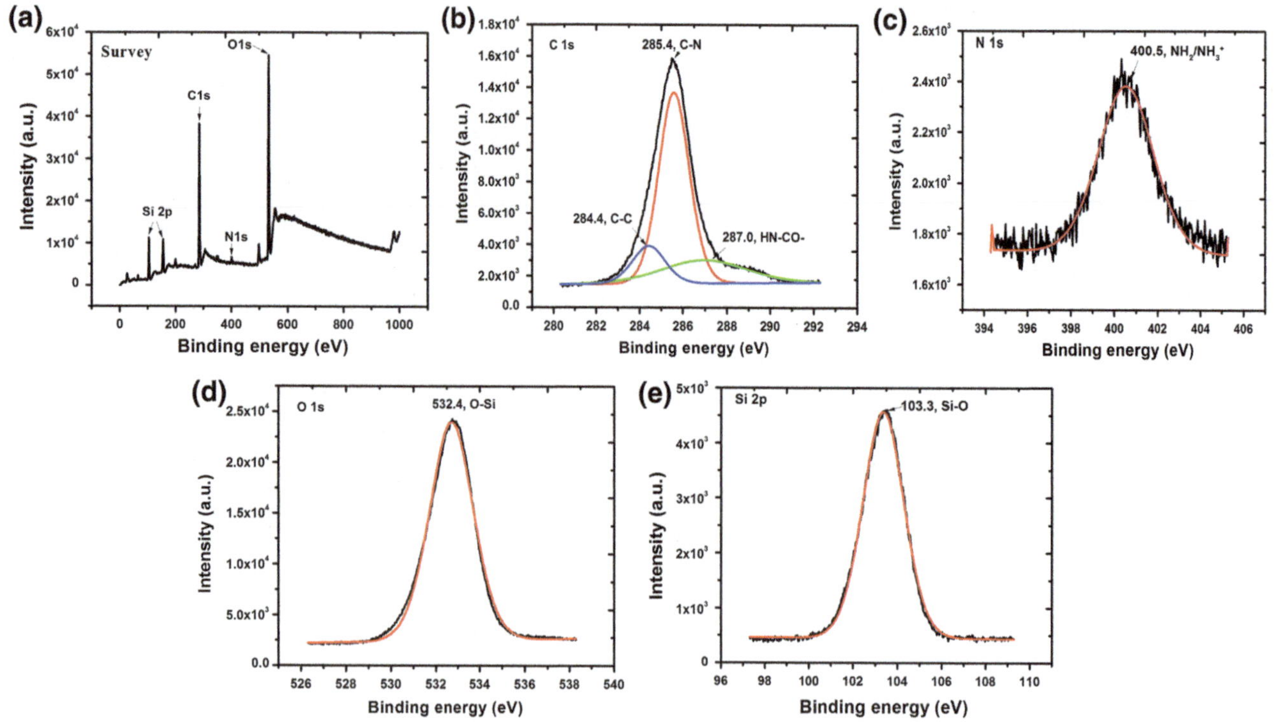

Fig. 10 XPS spectra of amine modified GS surface. **a** Survey scan, **b** high-resolution scan of C 1s region, **c** N 1s region, **d** O 1s region, and **e** Si 2p region. N 1s, Si 2p, and O 1s indicate that the presence of the amine functional group on the APTES treated GS surface

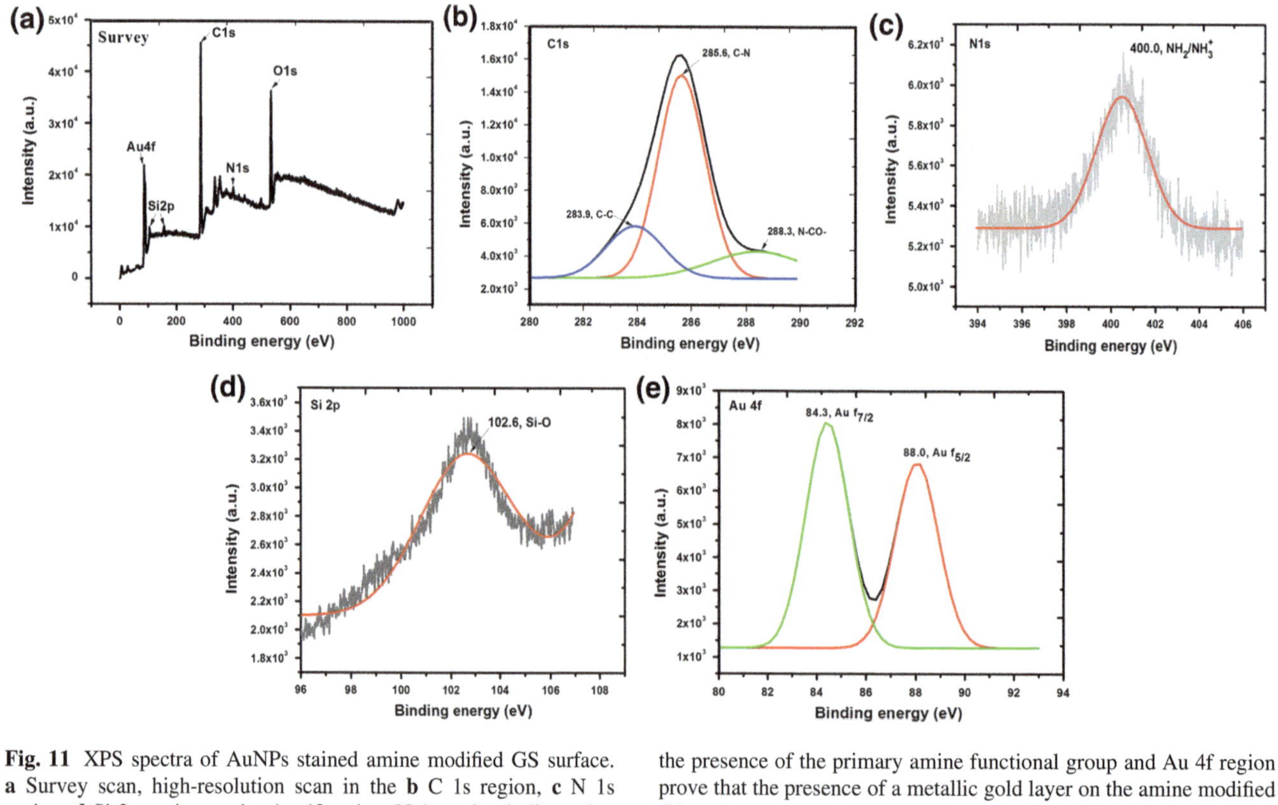

Fig. 11 XPS spectra of AuNPs stained amine modified GS surface. **a** Survey scan, high-resolution scan in the **b** C 1s region, **c** N 1s region, **d** Si 2p region, and **e** Au 4f region. N 1s region indicate that the presence of the primary amine functional group and Au 4f region prove that the presence of a metallic gold layer on the amine modified GS surface

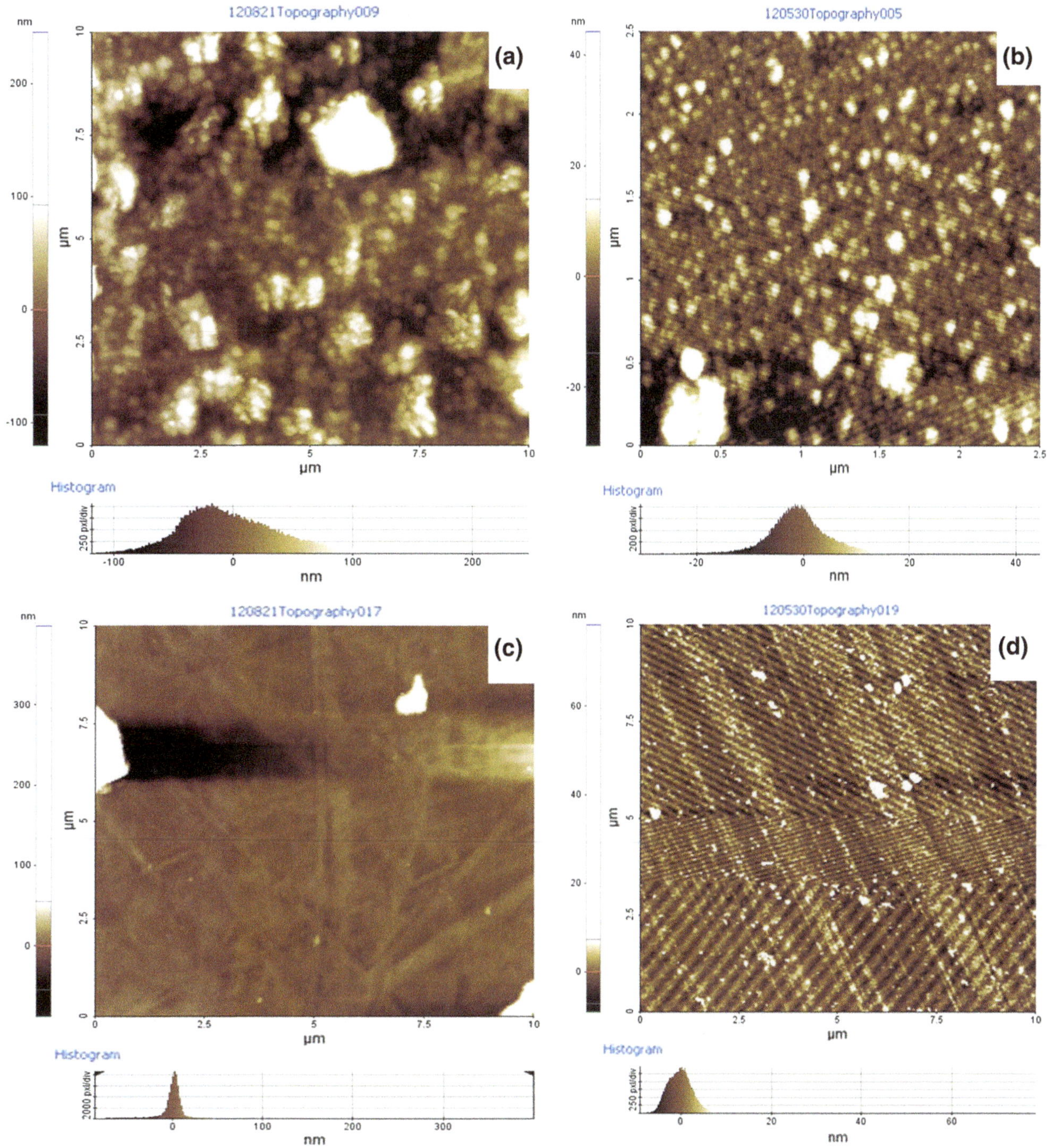

Fig. 12 AFM topographic images with histogram of AuNPs thin metallic film on the amine modified PS plate (**a**) and GS surfaces (**b**), respectively. **c**, **d** show AFM images with histogram of unmodified PS plate and GS surfaces, respectively

characteristics of C–C/C–H, C–N, and C–O bonding of APTES, respectively (Fig. 9b). Peak corresponding to the binding energy of primary/protonated amine (NH_2 or NH_3^+) was found at 400.5 eV (Fig. 9c). Peak for Si 2p was observed at 102.8 eV which was the characteristic feature of Si–O bonding nature (Fig. 9d). Figure 9e shows the binding energy peaks at 83.8 and 87.4 eV specific for Au $4f_{7/2}$ and Au $4f_{5/2}$, respectively. The characteristic features of these peaks indicate the presence of AuNPs in the metallic state on the modified PS plate surface (Young et al. 2011). Also, it indicates that the physical adsorption of AuNPs on the aminated PS plate surface through the electrostatic interactions and the hydrogen bonds (Carré and Lacarrière 2007). This study confirms that the AuNPs interact with modified PS plate surface through the weak interactions that lead to aggregation and the color changes of AuNPs from red to blue.

Figure 10a (XPS survey spectra) shows the presence of the C, N, Si, and O on the amine modified glass slide surface. The C 1s spectra show peaks at 284.4, 285.4, and 287.0 eV which correspond to C–C, C–N, and HN–CO, respectively (Fig. 10b). The N 1s and O 1s spectra show peaks at 400.5 and 532.4 eV, respectively, the characteristic features of protonated amine and O–Si (Fig. 10c, d). Also, Si–O corresponding binding energy peak was found at 103.3 eV in Si 2p spectra (Fig. 10e). AuNPs stained amine modified GS surface XPS survey spectra were shown in Fig. 11a and 1 eV of binding energy was subtracted for curve fitting with respect to carbon charge correction. Peaks at 283.9 eV (C–C), 285.6 eV (C–N), 288.3 eV (HN–CO), 400.0 eV (NH_2/NH_3^+), 102.6 eV (Si–O), 84.3 eV (Au $f_{7/2}$), and 88.0 eV (Au $f_{5/2}$) were found in C 1s, N 1s, Si 2p, and Au 4f of high-resolution spectra, respectively (Fig. 11b–e, respectively). These spectral data show that the AuNPs present in the metallic state even after the adsorption on the modified surface (Flores-Perez and Ivanisevic 2007; Durdureanu-Angheluta et al. 2012; Young et al. 2011).

AFM analysis of AuNPs stained PS plate and GS surfaces

Adsorpance of AuNPs on the aminated PS plates and GS surfaces was reconfirmed by AFM topographical analysis. Figure 12 represents AFM images of AuNPs on the amine modified and unmodified PS plate and GS surfaces, respectively. In the modified PS plate and GS surfaces, the size of AuNPs was found to be 10–100 nm (Fig. 12a, b, respectively) and they are not found on the unmodified PS plate and GS surfaces (Fig. 12c, d, respectively) which could be observed with the corresponding histograms. The size of AuNPs up to 100 nm also indicates the aggregation of AuNPs is due to the presence of amine groups on the PS plate and GS surfaces. This result also proves that AuNPs adsorption and aggregation take place only on amine modified surfaces (Carré and Lacarrière 2007; Fujiwara et al. 2006; Huang et al. 2012).

Conclusion

Citrate-capped AuNPs used to characterize the amine functionalized PS multiwell plate and GS surfaces without help of sophisticated instruments were demonstrated. Here, the surface modification could be observed by naked eyes based on the color changes of AuNPs from red to purple or blue. This will clearly discriminate the modified and unmodified surfaces by simply without the addition of salts. The effect of AuNPs concentration on the characterization of amine modified surfaces with AuNPs was also investigated and we found satisfactory results even in the very low concentration of AuNPs (0.2 ratio). This characterization explored that the interactions and aggregation of AuNPs on the modified PS and GS surfaces by colorimetric and simple scanning methods. These AuNPs assisted characterization method for amine modified surfaces is very simple, cost-effective, less time consuming, and required very low concentration of AuNPs (0.2 ratio) and we hope this study may help to characterize the amine modified solid surfaces rapidly in the routine process in commercial and research point of view.

Acknowledgments The authors would like to thank The Director and Mr. Thirumal, NCNS & NT, University of Madras (UNOM) for HR-TEM and XPS characterizations. The authors profusely thank The Director and Mr. Thangappan, Research Scholar, National Centre for Nanoscience and Technology, Anna University, A.C. Tech Campus for providing the AFM facility for topology analysis. And also thank to Dr. Abhijit P. Deshpande, Department of Chemical Engineering, IIT Madras and his students for their help in WCA measurement. This work was financially supported by the National Centre for Nanoscience and Nanotechnology (NCNS&NT) (F. No.: C-2/Res.Pro/NSNT/Proj.No.7/2011/192), UNOM.

References

Butler JE (2000) Solid supports in enzyme-linked immunosorbent assay and other solid- phase immunoassays. Methods 22:4–23

Carré Birch W, Lacarrière V (2007) Glass substrates modified with organosilanes for DNA immobilization. Silanes and Other Coupling Agents 4:1–14

Chu X, Xiang ZF, Fu X, Wang SP, Shen GL, Yu RQ (2005) Silver-enhanced colloidal gold metalloimmunoassay for Schistosoma japonicum antibody detection. J Immunol Methods 301:77–88

Dixit CK, Vashist SK, O'Neill FT, O'Reilly B, MacCraith BD, O'Kennedy R (2010) Development of a high sensitivity rapid sandwich elisa procedure and its comparison with the conventional approach. Anal Chem 82:7049–7052

Dixit CK, Vashist SK, MacCraith BD, O'Kennedy R (2011) Multisubstrate-compatible ELISA procedures for rapid and high-sensitivity immunoassays. Nat Protoc 6:439–445

Durdureanu-Angheluta A, Dascalu A, Fifere A, Coroaba A, Pricop L, Chiriac H, Tura V, Pinteala M, Simionescu BC (2012) Progress in the synthesis and characterization of magnetite nanoparticles with amino groups on the surface. J Magn Magn Mater 324:1679–1689

Flores-Perez R, Ivanisevic A (2007) Molecular recognition of chromophore molecules to amine terminated surfaces. Appl Surf Sci 253:4176–4181

Fujiwara K, Watarai H, Itoh H, Nakahama E, Ogawa N (2006) Measurement of antibody binding to protein immobilized on gold nanoparticles by localized surface plasmon spectroscopy. Anal Bioanal Chem 386:639–644

Goddard JM, Hotchkiss JH (2007) Polymer surface modification for the attachment of bioactive compounds. Prog Polym Sci 32:698–725

Grabar KC, Freeman RG, Hommer MB, Natan MJ (1995) Preparation and characterization of Au colloid monolayers. Anal Chem 67:735–743

Hirayama H, Tamaoka J, Horikoshi K (1996) Improved immobilization of DNA to microwell plates for DNA–DNA hybridization. Nucleic Acids Res 24:4098–4099

Huang KW, Hsieh CW, Kan HC, Hsieh ML, Hsieh S, Chau LK, Cheng TE, Lin WT (2012) Improved performance of aminopropylsilatrane over amino propyltriethoxysilane as a linker for nanoparticle-based plasmon resonance sensors. Sensor Actuat B-Chem 163:207–215

Idage SB, Badrinarayanan S (1998) Surface modification of polystyrene using nitrogen plasma: an X-ray photoelectron spectroscopy study. Langmuir 14:2780–2785

Ji H, Dong H, Yan F, Lei J, Ding L, Gao W, Ju H (2011) Visual scanometric detection of DNA through silver enhancement regulated by gold-nanoparticle aggregation with a molecular beacon as the trigger. Chem Eur J 17:11344–11349

Kannoujia DK, Ali S, Nahar P (2010) Single-step covalent immobilization of oligonucleotides onto solid surface. Anal Methods 2:212–216

Kaur J, Raman Suri C (2007) Direct hapten coated ELISA for immunosensing of low molecular weight analytes. Nat Protoc Exch.

Kim YP, Oh YH, Kim HS (2008) Protein kinase assay on peptide-conjugated gold nanoparticles. Biosens Bioelectron 23:980–986

Kobayashi Y, Tadaki Y, Nagao D, Konno M (2007) Deposition of gold nanoparticles on polystyrene spheres by electroless metal plating technique. J Phys: Conf Ser 61:582–586

Lin HY, Chen CT, Chen YC (2006) Detection of phosphopeptides by localized surface plasma resonance of titania-coated gold nanoparticles immobilized on glass substrates. Anal Chem 78:6873–6878

Moirangthem RS, Yaseen MT, Wei PK, Cheng JY, Chang YC (2012) Enhanced localized plasmonic detections using partially-embedded gold nanoparticles and ellipsometric measurements. Biomed Opt Express 3:899–910

Nikiforov TT, Rendle RB, Goelet P, Rogers YH, Kotewicz ML, Anderson S, Trainor GL, Knapp MR (1994) Genetic bit analysis: a solid phase method for typing single nucleotide polymorphisms. Nucleic Acids Res 22:4167–4175

North SH, Lock EH, Cooper CJ, Franek JB, Taitt CR, Walton SG (2010) Plasma-based surface modification of polystyrene microtiter plates for covalent immobilization of biomolecules. ACS Appl Mater Interfaces 2:2884–2891

Pack SP, Kamisetty NK, Nonogawa M, Devarayapalli KC, Ohtani K, Yamada K, Yoshida Y, Kodaki T, Makino K (2007) Direct immobilization of DNA oligomers onto the amine-functionalized glass surface for DNA microarray fabrication through the activation-free reaction of oxanine. Nucleic Acids Res 35:1–10

Qin M, Hou S, Wang L, Feng X, Wang R, Yang Y, Wang C, Yu L, Shao B, Qiao M (2007) Two methods for glass surface modification and their application in protein immobilization. Colloids Surf B 60:243–249

Rebeski DE, Winger EM, Shin YK, Lelenta M, Robinson MM, Varecka R, Crowther JR (1999) Identification of unacceptable background caused by non-specific protein adsorption to the plastic surface of 96-well immunoassay plates using a standardized enzyme-linked immunosorbent assay procedure. J Immunol Methods 226:85–92

Wang W, Vaughn MW (2008) Morphology and amine accessibility of (3-aminopropyl) triethoxysilane films on glass surfaces. Scanning 30:65–77

Yang Y, Hori M, Hayakawa T, Nogami M (2005) Self-assembled 3-dimensional arrays of Au@SiO2 core-shell nanoparticles for enhanced optical nonlinearities. Surf Sci 579:215–224

Young JK, Lewinski NA, Langsner RJ, Kennedy LC, Satyanarayan A, Nammalvar V, Lin AY, Drezek RA (2011) Size-controlled synthesis of monodispersed gold nanoparticles via carbon monoxide gas reduction. Nanoscale Res Lett 6:4281–42811

21

Influences of carbon nanofillers on mechanical performance of epoxy resin polymer

Shraddha Singh · V. K. Srivastava · Rajiv Prakash

Abstract The influence of multi-walled carbon nanotubes (MWCNTs) and graphene nanoplatelets (GnPs) on epoxy resin was investigated to compare their mechanical properties. MWCNT/epoxy resin and GnP/epoxy resin composites were compared with each other for their tensile strength, compressive strength, Charpy Impact and Izod impact energy with the variation of weight percentage ratio of nanofiller ranging from 0.5, 1.0, 2.0 and 3.0, respectively. The result shows that GnP/epoxy resin composite gave better tensile and compressive strength compared to MWCNT/epoxy resin composite whereas Izod impact energy, Charpy impact energy and dynamic fracture toughness of MWCNT/epoxy resin composite resulted in better impact resistance than the GnP/epoxy resin composite. Thermal stability and microstructural properties of composites were measured using Thermogravimetric analysis (TGA), transmission electron microscope (TEM) and scanning electron microscope (SEM).

Keywords Carbon · Composites · Mechanical properties · Thermogravimetric analysis · Scanning electron microscopy

Introduction

Carbon-reinforced epoxy resin composites are often used in aircraft. Because of their high specific mechanical properties, they are mainly used as structural components. Epoxy resin is sometimes used as a matrix material in these composites. Graphene nanoplatelets (GnPs) and multi-walled carbon nanotubes (MWCNTs) are new carbon materials that have recently been developed. Graphene has generated much more interest due to its high specific area and novel mechanical, electrical and thermal properties. (Higginbotham et al. 2010; Stankovich et al. 2006; Jiao et al. 2009; Kosynkin et al. 2009; Li et al. 2008; Wang et al. 2009; Rafiee et al. 2010; Han et al. 2007) combined high electrical conductivity with record of thermal conductivity and mechanical properties. Used as additives, they can impart their properties to the materials and coatings they are added to. They have countless potential applications, some of which we cannot even begin to imagine today (Chen et al. 2007; Iijima 1991; Sinnott and Andrews 2001).

Carbon nanotubes are the most promising 1D material for nanotechnology, optical, electronic and composite material applications. Advances in carbon nanotube growth technique led to considerable increases in both production rates. Nanotube synthesis by CVD process was chosen because it offers a promising method to bulk production of high-purity nanotube that can be carried to commercialization. Useful composite research requires a bulk supply of nanotube of high purity and in a useable (i.e. easily dispersible) form. Existing technologies for the production of single-walled carbon nanotube (SWCNT) do not yield sufficient quantities and lack the required purity. Purification of these materials is often tedious, low in yield, and damaging to the tubes' structure through oxidative shortening. For applications such as conductive filler or as reinforcing fibre, MWCNT is likely to be preferred over SWCNT on a cost basis. Also, technologies developed for MWCNT applications can be directly transferred to SWCNT composites should SWCNT synthesis reach industrial scale (Gojny et al. 2005).

S. Singh · R. Prakash
School of Materials Science and Technology, Indian Institute of Technology, (BHU), Varanasi 221005, India

V. K. Srivastava (✉)
Department of Mechanical Engineering, Indian Institute of Technology, (BHU), Varanasi 221005, India
e-mail: vijayks210@gmail.com

One of the most promising applications of MWCNT and GnP materials is in polymer composite, which incorporates nano-scale filler materials. Polymer composites show substantial property enhancements at much lower loadings than polymer composites with conventional microscale fillers (such as glass or carbon fibre), which ultimately results in lower component weight and can simplify processing; moreover, the multifunctional property enhancements made possible with composites may create new applications of polymers (Dassios et al. 2012; King et al. 2013; Singh et al. 2013).

The main objective of the present work was to compare the mechanical properties of GnP/epoxy resin and MWCNT/epoxy resin composites with the variation of weight percentage ratio of nanofillers ranging from 0.5, 1.0, 2.0 and 3.0.

Experimentation

Materials and specimen preparation

Graphene nanoplatelet and multi-walled carbon nanotube were used as filler materials. Graphene nanoplatelet was obtained from J. K. Impex Company, Mumbai with 99.5 % purities. Multi-walled carbon nanotube was available from TUHH, Harburg, Hamburg, Germany. Epoxy resin was obtained from Ciba-Geigy Ltd., Mumbai. The material properties are listed in Table 1.

First, araldite (LY-556) 55 %, hardener (HY-917) 49 % and accelerator (DY-070) 0.28 % were used as epoxy resin matrix. MWCNT filler was used as reinforcement in epoxy resin with weight percentage variation of MWCNT filler ranging from 0.5, 1.0, 2.0 and 3.0. MWCNT particle was manually mixed with the resin (LY-556). MWCNT filled resin was dispersed using a lab-scale three-roll-mill (Exakt 120 EXAKT Advanced Technology GmbH, Germany), which enables the introduction of very high shear forces (up to 200,000/s) throughout the suspension. The curing time 20 min is allowed to cure the specimen. The pre-dispersed suspension was then given batchwise onto the roll with dwell times of 2 min. The dispersive forces on the suspension were acting in the gap (5 μm) between the rolls. After dispersion of the MWCNT in the resin LY-556, the hardener and accelerator were usually added in a vacuum dissolver, to avoid trapped air in the suspension. Then the mixture was placed in a vacuum chamber for 20 min to eliminate the bubbles introduced during the rolling process. The dispersed mixtures of MWCNT and resin were subsequently diluted by adding an appropriate amount of hardener (HY-917) in the weight ratio of 10:1. Again, MWCNT/epoxy resin mixture was dispersed by the same method and air bubble removed completely from the mixture before curing.

Table 1 Material properties

Material	Geometry	Elastic modulus, GPa
MWCNT 99.98 %	Length = ±2 μm Average inner diameter = ±6.5 nm Average outer diameter = ±40 nm	1,000.0
Epoxy resin	–	2.0
Graphene nanoplatelet 99.5 %	Thickness = 8–10 nm, Diameter = 5.25 μm,	Elastic modulus = 1,006.0

In this paper, we were considered to differentiate between mechanical properties of MWCNT/epoxy resin composite and GnP/epoxy resin composite with the same percentage of fillers. For this objective, GnP composite was manually mixed with the resin and then mixer was placed in the mechanical stirrer at the speed of 1 m/min for 1 h. After proper dispersion of graphene nanoplatelet in the resin, hardener was added to the weight ratio of 10:1. Again, mixture of graphene nanoplatelet (weight percentage variation of ranging from 0.5, 1.0, 2.0 and 3.0) added to the resin and hardener was gently mixed with the stirrer and allowed removing the air bubble from the mixture. Finally, GnP/epoxy resin mixture was placed in vacuum for 20 min to escape out the bubbles introduced during the mixing process. The dispersed mixtures were poured into the mould and allowed curing the mixture for 24 h. After curing, tensile, compressive, Izod and Charpy impact test specimens were prepared as per the ASTM standard for the measurement of mechanical properties.

Mechanical tests

Mechanical properties of the MWCNT/epoxy resin and GnP/epoxy resin composites were measured under tensile, compression, Izod and Charpy impact tests. The shape and size of the specimens were prepared according to the ASTM standard (Dassios et al. 2012). A dog bone-shaped tensile specimen size of 165 × 5.3 × 3 mm was used. The compression specimen size (25 × 7 × 6.5 mm) was used in short length to avoid the buckling effect. These tests were performed on the universal testing machine (UTM-5T; SC Deys & Company, Calcutta) with the cross-head speed of 0.05 mm/min. Load versus displacement results were used to measure elastic modulus under the tensile and compressive loading conditions. At least five specimens were tested for each sample with the variation of 10 %. The tensile strength, compressive strength and elastic modulus are reported on average.

The impact test was performed with instrumented Izod and Charpy equipment (Model; Resil Impactor-50, CEAST, S. p. A., Italy). The impact hammer and vice lever with specimen adapter were used different in Izod and Charpy impact tests. The impact length and impact velocity were 0.327 m and 3.46 m/s. Izod and Charpy impact tests were performed on notched and un-notched sandwich specimens. The dimensions of Izod impact test were length 66 mm, width 20 mm, thickness 18 mm. Charpy impact test dimensions are length 123 mm, width 20 mm and thickness 18 mm. A V-notch of depth 2 mm was introduced in Izod and Charpy impact test specimens.

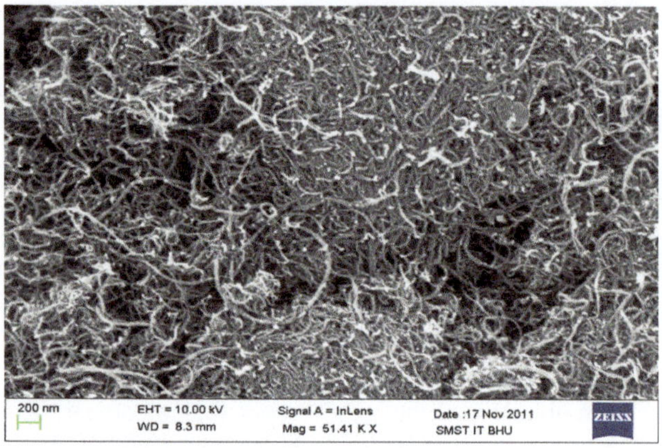

Fig. 2 SEM image of MWCNT particles

Microstructure observation and surface analysis

The morphologies of the fractured specimens of graphene nanoplatelet/epoxy resin and MWCNT/epoxy resin composites were observed by scanning electron microscope (High resolution SEM SUPRA 40, 5 kV, Zeiss, Germany). TEM (Jeol JEM-2100, HR) was carried out to characterize the structure of nanotubes. To prepare TEM samples, some alcohol was dropped on the nanotubes film, then, these films were transferred with a pair of tweezers to a carbon-coated copper grid.

Results and discussion

The SEM and TEM images of nanofiller are presented in Figs. 1, 2, 3. Figures 1, 2 clearly indicate that the nanofillers are well dispersed within the epoxy resin. It is obvious, from the images that all the nanofillers are hollow and tubular in shape as shown in Fig. 3a. In some of the images, catalyst particles can be seen inside the nanotubes. TEM images indicate that the nanotubes are high purity, with uniform diameter distribution and contain no deformity in the structure. While Fig. 3b shows the GnP particles interact well with each other at magnification of 20 nm. This indicates that the GnP particles are dispersed uniformly within the epoxy resin.

Fig. 1 SEM image of graphene nanoplatelet particles

To measure the influence of MWCNT and GnP nanofiller on mechanical properties of epoxy resin matrix, the samples with specific size of composite were prepared according to the requirements of tensile, compressive, Izod impact and Charpy impact tests. The tensile strength and compressive strength were obtained based on the maximum load. Figures 4, 5 show that the tensile strength and compressive strength increase with increase of weight percentage of nanofillers. The tensile strength and compressive strength of GnP/epoxy resin composite specimen give better strength compared to neat epoxy resin and MWCNT/epoxy resin composite specimens. However, the proper dispersion of nanofillers has great effect on the mechanical properties of the epoxy resin. The tensile strength and compressive strength of epoxy resin were increased about 20 and 31 % with the addition of GnP filler, whereas tensile strength and compressive strength of epoxy resin were increased about 18 and 19 % with the addition of MWCNT filler. These results clearly indicate that the nanofiller improve the tensile and compressive strength of epoxy resin. The results are parallel to other reported works (Dassios et al. 2012; King et al. 2013). It is now well known that the mechanical properties of nanofiller-filled polymer matrix, especially the strength and modulus depend to a great extent on filler dispersion and interfacial interaction. Modulus measurement by nanohardness followed the trends reported in the literature for polymers, that is, the modulus obtained was higher than the results obtained from macroscopic testing (King et al. 2013; Singh et al. 2013).

Izod impact and Charpy impact test results are shown in Figs. 6, 7. Also, dynamic fracture toughness (α_k) was calculated using the following equation (Gojny et al. 2005),

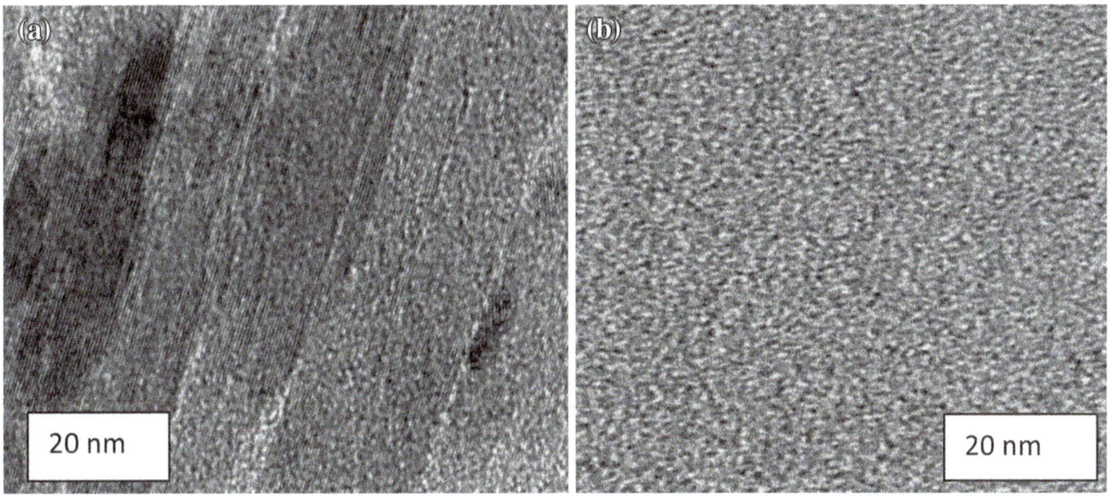

Fig. 3 TEM micrograph shows that interact with each other **a** MWCNT particles, **b** GnP particles

Fig. 4 Variation of tensile strength with weight percentage of nanofillers

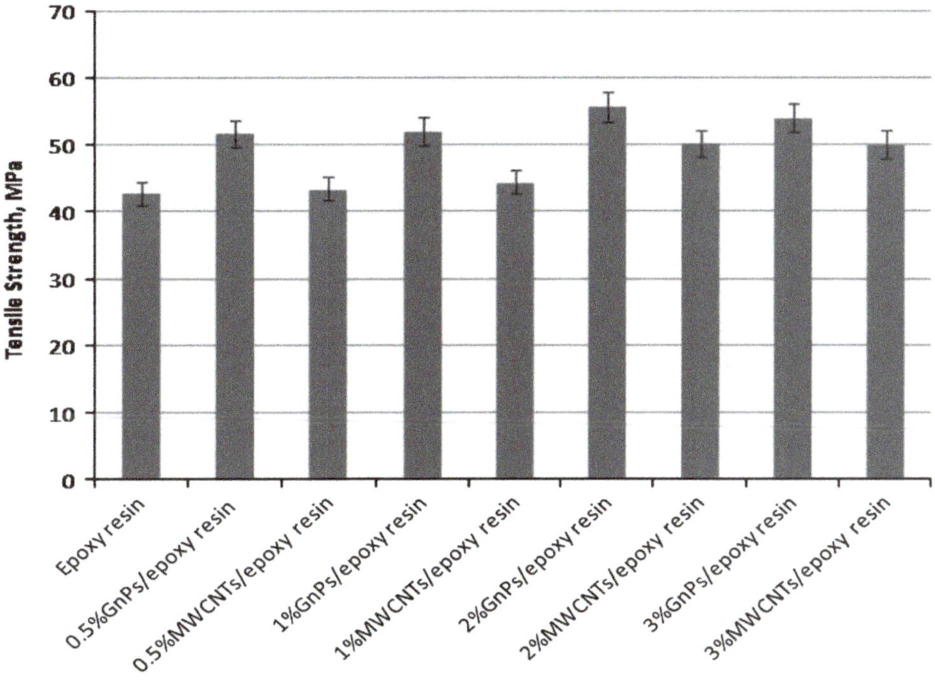

$$\alpha_k = \Delta E/wh, \quad (1)$$

where ΔE is the absorbing energy of material during impact processing, w and h are the width and thickness of specimen, respectively.

The impact tests result shows that the MWCNT/epoxy resin composite gave higher Izod impact energy, Charpy impact energy, dynamic fracture toughness (Izod) and dynamic fracture toughness (Charpy) compared to GnP/epoxy resin composite and neat epoxy resin specimen as indicated in Figs. 6, 7, 8, 9.

Figures 6, 7 indicate the variation of Izod impact energy and Charpy impact energy with the percentage of filler ratio. The higher impact resistance clearly represents that the multi-walled nanotube absorbed more energy and it is dissipated within the area of nanotubes and take more time to penetrate the more area of the specimen (Singh et al. 2013). But, in the case of graphene nanoplatelets, impact energy spread easily all over the specimen because graphene nanoplatelets dispersed in the plate form. The reduced in absorbed energy may be caused by stress concentration in the vicinity of the graphene nanoplatelet (Chen et al. 2007); this typically occurs when hard fillers are incorporated into a brittle matrix. Besides, agglomeration of MWCNT/epoxy resin or GnP/epoxy resin

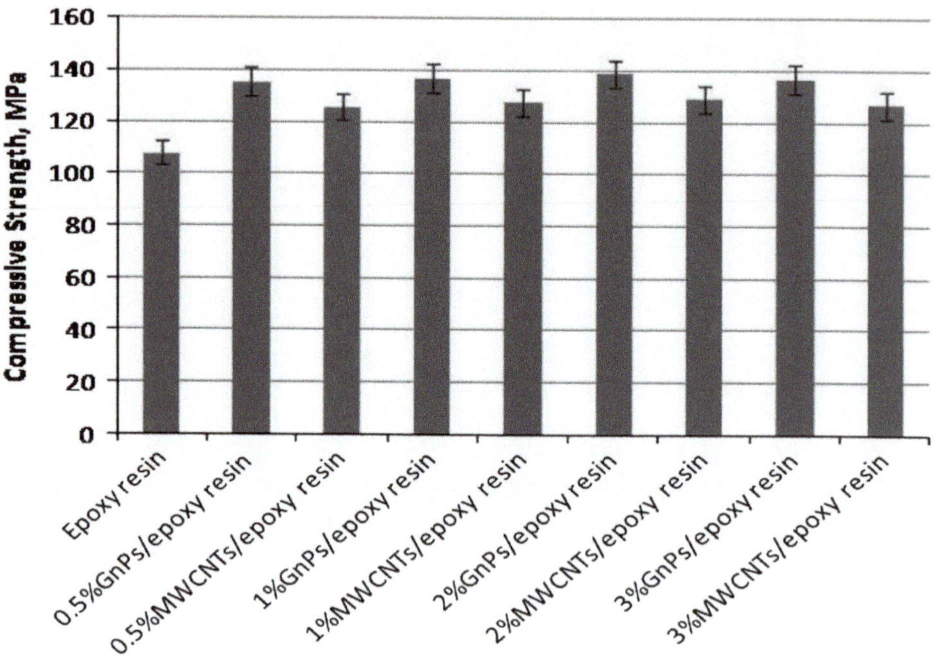

Fig. 5 Variation of compressive strength with weight percentage of nanofillers

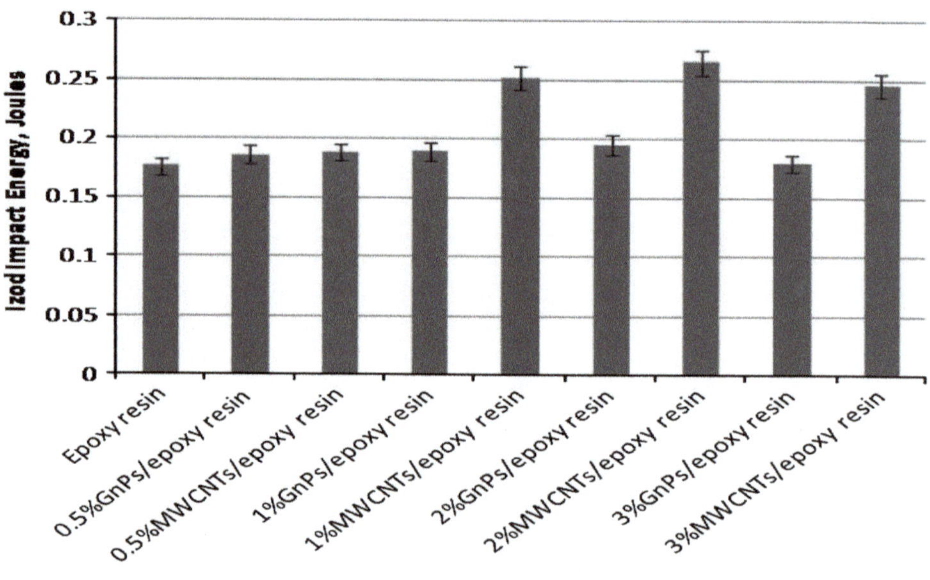

Fig. 6 Variation of Izod impact energy with weight percentage of nanofillers

composites can lead to defects in the matrix that can act as seed points for crack initiation and premature fracture (Dassios et al. 2012; King et al. 2013).

Figures 8, 9 show the variation in dynamic fracture toughness under Izod and Charpy impact tests, which indicates the material toughness of neat epoxy resin, MWCNT/epoxy resin and GnP/epoxy resin composites. MWCNT/epoxy resin composite shows higher dynamic fracture toughness than the neat epoxy resin and GnP/epoxy resin composite. This result is evident that the lower weight fractions of nanofillers begin to agglomerate in epoxy resin. Figure 10 shows that MWCNT particle is well dispersed in epoxy resin, which indicates that the three mill rolling process is most suitable for dispersion of MWCNT in epoxy resin (Sinnott and Andrews 2001).

However, GnP nanofiller is not fully dispersed because of stirrer mixing process, as shown in Fig. 11. This highlights the need for continued research to develop new methods to enhance GnP dispersion at higher weight fraction (Han et al. 2007).

To understand the underlying mechanisms that are responsible for the improved performance of GnPs, we compared the thermal and microstructure chemistry of GnPs/epoxy resin and MWCNT/epoxy resin by TGA and

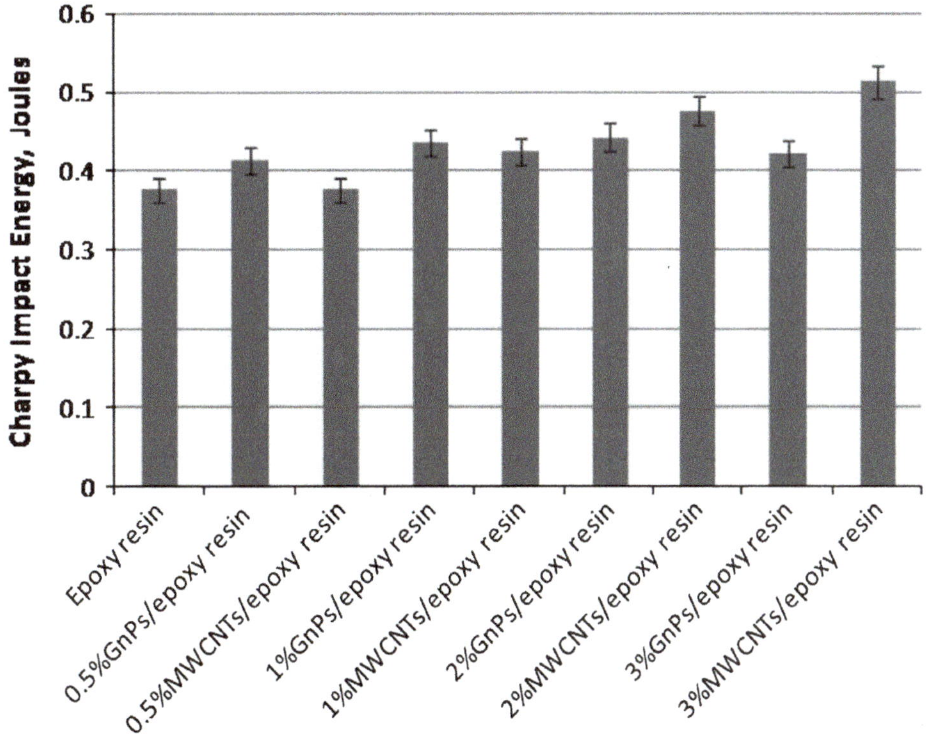

Fig. 7 Variation of Charpy impact energy with weight percentage of nanofillers

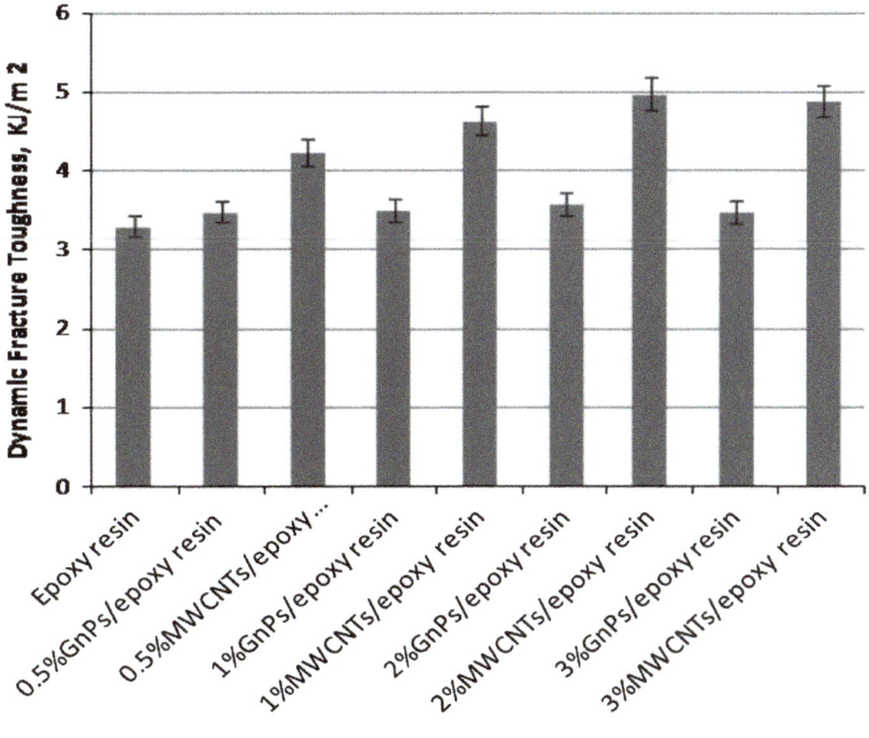

Fig. 8 Variation of dynamic fracture toughness with weight percentage of nanofillers under Izod impact energy

SEM analysis. We find that the MWCNT and GnP filler can be affected with the temperature due to variation in weight loss, as shown in Fig. 12. It is observed from the Fig. 12 that the nanocomposite made with MWCNT/epoxy resin can increase the initial decomposing temperature than the GnP-filled epoxy resin composite.

This means the percentage weight loss reduces abruptly with increase of temperature. MWCNT/epoxy resin composite could significantly enhance interfacial interaction between the MWCNT and epoxy resin matrix in composites (Singh et al. 2013). Where, GnP/epoxy resin composite reduces the stability of the composite at higher

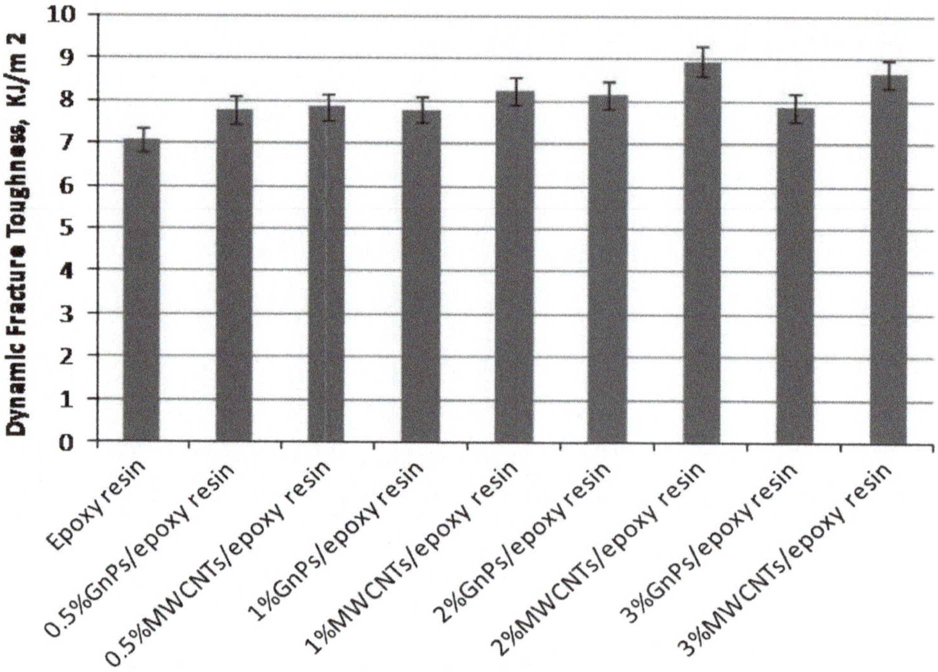

Fig. 9 Variation of dynamic fracture toughness with weight percentage of nanofillers under Charpy impact energy

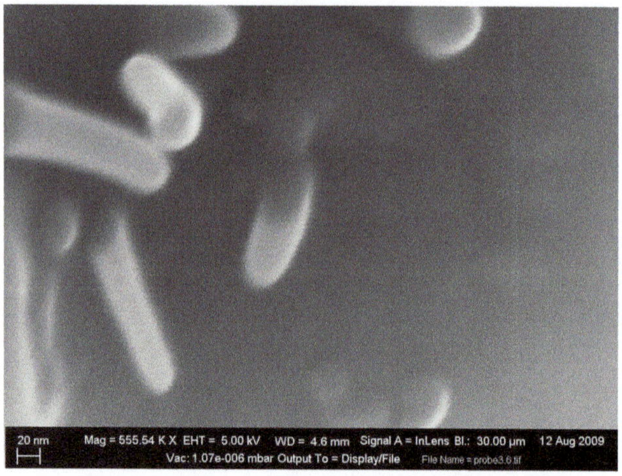

Fig. 10 SEM micrographs showing the dispersion of MWCNTs in epoxy resin

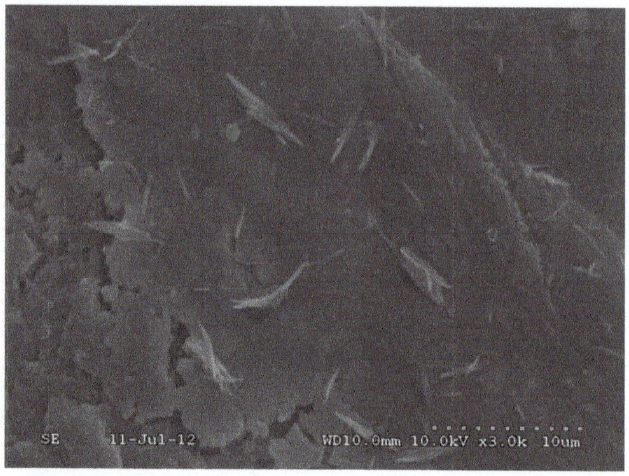

Fig. 11 SEM micrographs showing the dispersion of GnPs in epoxy resin

temperature, because GnP molecules can be retarded under high temperature. Therefore, dispersion of nanofillers can be improving the thermal stability of nanocomposites (Iijima 1991). But a noticeable increase in decomposition temperature was observed with incorporation of 2 % MWCNT and GnP filler in epoxy resin. This was solely attributed to the synergistic effect of the composite structure formed between MWCNT and epoxy resin, as shown in Fig. 10. The increase through the synergistic effect was more than the composite containing GnPs (0.5 %), but considerably smaller compared to the composite containing 2 % MWCNT. The likely possibility for this low increase compared to a binary system containing 2 % MWCNT may be due to the effective low concentration of GnPs (0.5 %) in the epoxy resin. Another possible reason may be due to the fact that the high thermal conductivity of the MWCNT/epoxy resin composites resulted in a greater flow of heat throughout the sample. GnP/epoxy resin and MWCNT/epoxy resin composites started to lose weight as temperature increases continuously (Dassios et al. 2012) as indicated in Fig. 12.

Figure 13 shows SEM images of the fracture surfaces of GnP/epoxy resin composites. These composite fails in catastrophic mode. Figure 13a, b shows the fracture surface of the polymer composites made with GnPs, this clearly indicate that the epoxy resin fractured as a catastrophic pattern, because epoxy resin becomes more brittle due to addition of nanofillers. This shows the fracture surface of

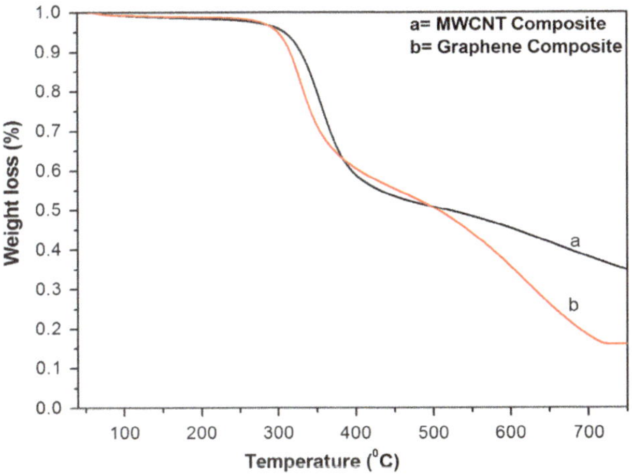

Fig. 12 Variation of weight loss with temperature

composite made with graphene nanoplatelet. Micrographs indicate that the cracks were propagated at different planes and continued all around the specimen. Once cracks started, they spread quickly along the crack direction and then developed into macroscopic failure. Also, dispersion and agglomeration graphene nanoplatelet were clearly identified from the Fig. 13c, d. In most cases, the crack damage started from the interface, and then nanofillers suffered from the external force and were pulled out, leaving the smooth resin matrix which exposed the weak interfacial bonding. It was discovered that there were many CNT existing in the surrounding agglomerates, which might have been caused by the shear force around the nanofillers during the fabrication process, or by poor dispersion (Singh et al. 2013). However, the morphology of fractured sur-

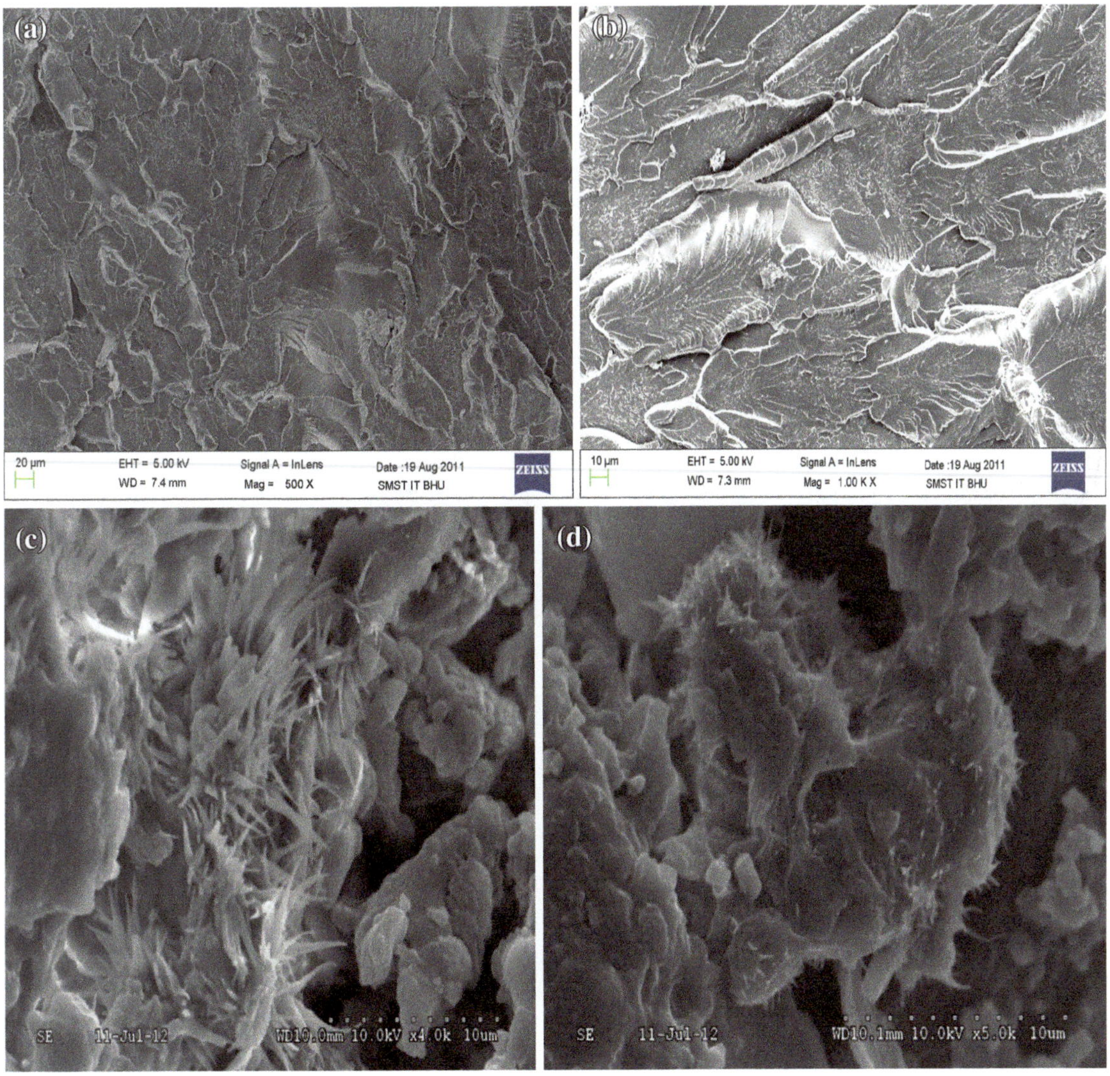

Fig. 13 SEM micrograph showing **a** microcracks formed in the resin area and **b** cracks formed around the graphene platelets, **c** graphene nanoplatelets agglomerated in the epoxy resin and **d** graphene nanoplatelets sticked with resin-like mushroom

faces depicts the intimate contact and high embedding with polymer matrices, indicating good interfacial bonding between nanofillers and the epoxy resin. This further confirms that the presence of hydroxyl groups on graphene nanoplatelet surfaces has a strong interaction with the epoxy resin and significantly improves adhesion at interfaces. In general, nanoplatelets can be easily attracted to each other due to their very high specific area and high surface energy, due to these properties, tensile and compressive strength of GnP/epoxy resin composite give more value compared to MWCNT/epoxy resin composite (King et al. 2013).

Conclusion

Based on the experimental observation, the results clearly indicate that the GnP and MWCNT nanofiller can improve the performance of epoxy resin. The tensile strength and compressive strength of GnP/epoxy resin composite are higher than those of the MWCNT/epoxy resin composite with the variation of weight percentage of nanofiller ranging from 0.5, 1, 2 and 3. Whereas, MWCNT/epoxy resin composite gives higher impact resistance compared to GnP/epoxy resin composite at the same variation of nanofiller. The results of this study suggest that MWCNT plays an important role of absorbing the impact resistance because nanotube dampens the more energy compared to graphene nanoplatelet. On the other side, tensile strength and compressive strength of GnPs/epoxy resin increase due to more surface area and significant increase in the interfacial contact area of graphene nanoplatelet compared to multi-walled carbon nanotube. The GnPs are more effective than MWCNT, which contributes to better interfacial binding. These results indicate that GnPs show significant potential as a structural reinforcement additive in polymer-based composite materials.

Acknowledgments The authors are thankful to School of Materials Science and Technology, Department of Mechanical Engineering, Indian Institute of Technology, (BHU), Varanasi, India and Department of Science and Technology, New Delhi, India for their supports.

References

Chen Z, Lin YM, Rooks MJ, Avouris P (2007) Graphene nanoribbons electronics. Phys E 40:228–232

Dassios KG, Musso S, Galiotis C (2012) Compressive behaviour of MWCNT/epoxy composite mats. Compos Sci Technol 72:1027–1033

Gojny FH, Wichmann MHG, Fiedler B, Schulte K (2005) Influence of different carbon nanotubes on the mechanical properties of epoxy matrix composites—a comparative study. Compos Sci Technol 65(15–16):2300–2313

Han MY, Oezyilmaz B, Zhang Y, Kim P (2007) Energy band-gap engineering of graphene nanoribbons. Phys Rev Lett 98:206805

Higginbotham AL, Kosynkin DV, Sinitskii A, Sun Z, Tour JM (2010) Low defect graphene oxide nanoribbons from multiwalled carbon nanotubes. ACS Nano 4:2059–2069

Iijima S (1991) Helical microtubules of graphitic carbon. Nature 354:56–58

Jiao L, Zhang L, Wang X, Diankov G, Dai H (2009) Graphene nanoribbons from carbon nanotubes. Nature 458:877–880

King JA, Klimek DR, Miskioglu I, Odegard GM (2013) Mechanical properties of graphene nanoplatelets/epoxy composites. J Appl Polym Sci 128:4217–4223

Kosynkin DV, Higginbotham AL, Sinitskii A, Lomeda JR, Dimiev A, Price BK, Tour JM (2009) Longitudinal unzipping of carbon nanotubes from graphene nanotubes. Nature 458:872–876

Li X, Wang X, Zhang L, Lee S, Dai H (2008) Chemically derived ultrasmooth graphene nanoribbon semiconductors. Science 319:1229–1232

Rafiee MA, Lu W, Thomas AV, Zandiatashbar A, Rafiee J, Tour JM, Koratkar NK (2010) Graphene nanoribbon composites. ACS Nano 4(12):7415–7420

Singh S, Srivastava VK, Prakash R (2013) Characterization of multi-walled carbon nanotube reinforced epoxy resin composites. Mater Sci Technol 29(9):1130–1134

Sinnott SB, Andrews R (2001) Carbon nanotubes: synthesis, properties, and applications. Crit Rev Solid State Mater Sci 26:145–249

Stankovich S, Dikin DA, Dommett GHB, Kohlhaas KM, Zinney EJ, Stach EA, Piner RD, Nguyen ST, Ruoff RS (2006) Graphene-based composite materials. Nature 442:282–286

Wang X, Li X, Zhang L, Yoon Y, Weber PK, Wang H, Guo J, Dai H (2009) N-dopping of graphene through electrothermal reactions with ammonia. Science 324:768–771

22

Preparation and electrochemical characterization of manganese dioxide-zirconia nanorods

R. R. Muthuchudarkodi · C. Vedhi

Abstract MnO_2–ZrO_2 nanorods were prepared by wet chemical method by mixing the solutions of $MnSO_4$ and $ZrOCl_2$ varying in the range (0.05–0.45 M) in aqueous NaOH at an elevated temperature. The morphologies of the synthesized products are characterized by scanning electron microscopy and transmission electron microscopy (TEM). X-ray diffraction (XRD) and energy-dispersive spectroscopic measurements were also employed for the characterization of the nanostructures. The synthesized nanoparticles were also characterized by ultraviolet visible spectroscopy, Fourier transform infrared spectroscopy, electrochemical impedance and cyclic voltammetric studies. The morphological studies of the nanoparticles revealed particle distribution with uniform rod-like structure. Energy-dispersive analysis indicated the presence of Mn, Zr and O. The nanostructures of the product were characterized by TEM studies and the mixed rod and granular structure that was found clearly indicated the presence of MnO_2–ZrO_2 mixed oxide. The size of the synthesized nanorod was found to be 20 nm. From XRD studies the size of the nanorods was found to be in the range 39–56 nm calculated by Debye–Scherrer's formula. Thermal stability of the nanorods was characterized by thermogravimetric and differential scanning colorimetric analysis. Cyclic voltammetric studies exhibit good adherent behavior on electrode surface and good electroactivity at a pH value of 1.0.

Keywords MnO_2–ZrO_2 nanorods · Impedance · SEM · TEM and cyclic voltammetry

R. R. Muthuchudarkodi · C. Vedhi (✉)
Department of Chemistry, V.O Chidambaram College,
Tuticorin 628008, Tamilnadu, India
e-mail: cvedhi23@gmail.com; cvedhi@rediffmail.com

R. R. Muthuchudarkodi
e-mail: muthu.rajaram@gmail.com

Introduction

One-dimensional (1D) nanostructured materials, including nanotubes, nanorods, and nanowires, have attracted intense research interest, owing to their novel physical and chemical properties (Chen et al. 2005). Various methods have been developed to prepare various 1D nanomaterials, such as arc discharge, laser ablation, chemical vapor deposition, templating, and hydrothermal methods, etc. MnO_2 is one of the most attractive materials due to its ion exchange, molecular adsorption, catalytic, electrochemical, and magnetic properties (Chen et al. 2005). They are widely used as catalysts (Chen et al. 2005; Brock et al. 1998; Shi et al. 2010) molecular-sieves, ion-sieves (Feng et al. 1999) and especially as electrode materials in Li/MnO_2 batteries because of its energetic compatibility in a reversible lithium electrochemical system, eco-friendliness, and low cost. The properties of MnO_2 depend not only on the manganese oxidation state, but also on the structure type of MnO_2 crystal. Great effort has been made to prepare nanocrystalline MnO_2 with different structures. Over the past few years, the synthesis of nanostructures with controllable size and shape has been increased attracting attention (Dong et al. 2006). Manganese compounds, such as various manganese oxides/oxyhydroxides have been noticed owing to their specialities. Manganese oxyhydroxides, MnOOH, are of considerable importance in many technological applications, e.g., electrochemical reaction, battery and electrochromic application (Dong et al. 2006). Mn_3O_4 is known to be an active catalyst in several oxidations and reductions, and can be used as a catalyst for the oxidation of methane and carbon monoxide of the selective reduction of nitrobenzene (Dong et al. 2006). Nanorods have potential applications in nanodevices (Xia et al. 2003; Lu and Lieber 2006). Manganese oxides have a wide range of applications such as battery materials (Hosono et al. 2009), catalysts and sensors. At present, a tremendous amount of comprehensive investigations are under way into the unique applications of nanorods and nanotubes, because they provide a great opportunity to investigate the dependence of optical/electrical properties, thermal transport, and mechanical performance on the nano-scaled dimensionality and size (Wang 2000). Nanomaterials, especially nanorods have enhanced performances in many fields such as sensors (Dietl et al. 2000), solar cells (Yu 2009) catalysts and battery materials. Of the various non-noble metals or transition-metal oxides studied, MnO_2 enjoys a place of pride because of its cost-effectiveness and eco-friendly nature. Beyond these advantageous properties, MnO_2 is a very promising material in a neutral electrolyte system. In the case of pseudo capacitors, various noble and transition-metal oxides such as RuO_2, IrO_2, NiO, CoOx, SnO_2 and MnO_2 were used as electrode materials (Subramanian et al. 2008). Zirconium dioxide is the most studied ceramic material. Pure ZrO_2 has a monoclinic structure at room temperature and transitions to tetragonal and cubic at increasing temperature. Zirconia supports are of particular interest as they can possess relatively high surface areas; they are mechanically and thermally stable and are solids with adsorbent properties, as well as catalysts themselves. Wambach et al. (1999) have recently reviewed the preparation of metal–zirconia catalysts along with their structural and chemical characteristics. Stabilized zirconia is used in oxygen sensors, fuel cell membranes and electroceramics. It will be used as insulators in transistors in future nanoelectronic devices. In this study an attempt has been made to synthesize and characterize novel, mixed MnO_2–ZrO_2 nanorods.

Experiment

Materials

The precursors zirconiumoxychloride ($ZrOCl_2$), manganese sulfate ($MnSO_4$) and the precipitant (NaOH) were purchased from Aldrich. All solutions were made up with deionised water.

Preparation of simple metal oxide nanoparticles

Fifty milliliters of 0.1 M $MnSO_4$ was added dropwise to an aqueous solution of NaOH (50 ml, 1.0 M), making a final volume of 100 ml. The mixture was stirred well and refluxed at an elevated temperature for 2.0 h. The sample was collected by centrifugation, washed with water and dried over for 4 days at room temperature. Similar procedure was carried out for the preparation of ZrO_2 nanoparticles using $ZrOCl_2$ as the precursor.

Preparation of MnO_2–ZrO_2 nanorods

MnO_2–ZrO_2 nanorod was prepared by wet chemical method. In this method, 25 ml of 0.45 M $ZrOCl_2$ was added to the aqueous solution of 50 ml of 1.0 M NaOH solution and stirred well. To this mixture, 25 ml of 0.45 M $MnSO_4$ was added making a final volume of 100 ml. The resulting mixture was stirred well and refluxed at an elevated temperature for 2.0 h. The sample was collected by centrifugation, washed with water and dried over for 4 days at room temperature. Similar procedure was carried out to prepare different concentrations of (0.05–0.35 M) MnO_2–ZrO_2 mixed oxides.

Characterizations

The solution of the metal oxide nanoparticles in DMSO was used for recording the ultraviolet visible spectroscopic (UV–Vis) spectra. For recording the UV–Visible absorption spectra, a computer-controlled JascoV-500 spectrophotometer was used. The FTIR spectra were recorded using a Shimadzu instrument. The X-ray diffraction (XRD) patterns were recorded for the powdered materials using a Bruker AXS (D8 Advance) X-ray diffractometer. Energy-dispersive spectroscopic (EDAX) and SEM measurements were carried out by JEOL JSM-6360F field emission scanning electron microscope. Transmission electron microscopy (TEM) images were recorded using Philips CM 200 model with the operating voltage range 20–200 and with a resolution of 2.4 Å.

The electrochemical studies were carried out in a three-electrode cell. Pt wire was used as a counter electrode, silver–silver chloride electrode as a reference electrode and metal oxide nanoparticle-coated GCE was used as working electrode. The impedance studies were carried out using electrochemical workstation (mode 650C), CH-Instrument Inc., TX, USA. The charge transfer resistance was obtained from the diameter of the semicircles of the Nyquist plots. TGA/DTA analysis was carried out using PerkinElmer model and DSC analysis using METTLER TOLEDO model.

Result and discussion

UV–Vis studies

Ultraviolet visible (UV–Vis) absorbtion spectra of nano MnO_2, nano ZrO_2 and mixed MnO_2–ZrO_2 were recorded in the range 200–800 nm as shown in (Fig. 1). The absorption

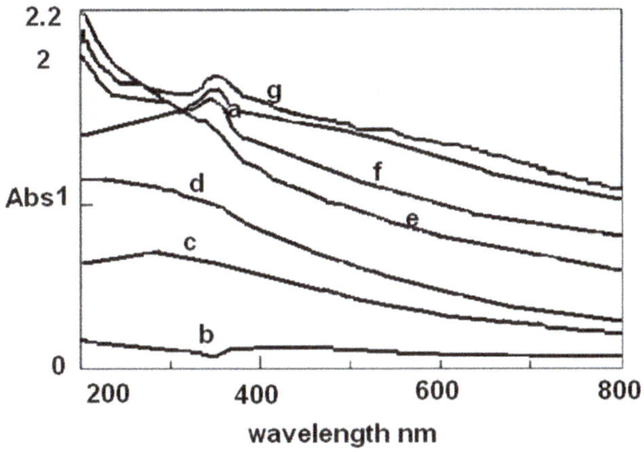

Fig. 1 UV-VIS spectra of (*a*) nano MnO_2, (*b*) nano ZrO_2, (*c*) 0.05 M MnO_2–ZrO_2, (*d*) 0.15 M MnO_2–ZrO_2, (*e*) 0.25 M MnO_2–ZrO_2, (*f*) 0.35 M MnO_2–ZrO_2, (*g*) 0.45 M MnO_2–ZrO_2

wavelength appears at 348 nm for nano MnO_2 (Luo et al. 2004) and 335 nm for nano ZrO_2. The absorption peaks for different concentrations (0.05–0.45 M) of MnO_2–ZrO_2 mixed oxides have been found to be in the range 281–348 nm, and are shown in Fig. 1. The variations in the absorption peaks for simple and mixed oxides are due to the smaller size of the nanoparticles (Singh et al. 2012). The mixed oxide nanoparticles exhibited blue shift compared with that of the simple oxides. The blue shift of the absorption peaks of the metal oxide nanoparticles result from certain unique effects of nanomaterials such as nanoscale effect and the blue shift reduces absorption of longer wavelength and is thus undesirable for UV protection (Yang et al. 2004). The absorption peaks of metal oxide nanoparticles appeared at a shorter wavelength region. Thus the metal oxide nanoparticles are found to be

Fig. 2 FTIR Spectra of (a) nano ZrO_2, b nano MnO_2, c 0.05 M MnO_2–ZrO_2, d 0.15 M MnO_2–ZrO_2, e 0.25 M MnO_2–ZrO_2, f 0.35 M MnO_2–ZrO_2, g 0.45 M MnO_2–ZrO_2

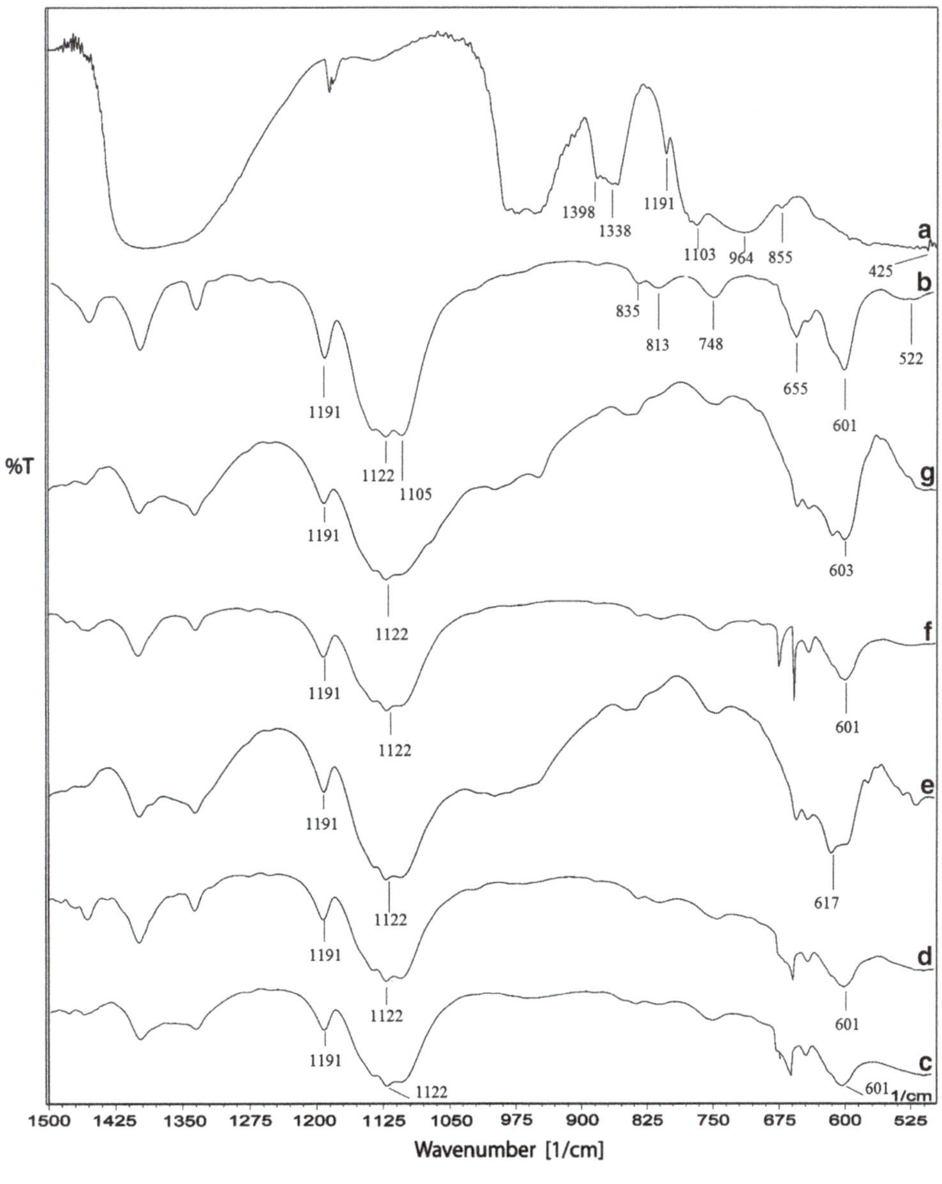

solar UV blockers (Yang et al. 2004). With the increase in concentrations of nano MnO_2–ZrO_2 mixed oxides there appeared increase in absorption values.

FTIR studies

Metal oxides generally give absorption bands below 1,000 cm^{-1} that arise from interatomic vibrations given in Fig. 2. The frequencies observed at 511–954 and 425–964 cm^{-1} correspond to Mn–O (Rema Devi et al. 2007) and Zr–O (Ṕerez-Maquedaa and Matijevicb 1997) bond vibrations, respectively. The peaks observed at 1,336 and 1,338 cm^{-1} are assigned to O_2 stretching frequency (Guedes et al. 2009). Symmetric frequencies of Mn–O and Zr–O were observed at 1,191 and 1,103 cm^{-1} (Guedes et al. 2009; Du et al. 2009). The peak observed at 3,401 cm^{-1} corresponds to O–H vibration of water. The characteristic peaks observed at 1,122 and 1,103 cm^{-1} are due to the presence of inorganic ions (Rema Devi et al. 2007).

Fourier transform infrared spectroscopic (FTIR) spectra of MnO_2–ZrO_2 mixed oxides synthesized at five different concentrations of both $MnSO_4$ and $ZrOCl_2$ (0.05, 0.15, 0.25, 0.35 and 0.45) are shown in Fig. 2c–g. For mixed oxide nanorods (0.45 M) MnO_2–ZrO_2, the combination of both Mn–O and Zr–O bonds appear in the range 511–999 cm^{-1} Fig. 2g. The frequencies observed at 1,336 and 1,396 cm^{-1} are assigned to O_2 stretching and bending frequencies, respectively (Partha Sarathi and Thilagavathi 2011). The characteristic peak at 1,122 cm^{-1} clearly indicates the presence of inorganic ions (Du et al. 2009). In the case of other concentrations of (0.05–0.35 M) MnO_2–ZrO_2 mixed oxides, the bands were almost similar to 0.45 M MnO_2–ZrO_2.

SEM and EDAX behaviors of nano MnO_2, nano ZrO_2 and MnO_2–ZrO_2

Scanning electron microscopy (SEM) was used to identify the morphology of the synthesized metal oxides and mixed metal oxides MnO_2, ZrO_2 and MnO_2–ZrO_2. From Fig. 3a, b the prepared MnO_2 and ZrO_2 display sponge-like structure, granular flakes and mixed granular appearance. When $MnSO_4$ and $ZrOCl_2$ were mixed, the surface morphologies and roughness of the particles are changed to rod-like structure and become homogeneous which confirmed the formation of MnO_2–ZrO_2 nanoparticles. SEM micrographs of nano MnO_2–ZrO_2 synthesized at five different concentrations of both $MnSO_4$ and $ZrOCl_2$ (0.05, 0.15, 0.25, 0.35, and 0.45) are shown in Fig. 3c–g. Particles synthesized in 0.05 M of $MnSO_4$ and $ZrOCl_2$ appear to be very uniform in spherical morphology. In the case of 0.15 M of $MnSO_4$ and $ZrOCl_2$ and 0.25 M of $MnSO_4$ and $ZrOCl_2$, the particles were modified from spherical into rod-like morphology. However, the nano MnO_2–ZrO_2 mixed oxides synthesized in the 0.35 M of $MnSO_4$ and $ZrOCl_2$; 0.45 M of $MnSO_4$ and $ZrOCl_2$ were completely changed into nanorods. By choosing the adequate concentration we would be able to form nanorods. EDAX analysis confirms the presence of Mn, Zr and O as shown in Fig. 3.

TEM behaviors of nano MnO_2, nano ZrO_2 and MnO_2–ZrO_2

Transmission electron microscopy (TEM)'s image of MnO_2 nanoparticles shows the size of the nanoparticles to be 20 nm and Fig. 4a shows the size of the ZrO_2 nanoparticles to be 100 nm. The morphology of the synthesized nano ZrO2 also appeared rod-like and that of nano MnO2, sphere-like (Fig. 4b). The morphology of mixed nano metal oxide (0.45 M of MnO_2–ZrO_2) was found to be a uniform rod and the size is observed to be 20 nm as shown in Fig. 4c.

XRD behaviors of nano MnO_2, nano ZrO_2 and MnO_2–ZrO_2

The X-ray powder diffraction patterns for the simple and mixed oxide nanoparticles as in Fig. 5 are typical of crystalline nanoparticles. The particle size was calculated using Scherrer's equation. In the case of nano MnO_2 the particle size was 45 nm and in the case of nano ZrO_2 the particle size was 8.3 nm. The average crystallite sizes of nanorods, MnO_2–ZrO_2 is found to be in the range of 39–56 nm. With the increase in concentrations of MnO_2–ZrO_2 from 0.05 to 0.45 M there is corresponding increase in size. The sharp peaks indicate more crystallinity and hence it shows more conductivity because high crystallinity materials usually show higher conductivity.

Thermal analysis

Figure 6 shows the thermograms of nano ZrO_2, nano MnO_2 and different concentrations of mixed MnO_2–ZrO_2 oxides. Figure 6a shows the TGA/DTA behavior of MnO_2. The first weight loss step from 42 to 90 °C corresponds to loss of moisture. The next step from 91 to 580 °C was due to the presence of extra bounded water molecules. The final weight loss step from 581 °C onwards corresponds to the degradation of MnO_2. The first weight loss step from 30 to 100 °C (Fig. 6b) corresponds to loss of moisture. The next step from 101 to 170 °C was due to the presence of extra bounded water molecules. The final weight loss step from 170 °C onwards corresponds to the degradation of ZrO_2. The DTA analysis also exhibited the same behavior.

Thermal stability of nano MnO_2–ZrO_2 mixed oxides synthesized at five different concentrations of both $MnSO_4$

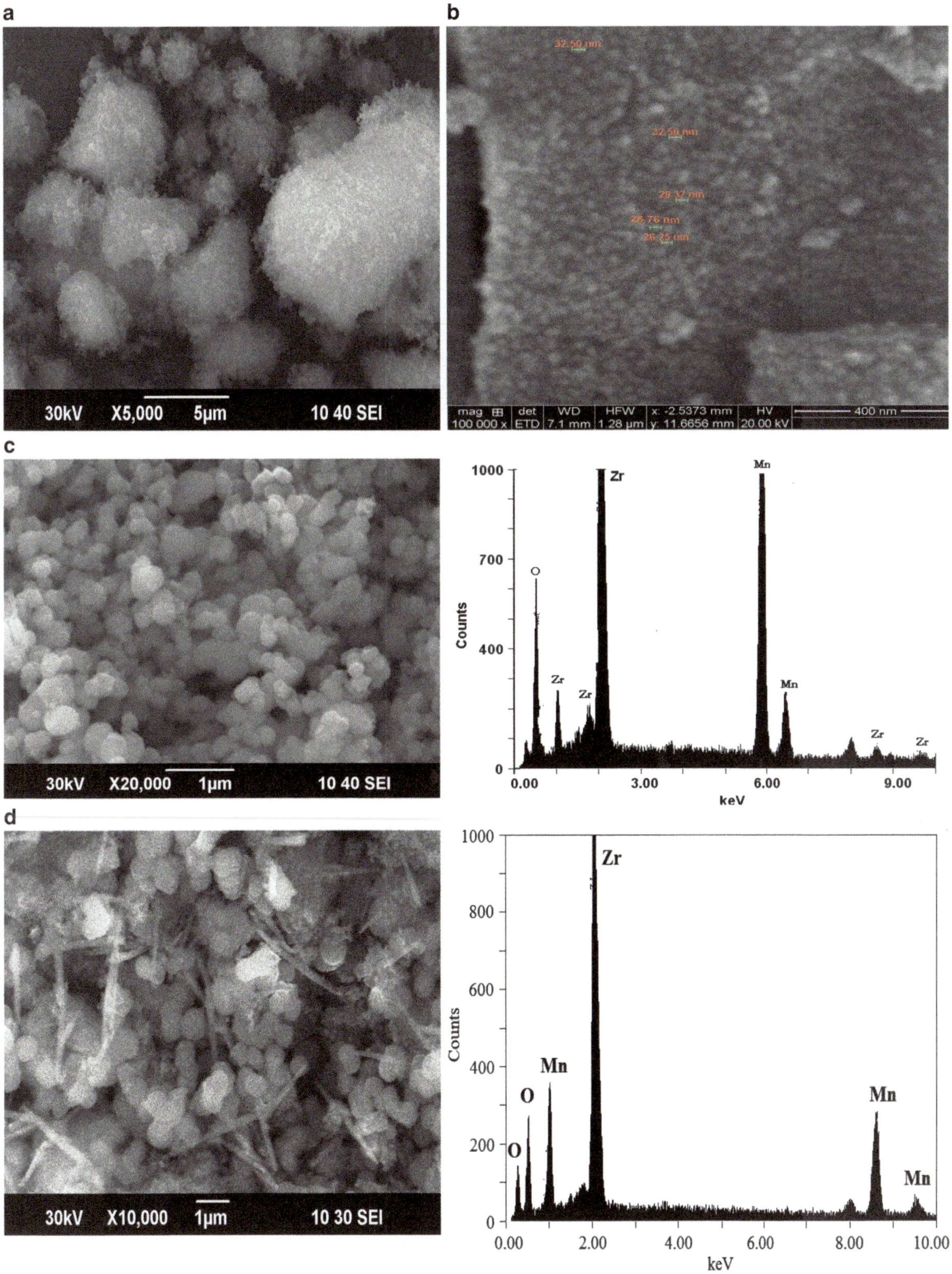

Fig. 3 SEM behavior of (**a**) nano MnO_2, **b** nano ZrO_2 and SEM image EDAX pattern of (**c**) MnO_2–ZrO_2, **d** MnO_2–ZrO_2, **e** MnO_2–ZrO_2, **f** MnO_2–ZrO_2, **g** 0.45 M MnO_2–ZrO_2

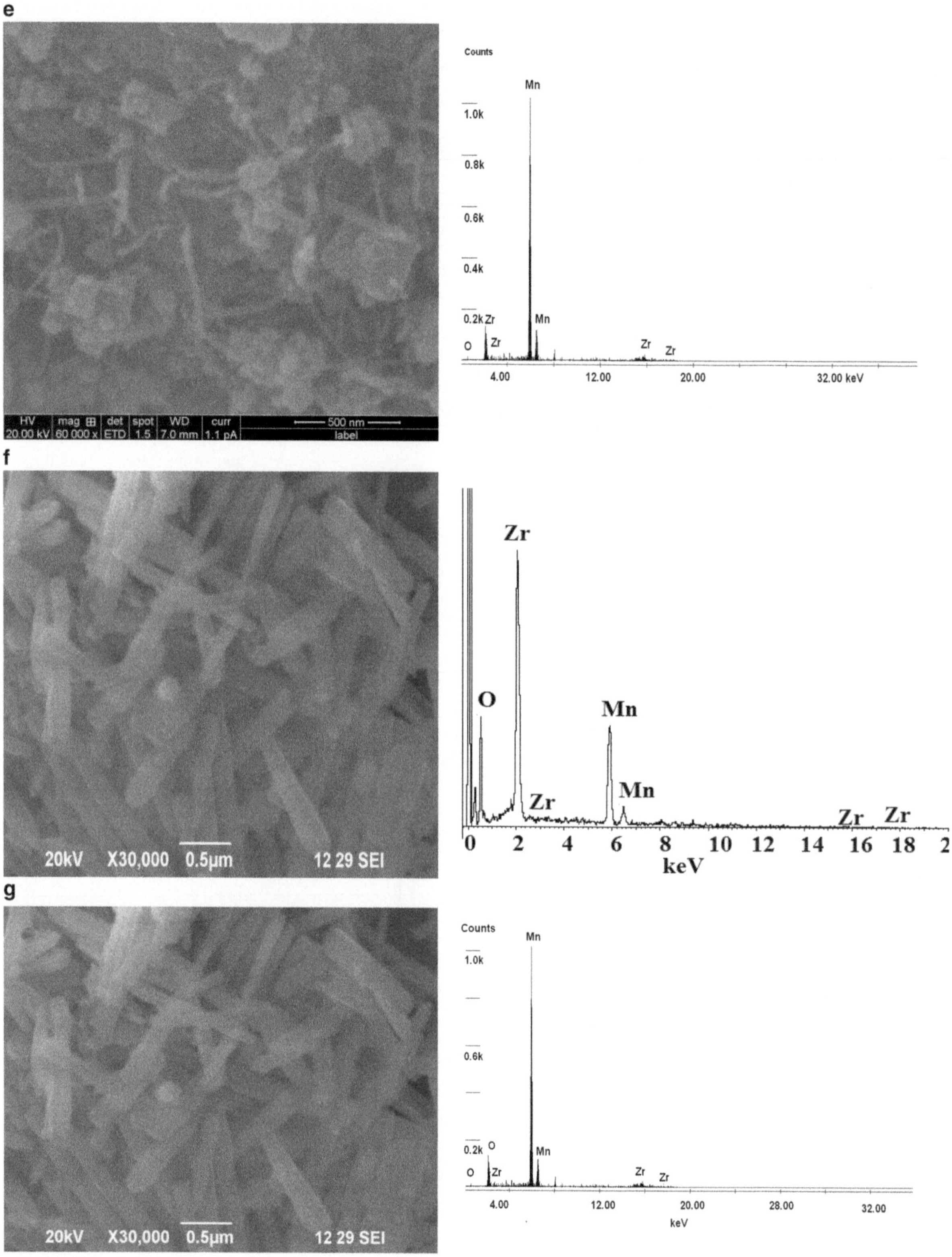

Fig. 3 continued

and $ZrOCl_2$ (0.05, 0.15, 0.25, 0.35, 0.45 M) is characterized by TGA/DTA analysis and the results are shown in Table 1. In these curves the first weight loss step from 42 to 170 °C corresponds to the loss of moisture. The next weight loss step from 91 to 390 °C corresponds to extra bounded water molecules. The final weight loss step from

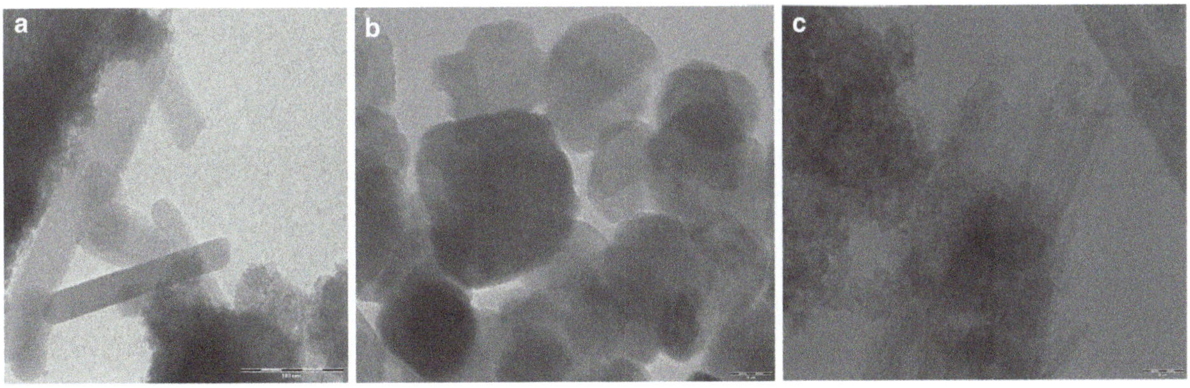

Fig. 4 TEM image of (**a**) nano ZrO_2, **b** MnO_2 nanoparticles, **c** MnO_2–ZrO_2 (0.45 M) nanorod

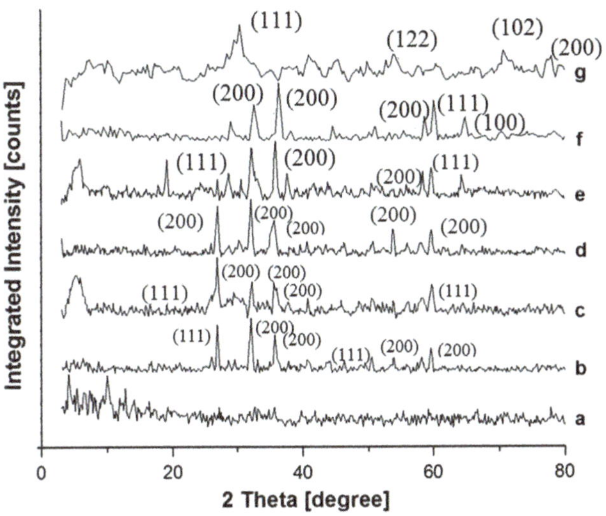

Fig. 5 XRD Pattern of (*a*) nano 0.05 M MnO_2–ZrO_2, (*b*) 0.15 M MnO_2–ZrO_2, (*c*) MnO_2–ZrO_2, (*d*) MnO_2–ZrO_2, (*e*) MnO_2–ZrO_2, (*f*) nano MnO_2, (*g*) nano ZrO_2

281 to 321 °C corresponds to the degradation of mixed oxides (MnO_2–ZrO_2). The weight loss values at the critical temperatures such as 280, 300, 320 and 390 °C reveal that the type of simple oxides employed affects the thermal properties of the resulting mixed oxides. The thermogram of 0.05 M MnO_2–ZrO_2 is shown in Fig. 6c.

DSC thermogram of MnO_2, ZrO_2 and different concentrations of mixed MnO_2–ZrO_2 are recorded at the heating rate 10 °C min^{-1}. The glass transition temperature (T_g) crystallization temperature (T_C) and melting point (T_M) of the mixed oxide are determined from the DSC curves (Fig. 7a) shows the melting point (T_M) of MnO_2 at a temperature of 237 °C and the (T_C) value at 220 °C. The melting point (T_M) of ZrO_2 was recorded at a temperature of 95 °C, the T_C value recorded at 10 °C and the T_g value recorded at −70 °C. These results can be seen in Fig. 7b. The T_C value of MnO_2–ZrO_2 mixed oxide is in the range of

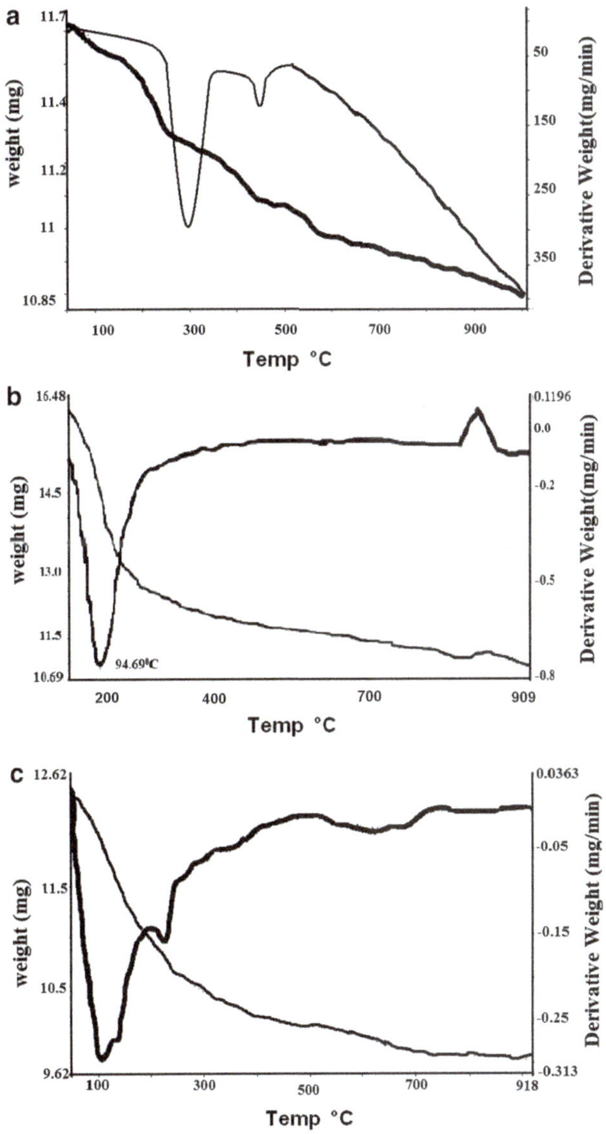

Fig. 6 TGA/DTA Curves of (**a**) nano MnO_2, **b** nano ZrO_2, **c** 0.05 M MnO_2–ZrO_2

Table 1 TGA analysis of nano ZrO_2, nano MnO_2 and MnO_2–ZrO_2 mixed oxides

Nanoparticles	Temperature due to loss of moisture (°C)	Temperature due to loss of water (°C)	Temperature due to degradation of mixed oxides (°C)
ZrO_2	30–100	101–170	From 171 onwards
MnO_2	42–90	91–580	From 581 onwards
0.05 M MnO_2–ZrO_2	55–170	171–390	From 391 onwards
0.15 M MnO_2–ZrO_2	42–100	101–320	From 321 onwards
0.25 M MnO_2–ZrO_2	45–180	181–320	From 321 onwards
0.35 M MnO_2–ZrO_2	42–90	91–300	From 301 onwards
0.45 M MnO_2–ZrO_2	45–120	121–280	From 281 onwards

29.93–36.35 °C. The mixed oxide was melted at the temperature range of 76.24–93.11 °C. According to DSC the melting temperature T_M of different concentrations of mixed oxide was decreased remarkably than the simple oxides. Table 2 shows the T_C and T_M values of 0.05–0.45 M MnO_2–ZrO_2 mixed oxides. DSC thermogram of 0.05 M MnO_2–ZrO_2 and is shown in Fig. 7c.

Cyclic voltammetric behavior of nano MnO_2, nano ZrO_2 and MnO_2–ZrO_2

Cyclic voltammetric studies of ZrO_2 exhibited one oxidation peak at 0.179 V with higher peak current at pH 1.0 which led to the selection of pH at 1.0 as optimum pH for further voltammetric studies. Cyclic voltammetric behavior of MnO_2 showed one anodic peak (Fig. 8a) at 0.415 V which is due to the presence of MnO_2 and it is reduced at 0.352 V. Cyclic voltammetric behavior of ZrO_2 showed one oxidation peak (Fig. 8b) observed at 0.179 V, which indicated the formation of ZrO_2, whereas for different concentrations of mixed oxides, two oxidation peaks were observed in the range 0.127–0.3238, 0.6,535–0.6571 V (Fig. 9), respectively, which were entirely different from the behavior, those obtained from nano MnO_2 and nano ZrO_2, confirmed the formation of mixed nano MnO_2–ZrO_2 oxide. Cyclic voltammetric behavior of nano MnO_2–ZrO_2 mixed oxides synthesized at five different concentrations of both $MnSO_4$ and $ZrOCl_2$ (0.05, 0.15, 0.25, 0.35, 0.45 M) are shown in Fig. 9a–e. With the increase in concentrations (0.05–0.45 M) of nano MnO_2–ZrO_2

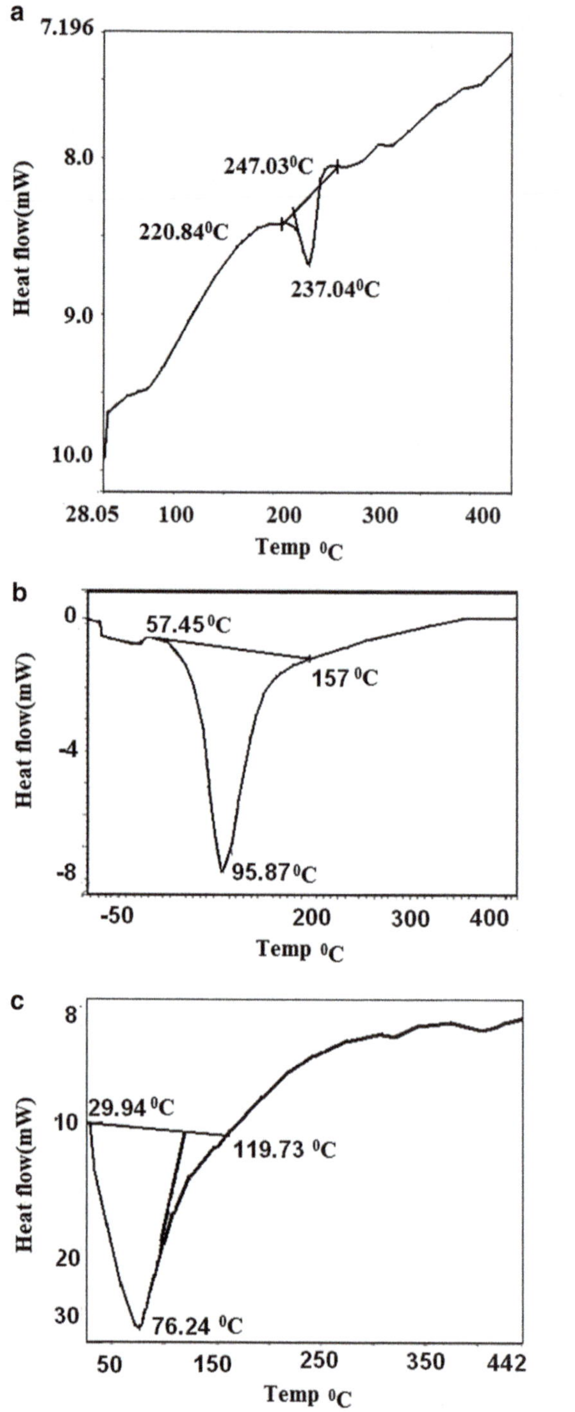

Fig. 7 DSC Thermograms of (**a**) nano MnO_2, **b** nano ZrO_2, **c** 0.05 M MnO_2–ZrO_2

there is corresponding increase in peak potentials and peak currents also.

The plot of peak current versus different scan rate for nano MnO_2–ZrO_2 mixed oxide exhibits a straight line (Fig. 10) indicating a good adherent behavior on electrode surface. When peak currents of nano MnO_2–ZrO_2 are correlated with the square root of scan rate, a straight line is

Table 2 T_C and T_M for 0.05–0.45 M MnO_2–ZrO_2 mixed oxides

Concentration of MnO_2–ZrO_2 nanoparticles (M)	T_C °C	T_M °C
0.05	29.94	76.24
0.15	36.35	77.20
0.25	32.60	81.90
0.35	29.93	82.41
0.45	34.59	93.11

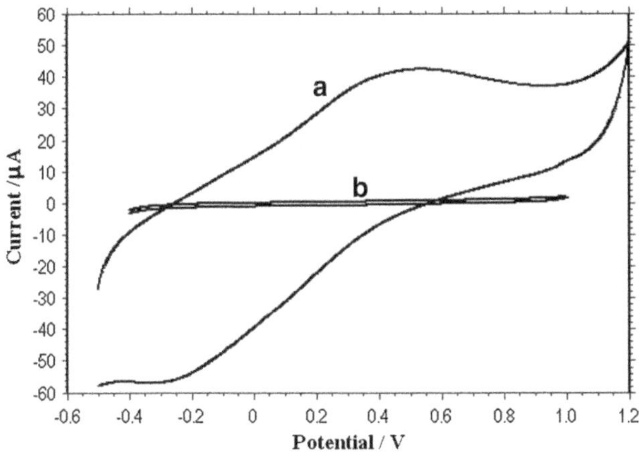

Fig. 8 Cyclic Voltammetric behavior of (a) nano MnO_2, b nano ZrO_2

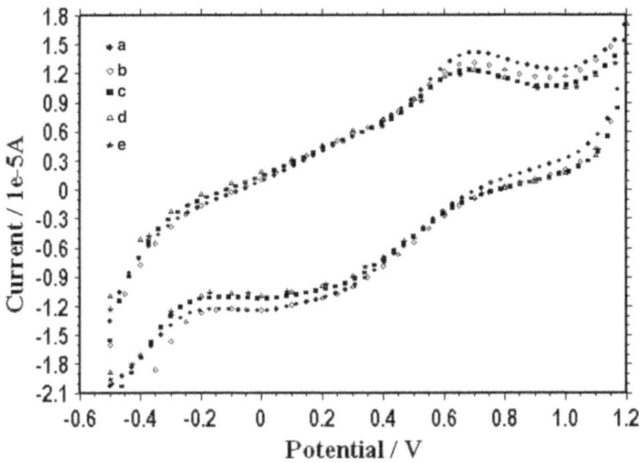

Fig. 9 Cyclic voltammetric behavior of (a) 0.05 M nano MnO_2–ZrO_2, b 0.15 M MnO_2–ZrO_2, c 0.25 M MnO_2–ZrO_2, d 0.35 M MnO_2–ZrO_2, e 0.45 M MnO_2–ZrO_2

observed (Fig. 11). These facts reveal that the voltammetric redox behavior of mixed metal oxide nanoparticles is controlled by adsorption process. Thus it proves to be an anticorrosive agent for paints.

When the metal salts of Mn and Zr were mixed during chemical co-precipitation, the mixed metal oxide nanoparticles were produced. The mixed metal oxide was coated on glassy carbon electrode and the cyclic voltammogram was recorded. The voltammogram also exhibited oxidation and reduction peaks with higher peak current.

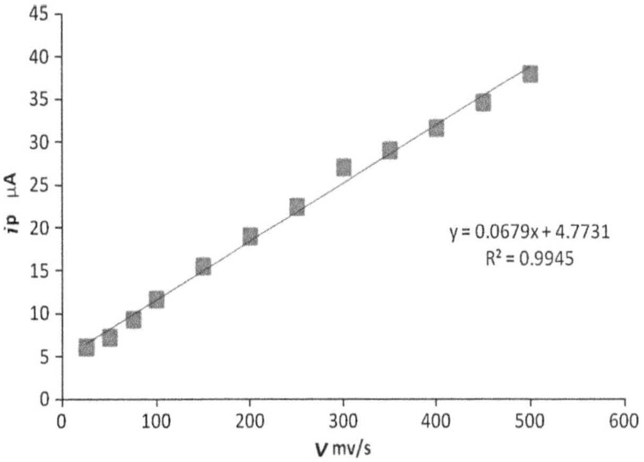

Fig. 10 Plot of peak current versus scan rate for 0.45 M nano MnO_2–ZrO_2

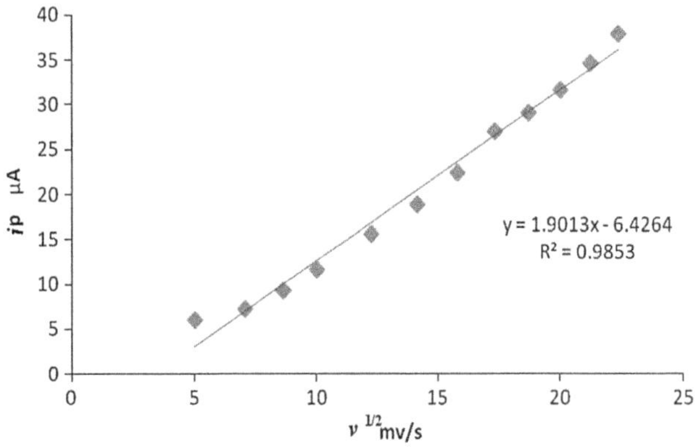

Fig. 11 Plot of peak current versus square root of scan rate for 0.45 M nano MnO_2–ZrO_2

This might be due to the formation of mixed oxide from metal salts of Mn and Zr.

The increase in the peak current of the mixed oxide from that of simple oxides and the increase in the peak current due to increase in the metal concentration confirm the increase in the conductivity of the mixed oxide. To temper the cutting edge of electrochemistry, it will be necessary to carry out capacitance measurements, as these are closer to bulk-like techniques.

For a simple parallel plate capacitor, charge on the capacitor, Q, is proportional to the voltage drop across the capacitor V, and C is the capacitance. Therefore the equation is

$$Q = CV$$

Capacitance is a crucial factor in electrochemical experiments because it gives rise to current during the charging of the capacitor. To calculate the magnitude of this current, with respect to time (t) and capacitance should be treated as constant recognizing that dQ/dt is an expression for current and dV/dt is the potential scan rate. The equation is

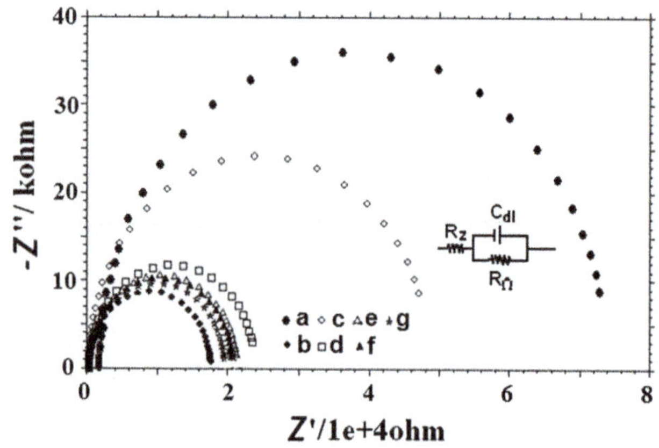

Fig. 12 Electrochemical impedance spectra of (a) nano MnO_2, b nano ZrO_2, c 0.05 M MnO_2–ZrO_2, d 0.15 M MnO_2–ZrO_2, e 0.25 M MnO_2–ZrO_2, f 0.35 M MnO_2–ZrO_2, g 0.45 M MnO_2–ZrO_2

$$dQ/dT = CdV/dt$$

Therefore $i = Cv$

From this very simple derivation, we have an expression for the charging current at the steady state when applying a ramping voltage. If there are no possibilities for electron transfer between the solution and the electrode this is the only current that we will observe. The cyclic voltammogram of mixed oxide shows the influence of various scan rates. The capacitance was calculated from cyclic voltammograms using the above said equation. The calculated values of MnO_2, ZrO_2 and 0.05 M MnO_2–ZrO_2 mixed oxide are 13, 30 and 39 μF respectively. The capacitance of chemically synthesized mixed oxide is suggested that, they can well be used as an electronic material.

EIS studies of nano MnO_2, nano ZrO_2 and MnO_2–ZrO_2

Electrochemical impedance measurements were carried out at a frequency range say from 1,000 to 0.01 Hz at open circuit potential. The simple equivalent Randle circuit for studies is given as an insert in (Fig. 12). The chemically synthesized simple oxide and mixed oxide nanoparticles coated on glassy carbon electrode are used (0.0314 cm^2) as working electrode. The cell is composed by a 1.0 cm^2 Pt electrode, Ag/AgCl serves as reference electrode. The measurements were made for simple oxides such as nano MnO_2, nano ZrO_2 (Fig. 12a, b) and nano MnO_2–ZrO_2 mixed oxides synthesized at five different concentrations of both $MnSO_4$ and $ZrOCl_2$ (0.05, 0.15, 0.25, 0.35, 0.45 M) as shown in Fig. 12c–g. To understand the electrical properties of the electrode/interfaces clearly the Randle's equivalent circuit was chosen to fit the obtained impedance data. In Randle's circuit, it is assumed that the resistance to charge transfer (R_{ct}) and the diffusion impedance (R_{ct}) are parallel to the interfacial capacity (C_{dl}). The parallel combination of R_{ct} and C_{dl} gives rise to a semicircle in the complex plane plot of Z'' against Z', the semicircle diameter equals charge transfer resistance (R_{ct}). This resistance exhibits the electron transfer kinetics of the redox probe at the electrode interface. From the Table 3, it is clear that the decrease in electrical resistivity with the increase in concentration is due to the improvement in crystallite and/or grain size, decrease in defects. This fact confirms the anticorrosive activity of the synthesized nano MnO_2–ZrO_2 mixed oxide samples. The lower the concentration of MnO_2–ZrO_2 mixed oxide nanoparticles, higher the resistance which is responsible for anti corrosive activity. The small observed variations might be from the nature of electrode conductivities (Chandrasekaran et al. 2008). More resistivity implies a physical adsorption of the corresponding electrodes during the diffusion path way whereas the lower value indicates the good conductivity behavior.

Table 3 R_{ct} and C_{dl} for 0.05–0.45 M MnO_2–ZrO_2 mixed oxides

Concentration of nano MnO_2–ZrO_2 M	R_{ct} (Ω cm^2)	C_{dl} (μF cm^{-2})
0.05	8.393	0.0107
0.15	2.353	0.0769
0.25	2.113	0.1000
0.35	2.053	0.1053
0.45	1.940	0.1205

Conclusion

To summarize, nano MnO_2, ZrO_2 and MnO_2–ZrO_2 were synthesized by wet chemical method. The characterizations of the chemically synthesized nanoparticles were done using UV–Visible spectroscopic and FTIR studies. The sizes of the synthesized oxides were in the nm range and they were found to be thermally stable. SEM and TEM studies show a uniform rod-like morphology in the case of mixed oxide. CV studies revealed that the mixed oxide has good adherent and electrochemical activity on GC and thus it is found to be corrosive protection agent for paints formulation. The capacitance of chemically synthesized mixed oxide was suggested that, they might well be used as an electronic material.

Acknowledgments The authors are extremely grateful to Department of Science and Technology (FAST TRACK and FIST) New Delhi, INDIA for using CHI Electrochemical workstation at V.O.C.College, Tuticorin-8 and Jasco UV–VISIBLE Spectrophotometer.

References

Brock SL, Duan NG, Tian ZR, Giraldo O, Zhou H, Suib SL (1998) A review of porous manganese oxide materials. Chem Mater 10:2619–2628

Chandrasekaran R, Soneda Y, Yamashita J, Kodama M, Hatori H (2008) Effect of poly(3,4-ethylenedioxythiophene) (PEDOT) in carbon-based composite electrodes for electrochemical supercapacitors. J Solid State Electrochem 12:1349–1352

Chen Y, Liu C, Li F, Chen H-M (2005) Preparation of single-crystal α-MnO_2 nanorods and nanoneedles from aqueous solution. J Alloy Compd 397:282–285

Dietl T, Matsukura H, Cibert F, Ferrand J, Zener D (2000) Model description of ferromagnetism in zinc-blende magnetic semiconductors. Science 287:1019–1022

Dong X, Zhang X, Liu B, Wang H, Li Y, Huang Y, Du Z (2006) Synthesis of manganese oxohydroxide (MnOOH) and Mn_3O_4 Nanorods using novel reverse micelles. J Nanosci Nanotechnol 6:818–822

Du F, Liu J, Guo Z, Shape J (2009) Controlled synthesis of CuO and its catalytic application to synthesise amorphous carbon nanofibres. J Mater Res Bull 44:25–29

Feng Q, Kanoh H, Ooi K (1999) Manganese oxide porous crystals. J Mater Chem 9:319–333

Guedes M, Ferreira JMF, Ferro AC (2009) Dispersion of CuO particles in aqueous suspensions containing 4,5-dihydroxy 1,3-benzenedisulphonic acid disodium salt. J Ceram Int 35:1939–1945

Hosono E, Kudo T, Honma I, Matsuda H, Zhou HS (2009) Synthesis of single crystalline spinel $LiMn_2O_4$ nanowires for a lithium ion battery with high power density. Nano Lett 9:1045–1051

Lu W, Lieber CM (2006) Semiconductor nanowires. J Phys D Appl Phys 39:R387–R406

Luo X-L, Xu J-J, Zhao W, Chen H-Y (2004) Cholesterol biosensor based on MWCNTs-MnO_2 Nanoparticle using FFT continuous cyclic voltammetry. Biosens Bioselectron 19:1295–1300

Partha Sarathi V, Thilagavathi G (2011) Synthesis and characterization of zinc oxide nanopartilce and its application on fabrics for microbe resistant defence clothing. Int J Pharm Pharm Sci 3:392–398

Perez-Maquedaa LA, Matijevicb E (1997) Preparation and characterization of nanosized zirconium (hydrous) oxide particles. J Mater Res 12:3286–3292

Rema Devi BS, Raveendren R, Vaidyan AV (2007) Synthesis and characterization of Mn^{2+}-doped ZnS nanoparticles. Pramana J Phys 68:679–687

Shi R, Wang YJ, Li D, Xu J, Zhu YF (2010) Synthesis of $ZnWO_4$ nanorods with [100] orientation and enhanced photocatalytic properties. Appl Catal B Environ 100:173–178

Singh DK, Pandey DK, Yadav RR, Singh D (2012) A study of nanosized zinc oxide and its nanofluid. Pramana J Phys 78:759–766

Subramanian V, Zhu H, Wei B (2008) Nanostructured manganese oxides and their composites with carbon nanotubes as electrode materials for energy storage devices. Pure Appl Chem 80:2327–2343

Wambach J, Baiker A, Wokaun A (1999) CO_2 hydrogenation over metal/zirconia catalysts. Phys Chem Chem Phys 1:5071–5080

Wang ZL (2000) Characterizing the structure and properties of individual wire like nano entities. Adv Mater 12:1295–1298

Xia YN, Yang PD, Sun YG, Wu XY, Mayers B, Gates B, Yin YD, Kim F, Yan HQ (2003) One-dimensional nanostructures: synthesis, characterization and applications. Adv Mater 15:353–389

Yang H, Zhu S, Pan N (2004) Studying the mechanisms of titanium dioxide as ultraviolet-blocking additive for films and fabrics by an improved scheme. J Appl Polym Sci 92:3201–3210

Yu KH, Chen JH (2009) Enhancing solar cell efficiencies through 1-D nanostructures. Nanoscale Res Lett 4:1–10

23

Synthesis and characterization of nickel–diamond nanocomposite layers

C. Gheorghies · D. E. Rusu · A. Bund ·
S. Condurache-Bota · L. P. Georgescu

Abstract Electrodeposition can be used in order to prepare nanocomposite coatings made of nanoparticles dispersed into a metallic matrix, onto a metallic substrate. These kinds of coatings can combine the properties of the matrix and of the dispersed nanoparticles, which can lead to some interesting properties of the mixture. Including diamond nanoparticles into a metallic matrix can take advantage of the diamond properties, especially of its hardness and inertness in many chemical media. Still, the proper control of the electrodeposition parameters is a must in order to obtain the best microhardness and corrosion behavior of such nanocomposite coating. This paper proposes the comparative analysis between Ni/diamond nanocomposite coatings, prepared by electrodeposition, by varying several deposition parameters and the corresponding pure nickel coatings onto the same type of substrate, namely copper. Microstructural and morphological analysis of the two kinds of coatings are also performed and studied, along with microhardness measurements and corrosion resistance tests. It is concluded that specific electrodeposition technique and corresponding parameters must be used in order to prepare those Ni/diamond nanocomposite coatings with the best microhardness and best corrosion resistance.

C. Gheorghies (✉) · S. Condurache-Bota · L. P. Georgescu
Dunarea de Jos University of Galati, 47 Domneasca Street,
800008 Galati, Romania
e-mail: cgheorg@ugal.ro

D. E. Rusu
S.C. Automobile DACIA S.A., (Renault Technologie
Roumanie), UMCD-Department ACU, Mioveni, Arges,
Uzinei Street, No.1-3, 115400 Mioveni, Romania

A. Bund
Fachgebiet Elektrochemie und Galvanotechnik II, Fakultät
für Elektrotechnik und Informationstechnik, Institut für
Werkstofftechnik, Technische Universität Ilmenau,
Ilmenau, Germany

Keywords Electrodepostion · Nickel · Diamond · Nanocomposite · Structure · Corrosion

Introduction

The co-deposition of particles from an electrolytic bath can be used in order to obtain nanocomposite layers within a metallic matrix, which contains inert particles spread as disperse phases (Watson 1993; Suzuki and Asai 1987). Recently, this technique has captured the attention of scientists, considering its interesting possibilities, especially from the point of view of the size of the inert particles, such as: ceramics, polymers and metallic powders, which can reach sizes as small as micro- and even nanometers. The electrodeposition of inert powders implies the initial dipping of the electrode into the plating solution and then, the nanocomposite layers are obtained by applying an electrical current. The dispersed phases incorporated into the nanocomposite layers also include carbide powders, nitrites, boron compounds and ceramic oxides. The crystals sizes, the surface morphology, the preferential orientation and the thickness of the nickel electrodepositions depend both on the electrolysis conditions and on the composition of the electrolyte (Feldstein 1981; Kollia and Spyrellis 1993; Pushpavanam et al. 2007; Chandrasekar and Pushpavanam 2008).

Studies were reported concerning the preparation of diamond nanocomposites, which are already produced at industrial scale by means of controlled detonation, through the plasma jet process and/or by electro-explosive wires.

These coatings can be used for various industrial purposes (Shen et al. 2009; Li et al. 2002, 2007). Moreover, the diamond nanoparticles included into the nickel matrix lead to fine polishing surfaces that can be used in dentistry (Sato et al. 1997). The various applications of such coatings are given by the unique properties of diamond, namely the high hardness, the low friction coefficient, the inertness to some acids and to the attack by other chemical substances (Feldstein and Lancsek 1984; Ranjan et al. 2010).

Experimental details

Pure Ni and Ni/diamond nanocomposite coatings were electrodeposited, according to the conditions given in Table 1. The electrodeposition of the nanocomposites was performed by using a Watts electrolyte, which is largely used in nickel electrodeposition solution (Subhash and Jitendra 2008).

The electrodeposition was done by means of an EG&G 263 A potentiostat/galvanostat, made in USA. Comparative experiments were performed between pure nickel electrodepositions and Ni/diamond nanocomposites in a Watts electrolyte, by employing either the direct current method or the pulse method.

The deposition conditions were chosen by carefully studying the similar works found in the scientific literature and by successive experimental trials, which allowed us to select such a combination of deposition parameters that would be expected to lead to diamond nanocomposite coatings with improved properties. Thus, the concentration of the diamond nanoparticles into the electrolyte was chosen as 5 g/l, as obtained by weighting. This value was also chosen according to Pushpavanam et al. (2007), is being lower than one used by Lee et al. (Ranjan et al. 2010). The diamond nanoparticles had sizes ranging from 70 nm till 80 nm and it were ultrasonically dispersed into the electrolyte for 10 min prior to the electrodeposition. The pH of the electrolyte was kept at 4.0 ± 0.5 [similarly to (Pushpavanam et al. 2007)], being adjusted with 98 % H_2SO_4 and 99.5 % $NaHCO_3$. The electrolyte solution was mechanically stirred, with 250-300 rpm.

The working temperature was 50 ± 1 °C, being controlled by means of a GD1-type of Haake thermostat, with ±1 °C precision. As far as studied in the literature, lower working temperatures were used by other scientists in similar type of experiments [30 °C in (Pushpavanam et al. 2007) and 40 °C in (Ranjan et al. 2010)]. We used such a higher working temperature since at higher temperatures one would expect a more efficient crystal growth and lower stress of the crystalline grains.

Small copper plates embedded into a rigid holder, with 2.26 cm^2 active surface were used as working electrode. The copper plates were vertically centered and they were surrounded by a cylinder-shape electrode, with opposite polarity, made of high-purity (99.99 %) nickel. The saturated calomel electrode (SCE) was used as reference. Prior to the experimental measurements, the copper substrate was mechanically polished with silicon carbide (SiC) abrasive paper, with several granulations (800, 1,200, 2,400 and 4,000 μm, respectively). Then, the polished copper substrate was electrochemically degreased at 3 V and 0.2 A/dm^2, for 20–30 s, into an alkaline solution made of UNAR EL 63 (Schering, Germany). After this, the substrates were activated at room temperature into 1 N hydrochloric acid solution, followed by washing with bi-distilled water and drying with air under pressure.

The diamond nanoparticles within the nanocomposite coatings obtained in nickel matrix were studied for their particular properties. Thus, the morphology, the hardness and the corrosion behavior of the nanocomposite coatings were studied. An X-ray diffractometer type DRON-3, with copper target, operating at the K_α line ($\lambda = 1.5416$ Å) was used for structural analysis, while the morphology was studied by means of a Quanta 200 scanning electron microscope (SEM). The EIS measurements were performed by using a VoltaLab 10 device, comprising a PGZ100 Potentiostat and a VoltaMaster 4 Electrochemical Software.

The influence of the particles

The diamond nanoparticles incorporated into nickel matrix used by electrodeposition had high purity (>96 %) and their distribution within the metallic matrix had a good uniformity.

Figure 1 presents the correlation between the incorporation percentage of the diamond nanoparticles into the Ni

Table 1 Electrodeposition conditions of Ni and Ni/diamond nanocomposite layers

Composition of solutions/nanoparticles	Concentration	Electrodepositionparameters
Nickel sulphate, $NiSO_4 \cdot 6H_2O$	0.9 M	Temperature 40–60 °C
Nickel chloride, $NiCl_2 \cdot 6H_2O$	0.21 M	Current intensity, $i = 1$–10 A/dm^2
Boric acid, H_3BO_3	0.48 M	pH 4
Sodium dodecil sulphate, $C_{12}H_{25}SO_4Na$	0.35 103^{-3} M	Stirring speed: 300 rpm
Diamond nanoparticles ($\phi = 70$–80 nm)	5 g/l	

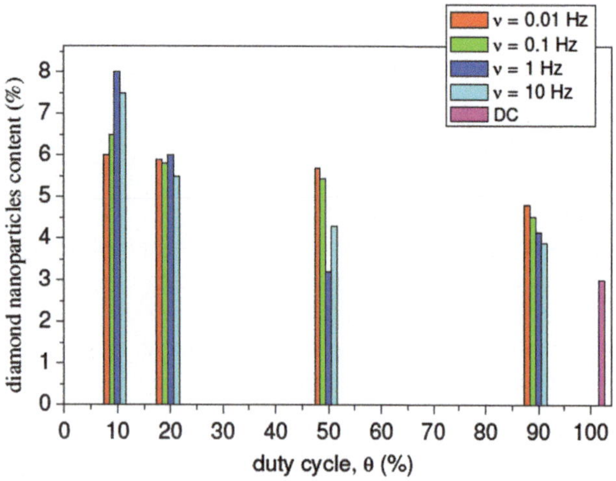

Fig. 1 Correlation between the diamond nanoparticles content within the coatings and the duty cycle, θ and the frequencies, υ of the applied pulses

matrix obtained by direct current (D.C.) and by pulsed current as a function of the duty rates, θ. The selected frequencies, υ of the pulsed current (P.C.) are written in the inset of Fig. 1. From this figure, it can be concluded that the nanocomposite electrodepositions done by P.C. present a higher incorporation percentage into the Ni matrix as compared to the ones obtained by D.C. The highest incorporation percentage was obtained at $\theta = 10$ % and at $\upsilon = 10$ Hz. The data for the incorporation percentage of the nanoparticles into the Ni matrix through D.C. are presented in Fig. 1 at $\theta = 100$ %.

Microstructural characteristics of the Ni/diamond nanocomposite coatings

The investigation of the surface of the nanocomposite coatings, having a thickness up to 5 μm, represents the most important aspect when studying the metallic morphology. The morphology and structure of the Ni/diamond nanocomposite coatings were examined by means of scanning electron microscopy (SEM) and by X-ray diffraction (XRD), respectively. The SEM micrographs of the Ni/diamond nanocomposite coatings as compared to pure Ni coatings, with similar electrodeposition conditions, obtained for several current frequencies, υ and different duty cycles, θ are presented in Fig. 2. These images reveal important structural changes within the nanocomposite coatings as compared to the pure nickel coatings.

Both the structural analysis and the morphology investigation are useful both for showing the aspect of the surfaces of the nickel coatings as compared to the Ni/diamond nanocomposite coatings and also for checking the growing pattern of the electrodeposited crystals. Thus, the uniformity of the nanoparticles included into the nickel matrix is revealed.

The sizes of the grains vary with the deposition conditions. Thus, for P. C. depositions at $\theta = 20$ % and $\upsilon = 0.01$ Hz (see Fig. 2a), the grains range between 0.1 and 3 μm for pure Ni coatings, while they decrease to 0.1–1 μm for Ni/diamond nanocomposites. At $\theta = 50$ % and $\upsilon = 0.1$ Hz (see Fig. 2b), the biggest grains of pure Ni do not exceed 2 μm, while the Ni/diamond nanocomposite deposited in similar conditions present smaller grains, not surpassing 0.5 μm.

One can notice form the SEM images that the dispersion of the diamond nanoparticles was uniform, especially at those surfaces obtained at high frequency of the applied pulses of current: $\upsilon = 1$ Hz (see Fig. 2c). This time, for pure Ni coatings, one can notice a predominance of big grains and fewer small grains, while the Ni/diamond nanocomposite deposited also at $\theta = 90$ % and $\upsilon = 1$ Hz have fragmented grains, reaching not more than 2 μm.

The coatings deposited by direct current present smaller grains (see Fig. 2d), in average than the coatings deposited by pulsed current. The pure Ni coating deposited by D. C. has a more uniform grain size distribution, while the Ni/diamond nanocomposite deposited by D. C. is similar to the one in Fig. 2a. Thus, as expected, after adding diamond to Ni, no more high grains are forming within the nanocomposites.

Most of the Ni/diamond nanocomposite electrodepositions, obtained in Watts electrolyte have a shining aspect. The morphology of the coatings obtained at lower current frequencies (0.01 and 0.1 Hz, respectively) exhibit differences as compared to the ones prepared at higher frequencies (1 and 10 Hz, respectively). Thus, the coatings with bigger crystals have a higher microhardness. At $\upsilon = 10$ Hz, the lowest microhardness for $\theta = 90$ % reaches 262 HV, while the highest microhardness for $\theta = 10$ % is 459 HV. This aspect proves the expected correlation between microhardness and granulation. Figure 3 reveals the difference between the morphologies of each of the Ni/diamond discussed coatings, having extreme values of the microhardness, at $\upsilon = 10$ Hz.

Figure 4 presents the XRD profiles for the P.C.-electrodeposited Ni/diamond coatings, at $\upsilon = 10$ Hz and at constant current density of 5 A/dm^2, for different duty cycles, θ. The relative intensities of the XRD peaks are those who are plotted, but they are now showed because of the waterfall-type of overlapped representation of multiple XRD patterns. Table 2 resumes the precise values of the relative intensities of the XRD peaks from all the graphs in Fig. 3. The crystalline planes are presented only for Nickel, the other peaks being merely assigned to the other chemical elements displaying XRD peaks (diamond—C and Copper—Cu).

Fig. 2 SEM micrographs of pure Ni coatings (*left-side images*) and of Ni/diamond nanocomposite coatings (*right-side images*), prepared by: **a** P.C. at $\theta = 20$ % and $\upsilon = 0.01$ Hz, **b**. P.C.at $\theta = 50$ % and $\upsilon = 0.1$ Hz, **c** P.C. at $\theta = 90$ % and $\upsilon = 1$ Hz, **d** D.C. at $\theta = 5$ A/dm^2

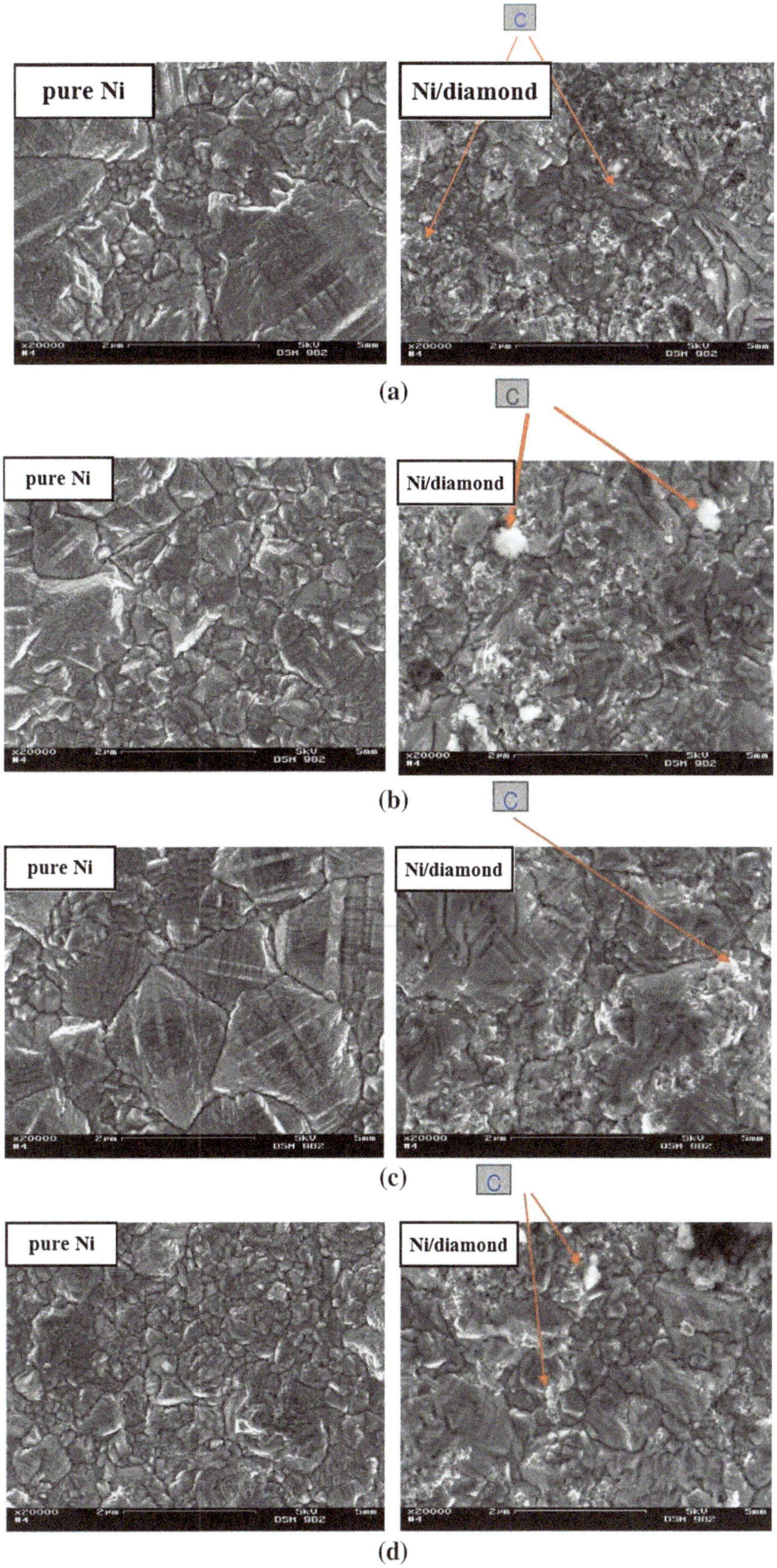

Fig. 3 SEM micrographs of the Ni/diamond coatings prepared in Watts electrolyte, coatings having **a** the lowest microhardness (262 HV); **b** the highest microhardness (459 HV)

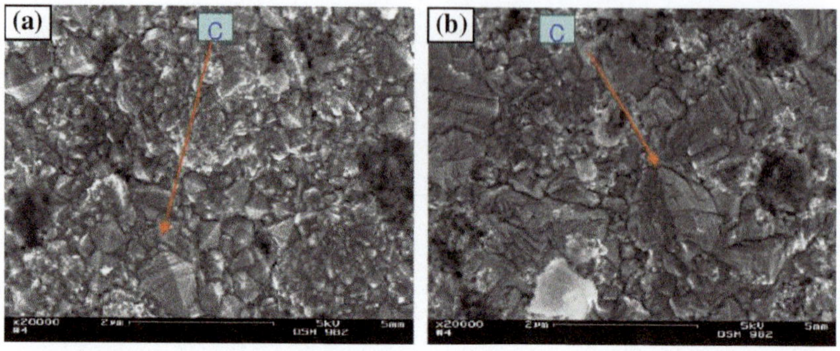

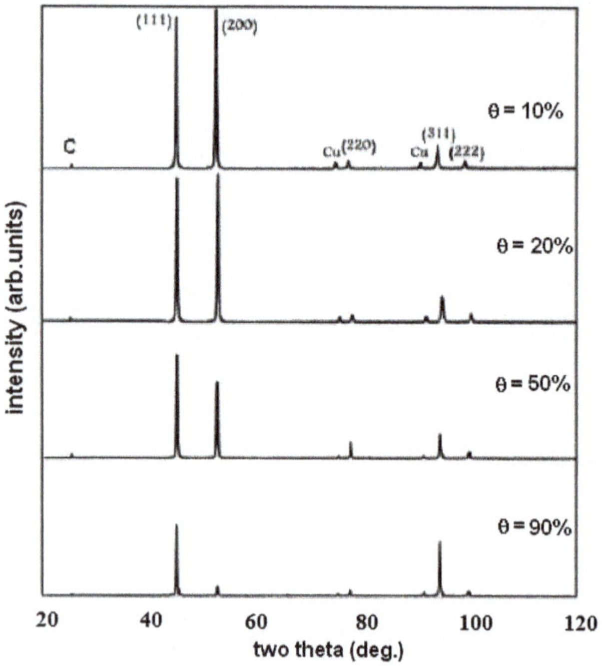

Fig. 4 The duty cycle effect on the structure of the dispersed phase of the Ni/diamond nanocomposite coatings, obtained at constant current density of 5 A/dm^2 and for a current frequency $\upsilon = 10$ Hz

Table 2 Typical data on XRD analysis of nanocomposite layers

Peak No	Element giving the XRD peak	Crystalline planes	θ (%)			
			10	20	50	90
1	C		2.38	1.30	1.85	0.53
2	Ni	(111)	94.05	97.66	100.00	100.00
3	Ni	(200)	100.00	100.00	73.33	11.76
4	Cu		3.81	3.90	0.74	1.07
5	Ni	(220)	4.76	5.19	16.67	6.95

From the XRD analysis, it can be inferred that the studied nanocomposite coatings have a face-centered cubic structure, with different orientations, which where influenced by the values of the P. C. duty cycles. Thus, if for $\theta = 10$ and 20 %, respectively, the highest peaks is asserted to the (200) planes, instead, for higher θ, namely for 50 and 90 %, respectively, the highest peak is associated to the (111) planes. The other XRD peaks are also changing from one set of deposition conditions to another. Thus, the (220) peak varies from 4.76 for $\theta = 10$ to 16.67 % for $\theta = 50$ %. Instead, the peak asserted to the (311) plane stays almost the same for $\theta = 10$, 20 and 50 %, respectively, but for 90 % jumps at a very high value, of 76.47 arb. units. The (222) peak does really change with changing deposition conditions.

Thus, a change of the intensity of the XRD peaks corresponding to nickel can be noticed when the duty cycle increases. When the duty cycle deposition parameter is changed to higher values (10 Hz), this gives rise to a decrease of the peak (200) in favor to the (111) peak and to an increase of (311) to (111) surface orientations ratio. This aspect is an indication of the grain refinement with increasing duty cycle.

Still, it remains to be established how the deposition frequencies of 0.01 and 10 Hz, respectively, for different duty cycles, influence the uniform particle distribution or if nucleation and subsequent aggregation of incorporated diamond particles alter the pure Ni matrix grain size and preferred surface orientations during its growth as a function of P.C. parameters and kinetics of each process (diamond adsorption and Ni plating).

Figure 5 reveals the influence of some deposition parameters, such as: current frequency and duty cycle on the microhardness of the Ni/diamond nanocomposites as compared to pure Ni coatings, prepared at constant current density of 5 A/dm^2.

The microhardness measurements revealed values between 210 and 334 HV for the pure nickel electrodeposited coatings, while for the Ni/diamond nanocomposites values between 262.61 and 459.21 HV were obtained. Increases and decreases of the microhardness were noticed for the studied coatings with changing pulse frequency. The standard deviation of the measurements were ±10–38 HV for the pure Ni coatings and ±21–47 HV for the Ni/diamond nanocomposite coatings.

As noticed from the SEM images, the pure Ni coatings present larger grains than the Ni/diamond nanocomposites. The decrease of Ni grains surfaces due to diamond addition

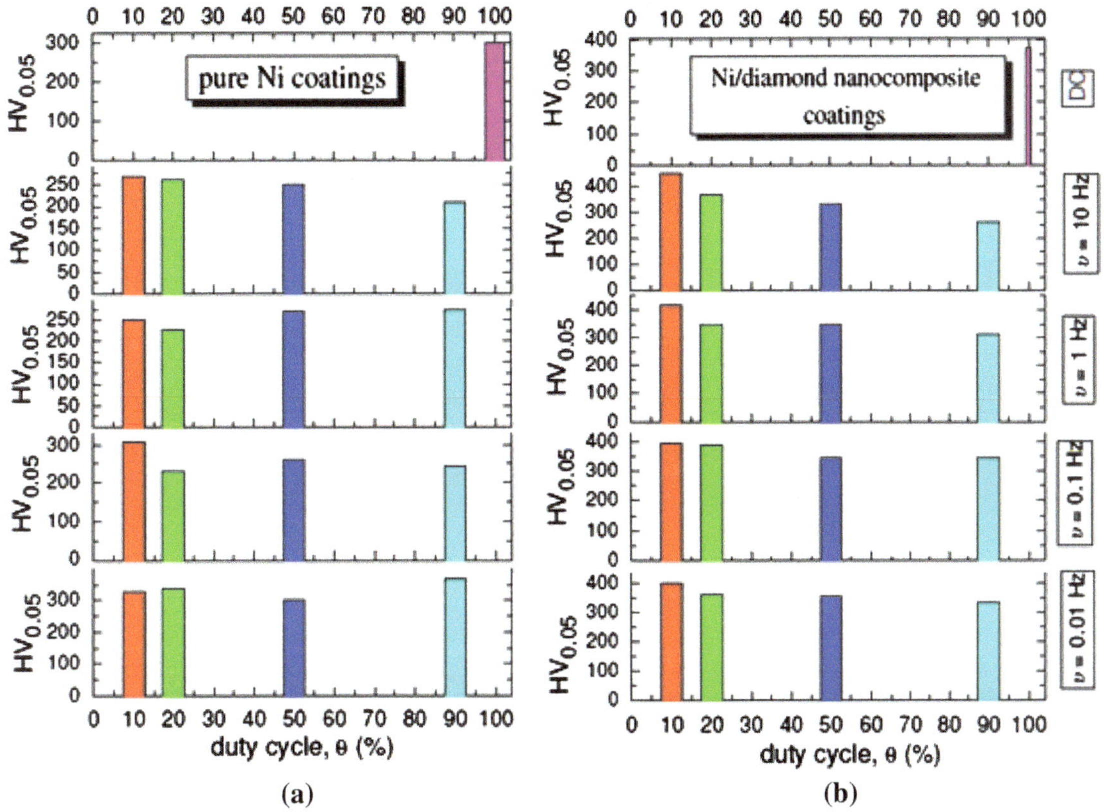

Fig. 5 Microhardness variation of the pure Ni coatings (**a**) as compared to the Ni/diamond nanocomposite coatings (**b**) with duty cycle and with current frequency, and also for D. C. electrodeposition

leads to an increase of the microhardness of the Ni/diamond nanocomposites as compared to the pure Ni ones. Also, the presence of the diamond powder itself is a cause of microhardness increase.

The crystallites growth is influenced by the frequency of the electrodes polarity changes during the depositions, such that the microstructure changes with frequency and, thus, the microhardness also changes.

Corrosion behavior

Potentiodynamic polarization tests

Corrosion studies were performed on the Ni/diamond nanocomposite coatings as compared to the pure Ni coatings, electrodeposited from Watts electrolyte. Two methods were employed: (1). the potentiodynamic polarization curves (LP); (2). The electrochemical impedance spectroscopy (EIS), by using 0.5 M Na_2SO_4 testing solution, with pH 2. Monitoring the corrosion potential of each sample was performed first. For the potentiodynamic polarization measurements, the initial potential were −0.5 V (SCE), while the final potential was 0.1 V, with 10 mV/s scanning speed. The values of the polarization resistance were obtained from the impedance diagrams performed at the corrosion potential for each sample. The diagrams of the potentiodynamic polarization curves of the studied coatings, obtained for 30 min and for 60 min immersion times, respectively are presented in Figs. 6, 7, 8 and the corresponding corrosion current, i_{cor}, polarization resistance, R_p and the Tafel constants, β_a and β_c are given in Tables 3 and 4. The corrosion diagrams are not represented with the same potential intervals on the horizontal axis, because of the differences between the potential where each analyzed coatings presents points of extreme for these kinds of dependences (log I versus E).

The potential of the current varies with increasing current frequency and duty cycle into the active area. The curves reveal a passive behavior between approximately −0.3 and 0.0 V for the depositions performed at different duty cycles, but with the same value for the current density, of 5 A/dm^2. The best corrosion resistance, of 20.3 $k\Omega\, cm^2$ was obtained for $\upsilon = 10$ Hz, $\theta = 20$ %, for 1 h immersion time into the testing solution. Thus, independently of the changes in the corrosion potential with pulse frequency and duty cycle, the experimental results are ascribed to the potential range from −0.3 to 0.0 V.

The corrosion resistance of the nanocomposite coatings is influenced both by the preparation of the

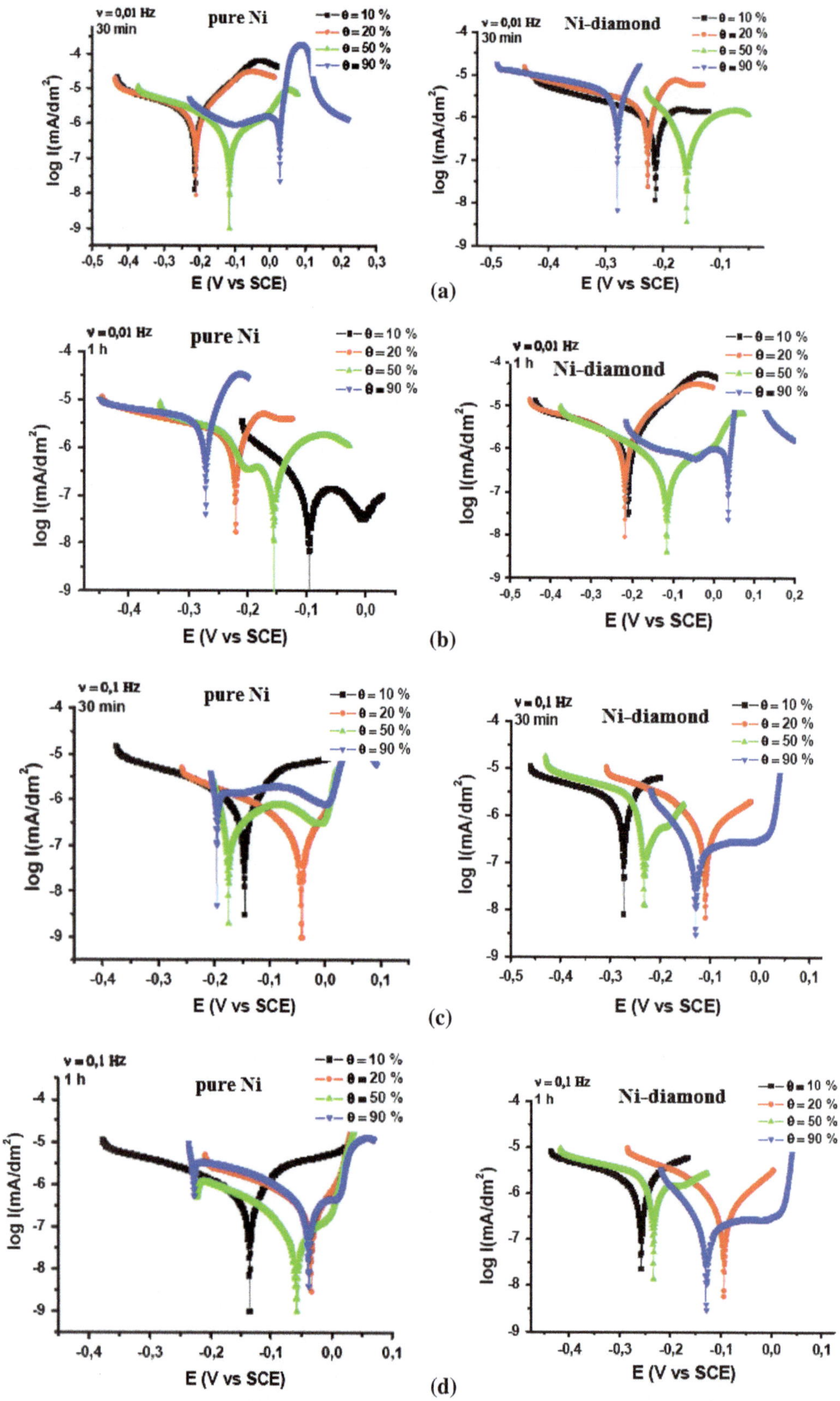

Fig. 6 Potentiodynamic polarization curves compared for pure Ni and for Ni/diamond coatings, electrodeposited at different duty cycles and evaluated after 30 min and also after 60 min immersion into 0.5 M Na_2SO_4 solution, at two impulse frequencies 0.01 and 0.1 Hz

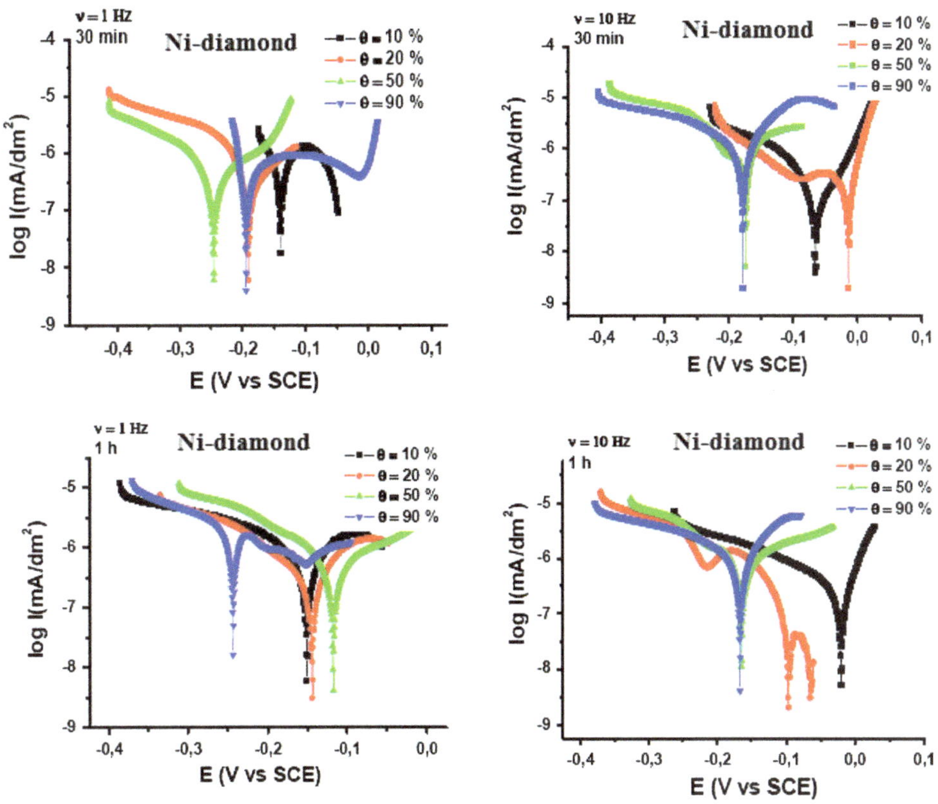

Fig. 7 Potentiodynamic polarization curves compared for Ni/diamond electrodeposited at different duty cycles, evaluated after 30 min and also after 60 min immersion into 0.5 M Na_2SO_4 solution, at two impulse frequencies: 1 and 10 Hz, respectively

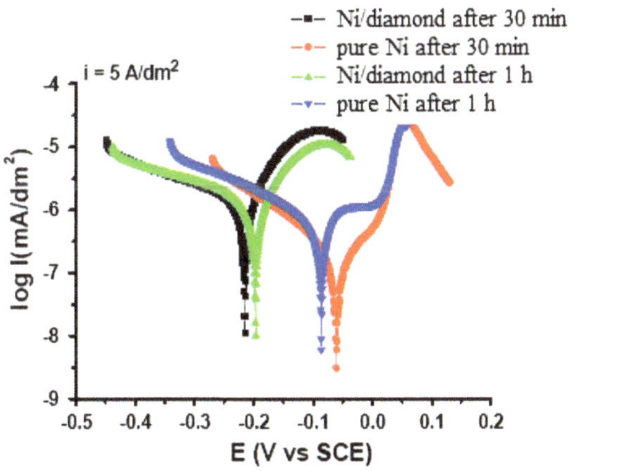

Fig. 8 Potentiodynamic polarization curves compared for pure Ni coatings and for Ni/diamond nanocomposites electrodeposited at 5 A/dm^2 and evaluated after 30 and 60 min immersion into 0.5 M Na_2SO_4 solution

electrodeposited Ni matrices and also by the incorporation of the diamond nanoparticles. It was reported that while the reverse ratio and the pulse frequency increased, more crystal nuclei were formed into the electrodeposited Ni coatings and that their granulation became smaller Lee et al. 1999. Wang et al., established that the corrosion resistance was improved by a decrease of the grain size onto the coating, since into finer grain coating the passivation was better (Lee et al. 1999).

For the Ni/diamond nanocomposite electrodepositions, the E_{cor} potential is displaced to more negative values than for pure nickel, which indicates an activation of those coatings in strong acid environment (see Fig. 8).

Analysis by electrochemical impedance spectroscopy (EIS)

Electrochemical impedance spectroscopy was also used for the analysis of the electrodeposited coatings, in order to have another representation of the corrosion behavior.

Differences between the polarization curves were noticed with increasing deposition duration. Almost each of the polarization curves presented a minimum, indicating the value of the polarization resistance. The Nyquist diagrams of the EIS spectra are presented in Fig. 9, for pulse current Ni/diamond nanocomposite coatings, electrodeposited for 1 h into the Watts nickel electrolyte, corresponding to 10 min, 1, 2, 4, 6, 24 and 30 h immersion time into 0.5 M Na_2SO_4 solution with pH 2.

By analyzing Fig. 9 one can notice that a 10 times increase in the frequency of the deposition leads to a drastic decrease, by four times, of the imaginary value Z'' of the impedance and to a more than 2 times decrease of the real values Z' of the impedance. Also, the immersion time strongly influences the aspect of the Nyquist diagrams.

Table 3 Corrosion parameters of electrodeposited layers for various values of (ν, θ)

υ (Hz)	θ (%)	*t* (min)	$E_{corr.}$ (V) SCE	$i_{corr.}$ (mA/dm^2)	β_a (V/dec)	β_c (V/dec)	R_p ($k\Omega\,cm^2$)	$V_{corr.}$ (μm/year)
0.01	10	30	−0.21	6.36	18.75	−13.58	5.4	68.51
		60	−0.10	7.49	25.06	−46.04	9.4	80.68
	20	30	−0.22	5.92	23.89	−8.29	4.5	63.77
		60	−0.22	6.21	38.05	−14.92	7.5	66.89
	50	30	−0.16	6.83	22.87	−30.17	8.3	73.57
		60	−0.15	6.98	42.59	−23.15	9.3	75.18
	90	30	−0.28	6.00	46.63	−23.58	11.3	64.63
		60	−0.27	5.96	55.13	−19.06	10.3	64.20
0.1	10	30	−0.27	6.18	31.46	−15.15	7.2	66.57
		60	−0.25	6.15	22.35	−10.81	5.1	66.25
	20	30	−0.11	6.68	16.38	−21.19	6.0	71.96
		60	−0.10	6.63	13.91	−18.83	5.2	71.42
	50	30	−0.23	6.73	13.93	−60.54	7.3	72.49
		60	−0.23	6.28	15.28	−32.12	7.1	67.65
	90	30	−0.13	7.34	21.72	−49.23	8.9	79.06
		60	−0.13	7.21	15.23	−41.12	6.7	77.67
1	10	30	−0.14	6.68	34.34	−42.61	12.3	71.96
		60	−0.15	6.58	19.40	−18.31	6.2	70.88
	20	30	−0.19	6.68	11.14	−30.99	5.3	71.95
		60	−0.14	6.79	23.36	−14.96	5.8	73.14
	50	30	−0.11	6.72	13.53	−27.10	5.8	72.38
		60	−0.12	6.74	21.47	−25.89	7.5	72.604
	90	30	−0.20	6.74	16.43	−93.54	9.0	72.60
		60	−0.24	6.32	40.03	−36.82	13.2	68.08
10	10	30	−0.06	7.20	20.19	−24.95	6.7	77.56
		60	−0.01	6.78	39.30	−10.69	5.4	73.03
	20	30	−0.02	7.08	75.65	−21.63	10.3	76.26
		60	−0.10	7.92	59.48	−97.81	20.3	85.31
	50	30	−0.17	6.47	43.03	−12.12	6.3	69.69
		60	−0.16	6.49	19.47	−28.00	77.0	69.91
	90	30	−0.18	6.48	35.46	−20.47	8.7	69.80
		60	−0.16	6.54	35.89	−22.80	9.2	70.45

Table 4 Corrosion parameters of electrodeposited layers for various values of (i, t)

$i = 5\ A/dm^2$	*t* (min)	$E_{corr.}$ (V) SCE	$i_{corr.}$ (mA/dm^2)	β_a (V/dec)	β_c (V/dec)	$V_{corr.}$ (μm/year)
Pure Ni	30	−0.06	7.18	18.92	−19.08	77.34
	60	−0.09	6.69	22.27	−17.34	72.06
Ni/diamond	30	−0.21	6.17	31.95	−7.79	66.46
	60	−0.19	6.33	25.42	−13.92	68.18

Thus, for υ = 0.01 Hz, the graphs for 4 and for 6 h of immersion into the Na_2SO_4 solution do not close to the horizontal axis, while the one reaching the fastest this axis corresponds to 30 h immersion time. Instead, for υ = 0.1 Hz, all the curves reach the horizontal axis, finally, the quickest corresponding to 2 h immersion time and the longest appearing for the coating which was immersed for 6 h.

Figure 10 presents the Nyquist diagrams for the pure nickel coatings (Fig. 10a) and for Ni/diamond nanocomposite coatings (Fig. 10b), prepared by direct current electrodeposition, corresponding to 10 min, 1, 2, 4, 6, 24 and 30 h immersion time into 0.5 M Na_2SO_4 solution with pH 2.

While the aspect of the Nyquist diagrams is irregular for pure Ni coatings, this is not the case for Ni/diamond

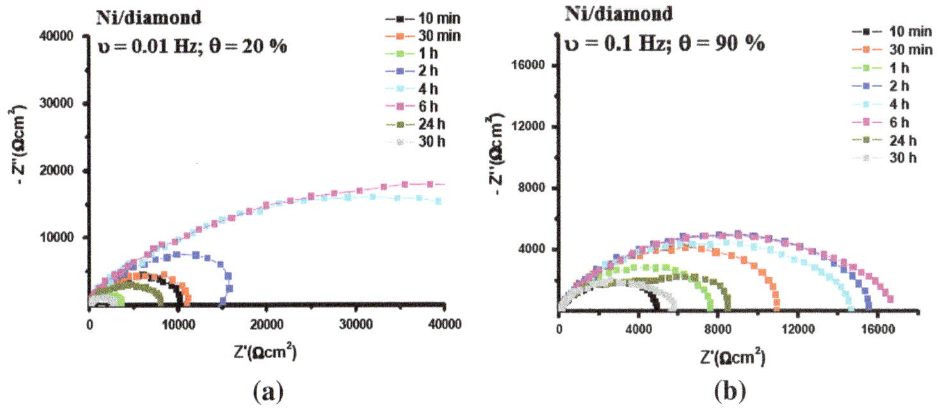

Fig. 9 Nyquist diagram of the impedance spectra for the Ni/diamond nanocomposite coatings, pulse current—electrodeposited, with 5 A/dm^2 and 3,600 s deposition duration, corresponding to testing in 0.5 M Na_2SO_4 solution: **a** $\upsilon = 0.01$ Hz and $\theta = 20$ %; **b** $\upsilon = 0.1$ Hz and $\theta = 90$ %

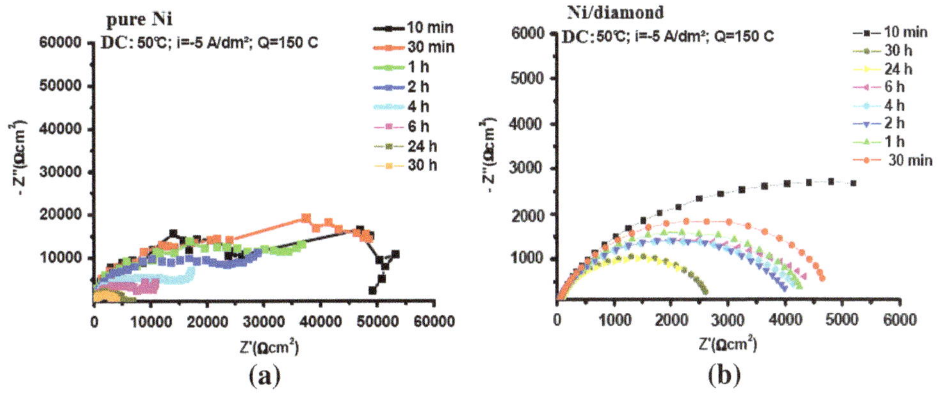

Fig. 10 Nyquist diagram of the impedance spectra for the pure Ni (**a**) and for the Ni/diamond nanocomposite (**b**) coatings, for direct current electrodeposition with 5 A/dm^2 and 3600 s deposition duration, corresponding to testing in 0.5 M Na_2SO_4 solution

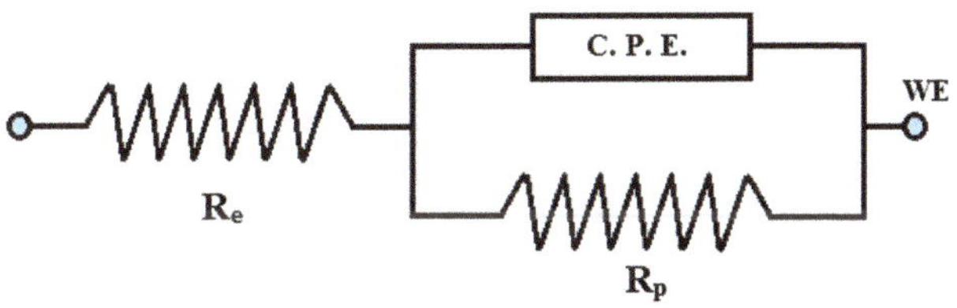

Fig. 11 The equivalent circuit used to fit the impedance data

nanocomposite coatings. Still, shorter the immersion time, higher the maximum reached values of Z' for the Ni/diamond nanocomposites.

It was noticed that both the Ni/diamond nanocomposite coatings and the pure nickel coatings presented a polarization resistance which decreased after 30 h of investigation. The corrosion potential was $E_{cor} = -0.21$ V and changed in time to more positive values. This behavior indicates an active dissolving with reduction of the oxidant (H^+).

As a reminder, the polarization resistance technique (sometimes referred to as linear polarization) is an experimental electrochemical technique that enables the estimation of the polarization resistance and from that the corrosion rate. The experiment requires polarizing an electrode between about −20 mV and +20 mV relative to the steady state corrosion potential. Some authors refer to the polarization resistance and to the capacitance terms as charge transfer resistance (RCT) and pseudo-capacitance.

Figure 11 presents the equivalent circuit for the Ni/diamond nanocomposite coatings, where the notations correspond to: R_e = the resistance of the electrolyte, measured between the working electrode and the reference electrode; R_p is the polarization resistance or the resistance transfer at the Ni/Na_2SO_4 interface; CPE is a constant phase element.

The charge transfer resistance determined by using the equivalent circuit model given in Fig. 11 should be taken as an overall or effective polarization resistance of the electrode. Effect of the electrode resistivity on the performance of the porous CEs is a subject of ongoing research (Haseko et al. 2006).

Figures 12 and 13 present the values of the polarization resistance, R_p and those of the capacitance computed with the equivalent circuit from Fig. 11, for the Ni/diamond nanocomposite coatings, P.C. and D.C. electrodeposited, respectively, for 3,600 s in Watts electrolyte, after different immersion duration into 0.5 M Na_2SO_4 solution, for all the tested electrodeposition conditions ($\upsilon = 0.01$, 0.1, 1 and 10 Hz, respectively).

The CPE parameter can be used to infer the surface roughness (Halme et al. 2006), but the present study didn't propose for roughness estimations in this phase of the

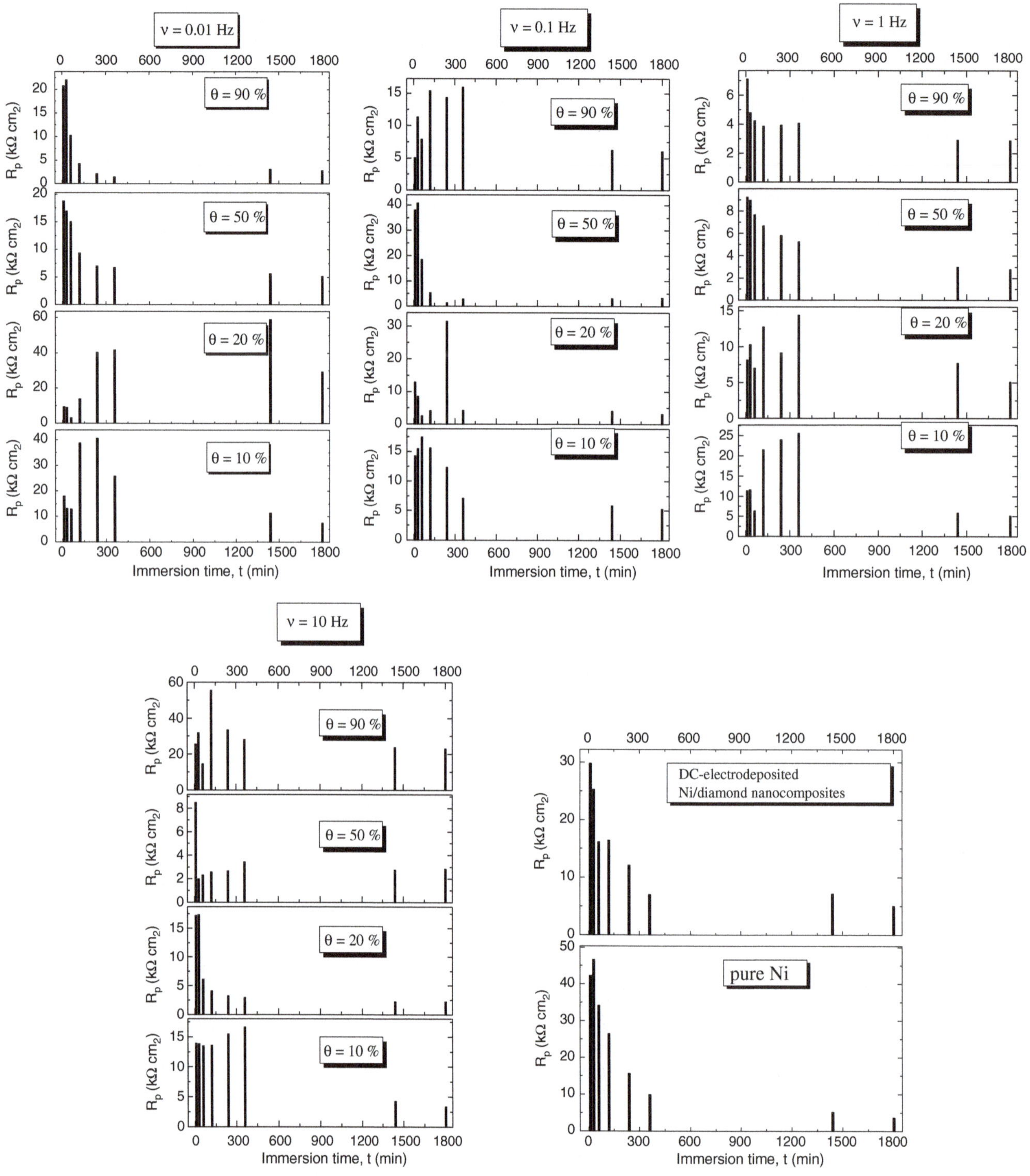

Fig. 12 The variation of the polarization resistance, R_p for the Ni/diamond nanocomposite coatings, P.C. and D.C. electrodeposited, respectively, for 3,600 s in Watts electrolyte, after different immersion duration into 0.5 M Na_2SO_4 solution, for all the tested electrodeposition conditions (υ = 0.01, 0.1, 1 and 10 Hz, respectively)

analysis. Still, there are some other factors influencing the CPR: surface reactivity, surface inhomogeneity, electrode porosity, etc. (Jorcin et al. 2006).

It can be noticed that the polarization resistance of the Ni/diamond nanocomposites had a smaller value than that of the pure nickel coating. The capacity of the double electric layer was around 19.40 μF/cm^2, which is somewhat smaller than that of pure nickel coating. The corrosion potential was driven to more negative values than pure electrodeposited nickel, having an average polarization

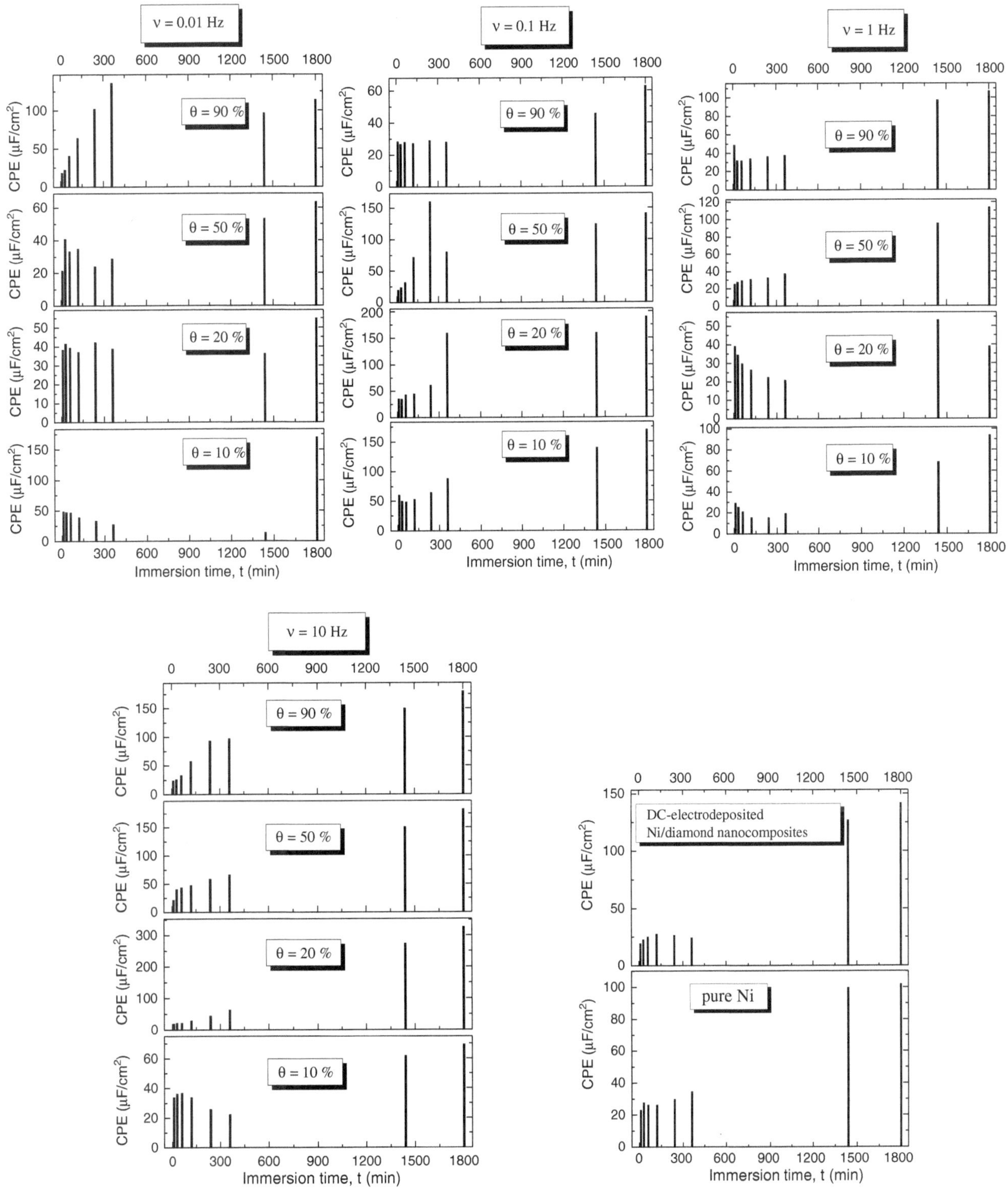

Fig. 13 The variation of the capacitance (constant phase element), CPE for the Ni/diamond nanocomposite coatings, P.C. and D.C. electrodeposited, respectively, for 3,600 s in Watts electrolyte, after different immersion duration into 0.5 M Na_2SO_4 solution, for all the tested electrodeposition conditions ($\upsilon = 0.01$, 0.1, 1 and 10 Hz, respectively)

resistance of 6.3 kΩ cm^2, but with a better stability as compared to the nanocomposite coatings. The values of the polarization resistance computed from the potentiodynamic diagrams, by using Stern-Geary formula are closer to the ones obtained from the electrochemical impedance Spectroscopy measurements.

The polarization resistance of the Ni/diamond nanocomposite coatings presents the highest value of 60.00 kΩ cm^2, after 30 h of immersion into the Na_2SO_4 solution, for the samples electrodeposited at $\upsilon = 0.01$ Hz and $\theta = 20$ %. The smallest value for the polarization resistance, namely 2.79 kΩ cm^2 was measured for the Ni/diamond nanocomposite coating electrodeposited at $\upsilon = 1$ Hz and $\theta = 90$ %, also after 30 h of immersion into the Na_2SO_4 solution.

The D.C. electrodeposition of/diamond nanocomposite led to somewhat smaller values for R_p, not exceeding 30 kΩ cm^2, values which are smaller than for pure Ni coatings.

As about the CPE, Fig. 13 reveals that it surpasses 300 μF/cm^2 for the nanocomposite coating P.C. deposited at the highest frequency, $\upsilon = 10$ Hz and for 20 % duty cycle. As for the D.C. electrodepositions, the pure Ni coatings present somewhat smaller values for CPE than the corresponding Ni/diamond nanocomposites.

From the obtained results one cannot make a complex evaluation of the kinetic reactions that take place and no unitary mechanism can be considered for the composite system. Vaster studies are needed. The nanocomposite system differentiates both by its forming mechanism and by its corrosion behavior in the studied electrolyte. The electrodeposition of the diamond nanoparticles into nickel matrix improves the corrosion behavior of the coatings. The obtained results are encouraging and suggest that the diamond nanoparticles coating can represent an attractive alternative for the protection of surfaces by introducing a barrier between the aggressive medium and the metallic substrate.

Conclusions

Ni/diamond nanocomposite coatings and pure Ni coatings were prepared by P.C. and D.C. electrodeposition onto copper substrates, with several values for the pulses frequencies and for different necessary durations. It was found out that the diamond nanoparticles content increases with increasing pulse frequency and decreasing duty cycle. The experimental results proved that the microhardness and the roughness of the Ni/diamond nanocomposite coatings were strongly influenced by the presence of the nanoparticles as compared to pure nickel coatings. By using the pulse current electrodeposition instead of direct current electrodeposition, the microhardness of the nanocomposite coatings increased, especially for low duty cycles and low current frequency. The corrosion behavior of the studied coatings represents a priority for the properties of new materials. Modern methods were employed for the study of the corrosion behavior, namely the potentiodynamic polarization curves and the electrochemical impedance spectra. These methods offer very useful information about the kinetics of the chemical reactions at the electrode. The corrosion studies in strongly corrosive electrolyte (0.5 M Na_2SO_4 solution, with pH 2) have revealed a better behavior of the Ni/diamond nanocomposite coatings as compared to the pure nickel coatings. Also, it was concluded that the corrosion resistance of the nanocomposite coatings increased with decreasing pulse frequency and reverse ratio. The best corrosion resistance $R_p = 20.3\,k\Omega\,cm^2$, was obtained for the nanocomposite coatings prepared in the following conditions: $\upsilon = 10$ Hz, $\theta = 20$ %, for 1 h of immersion into the testing solution. More complex studies are necessary in order to elucidate the phenomena and the kinetic reactions, such that the anticorrosive protection mechanism would be better understood and also its influence factors would be inferred for the Ni/diamond nanocomposite coatings.

Acknowledgments The research of this paper has been supported by POSDRU/88/1.5/S/61445 Project.

References

Chandrasekar MS, Pushpavanam M (2008) Electrochim Acta 53:3313–3322
Feldstein N (1981) Composite coatings. Materials Engineering, Cleveland
Feldstein N, Lancsek T (1984) J Electrochem Soc 131(12):3026–3027
Halme J, Toivola M, Tolvanen A, Lund P (2006) Sol Energ Mat Sol C 90(78):872886
Haseko Y, Shrestha NK, Teruyama S, Saji T (2006) Electrochim Acta 51(18):3652–3657
Jorcin JB, Orazem ME, Pebere N, Tribollet B (2006) Electrochim Acta 51:1473–1479
Kollia C, Spyrellis N (1993) Surf Coat Tech 57:101–105
Lee WH, Tang SC, Chung KC (1999) Surf Coat Tech 120–121:607–611
Li YD, Li GS, Jiang H, He YP (2002) Ind Diam Rev 62:259–262
Li YD, Jiang H, Pang L, Wang B (2007) Surf Coat Tech 201(12):5925–5930
Pushpavanam M, Manikandan H, Ramanathan K (2007) Surf Coat Tech 201(14):6372–6379
Ranjan S, Ashutosh S, Sumit B, Siddhartha D, Karabi D (2010) J Nanosci Nanotechnol 10:4998–5003
Sato K, Suzuki K, Yokoyama T, Kagawa H (1997) Mater J Proc Tech 63:829–832
Shen B, Sun FH, Zhang ZM (2009) J Nanosci Nanotechnol 9:788–792
Subhash T, Jitendra K (2008) J Nanosci Nanotechnol 8:4073–4080
Suzuki Y, Asai O (1987) J Electrochem Soc 134:1905–1910
Watson SW (1993) J Electrochem Soc 140(8):2235–2238

Permissions

All chapters in this book were first published in AN, by Springer; hereby published with permission under the Creative Commons Attribution License or equivalent. Every chapter published in this book has been scrutinized by our experts. Their significance has been extensively debated. The topics covered herein carry significant findings which will fuel the growth of the discipline. They may even be implemented as practical applications or may be referred to as a beginning point for another development.

The contributors of this book come from diverse backgrounds, making this book a truly international effort. This book will bring forth new frontiers with its revolutionizing research information and detailed analysis of the nascent developments around the world.

We would like to thank all the contributing authors for lending their expertise to make the book truly unique. They have played a crucial role in the development of this book. Without their invaluable contributions this book wouldn't have been possible. They have made vital efforts to compile up to date information on the varied aspects of this subject to make this book a valuable addition to the collection of many professionals and students.

This book was conceptualized with the vision of imparting up-to-date information and advanced data in this field. To ensure the same, a matchless editorial board was set up. Every individual on the board went through rigorous rounds of assessment to prove their worth. After which they invested a large part of their time researching and compiling the most relevant data for our readers.

The editorial board has been involved in producing this book since its inception. They have spent rigorous hours researching and exploring the diverse topics which have resulted in the successful publishing of this book. They have passed on their knowledge of decades through this book. To expedite this challenging task, the publisher supported the team at every step. A small team of assistant editors was also appointed to further simplify the editing procedure and attain best results for the readers.

Apart from the editorial board, the designing team has also invested a significant amount of their time in understanding the subject and creating the most relevant covers. They scrutinized every image to scout for the most suitable representation of the subject and create an appropriate cover for the book.

The publishing team has been an ardent support to the editorial, designing and production team. Their endless efforts to recruit the best for this project, has resulted in the accomplishment of this book. They are a veteran in the field of academics and their pool of knowledge is as vast as their experience in printing. Their expertise and guidance has proved useful at every step. Their uncompromising quality standards have made this book an exceptional effort. Their encouragement from time to time has been an inspiration for everyone.

The publisher and the editorial board hope that this book will prove to be a valuable piece of knowledge for researchers, students, practitioners and scholars across the globe.

List of Contributors

R. Augustine
International and Interuniversity Centre for Nanoscience and Nanotechnology, Mahatma Gandhi University, Priyadarshini Hills P O, Kottayam 686 560, Kerala, India

N. Kalarikkal
International and Interuniversity Centre for Nanoscience and Nanotechnology, Mahatma Gandhi University, Priyadarshini Hills P O, Kottayam 686 560, Kerala, India
School of Pure and Applied Physics, Mahatma Gandhi University, Priyadarshini Hills P O, Kottayam 686 560 Kerala, India

S. Thomas
International and Interuniversity Centre for Nanoscience and Nanotechnology, Mahatma Gandhi University, Priyadarshini Hills P O, Kottayam 686 560, Kerala, India
School of Chemical Sciences, Mahatma Gandhi University, Priyadarshini Hills P O, Kottayam 686 560, Kerala, India

Maryam N. Shaker
Physics Department, Faculty of Science, Alexandria University, Alexandria, Egypt

Heba S. Ramadan
Biophysics Department, Medical Research Institute, Alexandria University, Alexandria, Egypt

Moustafa M. Mohamed
Biophysics Department, Medical Research Institute, Alexandria University, Alexandria, Egypt

Ahmed M. El khatib
Physics Department, Faculty of Science, Alexandria University, Alexandria, Egypt

Gamal D. Roston
Physics Department, Faculty of Science, Alexandria University, Alexandria, Egypt

N. Shanmugam
Department of Physics, Annamalai University, Annamalai Nagar, Chidambaram 608 002, Tamilnadu, India

P. Rajkamal
Department of Physics, Annamalai University, Annamalai Nagar, Chidambaram 608 002, Tamilnadu, India

S. Cholan
Department of Physics, Annamalai University, Annamalai Nagar, Chidambaram 608 002, Tamilnadu, India

N. Kannadasan
Department of Physics, Annamalai University, Annamalai Nagar, Chidambaram 608 002, Tamilnadu, India

K. Sathishkumar
Department of Physics, Annamalai University, Annamalai Nagar, Chidambaram 608 002, Tamilnadu, India

G. Viruthagiri
Department of Physics, Annamalai University, Annamalai Nagar, Chidambaram 608 002, Tamilnadu, India

A. Sundaramanickam
CAS in Marine Biology, Faculty of Marine Sciences, Annamalai University, Parangipettai, Tamil Nadu, India

Susmita Paul
Department of Physics, Tezpur University, Napaam 784028, India

Amarjyoti Choudhury
Department of Physics, Tezpur University, Napaam 784028, India

Jolly Pal
School of Studies in Chemistry, Pandit Ravishankar Shukla University, Raipur 492 010, Chhattisgarh, India

Manas Kanti Deb
School of Studies in Chemistry, Pandit Ravishankar Shukla University, Raipur 492 010, Chhattisgarh, India

O. Anwar Bég
Gort Engovation Research (Propulsion/Biomechanics), 15 Southmere Avenue, Bradford BD73NU, UK

M. S. Khan
Mathematics Discipline, Khulna University, Khulna 9208, Bangladesh

Ifsana Karim
Mathematics Discipline, Khulna University, Khulna 9208, Bangladesh

Md. M. Alam
Mathematics Discipline, Khulna University, Khulna 9208, Bangladesh

M. Ferdows
Department of Mathematics, Dhaka University, Dhaka, Bangladesh
Quantum Beam Science Directorate, Japan Atomic Energy Authority, Tokyo, Japan

M. J. Uddin
American International University-Bangladesh, Dhaka 1206, Bangladesh

M. Ferdows
Department of Mathematics, University of Dhaka, Dhaka 1000, Bangladesh

N. A. Shah
Department of Electronics, Saurashtra University, Rajkot 360 005, India

Polina A. Demina
Lomonosov University of Fine Chemical Technology, 119571 Moscow, Russia

Alexey N. Kuz'michev
Bonifatius College, Burg. Fockema Andreaelaan 7, 3582 Utrecht, The Netherlands

Andrei M. Tsybinsky
Lomonosov University of Fine Chemical Technology, 119571 Moscow, Russia

Lubov N. Obolenskaya
Lomonosov University of Fine Chemical Technology, 119571 Moscow, Russia

Galina M. Kuz'micheva
Lomonosov University of Fine Chemical Technology, 119571 Moscow, Russia

Elena N. Domoroshchina
Lomonosov University of Fine Chemical Technology, 119571 Moscow, Russia

Elena V. Savinkina
Lomonosov University of Fine Chemical Technology, 119571 Moscow, Russia

S. Omeiri
Laboratory of Storage and Valorization of Renewable Energies, Faculty of Chemistry, University of Science and Technology (USTHB), B P 32, El Alia, 16111 Algiers, Algeria

N. Allalou
Laboratory of Storage and Valorization of Renewable Energies, Faculty of Chemistry, University of Science and Technology (USTHB), B P 32, El Alia, 16111 Algiers, Algeria

G. Rekhila
Laboratory of Storage and Valorization of Renewable Energies, Faculty of Chemistry, University of Science and Technology (USTHB), B P 32, El Alia, 16111 Algiers, Algeria

Y. Bessekhouad
Laboratory of Storage and Valorization of Renewable Energies, Faculty of Chemistry, University of Science and Technology (USTHB), B P 32, El Alia, 16111 Algiers, Algeria

M. Trari
Laboratory of Storage and Valorization of Renewable Energies, Faculty of Chemistry, University of Science and Technology (USTHB), B P 32, El Alia, 16111 Algiers, Algeria

D. Raju
CSIR-National Chemical Laboratory, Plant Tissue Culture Division, Pune 411008, Maharashtra, India

Nikita Paneliya
CSIR-National Chemical Laboratory, Plant Tissue Culture Division, Pune 411008, Maharashtra, India

Urmil J. Mehta
CSIR-National Chemical Laboratory, Plant Tissue Culture Division, Pune 411008, Maharashtra, India

E. Chigo Anota
Facultad de Ingeniería Química, Benemérita Universidad Autónoma de Puebla, Ciudad Universitaria, San Manuel, Puebla 72570, Mexico

A. Torres Soto
Facultad de Ingeniería Química, Benemérita Universidad Autónoma de Puebla, Ciudad Universitaria, San Manuel, Puebla 72570, Mexico

Gregorio H. Cocoletzi
Instituto de Física 'Luís Rivera Terrazas', Benemérita Universidad Autónoma de Puebla, Apartado Postal J-48, Puebla 72570, Mexico

B. P. Singh
Physics and Engineering of Carbon, Division of Materials Physics and Engineering, CSIR-National Physical Laboratory, New Delhi, India
Centre for Polymer Science and Engineering, Indian Institute of Technology Delhi, Delhi 16, India

Veena Choudhary
Centre for Polymer Science and Engineering, Indian Institute of Technology Delhi, Delhi 16, India

V. N. Singh
Electron and Ion Microscopy Section, CSIR-National Physical Laboratory, New Delhi 110012, India

R. B. Mathur
Physics and Engineering of Carbon, Division of Materials Physics and Engineering, CSIR-National Physical Laboratory, New Delhi, India

Pooja Rani
Department of Physics, Panjab University, Chandigarh 160014, India

V. K. Jindal
Department of Physics, Panjab University, Chandigarh 160014, India

S. Nadeem
Department of Mathematics, Quaid-I-Azam University 45320, Islamabad 44000, Pakistan

S. T. Hussain
Department of Mathematics, Quaid-I-Azam University 45320, Islamabad 44000, Pakistan

K. Sri Sindhura
Department of Physics, Sri Venkateswara University, Tirupati 517 502, Andhra Pradesh, India

T. N. V. K. V. Prasad
Department of Nanotechnology, Institute of Frontier Technology, Regional Agricultural Research Station, Acharya Nagarjuna Ranga Agricultural University, Tirupati 517 502, Andhra Pradesh, India

P. Panner Selvam
Department of Soil Science and Agricultural Chemistry, Indian Institute of Horticultural Research, Bangalore 560 089, India

O. M. Hussain
Department of Physics, Sri Venkateswara University, Tirupati 517 502, Andhra Pradesh, India

S. Muruganandam
Department of Physics, Presidency College, Chennai 600 005, Tamilnadu, India

G. Anbalagan
Department of Physics, Presidency College, Chennai 600 005, Tamilnadu, India

G. Murugadoss
Centre for Nanoscience and Technology, Anna University, Chennai 600 025, Tamilnadu, India

Luky Hendraningrat
Department of Petroleum Engineering and Applied Geophysics, Norwegian University of Science and Technology, NTNU, 7491 Trondheim, Norway

Torsæter
Department of Petroleum Engineering and Applied Geophysics, Norwegian University of Science and Technology, NTNU, 7491 Trondheim, Norway

H. F. Aly
Hot Laboratories Center, Atomic Energy Authority, Nasr 13759, Egypt

M. A. Akl
Faculty of Science, Mansoura University, Mansoura, Egypt

Hesham M. A. Soliman
Advanced Technology and New Materials Research Institute, City for Scientific Research and Technology Applications, P. O. Box 21934, SRTA, Egypt

Aref M. E. AbdEl-Rahman
Advanced Technology and New Materials Research Institute, City for Scientific Research and Technology Applications, P. O. Box 21934, SRTA, Egypt

A. I. Abd-Elhamid
Advanced Technology and New Materials Research Institute, City for Scientific Research and Technology Applications, P. O. Box 21934, SRTA, Egypt

Gunasekaran Dharanivasan
Department of Biotechnology, University of Madras, Guindy Campus, Chennai 600 025, Tamil Nadu, India

Thangavelu Rajamuthuramalingam
Department of Biotechnology, University of Madras, Guindy Campus, Chennai 600 025, Tamil Nadu, India

Denison Michael Immanuel Jesse
Department of Biotechnology, University of Madras, Guindy Campus, Chennai 600 025, Tamil Nadu, India

Nagappan Rajendiran
Department of Polymer Science, University of Madras, Guindy Campus, Chennai 600 025, Tamil Nadu, India

Krishnan Kathiravan
Department of Biotechnology, University of Madras, Guindy Campus, Chennai 600 025, Tamil Nadu, India

Shraddha Singh
School of Materials Science and Technology, Indian Institute of Technology, (BHU), Varanasi 221005, India

V. K. Srivastava
Department of Mechanical Engineering, Indian Institute of Technology, (BHU), Varanasi 221005, India

Rajiv Prakash
School of Materials Science and Technology, Indian Institute of Technology, (BHU), Varanasi 221005, India

R. R. Muthuchudarkodi
Department of Chemistry, V.O Chidambaram College, Tuticorin 628008, Tamilnadu, India

C. Vedhi
Department of Chemistry, V.O Chidambaram College, Tuticorin 628008, Tamilnadu, India

C. Gheorghies
Dunarea de Jos University of Galati, 47 Domneasca Street, 800008 Galati, Romania

D. E. Rusu
S.C. Automobile DACIA S.A., (Renault Technologie Roumanie), UMCD-Department ACU, Mioveni, Arges, Uzinei Street, No.1-3, 115400 Mioveni, Romania

A. Bund
Fachgebiet Elektrochemie und Galvanotechnik II, Fakultät für Elektrotechnik und Informationstechnik, Institut für Werkstofftechnik, Technische Universität Ilmenau, Ilmenau, Germany

S. Condurache-Bota
Dunarea de Jos University of Galati, 47 Domneasca Street, 800008 Galati, Romania

L. P. Georgescu
Dunarea de Jos University of Galati, 47 Domneasca Street, 800008 Galati, Romania